ABIOTIC STRESS RESPONSE IN PLANTS – PHYSIOLOGICAL, BIOCHEMICAL AND GENETIC PERSPECTIVES

Edited by **Arun Kumar Shanker**
and **B. Venkateswarlu**

INTECHOPEN.COM

Abiotic Stress Response in Plants – Physiological, Biochemical and Genetic Perspectives
Edited by Arun Kumar Shanker and B. Venkateswarlu

Published by InTech

Janeza Trdine 9, 51000 Rijeka, Croatia

Edition 2014

Publishing Process Manager Dragana Manestar

Technical Editor Teodora Smiljanic

Cover Designer InTech Design Team

Image Copyright oriontrail, 2010. Used under license from Shutterstock.com

Additional hard copies can be obtained from orders@intechopen.com

Abiotic Stress Response in Plants – Physiological, Biochemical and Genetic Perspectives,
Edited by Arun Kumar Shanker and B. Venkateswarlu
p. cm.
ISBN 978-953-307-672-0

INTECH

open science | open minds

Contents

Preface IX

Part 1 Signalling in Abiotic Stress 1

Chapter 1 **Abiotic and Biotic Stress Response Crosstalk in Plants 3**
Saúl Fraire-Velázquez, Raúl Rodríguez-Guerra and Lenin Sánchez-Calderón

Chapter 2 **Reactive Oxygen in Abiotic Stress Perception - From Genes to Proteins 27**
Michael Wrzaczek, Julia P. Vainonen, Adrien Gauthier, Kirk Overmyer and Jaakko Kangasjärvi

Chapter 3 **Plant Organelles-to-Nucleus Retrograde Signaling 55**
Nadezhda Yurina and Margarita Odintsova

Part 2 Nucleic Acids, Proteins and Enzymes 75

Chapter 4 **Post-Translational Modifications of Nuclear Proteins in the Response of Plant Cells to Abiotic Stresses 77**
Jennifer Dahan, Emmanuel Koen, Agnès Dutartre, Olivier Lamotte and Stéphane Bourque

Chapter 5 **Facing the Environment: Small RNAs and the Regulation of Gene Expression Under Abiotic Stress in Plants 113**
Inês Trindade, Dulce Santos, Tamas Dalmay and Pedro Fevereiro

Chapter 6 **Cyclic Nucleotides and Nucleotide Cyclases in Plant Stress Responses 137**
Fouad Lemtiri-Chlieh, Ludivine Thomas, Claudius Marondedze, Helen Irving and Chris Gehring

Chapter 7 **Abiotic Stress-Induced Programmed Cell Death in Plants: A Phytaspase Connection 183**
Alexander I. Tuzhikov, Boris B. Vartapetian Andrey B. Vartapetian and Nina V. Chichkova

Chapter 8 **Plant Plasma Membrane H^{+}-ATPase in Adaptation of Plants to Abiotic Stresses 197**
Małgorzata Janicka-Russak

Part 3 Genes and Genomes 219

Chapter 9 **Plant Abiotic Stress: Insights from the Genomics Era 221**
Erik R. Rowley and Todd C. Mockler

Chapter 10 **Role of Plant Transcription Factors in Abiotic Stress Tolerance 269**
Charu Lata, Amita Yadav and Manoj Prasad

Chapter 11 **The Roles of Germin Gene Products in Plants Under Salt Stress 297**
Mahmut Caliskan

Part 4 Adaptation and Tolerance 321

Chapter 12 **Does Environmentally Contingent Variation in the Level of Molecular Chaperones Mirror a Biochemical Adaptation to Abiotic Stress? 323**
Branka Tucić, Sanja Manitašević Jovanović and Ana Vuleta

Preface

Plants, unlike animals, are sessile. This demands that adverse changes in their environment are quickly recognized, distinguished and responded to with suitable reactions. Drought, heat, cold and salinity are among the major abiotic stresses that adversely affect plant growth and productivity. Abiotic stress is the principal cause of crop yield loss worldwide, reducing normal yields of major food and cash crops by more than 50 percent and thereby causing enormous economic loss as well. Water availability and water use efficiency are among the important abiotic factors that have had and continue to have a decisive influence on plant evolution. Water stress in its broadest sense encompasses both drought and flooding stress. Salinity usually accompanies water stress and may occur concurrently. Drought and salinity are becoming particularly widespread in many regions, and may cause serious salinization of more than 50% of all arable lands by the year 2050. In general, abiotic stress often causes a series of morphological, physiological, biochemical and molecular changes that unfavorably affect plant growth, development and productivity. Drought, salinity, extreme temperatures (cold and heat) and oxidative stress are often interrelated; these conditions singularly or in combination induce cellular damage. These stress stimuli are complex in nature and may induce responses that are equally, if not more, complex in nature. For example severe drought during critical growth phases may directly result in mechanical damage, changes in the synthesis of macromolecules, and low osmotic potential in the cellular settings. In addition it should be noted that almost all of these abiotic stresses lead to oxidative stress and involve the formation of reactive oxygen species (ROS) in plant cells. Usually, plants have mechanisms to reduce their oxidative damage by the activation of antioxidant enzymes and the accumulation of compatible solutes that effectively scavenge ROS. However, if the production of activated oxygen exceeds the plant's capacity to detoxify it, deleterious degenerative reactions do occur, the typical symptoms being loss of osmotic responsiveness, wilting and necrosis. Therefore, it is the balance between the production and the scavenging of activated oxygen that is critical to the maintenance of active growth and metabolism of the plant and overall environmental stress tolerance.

There has been considerable progress in the area of abiotic stress research, especially in the direction of producing improved crop varieties that counter these stresses effectively. Plant engineering strategies for abiotic stress tolerance has been focused largely on the expression of genes that are involved in osmolyte biosynthesis (glycine betaine, mannitol, proline, trehalose etc.); genes encoding enzymes for scavenging ROS (super oxide dismutase (SOD), glutathione S- transferase, Glutathione reductase, glyoxylases etc); genes encoding late embryogenesis protein (LEA) (LEA, HVA1, LE25, Dehydrin etc); genes encoding heterologous enzymes with different temperature optima; genes for molecular chaperons (Heat Shock Proteins (HSPs));genes encoding transcription factors (DREB 1A,CBF 1, Alfin 1); engineering of cell membranes; proteins involved in ion homeostasis. These aspects have undoubtedly opened up the avenue to produce transgenics with improved tolerance. To cope with abiotic stresses it is of paramount significance to understand plant responses to abiotic stresses that disturb the homeostatic equilibrium at cellular and molecular level in order to identify a common mechanism for multiple stress tolerance. A very crucial and highly productive role is envisaged here for biotechnology in identifying metabolic alterations and stress signaling pathways, metabolites and the genes controlling these tolerance responses to stresses and in engineering and breeding more efficient and better adapted new crop cultivars.

This book is broadly divided into sections on signaling in abiotic stress, nucleic acids, proteins and enzymes, genes and genomes and adaptation and tolerance. It focuses on in depth molecular mechanism of abiotic stress effects on plants. In addition, insights from the genomics area are highlighted in one of the chapters of the book. Of special significance in the book is the comprehensive state of the art understanding of stress and its relationship with cyclic nucleotides in plants.

This multi authored edited compilation attempts to put forth an all-inclusive biochemical and molecular picture in a systems approach wherein mechanism and adaptation aspects of abiotic stress will be dealt with. The chief objective of the book hence is to deliver state of the art information for comprehending the effects of abiotic stress in plants at the cellular level. Our attempt here was to put forth a thoughtful mixture of viewpoints which would be useful to workers in all areas of plant sciences. We trust that the material covered in this book will be valuable in building strategies to counter abiotic stress in plants.

Arun K. Shanker and B. Venkateswarlu
Central Research Institute for Dryland Agriculture (CRIDA)
Indian Council of Agricultural Research (ICAR),
Santoshnagar, Saidabad P.O, Hyderabad - 500 059
Andhra Pradesh,
India

Part 1

Signalling in Abiotic Stress

1

Abiotic and Biotic Stress Response Crosstalk in Plants

Saúl Fraire-Velázquez[1], Raúl Rodríguez-Guerra[2] and Lenin Sánchez-Calderón[1]
[1]*Unidad de Biología Experimental, Universidad Autónoma de Zacatecas Av. Revolución S/N, Guadalupe, Zacatecas*
[2]*INIFAP Campo Experimental General Terán. Carretera Montemorelos-China km. 31 General Terán, Nuevo León*
[1,2]*México*

1. Introduction

In the course of its evolution, plants have developed mechanisms to cope with and adapt to different types of abiotic and biotic stress imposed by the frequently adverse environment. The biology of a cell or cells in tissues is so complicated that with any given stimulus from the environment, multiple pathways of cellular signaling that have complex interactions or crosstalk are activated; these interactions probably evolved as mechanisms to enable the live systems to respond to stress with minimal and appropriate biological processes. The sensing of biotic and abiotic stress induces signaling cascades that activate ion channels, kinase cascades, production of reactive oxygen species (ROS), accumulation of hormones such as salicylic acid (SA), ethylene (ET), jasmonic acid (JA) and abscisic acid (ABA). These signals ultimately induce expression of specific sub-sets of defense genes that lead to the assembly of the overall defense reaction.

In plants, defense response genes are transcriptionally activated by different forms of environmental stress or by pathogens. The induction of expression of defense genes in the response against certain pathogens is further dependent on temperature and humidity, suggesting the existence of a complex signaling network that allows the plant to recognize and protect itself against pathogens and environmental stress. A body of research has shown that calcium and reactive oxygen species are second messengers in the early response to abiotic and biotic stress. For example, cytosolic calcium (Ca^{2+}) levels increase in plant cells in response to various harsh environmental conditions, including pathogen attack, osmotic stress, water stress, cold and wounding. After the increase of Ca^{2+} concentration in the intracellular space, several simultaneous pathways are activated by calcium-interacting proteins such as Ca^{2+}-dependent protein kinases (CDPKs), calmodulin and calcineurin B-like proteins (CBLs), all proteins with the structural 'EF-hand' calcium-binding motif.

It is also known that plants respond with an oxidative burst to avirulent microbial intruders or to the previously mentioned abiotic stress factors. In this response, NADPH oxidases generate O_2^- that is rapidly converted to H_2O_2. Recent evidence demonstrated that the NADPH oxidases are activated by Ca^{2+} signatures. ROS are generated by NADPH oxidases in the plasma membrane and increase in concentration in the cytoplasm; these species are also formed in mitochondrion and chloroplast.

The intricate and finely tuned molecular mechanisms activated in plants in response to abiotic and biotic environmental factors are not well understood, and less is known about the integrative signals and convergence points in different sets of partially overlapping reactions. It is now recognized that crosstalk between the second messengers Ca^{2+} and ROS modulates the activity of specific proteins that act at the nuclear level to control the expression of determinate defense genes. Recent studies exploring molecular players have identified and characterized several new genes, including kinases and transcription factors, that are involved in the crosstalk between signaling cascades involved in the responses against two or more types of stress.

Phytohormones also play central roles in abiotic and biotic stress signaling. SA, JA and ET have central roles in biotic stress signaling. ABA is involved in the response to abiotic stress as low temperature drought and osmotic stress. ABA appears to function as a negative regulator in disease resistance, in opposite action to SA, ET and JA. Several transcription factors including AtMYC2, BOS1 and RD26 are mediators in multiple hormone signaling pathways.

In our recent studies of a *Phaseolus vulgaris/Colletotrichum lindemuthianum* pathosystem, genes such as *SUMO* (Small Ubiquitin-like MOdifier) and a calcium-binding like protein (*CaM*) were induced to different levels during the time course of the response to avirulent pathogen inoculation, ultraviolet (A-B) light or extreme temperatures. These findings indicate that these two molecules should be included in the category of integrative signals in abiotic and biotic stress response in plants.

Other well known players in plant response to abiotic and biotic stress are members of the WRKY transcription factor family. Expression patterns of *VvWRKY11*, *AtWRKY39* and *AtWRKY53* genes indicate that protein products of these genes are co-regulators of the plant response against pathogens, hydric stress and heat stress. In addition, some WRKY transcription factors (OsWRKY24 and OsWRKY45) antagonize ABA function by repression of ABA-inducible promoters, indicating that these molecules operate with versatile capabilities. Clearly, the signaling components in plant responses to different abiotic and biotic stress often overlap. Commonly the activated signaling cascades act via synergistic and antagonistic actions.

Powerful molecular tools, including transcriptome and proteome analysis, sequencing of entire genomes in plants, bioinformatic analysis and functional studies, are enabling the disection of networks and identification of key factors in abiotic and biotic signaling cascade crosstalk, and will reveal novel interplays between parallel signaling pathways in the plant responses to pathogens and abiotic stress.

2. Calcium (Ca^{2+}) and reactive oxygen species (ROS) as second messengers common to abiotic and biotic stress responses

In plants, Ca^{2+} and ROS constitute important and common signaling molecules in the early response to abiotic and biotic stress. Levels of Ca^{2+} and ROS rapidly increase in cells of local tissue soon after pathogen attack or stress exerted by environmental conditions. Calcium is perhaps the main signal transducer in the signaling cascades activated in plant response to any stimulus or stress, and the ubiquitous characteristic of this molecule in stress signaling justifies the role of the Ca^{2+} cation as an important node at which crosstalk between pathways can occur. Cytosolic Ca^{2+} levels increase in plant cells in response to various harsh environmental conditions, including pathogen challenge, osmotic stress, water stress, cold and wounding (Dey et al., 2010; Takahashi et al., 2011). For example, plant Ca^{2+} signals are involved in an array of intracellular signaling pathways after pest invasion. Upon herbivore feeding there is a dramatic Ca^{2+} influx, followed by the activation of Ca^{2+}-dependent signal transduction pathways that include interacting downstream networks of kinases (Arimura and Maffei, 2010).

In the last three decades, it has become clear that Ca^{2+} is a universal message transducer that acts on sub-cellular and spatio-temporal patterns of accumulation and protein interaction. Ca^{2+} influx through membrane Ca^{2+} ion channels or carriers yields specific spatial and temporal sub-cellular calcium ion elevations (Errakhi et al., 2008). These signals are then transduced downstream through several simultaneous pathways by calcium-interacting proteins such as CDPKs and CBLs; these Ca^{2+}-binding proteins all contain the 'EF-hand' calcium-binding motif (Kim et al., 2009). An example of Ca^{2+} concentration signatures related to specific signaling pathways is observed in tobacco stressed by wounding: Three calmodulin (CaM) isoforms (wound-inducible type I, hypersensitive response-inducible type III, and constitutive type II) are enabled at different cytosolic Ca^{2+} concentrations to activate the target enzymes NO synthase and NAD kinase (Karita et al., 2004).

There is ample evidence that ROS are also crucial second messengers involved in the response to diverse abiotic and biotic forms of stress. An oxidative burst takes place in response to avirulent microbial intruders (Lamb and Dixon, 1997) or to the previously mentioned abiotic stress factors including heat (Wahid et al., 2007), cold (Kwon et al., 2007), drought, salinity (Miller et al., 2010) and others. ROS production in plants by plasma membrane NADPH oxidases and apoplastic oxidases following pathogen recognition is well documented process (Allan and Fluhr, 1997; Lamb and Dixon, 1997; Bolwell et al., 2002; Torres et al., 2006; Galletti et al., 2008). Indeed, in plants a positive feedback mechanism involving NADPH oxidase, ROS and Ca^{2+} has been reported. Reduced levels of ROS stimulate Ca^{2+} influx into the cytoplasm and Ca^{2+} in turn activates NADPH oxidase to produce ROS (Takeda et al., 2008). Plant NADPH oxidases generate O_2^- that is converted to H_2O_2 by superoxide dismutase (SOD) and the peroxide diffuses through the cell wall to the extracellular medium and enters into the cell (Hammond-Kosack and Jones, 1996).

Reactive oxygen species are usually generated by NADPH oxidases in the plasma membrane, but in tobacco cells in response to abiotic stress as cadmium heavy metal, the anion superoxide is generated in mitochondria (Garnier et al., 2006). Mitochondria also serve as the site of ROS production upon abiotic stress exerted by copper in the marine alga *Ulva compressa* (Gonzalez et al., 2011). The NADPH oxidase is a multicomponent complex known as respiratory burst oxidase (RBO), initially described in mammals (Lambeth, 2004). The RBO enzymatic subunit is the transmembrane gp91^{phox} protein that transfers electrons to molecular oxygen to generate superoxide (Lherminier et al., 2009). In *Arabidopsis thaliana*, ten gp91phox homologs have been reported (Torres and Dangl, 2005). It has been shown that members of the Rboh family mediate the ROS production in defense responses to microorganisms, as well as in response to wounding or mechanical stress (Yoshioka et al., 2003; Torres and Dangl, 2005). In *Arabidopsis*, the NADPH oxidase *AtrbohD*, which contains two EF-hand calcium binding motifs, is synergistically activated by Ca^{2+} and phosphorylation. Phosphorylation levels are correlated with ROS production (Ogasawara et al., 2008).

In the early signaling pathways in the plant defense response to pathogens, the opening of Ca^{2+}-associated of plasma membrane anion channels concomitant with the reactive oxygen species potential response have been described (Jurkowski et al., 2004; Dey et al., 2010). Crosstalk between these two signals in the plant response to abiotic stress has also been reported. In pea plants, the cellular response to long-term cadminum exposure consists of crosstalk between Ca^{2+}- and ROS- activated pathways and signaling mediated by nitric oxide (NO) (Rodriguez-Serrano et al., 2009). In roots in *Arabidopsis thaliana*, mechanical stimulation triggers rapid and transient cytoplasmic Ca^{2+} concentration increases; this mechanical stimulation likewise elicites apoplastic ROS production with the same kinetics (Monshausen et al., 2009). Certainly, the ROS (specifically H_2O_2) production in a Ca^{2+}-dependent manner and then the Ca^{2+}concentration regulation in cytoplasm by ROS through the activation of Ca^{2+} channels in the plasma membrane have been established (Takeda et al., 2008; Mazars et al., 2010).

The co-occurrence and the levels of the induction of Ca^{2+} and ROS signatures vary greatly and is dependent on pathosystem and environmental situation. For example, in callose deposition in *Arabidopsis* in response to the flagelin epitope Flg22 and the polysaccharide chitosan, environmental variability that imposes differential growth conditions is correlated with levels of hydrogen peroxide production. This demonstrates that callose deposition is a multifaceted response controlled by multiple signaling pathways, depending of the environmental conditions and the challenging pathogen-associated molecular pattern (Luna et al., 2011). In another example, pharmacological studies indicate that acclimation to low temperatures requires Ca^{2+} influx across the plasma membrane and a transient increase of Ca^{2+} in the cytoplasm (White and Broadley, 2003), and in *Arabidopsis* mesophyll cells, cold transiently activates Ca^{2+}-permeable channels (Carpaneto et al., 2007). The plant response to low temperature stress also includes production of reactive oxygen species (Heidarvand and Amiri, 2010).

Taking in account the aforementioned antecedents it is clear that responses to two or more forms of stress (biotic or abiotic) may overlap or converge in a common signaling element, for instance, Ca^{2+} or ROS or both, leading to similar downstream events. Calcium and ROS are ubiquitous second messengers in the abiotic and biotic stress signaling pathways and are in variable ways interconnected elements. There is strong evidence that Ca^{2+}-dependent ROS production through respiratory burst oxidase homolog (RBOH) enzyme activation is the first link. Induction of Ca^{2+} plasma membrane channels through the increase of cytoplasmic ROS is a second connection. Although these signals co-occur, their magnitudes, spatial location and timing depend on the biological system. The fine signatures in Ca^{2+} and the recently introduced concept of signatures in ROS (sub-cellular and spatiotemporal patterns of ROS) (Mazars et al., 2010) explain the downstream signaling independence that results in unique molecular responses in plant systems to the environment constraints with specific and adaptive responses.

3. Calcium-dependent protein kinases (CDPKs) and mitogen-activated protein kinases (MAPKs) crosstalk in response to abiotic and biotic stress

The transient changes in cytosolic calcium content with their diverse spatio-temporal signatures observed under biotic or abiotic stress conditions require different calcium sensors. A larger and defined group of calcium sensors are the calcium-dependent protein kinases (CDPKs) which in turn have many different substrates. CDPKs possess a carboxyterminal calmodulin-like domain containing EF-hand calcium-binding sites plus a N-terminal protein kinase domain (Cheng et al., 2002). Thus, the signaling pathways activated in response to stress stand in part on CDPKs. The *Arabidopsis* genome encodes 34 CDPKs, but few substrates of these enzymes have been identified (Uno et al., 2009). Mitogen-activated protein kinases (MAPKs) are a family of Ser/Thr protein kinases widely conserved among eukaryotes. Them respond to extracellular stimuli and regulate various cellular activities, such as gene expression, mitosis, differentiation, proliferation, and cell survival/apoptosis. They work downstream of sensors/receptors and transmit extracellular stimuli into intracellular responses and at the same time amplifying the transducing signal (Ichimura et al., 2002). Amplification is accomplished by a MPK cascade of three hierarchically arranged, interacting types of kinases. MPK activity is induced upon phosphorylation by MPK kinases (MPKKs, MAPKKs, or MEKs), which are in turn phosphorylation activated by MPKK kinases (MPKKKs, MAPKKKs, or MEKKs). In *Arabidopsis*, there are 20 MPKs, 10 MPKKs, and 80 MPKKKs (Colcombet and Hirt, 2008). MAPKs act as last component in a protein kinase cascade, and one of their major tasks is to transducer an extracellular stimulus into a transcriptional response in the nucleus (Wurzinger et al., 2010).

In eukaryotes, CDPKs together with MAPKs are two signaling cascades widely activated in response to changing environmental abiotic and biotic stresses. In several pathosystems both cascades could be activated in response to the same stressing factor suggesting a crosstalk between those pathways (Wurzinger et al., 2010), or a specific CDPK or MAPK could be induced or activated in response to different biotic and abiotic stresses. Several studies in *Arabidopsis* demonstrate that: a) upon challenge exposure to biotic (bacterial pathogens) or abiotic (BTH, SA, and 4-chloro-SA) stress, MPK3 and MPK6 are activated and their respective mRNAs accumulate (Gerold et al., 2009); b) MKK2 is a key regulator of the cold- and salt-stress response (Teige et al., 2004) but, it was similarly involved in disease resistance to *Pseudomonas syringae* (Brader et al., 2007); c) the actived MKK9 protein in transgenic plants, induces the synthesis of ethylene and camalexin through the activation of the endogenous MPK3 and MPK6 kinases, moreover enhances the sensitivity to salt stress (Xu et al., 2008). In other hand, CDPKs CPK6 and CPK3 operate in ABA regulation of guard cell S-type anion- and Ca2+- permeable channels and stomatal closure (Mori et al., 2006), but besides its well-established role in abiotic stress adaptation, recent results in rice plants indicate that ABA is also involved in the regulation of pathogen defense responses, and mediates the repression of pathogen-induced ethylene signaling pathway in an MPK5-dependent manner (De Vleesschauwer et al., 2010).

From the accumulated data, the biological significance of crosstalk among signaling pathways under stress conditions that operate by CDPKs alone or together with MAPKs and viceversa, demonstrate that these two groups of calcium-dependent enzymes and the mitogen-activated protein kinases are involved in signaling pathways that in plants, in some cases signify the establishment of cellular mechanisms that lead to the simultaneous reinforcement of the defense responses to pathogens as well to other forms of abiotic stress. We are just to begin to uncover convergence points that command the crosstalk between these signaling pathways under various stress conditions.

4. Genetic pathways crosstalk in response to abiotic and biotic stress

A body of research demonstrates that plant defense response genes are transcriptionally activated by pathogens, as well by different forms of abiotic stress, or even more, the induction of specific defense genes in the response against certain pathogens, are dependent on specific environmental conditions, suggesting the existence of a complex signaling network that allows the plant to recognize and protect itself against pathogens and environmental stress. Similar induction patterns of members of the *14.3.3* gene family (*GF14b* and *GF14c*) by abiotic and biotic stresses such as salinity, drought, ABA and fungal inoculation have been documented in rice. The rice *GF14* genes contain *cis*-elements in their promoter regions that are responsive to abiotic stress and pathogen attack. The *14-3-3s* family genes are also subject to the regulation by certain transcript factors (Chen et al., 2006). In rice, the *RO-292* gene is up-regulated in roots by salt or drought stresses and by blast fungus infection (Hashimoto et al., 2004). Similarly, the *Mlo* gene in barley (*Hordeum vulgare*) act as modulators of defense and cell death in response to *Blumeria graminis* f. sp. *tritici* or *Magnaporte grisea* inoculation, and to wounding or the herbicide paraquat (Piffanelli et al., 2002). In *Arabidopsis*, at least five of the 29 *cytochrome P450* genes are induced by abiotic and biotic stress including *Alternaria brassicicola* or *Alternaria alternata*, paraquat, rose bengal, UV stress (UV-C), heavy metal stress (CuSO4), mechanical wounding, drought, high salinity, low temperature or hormones (salicylic acid, jasmonic acid, ethylene and abscisic acid). These five *cytochrome P450* genes (*CYP81D11*, *CYP710A1*, *CYP81D8*, *Cyp71B6* and *CYP76C2*) are co-induced by metal stress ($CuSO_4$), paraquat, salinity, ABA and pathogen inoculation. A common characteristic shared by all of these induced genes, as in the 14.3.3 genes family, is the presence of *cis*-acting elements in regulatory regions of the gene; W-box (DNA binding sites for WRKY transcription factors), P-box (a positive *cis*-acting regulator of pathogen defense) and MYB recognition sites are common (Narusaka et al., 2004). A collection of genes, including transcription factors are co-activated by pathogen challenge and abiotic stress, examples of these genes mediating crosstalk between signaling pathways for biotic and abiotic stress responses are *DEAR1*, *BOS1* and *SlERF5*. *DEAR1* is a transcriptional repressor of DREB protein that mediates plant defense and freezing stress responses in *Arabidopsis*; the *DEAR1* mRNA accumulates in response to both pathogen infection (*Pseudomonas syringae*) and cold treatment (Tsutsui et al., 2009). *BOS1* codes for a R2R3MYB protein that acts as transcription factor that in *Arabidopsis* regulates responses to *Botrytis cinerea* infection and to water deficit, increased salinity and oxidative stress (Mengiste et al., 2003). *SlERF5* is highly expressed in response to the harpin protein coded in the *hrp* gene clusters in many Gram-negative phytopathogens; the over-expression of *SlERF5* is involved in the induction of the dehydration-responsive genes through the ABA-mediated abiotic stress response (Chuang et al., 2010).

Studies in our laboratory in common bean (*Phaseolus vulgaris*) leaves detached, inoculated with fungal pathogen, and maintained in humid chamber demonstrate that *chalcone synthase* (*CHS*), a gene implicated in the biosynthesis of phytoalexins in response to pathogen challenge (Ferrer et al., 1999), is also responsive to wounding at early times after stress. As shown in Figure 1, *CHS* mRNA is detected 6 hours post-wounding of leaves or at latter times post-inoculation with *Colletotrichum lindemuthianum*; the mRNA disappears by 12 hours post-wounding stress. In plants, following exposure to environmental stresses including pathogen attack and wounding, the phenylpropanoid pathway has important functions in the production of compounds including lignin, flavonoids and phytoalexins. Chalcone synthase (CHS) is a key enzyme in this pathway, catalyzing the first step in flavonoid biosynthesis, whose expression can be induced in response to environmental stress (Richard et al., 2000). This evidence exhibits the importance of molecular events in downstream levels from the initial key factors (transcription factors), where secondary genes as chalcone synthase and phenylalanine ammonia-lyase (PAL) are able also to respond to abiotic and biotic stress, and are committed to achieve the relevant functions of biosynthesis of compounds with more direct actions toward microorganisms intruders, through phytoalexins; or the reinforcement of the cell wall with lignin, a macromolecule composed of highly cross-linked phenolic molecules, as a major component of secondary walls.

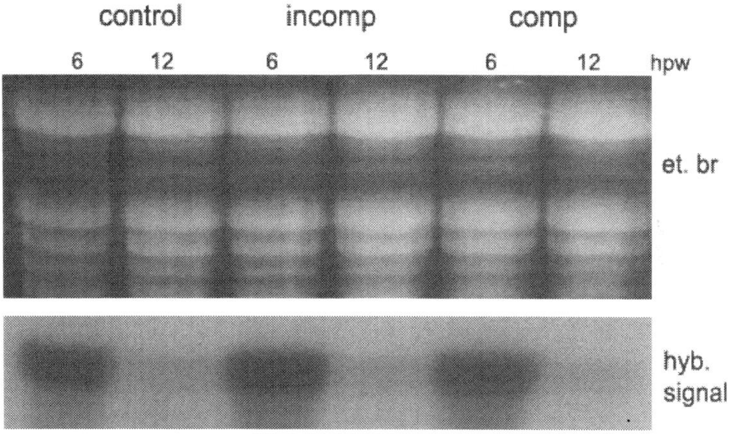

Fig. 1. *Chalcone synthase* mRNA levels in *Phaseolus vulgaris* leaves 6 and 12 hours after wounding besides pathogen interaction. Other leaves were inoculated with conidial suspensions of avirulent (incompatible) and virulent (compatible) *Colletotrichum lindemuthianum* pathotypes and total RNA was isolated after 6 and 12 hours. In the upper panel, total RNA (14 µg per lane) was stained with ethidium bromide; in bottom panel, the hybridization signal with a radiolabeled chalcone synthase cDNA probe is shown.

The signaling pathways in plants in response to microorganism intruders and to wound could be with a relevant level of crosstalk. In both cases, cytoplasmic Ca^{2+} increase and the reactive oxygen species production occur (Jurkowski et al., 2004; Karita et al., 2004; Dey et al., 2010), moreover the induction of *WRKY* and pathogenesis related (*PR*) gene expression (Leon et al., 2001; Takemoto et al., 2003; Huang et al., 2010). The level of crosstalk between different genetic pathways in the plant response to abiotic and biotic stress often vary, as expected, in accordance with the specificity of the stressors. On the biotic side, the response depends on the pathogen identity; on the abiotic side, it depends on the level of the stressing factor and the general environmental conditions. The commonality between different genetic pathways vary greater in relation with the species and the genotype in the plant species. In chickpea, the batteries of expressed genes identified in response to high salinity, drought, cold or pathogen inoculation show marked differential coincidences. It was found that the genes up-regulated in response to pathogens were more similar to these induced by high salinity than those up-regulated in response to cold or drought conditions. In 51 transcripts differentially expressed in plants inoculated with pathogen, 21 were common among *Ascochyta rabiei* inoculation and one or more of the other three abiotic conditions. It is noteworthy that no transcript was commonly differentially expressed across all the four stresses assessed. Conversely, other sets of genes were found to be specifically induced by only one treatment, indicating the existence of specific signaling routes in addition to shared pathways (Mantri et al., 2010). A similar convergence of signaling pathways was reported for systemin, oligosaccharide elicitors and UV-B radiation at the level of mitogen-activated protein kinases (MAPKs) in *Lycopersicon peruvianum* suspension-cultured cells. LeMPK1 and LeMPK2, were activated in response to systemin, four different oligosaccharide elicitors, and UV-B radiation, whereas LeMPK3, was only activated by UV-B radiation. The common activation of LeMPK1 and LeMPK2 by many stress signals is consistent with a substantial overlap among stress responses; while UV-B induces specific responses (Holley et al., 2003). In our studies, in a *Phaseolus vulgaris/Colletotrichum lindemuthianum* pathosystem, the *SUMO* gene and the *EF-hand calcium-binding protein* gene were responsive to pathogen as well to the abiotic stresses UV light (UV-A and UV-B), and extreme temperatures (8° and 38°C). These two genes are induced to different levels by UV light and extreme temperatures conditions. The highest expression for the *SUMO* mRNA upon UV treatment was lower than of the *EF-hand calcium-binding protein* mRNA: After 4 hours of heat (38°C) treatment, the *EF-hand calcium-binding protein* mRNA levels surpass the *SUMO* mRNA levels (Fig. 2) (Alvarado-Gutiérrez et al., 2008). Thus, clearly the levels of individual defense genes are differentially regulated transcriptionally by abiotic and biotic forms of stress. In relation to *SUMO*, five WRKY transcription factors are SUMO1 targets (WRKY3, WRKY4, WRKY6, WRKY33, WRKY72); many WRKY transcription factors are commonly involved in plant defense reaction to pathogens, moreover several forms of abiotic stresses. Therefore, resistance protein signaling and SUMO conjugation also converge at transcription complexes. It is known that SUMO conjugation is essential to suppress defense signaling in non-infected plants, and recently was suggested a model in which SUMO conjugation can transform transcription activators into repressors, thereby preventing defense induction in the absence of a pathogen (Burg and Takken, 2010).

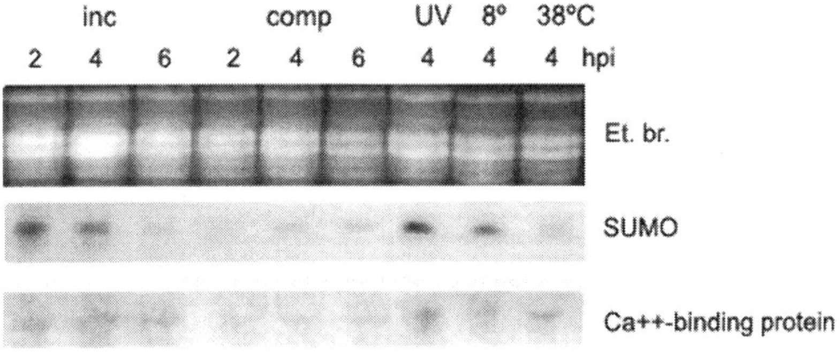

Fig. 2. *SUMO* and *EF-hand calcium-binding protein* mRNA levels in common bean plants after fungal infection or treatment with abiotic stresses UV light or extreme temperatures. Shown are northern blot assays of 12 µg of total RNA each lane. Radiolabeled probes for the two mRNAs were used. In the upper panel, total RNA was stained with ethidium bromide; in middle and bottom panels, signals for each gene are shown. Similar results were obtained from three independent experiments. Figure from previously reported results (Alvarado-Gutiérrez et al., 2008).

A complexity of the stress response in plants is evident when it is considered the natural fluctuating environmental conditions within a day or over longer periods of time. In the environment, changing states in light intensities, temperatures and pressures exerted by wind are normal. The dynamism inherent to factors that compose the environment impacts in changes in the profile of expression of some plant defense genes. As previously we reported, the *SUMO* and the *EF-hand calcium-binding protein* genes in the plant-pathogen interaction exhibit similar kinetics in the dark period, but not in the light period. For the *EF-hand calcium-binding protein* gene, the transcript levels in light in the control treated (H_2O sprinkled) leaves surpass those in the pathogen-treated leaves (Fig. 3) (Alvarado-Gutiérrez et al., 2008). Thus, these two genes, which are co-induced by two or more types of biotic and abiotic stresses, are also differentially regulated by the daily photoperiod advance and possibly by the circadian rhythm. These findings indicate that these two molecules should be included in the category of integrative signals in abiotic and biotic stress response in plants.

A number of *Arabidopsis thaliana* lesion-mimic mutants that show alterations in the responses to abiotic and biotic stresses have been reported. One class of these mutants exhibits constitutively increased *PR* gene expression, SA levels and heightened resistance to pathogen infection (Yoshioka et al., 2001; Jambunathan and McNellis, 2003; Jurkowski et al., 2004; Mosher et al., 2010); this class includes the *cpr22* mutant, which has mutations in two cyclic nucleotide-gated ion channels that impart the phenotype of spontaneous lesion formation, SA accumulation, constitutive *PR-1*, *PR-2* and *PR-5*gene expression and enhanced resistance to various pathogens (Mosher et al., 2010). Noteworthy, in the aforementioned mutants, the phenotypes exhibited are suppressed under high relative humidity and high temperature and are enhanced by low humidity and cold temperatures (Yoshioka et al., 2001; Mosher et al., 2010). Similarly, the effects on basal and resistance (*R*) gene-mediated resistance in *A. thaliana* and *Nicotiana benthamiana*/*Pseudomonas syringae* pathosystems are reduced at moderately elevated temperatures (Wang et al., 2009). In accordance with this data, a number of mutants in plants with de-regulated expression of R proteins have been shown temperature-dependent defense responses (Alcazar et al., 2009; Huang et al., 2010; Zhou et al., 2010). These data indicates that in these mutants, the resistance phenotypes are dependent on environmental conditions or that, at least, there are humidity and temperature sensitive steps (Mosher et al., 2010). Indeed, the resistance response mediated by *R* genes as well the basal resistance is attenuated when the temperature increases.

Collectively, these data suggest that specific batteries of defense genes are involved in different signaling cascades that converge with a degree of overlap in the response programs for pathogen defense and abiotic stress protection. There is a balanced interplay with fine-tuning between parallel signaling branches by different sets of partially overlapping reactions. Moreover, the genes that are the convergence points between different genetic pathways are differentially regulated, more evidently, when these genes are analyzed in the time scale, and definitely, the genetic pathways activated by *R* genes are modulated in variable levels by environmental factors. There are common factors in the defense signaling pathways to abiotic (humidity and temperatures variable conditions) and biotic (pathogen infection) stresses. These convergence points expose the superimposed complexity levels in the response to environmental changes. A pending task is the deciphering of the specificity of the signal transduction processes that conduit to the establishment of the commonality among different stress responses.

5. Phytohormones have central roles in abiotic and biotic stress signaling

Plant hormones, also called phytohormones, were first defined as "a substance which, being produced in any one part of the organism, is transferred to another part and there influences a specific physiological process" in the classical book *Phytohormones* written by Frits Went and Kenneth in 1937. The five classical phytohormones: auxin, cytokinin, ET, gibberellins, ABA and the recently identified brassinosteroids, JA and SA, are chemical messengers present in trace quantities; their synthesis and accumulation are tightly regulated. Depending on the context, they are subject to positive or negative feedback control and often are affected by crosstalk due to environmental inputs. Phytohormones move throughout the plant body via the xylem or phloem transport stream, move short distances between cells or are maintained in their site of synthesis to exert their influence on target cells where they bind transmembrane receptors located at the plasma membrane or endoplasmic reticulum or interact with intracellular receptors. The downstream effects of hormonal signaling include alterations in gene expression patterns and in some cases non-genomic responses. Changes in plant hormones concentrations and tissue sensitivity to them regulate a whole range of physiological process that have profound effects on growth and development. The phytohormomes affect all phases of the plant life cycle and their responses to environmental stresses, both biotic and abiotic. Hormonal signalling is critical for plant defenses against abiotic and biotic stresses (Crozier et al., 2000; Taiz and Zeiger, 2010; Williams, 2010).

Typically the phytohormones that regulate the responses against adverse cues are grouped into two types: those that play a major role in response to biotic stress (ET, JA and SA) and those that have pivotal roles regulating the abiotic stress responses (mainly ABA). Commonly the biotic defense signaling networks mediated by phytohormones are dependent on the nature of the pathogen and its mode of pathogenicity. SA plays a central role in the activation of defense responses against biotrophic and hemi-biotrophic pathogens as well as the establishment of systemic acquired resistance. By contrast, JA and ET are usually associated with defense against necrotrophic pathogens and herbivorous insects. Concerning to abiotic stress, ABA is the most studied stress-responsive hormone; it is involved in the responses to drought, osmotic and cold stress (Peleg and Blumwald, 2011; Wasilewska et al., 2008; Bari and Jones, 2009; Vlot et al., 2009).

5.1.2 Salicylic acid, ethylene, jasmonic acid and abscisic acid: are they working alone?

In addition to roles in activation of defense responses against biotrophic and hemi-biotrophic pathogens, SA is also important to the establishment of systemic acquired resistance (SAR) (Grant and Lamb, 2006; Vlot et al., 2009). When resistant tobacco and cucumber plants are inoculated with pathogens, the levels of SA increase (Malamy et al., 1990; Rasmussen et al., 1991). Exogenous applications of this chemical messenger result in the induction of *PR* genes increasing resistance to a broad range of pathogens (Vlot et al., 2009). In addition, transgenic plants and mutants of tobacco and *Arabidopsis* in which endogenous SA levels are reduced, fail to develop SAR or express *PR* genes; instead, they displayed heightened susceptibility to both virulent and avirulent pathogens. When these plants are treated with the SA synthetic analog, 2,6-dichloro-isonicotinic acid, resistance and *PR* genes expression are restored (Gaffney et al., 1993; Delaney et al., 1994; Vernooij et al., 1995.; Nawrath and Métraux, 1999; Nawrath et al., 2002; Genger et al., 2008; Vlot et al., 2009).

By contrast, over-expression of bacterial SA biosynthetic genes in transgenic tobacco confers highly elevated SA levels, *PR* gene expression, and enhanced resistance (Verberne et al., 2000). The SAR is induced systemically by a signal generated in the inoculated leaf; this signal is transmitted via the phloem to the uninfected portions of the plant (Grant and Lamb, 2006; Parker, 2009). SA levels rise coincidently with or just prior to SAR and systemic *PR* gene expression or peroxidase activation in pathogen-infected tobacco or cucumber, also was detected in the phloem of pathogen-infected cucumber and tobacco, and radio-tracer studies suggest that a significant amount of SA in the systemic leaves of pathogen-infected tobacco and cucumber is transported from the inoculated leaf. This was initially proposed to serve as signal in systemic acquired resistance; however, leaf detachment assays show that the mobile signal moves out of the infected leaf before increased SA levels are detected in petiole exudates from that leaf (Malamy et al., 1990; Rasmussen et al., 1991; Vlot et al., 2009). SA can be methylated to form methyl salicylate, in tobacco by the esterase SABP2 (an SA-binding protein). Recently, it has been shown that, methyl salicylate, which is induced upon pathogen infection, acts as an internal plant signal and also as an airborne defense signal (Forouhar et al., 2005; Park et al., 2007).

In plant defense responses against insects and microbial pathogens, JA is a crucial component. In *Arabidopsis* leaves, jasmonates control the expression of an estimated 67-85% of wound- and insect-regulated genes. Treatment of plants with JA results in enhanced resistance to herbivore challenge. Mutants defective in the biosynthesis or perception of JA show compromised resistance to herbivore attackers (Bari and Jones, 2009). Attack of herbivores such as *Manduca sexta* in tobacco induces the JA signaling activity (Paschold et al., 2007). Similarly, JA signaling is induced in tomato and *Arabidopsis* by *Tetranychus urticae* and *Pieris rapae*, respectably (Li et al., 2002; Reymond et al., 2004; De Vos et al., 2005). However, not all herbivores activate JA signaling in plants (Bari and Jones, 2009). The production of proteinase inhibitors (PIs) and other anti-nutritive compounds such as polyphenol oxidase (PPO), threonine deaminase (TD), leucine amino peptidase and acid phosphatase (VSP2) are mediated by JA in order to deter, sicken or kill the attacking insect (Howe and Jander, 2008). Also terpenoids and other volatile compounds produced by an herbivore-attacked plant are recognized by other carnivorous and parasitoid insects. The blends of compounds are specific to the particular plant/herbivore interaction, and the discerning carnivore uses this information to find its favorite meal (Howe and Jander, 2008; Williams, 2011).

5.1.2.1 Phytohormone signaling networks act together

Necrotrophic pathogens include most fungi and oomycetes as well as some bacteria. Defenses to these types of pathogens are often mediated by JA and ET. JA and ethylene operate synergistically to activate the expression of a subset of defense genes following pathogen inoculation in *Arabidopsis* (Thomma et al., 2001; Glazebrook, 2005). Experimental data confirm that JA and ethylene signaling pathways act together. Analysis of the mutants *coi1* (jasmonate insensitive) and *ein2* (ethylene insensitive) revealed that the induction of JA response marker gene *PDF1.2* by *Alternaria brassicicola* requires both JA and ethylene signaling pathways (Penninckx et al., 1998; Thomma et al., 2001). Genes acting as point controls between these two pathways have been described: *CEV1* acts as a negative regulator and *ERF1* (ethylene response factor 1) is a positive regulator (Ellis et al., 2002; Lorenzo et al., 2003).

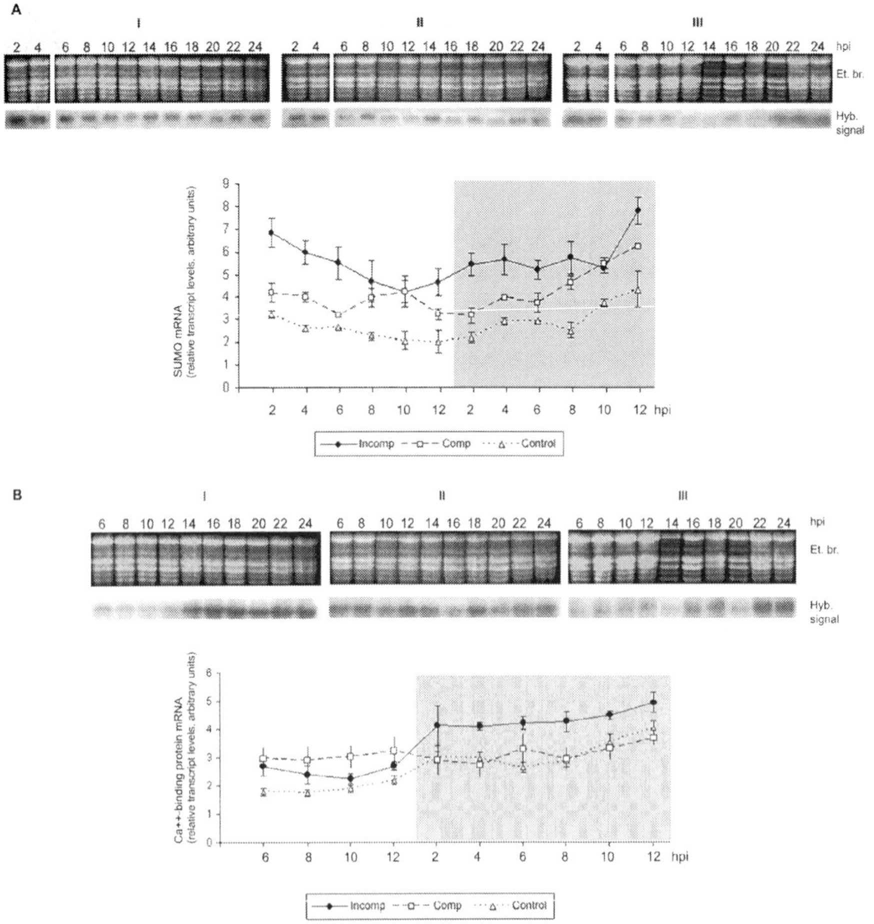

Fig. 3. *SUMO* and *EF-hand calcium-binding protein* mRNA levels in common bean plants infected with fungus through 24 hours with normal light and dark periods. Shown are northern blot assays with 14 µg of total RNA each lane. In panel A, RNA was hybridized with radiolabeled probe for *SUMO* mRNA. In panel B, a radiolabeled probe for the *EF-hand calcium-binding protein* mRNA was used. In A and B, from left to right: I, resistant interaction; II, susceptible interaction; and III, control plants in white background for day (light) period and in gray background for night (dark) period. The level of expression in the plot indicates transcript abundance relative to the 28S rRNA. Values are expressed as means of three independent experiments (± SE). Figure from previously reported results (Alvarado-Gutiérrez et al., 2008).

SA and JA are mutually antagonistic. Mutations that disrupt JA signaling (*coi1*) lead to the enhanced basal and inducible expression of the SA marker gene *PR1*, whereas mutations that disrupt SA signaling (*npr1*) lead to concomitant increases in the basal or induced levels of the JA marker gene *PDF1.2* (Kazan and Manners, 2008). Plants inoculated with virulent strains of *Pseudomonas syringae* pv. tomato treated with SA show compromised resistance to necrotrophic pathogen *Alternaria brassicicola*, which is sensitive to JA-dependent defenses (Spoel et al., 2007). The non-expresser of PR genes 1 (NPR1) is a master regulator of SA signaling. *Arabidopsis npr1* mutants fail in SA-mediated suppression of JA responsive genes suggesting that NPR1 plays an important role in the SA-JA interaction (Spoel et al., 2007). Acting downstream from NPR1, WRKY70, a transcription factor (TF) acts as a positive regulator of SA-dependent defenses and a negative regulator of JA-dependent defenses and plays central role in determining the balance between these two pathways. Suppression of *WRKY70* expression allows increased expression from JA-responsive genes and increased resistance to a pathogen sensitive to JA-dependent defenses. In contrast, over-expression of WRKY70 results in the constitutive expression of SA-responsive *PR* genes and enhanced resistance to SA-sensitive pathogens but reduces resistance to JA-sensitive pathogens (Li et al., 2004). Recently, WRKY6, WRKY53, mitogen activated protein kinase 4 (MPK4) and GRX480 (glutaredoxin) were reported to affect antagonism between SA- and JA-mediated signaling (Petersen et al., 2000; Brodersen et al., 2006; Mao et al., 2007; Miao and Zentgraf, 2007; Ndamukong et al., 2007). As we explained, plant hormone signaling pathways extensively interact during plant defense again pathogens and herbivores. Lifestyles of different pathogens are not often readily classifiable as purely biotrophic or necrotrophic. Therefore, those interacting points or crosstalk between SA and JA/ET pathways may be regulated in a pathogen-specific manner (Adie et al., 2007; Bari and Jones, 2009).

5.1.2.2 Abscisic acid in abiotic and biotic responses cross talk in plants

As sessile organisms, plants often have to cope with multiple environmental stresses; therefore most plants employ complex regulatory mechanisms to trigger effective responses against various biotic and abiotic stresses. In this scenario, phytohormones are the main players regulating these responses. To coordinate the complex interactions, an intense crosstalk among the regulatory networks is necessary. ABA is involved in the regulation of many aspects of plant growth and development and also is the major hormone that controls plant responses to abiotic stresses (Wasilewska et al., 2008).

In the last decade, our understanding of ABA involvement to pathogen susceptibility and its relationship to other phytohormones involved in biotic stress response have increased. Exogeneous ABA treatment increases the susceptibility of various plant species to bacterial and fungal pathogens (Heinfling et al., 1980; McDonald and Cahill, 1999; Thaler and Bostock, 2004; Mohr and Cahill, 2007)(Henfling et al., 1980; McDonald & Cahill., 1999; Mohr & Cahill, 2003; Thaler & Bostock, 2004; Ward et al., 1989). ABA-deficient tomato mutants show a reduction in susceptibility to the necrotroph *Botrytis cinerea* (Audenaert et al., 2002) and virulent isolates of *Pseudomonas syringae* pv tomato DC3000 (Thaler and Bostock, 2004; de Torres-Zabala et al., 2007), and ABA-deficient *Arabidopsis* has reduced susceptibility to the oomycete *Hyaloperonospora parasitica* (Mohr and Cahill, 2003). In general, ABA is involved in the negative regulation of plant defenses against various biotrophic and necrotrophic pathogens. However, the role of ABA appears to be complex and may vary depending on the pathosystem. The role of ABA as a positive regulator of defense has also been reported (Mauch-Mani and Mauch, 2005). ABA activates stomatal closure that acts as a barrier against bacterial infection (Melotto et al., 2006). As a result, ABA-deficient mutants show more susceptibility to *Pseudomonas syringae* pv. tomato. In addition, treatment with ABA protects plants against *Alternaria brassicicola* and *Plectosphaerella cucumerina* indicating that ABA acts as a positive signal for defense against some necrotrophs (Ton and Mauch-Mani, 2004). Pathogen challenge results in the alteration of ABA levels in plants. For example, tobacco plants infected with tobacco mosaic virus (TMV) have increased ABA levels, and treatment with ABA enhances TMV resistance in tobacco (Whenham et al., 1986). Similarly, *Arabidopsis* plants challenged with virulent isolates of *Pseudomonas syringae* pv tomato DC3000, accumulate higher levels of ABA and JA than unchallenged plants (de Torres-Zabala et al., 2007). Additionally, mutants deficient in ABA are more sensitive to infection by the fungal pathogens *Pythium irregulare* (Adie et al., 2007) and *Leptosphaeria maculans* (Kaliff et al., 2007). The situation becomes even more complicated when pathogens are tested on ABA signaling mutants, such as *abi4,* which displays opposite resistance responses towards these two fungi. Along the same line, the mutations *abi1-1* and *abi2-1* actually foster differential resistance responses against *Leptosphaeria maucans* (Kaliff et al., 2007; Wasilewska et al., 2008). Transcriptome and meta analyses of expression profiles altered by infection with the necrotroph *Pythium irregulare* identified many JA-induced genes but also highlighted the importance of ABA as a regulator, as the ABA responsive element (ABRE) appears in the promoters of many of the defense genes (Adie et al., 2007; Wasilewska et al., 2008). This indicates that ABA plays an important role in the activation of plant defense through transcriptional reprogramming of plant cell metabolism. Moreover, ABA is required for JA biosynthesis and the expression of JA responsive genes after *Phytium irregular* infection (Adie et al., 2007). Recently, it has been identified the first molecular component in crosstalk between biotic and abiotic stress, the rice *MAP* gene *OsMPK5*. ABA antagonize pathogen-activated ET signaling via OsMPK5 (De Vleesschauwer et al., 2010). The exact molecular mechanism of ABA action on plant defense responses against diverse pathogens started to be elucidated. Identification of more factors involved in ABA-mediated crosstalk between biotic and abiotic stress signaling merits extensive future study.

6. WRKY and other transcription factors as players in plant response to abiotic and biotic stress

Plant responses to environmental stimuli involve a network of molecular mechanisms that vary depending on the nature of environmental signal. In the signal transduction network that leads from the perception of stress signals to the expression of stress-responsive genes, transcription factors play an essential role. TFs are a group of master proteins that interact with *cis*-elements present in promoter regions upstream of genes and regulate their expression. Most TFs impact multiple physiologic processes such as metabolism, cell cycle progression, growth, development and reproduction (Fujita et al., 2009; Zhou et al., 2010; Hussain et al., 2011). Several transcription factors are mediators of multiple phytohormone signaling networks.

6.1 Transcription factors in crosstalk stress responses

The TFs are involved in responses against biotic and abiotic stress, and they play an esential role in regulation of plant adaptation to environmental changes. A few TFs have been reported to take part in the crosstalk between abiotic and biotic stress signaling nethworks. The basic helix-loop-helix (bHLH) domain-containing transcription factor AtMYC2 is a positive regulator of ABA signaling. The genetic lession of *AtMYC2* results in elevated levels of basal and activated transcription from JA-ethylene responsive defense genes (Abe et al., 2003; Anderson et al., 2004). MYC2 differentially regulates two branches of JA-mediated responses; it positively regulates wound-responsive genes, including *VSP2, LOX3,* and *TAT,* but represses the expression of pathogen-responsive genes such as *PR4, PR1,* and *PDF1.2.* These complex interactions are co-mediated by the ethylene-responsive transcription factor ERF1 (Lorenzo et al., 2003; Lorenzo et al., 2004). The botrytis susceptible 1 (*BOS1*) gene of *Arabidopsis* encodes an R2R3MYB transcription factor that mediates responses to certain signals, possibly through ROS intermediates from both biotic and abiotic stress agents (Mengiste et al., 2003). There are also four members of the *NAC* family of genes that encode plant-specific transcription factors involved in diverse biological processes. *OsNAC6, Arabidopsis transcription activation factor 1* (*ATAF1*), *ATAF2* and *dehydration 26* (*RD26*) are potentially involved in regulation of responses to abiotic and biotic stresses (Wu et al., 2009).

6.2 WRKY transcription factors

WRKY proteins are a recently identified class of DNA-binding proteins that recognize the TTGAC(C/T) W-box elements found in the promoters of a large number of plant defense-related genes (Dong et al., 2003). These TFs contain WRKY domains that appear to be unique to plants (Eulgem and Somssich, 2007). The name of the WRKY family is derived from its highly conserved 60 amino acid long WRKY domain, comprising highly conserved WRKYGQK at N-terminus and a novel metal chelating zinc finger signature at C-terminus. *WRKY* genes thought to be plant-specific TFs that have been subject to a large plant-specific diversification. Phylogenetic analysis shows that the *WRKY* genes are clustered into several different groups on the basis of their amino acid sequences (Yamasaki et al., 2005; Eulgem and Somssich, 2007). *WRKY* genes probably originated concurrently with the major plant phyla.

Current information suggests that WRKY factors play a key role in regulating the pathogen-induced defense program. From the beginning of research into WRKY transcription factors, it was evident that they play roles in regulating several different plant processes. It is common for a single WRKY transcription factor to regulate transcriptional reprogramming associated with multiple plant processes. The dynamic web of signaling in which WRKY factors operate has multiple inputs and outputs (Rushton et al., 2010). It is expected that a single WRKY transcription factor has activity on both abiotic and biotic stress pathways and cross talks with different signal transduction pathways. The rice *WRKY45 (OsWRKY45)* gene expression is markedly induced in response to ABA and various abiotic stress factors such as NaCl, dehydration; in addition expression is induced by pathogens such as *Pyricularia oryzae Cav.* and *Xanthomonas oryzae* pv. oryzae. Moreover, *OsWRKY45*-over-expressing plants exhibited several changes: a) the constitutive expression of ABA-induced responses and abiotic-related stress factors, b) markedly enhanced drought resistance and c) increased expression of *PR* genes and resistance to the bacterial pathogen *Pseudomonas syringae*. Thus, OsWRKY45 shows a dual role, acting as a regulator and as a protective molecule upon water deficit and pathogen attack (Qiu and Yu, 2009). *VvWRKY11* from *Vitis vinnifera* is a nuclear protein that is expressed rapidly and transiently in response to treatment with SA or pathogen *Plasmopara viticola.* Transgenic *Arabidopsis* seedlings over-expressing *VvWRKY11* have higher tolerance to water stress induced by mannitol than wild-type plants. These results demonstrate that the *VvWRKY11* gene is involved in the response to dehydration and biotic stress (Liu et al., 2011). Other well known players in plant responses to abiotic and biotic stresses are members of the WRKY transcription factor family. Expression patterns of *VvWRKY11, AtWRKY39* and *AtWRKY53* indicate that these genes are co-regulator of the plant response against pathogens and hydric and heat stress. In addition, some WRKY transcription factors (OsWRKY24 and OsWRKY45) antagonize ABA function, repressing an ABA-inducible promoter, indicating that these molecules operate with versatile capabilities.

7. Conclusion

Crop growth and crop yield are affected by environmental cues. There is a need of greater understanding of plant physiological responses to the abiotic and biotic stresses. We can understand stress as a stimulus or influence that is outside the normal range of homeostatic control in a given organism: If a stress tolerance is exceeded, mechanisms are activated at molecular, biochemical, physiological and morphological levels; once stress is controlled, a new physiological state is established, and homeostasis is re-established. When the stress is retired the plant may return to the original state or a new physiological state.

Plants continually encounter stress even under environmental conditions that we think of as normal. The environment changes during the day, day to day and throughout the year, thus plants must respond to stress over the course of each day and often must respond to several stresses at the same time. Study of stress responses show that there is much crosstalk among signaling networks during specific stress responses. Thus, plants may respond to stress perception by an initial global response and follow with specific stress responses.

As we discussed in this chapter, convergence points between biotic and abiotic stress signaling pathways have begun to be analyzed. Specific factors including transcription factors such as *WRKYs, ATAF1 and 2, MYC2, RD2, BOS1, OsNAC6* and OsMPK5 kinase are molecular player, common to multiple networks or involved in crosstalk between stress signaling pathways regulated by abscisic acid, salicylic acid, jasmonic acid and ethylene as well as ROS signaling. Powerful molecular tools, including transcriptome and proteome analyses, sequencing of entire genomes in plants, bioinformatic analyses and functional studies, will enable the dissection of networks and identification of key factors in abiotic and biotic signaling cascade crosstalk, which will reveal novel interplays between parallel signaling pathways in the plant responses to biotic and abiotic stress.

8. References

Abe H, Urao T, Ito T, Seki M, Shinozaki K, Yamaguchi-Shinozaki K (2003) Arabidopsis AtMYC2 (bHLH) and AtMYB2 (MYB) Function as Transcriptional Activators in Abscisic Acid Signaling. The Plant Cell Online 15: 63-78

Adie BAT, Pérez-Pérez J, Pérez-Pérez MM, Godoy M, Sánchez-Serrano J-J, Schmelz EA, Solano R (2007) ABA Is an Essential Signal for Plant Resistance to Pathogens Affecting JA Biosynthesis and the Activation of Defenses in Arabidopsis. The Plant Cell Online 19: 1665-1681

Alcazar R, Garcia AV, Parker JE, Reymond M (2009) Incremental steps toward incompatibility revealed by Arabidopsis epistatic interactions modulating salicylic acid pathway activation. Proc Natl Acad Sci U S A 106: 334-339

Alvarado-Gutiérrez A, Del Real-Monroy M, Rodríguez-Guerra R, Almanza-Sánchez L, Lozoya-Gloria E, Fraire-Velázquez S (2008) A Phaseolus vulgaris EF-hand calcium-binding domain is induced early in the defense response against Colletotrichum lindemuthianum and by abiotic stress: Sequences shared between interacting partners Physiol Mol Plant Pathol 72: 111-121

Allan AC, Fluhr R (1997) Two Distinct Sources of Elicited Reactive Oxygen Species in Tobacco Epidermal Cells. Plant Cell 9: 1559-1572

Anderson JP, Badruzsaufari E, Schenk PM, Manners JM, Desmond OJ, Ehlert C, Maclean DJ, Ebert PR, Kazan K (2004) Antagonistic Interaction between Abscisic Acid and Jasmonate-Ethylene Signaling Pathways Modulates Defense Gene Expression and Disease Resistance in Arabidopsis. The Plant Cell Online 16: 3460-3479

Arimura G-i, Maffei ME (2010) Calcium and secondary CPK signaling in plants in response to herbivore attack. Biochemical and Biophysical Research Communications 400: 455-460

Audenaert K, De Meyer GB, Höfte MM (2002) Abscisic Acid Determines Basal Susceptibility of Tomato toBotrytis cinerea and Suppresses Salicylic Acid-Dependent Signaling Mechanisms. Plant Physiology 128: 491-501

Bolwell GP, Bindschedler LV, Blee KA, Butt VS, Davies DR, Gardner SL, Gerrish C, Minibayeva F (2002) The apoplastic oxidative burst in response to biotic stress in plants: a three-component system. J Exp Bot 53: 1367-1376

Brader G, Djamei A, Teige M, Palva ET, Hirt H (2007) The MAP kinase kinase MKK2 affects disease resistance in Arabidopsis. Mol Plant Microbe Interact 20: 589-596

Brodersen P, Petersen M, Bjørn Nielsen H, Zhu S, Newman M-A, Shokat KM, Rietz S, Parker J, Mundy J (2006) Arabidopsis MAP kinase 4 regulates salicylic acid- and jasmonic acid/ethylene-dependent responses via EDS1 and PAD4. The Plant Journal 47: 532-546

Burg HA, Takken FL (2010) SUMO-, MAPK-, and resistance protein-signaling converge at transcription complexes that regulate plant innate immunity. Plant Signal Behav 5

Carpaneto A, Ivashikina N, Levchenko V, Krol E, Jeworutzki E, Zhu JK, Hedrich R (2007) Cold transiently activates calcium-permeable channels in Arabidopsis mesophyll cells. Plant Physiol 143: 487-494

Colcombet J, Hirt H (2008) Arabidopsis MAPKs: a complex signalling network involved in multiple biological processes. Biochem J 413: 217-226

Crozier A, Kamiya Y, Bishop G, Yokota T (2000) Biosynthesis of Hormones and Elicitor Molecules. *In* BB Buchanan, W Gruissem, RL Jones, eds, Biochemistry & Molecular Biology of Plants. American Society of Plant Biologist Rockville, Maryland. U.S.A., pp 850-929

Chen F, Li Q, Sun L, He Z (2006) The rice 14-3-3 gene family and its involvement in responses to biotic and abiotic stress. DNA Res 13: 53-63

Cheng S-H, Willmann MR, Chen H-C, Sheen J (2002) Calcium Signaling through Protein Kinases. The Arabidopsis Calcium-Dependent Protein Kinase Gene Family. Plant Physiol 129: 469-485

Chuang H, Harnrak A, Chen Y, Hsu C (2010) A harpin-induced ethylene-responsive factor regulates plant growth and responses to biotic and abiotic stresses. Biophysical Res Communic 402: 410-420 de Torres-Zabala M, Truman W, Bennett MH, Lafforgue G, Mansfield JW, Rodriguez Egea P, Bogre L, Grant M (2007) Pseudomonas syringae pv. tomato hijacks the Arabidopsis abscisic acid signalling pathway to cause disease. EMBO J 26: 1434-1443

De Vleesschauwer D, Yang Y, Vera Cruz C, Höfte M (2010) Abscisic Acid-Induced Resistance against the Brown Spot Pathogen Cochliobolus miyabeanus in Rice Involves MAP Kinase-Mediated Repression of Ethylene Signaling. Plant Physiology 152: 2036-2052

De Vos M, Van Oosten VR, Van Poecke RMP, Van Pelt JA, Pozo MJ, Mueller MJ, Buchala AJ, Métraux J-P, Van Loon LC, Dicke M, Pieterse CMJ (2005) Signal Signature and Transcriptome Changes of Arabidopsis During Pathogen and Insect Attack. Molecular Plant-Microbe Interactions 18: 923-937

Delaney TP, Uknes S, Vernooij B, Friedrich L, Weymann K, Negrotto D, Gaffney T, Gut-Rella M, Kessmann H, Ward E, Ryals J (1994) A Central Role of Salicylic Acid in Plant Disease Resistance. Science 266: 1247-1250

Dey S, Ghose K, Basu D (2010) Fusarium elicitor-dependent calcium influx and associated ros generation in tomato is independent of cell death. Eur J Plant Pathol 126: 217-228

Dong J, Chen C, Chen Z (2003) Expression profiles of the <i>Arabidopsis</i> WRKY gene superfamily during plant defense response. Plant molecular biology 51: 21-37

Ellis C, Karafyllidis I, Wasternack C, Turner JG (2002) The Arabidopsis Mutant cev1 Links Cell Wall Signaling to Jasmonate and Ethylene Responses. The Plant Cell Online 14: 1557-1566

Errakhi R, Dauphin A, Meimoun P, Lehner A, Reboutier D, Vatsa P, Briand J, Madiona K, Rona J, Barakate M, Wendehenne D, Beaulieu C, Bouteau F (2008) An early Ca2+ influx is a prerequisite to thaxtomin A-induced cell death in Arabidopsis thaliana cells. J Exp Botany 59: 4259-4270

Eulgem T, Somssich IE (2007) Networks of WRKY transcription factors in defense signaling. Current Opinion in Plant Biology 10: 366-371

Ferrer J-L, Jez JM, Bowman ME, Dixon RA, Noel JP (1999) Structure of chalcone synthase and the molecular basis of plant polyketide biosynthesis. Nat Struct Mol Biol 6: 775-784

Forouhar F, Yang Y, Kumar D, Chen Y, Fridman E, Park SW, Chiang Y, Acton TB, Montelione GT, Pichersky E, Klessig DF, Tong L (2005) Structural and biochemical studies identify tobacco SABP2 as a methyl salicylate esterase and implicate it in plant innate immunity. Proceedings of the National Academy of Sciences of the United States of America 102: 1773-1778

Fujita Y, Fujita M, Yamaguchi-Shinozaki K, Shinozaki K (2009) Transcription Factors Involved in the Crosstalk between Abiotic and Biotic Stress-Signaling Networks. Wiley-Blackwell

Gaffney T, Friedrich L, Vernooij B, Negrotto D, Nye G, Uknes S, Ward E, Kessmann H, Ryals J (1993) Requirement of Salicylic Acid for the Induction of Systemic Acquired Resistance. Science 261: 754-756

Galletti R, Denoux C, Gambetta S, Dewdney J, Ausubel FM, De Lorenzo G, Ferrari S (2008) The AtrbohD-Mediated Oxidative Burst Elicited by Oligogalacturonides in

Arabidopsis Is Dispensable for the Activation of Defense Responses Effective against Botrytis cinerea Plant Physiol 148: 1695-1706

Garnier L, Simon-Plas F, Thuleau P, Agnel J-P, Blein J-P, Ranjeva R, Montillet J-L (2006) Cadmium affects tobacco cells by a series of three waves of reactive oxygen species that contribute to cytotoxicity. Plant Cell Environ 29: 1956-1969

Genger RK, Jurkowski GI, McDowell JM, Lu H, Jung HW, Greenberg JT, Bent AF (2008) Signaling Pathways That Regulate the Enhanced Disease Resistance of Arabidopsis "Defense, No Death" Mutants. Molecular Plant-Microbe Interactions 21: 1285-1296

Gerold JM, Beckers, Michal J, Yidong L, William R, Underwood, Sheng Y, Shuqun Z, Uwe C (2009) Mitogen-Activated Protein Kinases 3 and 6 Are Required for Full Priming of Stress Responses in Arabidopsis thaliana Plant Cell 21: 944–953.

Glazebrook J (2005) Contrasting Mechanisms of Defense Agains Biotrophic and Necrotrophic Pathogens. Annual Review of Phytopathology 43: 205-227

Gonzalez A, Vera J, Castro J, Dennett G, Mellado M, Morales B, Correa JA, Moenne A (2011) Co-occurring increases of calcium and organellar reactive oxygen species determine differential activation of antioxidant and defense enzymes in Ulva compressa (Chlorophyta) exposed to copper excess. Plant Cell Environ 33: 1627-1640

Grant M, Lamb C (2006) Systemic immunity. Current Opinion in Plant Biology 9: 414-420

Hammond-Kosack KE, Jones JD (1996) Resistance gene-dependent plant defense responses. Plant Cell 8: 1773-1791

Hashimoto M, Kisseleva L, Sawa S, Furukawa T, Komatsu S, Koshiba T (2004) A novel rice PR10 protein, RSOsPR10, specifically induced in roots by biotic and abiotic stresses, possibly via the jasmonic acid signaling pathway. Plant Cell Physiol 45: 550-559

Heidarvand L, Amiri R (2010) What happens in plant molecular responses to cold stress? . Acta Physiol Plantarum 32: 419-431

Heinfling JWDM, Bostock RM, Kue J (1980) Effect of Abscisic Acid on Rishtin and Lubimin Accumulation and Resistance to Phytophthora infestans and Cladiosporimu cucumerinum in Potto Tuber Tissue Slices Physioloy and Biochemistry: 1074-1078

Holley S, Yalamanchili R, Moura D, Ryan C, Stratmann J (2003) Convergence of signaling pathways induced by systemin, oligosaccharyde elicitors, and ultraviolet-B radiation at the level of mitogen-activated protein kinases in Lycopersicon peruvianum suspension-cultured cells. Plant Physiol 132: 1728-1732

Howe GA, Jander G (2008) Plant Immunity to Insect Herbivores. Annual Review of Plant Biology 59: 41-66

Huang X, Li J, Bao F, Zhang X, Yang S (2010) A gain-of-function mutation in the Arabidopsis disease resistance gene RPP4 confers sensitivity to low temperature. Plant Physiol 154: 796-809

Hussain SS, Kayani MA, Amjad M (2011) Transcription factors as tools to engineer enhanced drought stress tolerance in plants. Biotechnology Progress 27: 297-306

Ichimura K, Shinozaki K, Tena G, Sheen J, Henry Y, Champion, A. K, M. , Zhang S, Hirt H (2002) Mitogen-activated protein kinase cascades in plants: A new nomenclature. Trends Plant Sci. 7: 301-308

Jambunathan N, McNellis TW (2003) Regulation of Arabidopsis COPINE 1 gene expression in response to pathogens and abiotic stimuli. Plant Physiol 132: 1370-1381

Jurkowski G, Smith R, Yu I, Ham J, Sharma S, Klessig D, Fengler K, Bent A (2004) Arabidopsis DND2, a second cyclic nucleotidegated ion channel gene for which mutation causes the "defense, no death" phenotype. Mol Plant-Microbe Interact 17: 511-520

Kaliff M, Staal J, Myrenås M, Dixelius C (2007) ABA Is Required for Leptosphaeria maculans Resistance via ABI1- and ABI4-Dependent Signaling. Molecular Plant-Microbe Interactions 20: 335-345

Karita E, Yamakawa H, Mitsuhara I, Kuchitsu K, Ohashi Y (2004) Three types of tobacco calmodulins characteristically activate plant NAD kinase at different Ca2+ concentrations and pHs. Plant Cell Physiol 45: 1371-1379

Kazan K, Manners JM (2008) Jasmonate Signaling: Toward an Integrated View. Plant Physiology 146: 1459-1468

Kim MC, Chung WS, Yun D-J, Cho MJ (2009) Calcium and Calmodulin-Mediated Regulation of Gene Expression in Plants. Molecular Plant 2: 13-21

Kwon SJ, Kwon SI, Bae MS, Cho EJ, Park OK (2007) Role of the methionine sulfoxide reductase MsrB3 in cold acclimation in Arabidopsis. Plant Cell Physiol 48: 1713-1723

Lamb C, Dixon RA (1997) The Oxidative Burst in Plant Disease Resistance. Annu Rev Plant Physiol Plant Mol Biol 48: 251-275

Lambeth JD (2004) NOX enzymes and the biology of reactive oxygen. Nat Rev Immunol 4: 181-189

Leon J, Rojo E, Sanchez-Serrano JJ (2001) Wound signalling in plants. J Exp Bot 52: 1-9

Lherminier J, Elmayan T, Fromentin J, Elaraqui KT, Vesa S, Morel J, Verrier JL, Cailleteau B, Blein JP, Simon-Plas F (2009) NADPH oxidase-mediated reactive oxygen species production: subcellular localization and reassessment of its role in plant defense. Mol Plant Microbe Interact 22: 868-881

Li J, Brader G, Palva ET (2004) The WRKY70 Transcription Factor: A Node of Convergence for Jasmonate-Mediated and Salicylate-Mediated Signals in Plant Defense. The Plant Cell Online 16: 319-331

Li L, Li C, Lee GI, Howe GA (2002) Distinct roles for jasmonate synthesis and action in the systemic wound response of tomato. Proceedings of the National Academy of Sciences 99: 6416-6421

Liu H, Yang W, Liu D, Han Y, Zhang A, Li S (2011) Ectopic expression of a grapevine transcription factor <i>VvWRKY11</i> contributes to osmotic stress tolerance in <i>Arabidopsis</i&gt. Molecular Biology Reports 38: 417-427

Lorenzo O, Chico JM, Sánchez-Serrano JJ, Solano R (2004) JASMONATE-INSENSITIVE1 Encodes a MYC Transcription Factor Essential to Discriminate between Different Jasmonate-Regulated Defense Responses in Arabidopsis. The Plant Cell Online 16: 1938-1950

Lorenzo O, Piqueras R, Sánchez-Serrano JJ, Solano R (2003) ETHYLENE RESPONSE FACTOR1 Integrates Signals from Ethylene and Jasmonate Pathways in Plant Defense. The Plant Cell Online 15: 165-178

Luna E, Pastor V, Robert J, Flors V, Mauch-Mani B, Ton J (2011) Callose Deposition: A Multifaceted Plant Defense Response. Molecular Plant-Microbe Interactions 24: 183-193

Malamy J, Carr JP, Klessig DF, Raskin I (1990) Salicylic Acid: A Likely Endogenous Signal in the Resistance Response of Tobacco to Viral Infection. Science 250: 1002-1004

Mantri NL, Ford R, Coram TE, Pang ECK (2010) Evidence of unique and shared responses to major biotic and abiotic stresses in chickpea. Env Exp Botany 69: 268-292

Mao P, Duan M, Wei C, Li Y (2007) WRKY62 Transcription Factor Acts Downstream of Cytosolic NPR1 and Negatively Regulates Jasmonate-Responsive Gene Expression. Plant and Cell Physiology 48: 833-842

Mauch-Mani B, Mauch F (2005) The role of abscisic acid in plant-pathogen interactions. Current Opinion in Plant Biology 8: 409-414

Mazars C, Thuleau P, Lamotte O, Bourque S (2010) Cross-Talk between ROS and Calcium in Regulation of Nuclear Activities. Molecular Plant 3: 706-718

McDonald KL, Cahill DM (1999) Influence of Abscisic Acid and the Abscisic Acid Biosynthesis Inhibitor, Norflurazon, on Interactions Between Phytophthora sojae and Soybean (Glycine max). European Journal of Plant Pathology 105: 651-658

Melotto M, Underwood W, Koczan J, Nomura K, He SY (2006) Plant Stomata Function in Innate Immunity against Bacterial Invasion. Cell 126: 969-980

Mengiste T, Chen X, Salmeron J, Dietrich R (2003) The BOS1 Gene Encodes an R2R3MYB Transcription Factor Protein That Is Required for Biotic and Abiotic Stress Responses in Arabidopsis. The Plant Cell Online

Mengiste T, Chen X, Salmeron J, Dietrich R (2003) The BOTRYTIS SUSCEPTIBLE1 Gene Encodes an R2R3MYB Transcription Factor Protein That Is Required for Biotic and Abiotic Stress Responses in Arabidopsis. Plant Cell 15: 2551-2565

Miao Y, Zentgraf U (2007) The Antagonist Function of Arabidopsis WRKY53 and ESR/ESP in Leaf Senescence Is Modulated by the Jasmonic and Salicylic Acid Equilibrium. The Plant Cell Online 19: 819-830

Miller G, Suzuki N, Ciftci-Yilmaz S, Mittler R (2010) Reactive oxygen species homeostasis and signalling during drought and salinity stresses. Plant Cell Environ 33: 453-467

Mohr P, Cahill D (2007) Suppression by ABA of salicylic acid and lignin accumulation and the expression of multiple genes, in <i>Arabidopsis</i> infected with <i>Pseudomonas syringae</i> pv. <i>tomato</i>. Functional & Integrative Genomics 7: 181-191

Mohr PG, Cahill DM (2003) Abscisic acid influences the susceptibility of Arabidopsis thaliana to Pseudomonas syringae pv. tomato and Peronospora parasitica. . Functional Plant Biology

Monshausen GB, Bibikova TN, Weisenseel MH, Gilroy S (2009) Ca2+ regulates reactive oxygen species production and pH during mechanosensing in Arabidopsis roots. Plant Cell 21: 2341-2356

Mori I, Murata Y, Yang Y, Munemasa S, Wang Y, Andreoli S, al. e (2006) CDPKs CPK6 and CPK3 function in ABA regulation of guard cell S-type anion- and Ca2+- permeable channels and stomatal closure. PLoS Biol 4: 327-332

Mosher S, Moeder W, Nishimura N, Jikumaru Y, Se-Hwan J, Urquhart W, Klessig D, Kim S, Nambara E, Yoshioka K (2010) The Lesion-Mimic Mutant cpr22 Shows Alterations in Abscisic Acid Signaling and Abscisic Acid Insensitivity in a Salicylic Acid-Dependent Manner. Plant Physiol 152: 1901-1913

Nawrath C, Heck S, Parinthawong N, Métraux J-P (2002) EDS5, an Essential Component of Salicylic Acid-Dependent Signaling for Disease Resistance in Arabidopsis, Is a Member of the MATE Transporter Family. The Plant Cell Online 14: 275-286

Nawrath C, Métraux J-P (1999) Salicylic Acid Induction-Deficient Mutants of Arabidopsis Express PR-2 and PR-5 and Accumulate High Levels of Camalexin after Pathogen Inoculation. The Plant Cell Online 11: 1393-1404

Ndamukong I, Abdallat AA, Thurow C, Fode B, Zander M, Weigel R, Gatz C (2007) SA-inducible Arabidopsis glutaredoxin interacts with TGA factors and suppresses JA-responsive PDF1.2 transcription. The Plant Journal 50: 128-139

Ogasawara Y, Kaya H, Hiraoka G, Yumoto F, Kimura S, Kadota Y, Hishinuma H, Senzaki E, Yamagoe S, Nagata K, Nara M, Suzuki K, Tanokura M, Kuchitsu K (2008) Synergistic activation of the Arabidopsis NADPH oxidase AtrbohD by Ca2+ and phosphorylation. J Biol Chem 283: 8885-8892

Park S-W, Kaimoyo E, Kumar D, Mosher S, Klessig DF (2007) Methyl Salicylate Is a Critical Mobile Signal for Plant Systemic Acquired Resistance. Science 318: 113-116

Parker JE (2009) The Quest for Long-Distance Signals in Plant Systemic Immunity. Sci. Signal. 2: pe31-

Paschold A, Halitschke R, Baldwin IT (2007) Co(i)-ordinating defenses: NaCOI1 mediates herbivore- induced resistance in Nicotiana attenuata and reveals the role of herbivore movement in avoiding defenses. The Plant Journal 51: 79-91

Peleg Z, Blumwald E (2011) Hormone balance and abiotic stress tolerance in crop plants. Current Opinion in Plant Biology In Press, Corrected Proof

Penninckx IAMA, Thomma BPHJ, Buchala A, Métraux J-P, Broekaert WF (1998) Concomitant Activation of Jasmonate and Ethylene Response Pathways Is Required for Induction of a Plant Defensin Gene in Arabidopsis. The Plant Cell Online 10: 2103-2114

Petersen M, Brodersen P, Naested H, Andreasson E, Lindhart U, Johansen B, Nielsen HB, Lacy M, Austin MJ, Parker JE, Sharma SB, Klessig DF, Martienssen R, Mattsson O, Jensen AB, Mundy J (2000) Arabidopsis MAP Kinase 4 Negatively Regulates Systemic Acquired Resistance. Cell 103: 1111-1120

Piffanelli P, Zhou F, Casais C, Orme J, Jarosch B, Schaffrath U, Collins N, Panstruga R, Schulze-Lefert P (2002) The Barley MLO Modulator of Defense and Cell Death Is Responsive to Biotic and Abiotic Stress Stimuli. Plant Physiol 129: 1076-1085

Qiu Y, Yu D (2009) Over-expression of the stress-induced OsWRKY45 enhances disease resistance and drought tolerance in Arabidopsis. Environmental and Experimental Botany 65: 35-47

Rasmussen JB, Hammerschmidt R, Zook MN (1991) Systemic Induction of Salicylic Acid Accumulation in Cucumber after Inoculation with Pseudomonas syringae pv syringae. Plant Physiology 97: 1342-1347

Reymond P, Bodenhausen N, Van Poecke RMP, Krishnamurthy V, Dicke M, Farmer EE (2004) A Conserved Transcript Pattern in Response to a Specialist and a Generalist Herbivore. The Plant Cell Online 16: 3132-3147

Richard S, Lapointe G, Rutledge RG, Seguin A (2000) Induction of chalcone synthase expression in white spruce by wounding and jasmonate. Plant Cell Physiol 41: 982-987

Rodriguez-Serrano M, Romero-Puertas M, Pazmino D, Testillano P, Risueno M, del Rio L, Sandalio L (2009) Cellular Rresponse of Pea Plants to Cadmium Toxicity: Cross Talk between Reactive Oxygen Species, Nitric Oxide, and Calcium. Plant Physiol 150: 229-243

Rushton PJ, Somssich IE, Ringler P, Shen QJ (2010) WRKY transcription factors. Trends in Plant Science 15: 247-258

Spoel SH, Johnson JS, Dong X (2007) Regulation of tradeoffs between plant defenses against pathogens with different lifestyles. Proceedings of the National Academy of Sciences 104: 18842-18847

Taiz L, Zeiger E (2010) Plant Physiology, Ed Fifth. Sinauer Associates, U.S.A

Takahashi F, Mizoguchi T, Yoshida R, Ichimura K, Shinozaki K (2011) Calmodulin-Dependent Activation of MAP Kinase for ROS Homeostasis in Arabidopsis. Mol Cell 41: 649-660

Takeda S, Gapper C, Kaya H, Bell E, Kuchitsu K, Dolan L (2008) Local positive feedback regulation determines cell shape in root hair cells. Science 319: 1241-1244

Takemoto D, Yoshioka H, Doke N, Kawakita K (2003) Disease stress-inducible genes of tobacco: expression profile of elicitor-responsive genes isolated by subtractive hybridization. Physiol Plantarum 118: 545-553

Teige M, Scheikl E, Eulgem T, Doczi R, Ichimura K, Shinozaki K, Dangl JL, Hirt H (2004) The MKK2 pathway mediates cold and salt stress signaling in Arabidopsis. Mol Cell 15: 141-152

Thaler JS, Bostock RM (2004) INTERACTIONS BETWEEN ABSCISIC-ACID-MEDIATED RESPONSES AND PLANT RESISTANCE TO PATHOGENS AND INSECTS. Ecology 85: 48-58

Thomma BPHJ, Penninckx IAMA, Cammue BPA, Broekaert WF (2001) The complexity of disease signaling in Arabidopsis. Current Opinion in Immunology 13: 63-68

Ton J, Mauch-Mani B (2004) β-amino-butyric acid-induced resistance against necrotrophic pathogens is based on ABA-dependent priming for callose. The Plant Journal 38: 119-130

Torres MA, Dangl JL (2005) Functions of the respiratory burst oxidase in biotic interactions, abiotic stress and development. Curr Opin Plant Biol 8: 397-403

Torres MA, Jones JD, Dangl JL (2006) Reactive oxygen species signaling in response to pathogens. Plant Physiol 141: 373-378

Tsutsui T, Kato W, Asada Y, Sako K, Sato T, Sonoda Y, Kidokoro S, Yamaguchi-Shinozaki K, Tamaoki M, Arakawa K, Ichikawa T, Nakazawa M, Seki M, Shinozaki K, Matsui M, Ikeda A, Yamaguchi J (2009) DEAR1, a transcriptional repressor of DREB protein that mediates plant defense and freezing stress responses in Arabidopsis. J Plant Res 122: 633-643

Uno Y, Rodriguez Milla MA, Maher E, Cushman JC (2009) Identification of proteins that interact with catalytically active calcium-dependent protein kinases from Arabidopsis. Mol Genet Genomics 281: 375-390

Verberne MC, Verpoorte R, Bol JF, Mercado-Blanco J, Linthorst HJM (2000) Overproduction of salicylic acid in plants by bacterial transgenes enhances pathogen resistance. Nat Biotech 18: 779-783

Vernooij B, Friedrich L, Goy P, Saub T, Kessmann H, Ryals J (1995.) 2,6-Dichloroisonicotinic acidinduced resistance to pathogens without the accumulation of salicylic acid. Mol. Plant-Microbe Interact. 8: 228-234

Vlot AC, Dempsey DMA, Klessig DF (2009) Salicylic Acid, a Multifaceted Hormone to Combat Disease. Annual Review of Phytopathology 47: 177-206

Wahid A, Gelani S, Ashraf M, Foolad M (2007) Heat tolerance in plants: An overview. Envir Exp Botany 61: 199-223

Wang Y, Bao Z, Zhu Y, Hua J (2009) Analysis of temperature modulation of plant defense against biotrophic microbes. Mol Plant-Microbe Interact 22: 498-506

Wasilewska A, Vlad F, Sirichandra C, Redko Y, Jammes F, Valon C, Frey NFd, Leung J (2008) An Update on Abscisic Acid Signaling in Plants and More .... Molecular Plant 1: 198-217

Whenham RJ, Fraser RSS, Brown LP, Payne JA (1986) Tobacco-mosaic-virus-induced increase in abscisic-acid concentration in tobacco leaves. Planta 168: 592-598

White PJ, Broadley MR (2003) Calcium in Plants. Annals of Botany 92: 487-511

Williams ME (2010) Introduction to phytohormones. The Plant Cell 22: 1-9

Williams ME (2011) Jasmonates: Defense andMore. The Plant Cell: 1-11

Wu Y, Deng Z, Lai J, Zhang Y, Yang C, Yin B, Zhao Q, Zhang L, Li Y, Yang C, Xie Q (2009) Dual function of Arabidopsis ATAF1 in abiotic and biotic stress responses. Cell Res 19: 1279-1290

Wurzinger B, Mair A, Pfister B, Teige M (2010) Cross-talk of calcium-dependent protein kinase and MAP kinase signalling. Plant Signal Behav 6

Xu J, Li Y, Wang Y, Liu H, Lei L, Yang H, al. e (2008) Activation of MAPK kinase 9 induces ethylene and camalexin biosynthesis and enhances sensitivityto salt stress in Arabidopsis. J Biol Chem 283: 26996-27006

Yamasaki K, Kigawa T, Inoue M, Tateno M, Yamasaki T, Yabuki T, Aoki M, Seki E, Matsuda T, Tomo Y, Hayami N, Terada T, Shirouzu M, Tanaka A, Seki M, Shinozaki K, Yokoyama S (2005) Solution Structure of an Arabidopsis WRKY DNA Binding Domain. The Plant Cell Online 17: 944-956

Yoshioka H, Numata N, Nakajima K, Katou S, Kawakita K, Rowland O, Jones J, Doke N (2003) Nicotiana benthamiana gp91(phox) homologs NbrbohA and NbrbohB participate in H2O2 accumulation and resistance to Phytophthora infestans. Plant Cell 15: 706-718

Yoshioka K, Kachroo P, Tsui F, Sharma S, Shah J, Klessig D (2001) Environmentally-sensitive, SA-dependent defense response in the cpr22 mutant of Arabidopsis. Plant Journal 26: 447-459

Zhou M-L, Ma J-T, Pang J-F, Zhang Z-L, Tang Y-X, Wu Y-M (2010) Regulation of plant stress response by dehydration responsive element binding (DREB) transcription factors. African Journal of Biotechnology 9: 9255-9279

2

Reactive Oxygen in Abiotic Stress Perception - From Genes to Proteins

Michael Wrzaczek, Julia P. Vainonen, Adrien Gauthier,
Kirk Overmyer and Jaakko Kangasjärvi
Plant Biology, Department of Biosciences, University of Helsinki
Finland

1. Introduction

Throughout their life plants have to adapt to variable environmental conditions. Changes in photoperiod, light intensity and quality, nutrient abundance and starvation, drought and flooding, variation in temperature, air and soil pollution and osmotic changes are among the abiotic factors that can cause stress (Apel & Hirt, 2004). To ensure constant monitoring of environmental conditions and a quick and appropriate response, plants have developed elaborate and robust perception and signal transduction mechanisms. The importance of the ability to adapt to a changing environment has been described in numerous research articles and reviews (Hirayama & Shinozaki, 2010). Recent years have seen tremendous progress in our understanding of the mechanisms and processes underlying abiotic stress adaptation and defence in different plant species (Hirayama & Shinozaki, 2010; Jaspers & Kangasjärvi, 2010). Importantly, the analysis of abiotic stress tolerant varieties of *Arabidopsis* and also rice has led to novel ideas for improving the stress resistance of crop species. The diversity of abiotic stresses implies that there should be a strong specific component in the individual stress responses (Jaspers & Kangasjärvi, 2010). However, there is a striking common component in the general response to all abiotic stresses (Vaahtera & Brosché, 2011). Essentially all abiotic stresses lead to the production of reactive oxygen species (ROS) albeit different forms and in different subcellular compartments (Jaspers & Kangasjärvi, 2010). In contrast to their presumed role as simply damaging agents in cells ROS act as signalling molecules in the regulation of stress adaptation but also in developmental regulation (Apel & Hirt, 2004; Jaspers & Kangasjärvi, 2010; Møller et al., 2007). For reviews on other aspects of abiotic stress we refer to reviews (Jaspers & Kangasjärvi, 2010; Miller et al., 2008; Munns & Tester, 2008; Vaahtera & Brosché, 2011; Zhu, 2002).

Despite the wealth of information on abiotic stress defence in plants the mechanisms of stress sensing have remained relatively elusive. In this review we turn our attention to the mechanisms of abiotic stress perception. Generally, stresses, as well as other stimuli, can be perceived in a direct or an indirect manner. In direct perception, the agent causing the stress is perceived through a receptor. Alternatively, in indirect perception, specific effects leading to stress caused by an agent are perceived. Evidence suggests that in abiotic stress perception plants use both modes in parallel. In indirect stress perception ROS are components frequently used as signalling molecules. However, ROS themselves can be subject to direct or indirect perception mechanisms (Figure 1). Since ROS are a response to common to many abiotic stresses particular emphasis will be placed on their role in stress perception. For general reviews on ROS signalling we refer the reader to recent reviews (Foyer et al., 2009; Foyer & Noctor, 2005; Foyer & Noctor, 2009; Foyer & Noctor, 2011; Jaspers & Kangasjärvi, 2010; Møller et al., 2007).

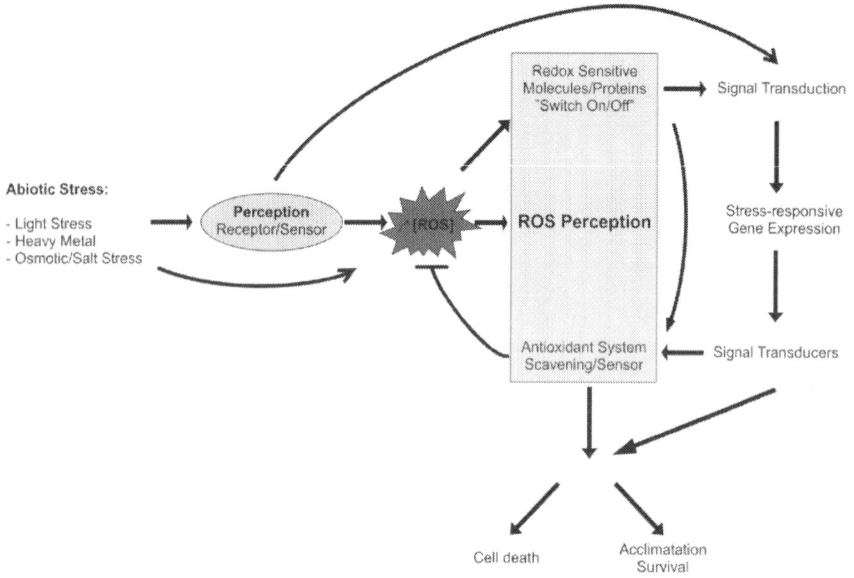

Fig. 1. Hypothetical model of the pathways involved in ROS perception in plants. Abiotic stress or its perception through transmembrane or intracellular receptors results in the overproduction of ROS, including singlet oxygen ($^1O_2$), superoxide ($O_2\cdot^-$), hydrogen peroxide (H_2O_2), or hydroxyl radicals ($HO\cdot$). The increase of ROS levels can be directly sensed by cellular sensors such as redox-sensitive proteins, e.g. transcription factors which can activate signal transduction to induce expression of stress-responsive genes. This results in differential regulation of gene expression that will affect pathways including the induction of the ROS scavenging system and repression of the ROS-producing mechanisms. Ultimately this sensing, signalling and transcriptional reprogramming will be critical for the future fate of the cell leading to adaptation to the stress or to cell death.

2. Perception of salt and osmotic stresses

Salt stress is an abiotic stress, for which some perception components have been identified. Salt stress, as induced by elevated concentrations of NaCl, can be separated into two components: an osmotic stress component and an ionic stress component, i.e., Na toxicity (Munns & Tester, 2008). Osmotic stress can also be caused by other osmotically active substances; mannitol is a frequently used chemical to analyze osmotic stress perception and regulation under laboratory conditions. To sense osmotic stress, a cell could employ either direct or indirect perception mechanisms. A direct sensing of osmotic stress would be difficult to imagine. If a putative direct osmosensor would act like a classical ligand-specific receptor, it would have to detect water activity (Wood et al., 2001). Alternatively, through indirect perception mechanisms, an osmosensor could sense other cellular properties that are affected by osmotic changes including cell volume, turgor pressure, membrane strain, individual solute concentrations, ionic strength and crowding of macromolecules in the cytoplasm (Wood et al., 2001).

Promising candidates as receptors for salt and osmotic stress are histidine kinases and their cognate response regulators. Histidine kinases are well known for their roles in hormone responses, e.g. cytokinin and ethylene perception (Urao et al., 2000; Urao et al., 2001). In prokaryotes and in yeast histidine kinases have been identified as osmosensors but the system is also conserved in other organisms (Heermann et al., 2009; Heermann & Jung, 2010; Maeda et al., 2006). In the photosynthetic cyanobacterium *Synechocystis* sp. PCC 6803 several different histidine kinases together with their response regulators (Hik33-Rre31, Hik34-Rre1, Hik10-Rre3, Hik16-Hik41-Rre17, Hik2-Rre1) regulate the transcription of distinct gene sets in response to osmotic and salt stress and are likely involved in stress perception (Paithoonrangsarid et al., 2004; Shoumskaya et al., 2005). In the model plant species *Arabidopsis thaliana* a hybrid-type histidine kinase *ARABIDOPSIS THALIANA HISTIDINE KINASE 1 (ATHK1)* has been proposed as an osmosensor as early as the late Nineteen-Nineties (Tran et al., 2007; Urao et al., 1999). Interestingly, ATHK1 is also involved in the regulation of the desiccation process during seed maturation, a process which seems to be connected to drought tolerance (Wohlbach et al., 2008). Similar osmosensors have also been identified in other plant species including the woody plant *Populus deltoides* (Chefdor et al., 2006). While the evidence supports a role for histidine kinases as sensors for salt and osmotic stress, the mechanism of this stress perception is still unclear. Much information originates from gene expression analysis of stress responses and from phenotypic analysis of mutant plants; direct biochemical evidence for salt or osmotic stress sensing is however lacking. Investigation of the homologous histidine kinase-response regulator system KpdD/KpdE from *Escherichia coli* identified different parts of the proteins in the perception of low potassium conditions compared to osmotic stress (Heermann et al., 2009). It has been proposed that histidine kinases as transmembrane proteins might perceive changes in turgor or some associated effect (Epstein, 1992; Laimins et al., 1981). However, for the bacterial histidine kinase KpdD turgor reduction appears not to be the stimulus; and similarity of the histidine kinases suggests that this might be the case also for other species (Hamann et al., 2008). It is abundantly clear, that histidine kinases have important roles during the regulation of salt and osmotic stress responses in addition to their roles in hormone signalling and they are probably very early elements in the perception machinery. However, their precise biochemical roles have yet to be clarified.

Histidine kinases are not the only candidates for sensors for osmotic and ionic stress. There is evidence linking several transporters to osmosensing thereby unifying osmoperception and osmoregulation in bacteria (Wood et al., 2001). Receptor-transporters in *E. coli* seem to respond specifically to internal solvents (Culham et al., 2003a; Culham et al., 2003b; Rübenhagen et al., 2000; van der Heide et al., 2001; Wood et al., 2001). Several transmembrane transporter proteins have been identified in plants including the plasma membrane Na^+/H^+ antiporter SALT-OVERLY-SENSITIVE 1 (SOS1) and the HIGH-AFFINITY K^+ TRANSPORTER 1 (HKT1), which mediates Na^+-K^+ co-transport (Horie et al.,

2009; Mäser et al., 2002a; Mäser et al., 2002b; Schachtman & Schroeder, 1994; Shi et al., 2000; Shi et al., 2002). SOS1 interacts with an important regulator of abiotic stress responses, RADICAL-INDUCED CELL DEATH 1 (RCD1; Katiyar-Agarwal et al., 2006). Another group of transporters, the Na^+/H^+ EXCHANGER AtNHX1, and the close orthologues AtNHX2, 3, 4, 5 and 6 are involved in compartmentalization of Na^+ to the vacuole (Horie et al., 2009; Qiu et al., 2004; Yokoi et al., 2002). While these transporters are critical determinants of salt tolerance there is only little evidence linking them to salt stress perception (Horie et al., 2009). SOS1 is regulated through phosphorylation of its autoinhibitory C-terminal domain by the SALT-OVERLY-SENSITIVE 2 (SOS2) protein kinase, a member of the Sucrose non-fermenting 1-related protein family (Quintero et al., 2011). SOS2 is regulated through interaction with the Ca^{2+}-sensing protein SALT-OVERLY-SENSITIVE 3 (SOS3). This suggests that calcium is a critical regulator of Na^+-transport and osmotic/ionic stress regulation. It is still unclear however, which sensory mechanism and which signaling events lead to the required changes in cellular Ca^{2+} concentrations that subsequently result in differential regulation of SOS1. Other Ca^{2+} signalling proteins seem to be involved in the acclimation to abiotic stresses as well. The Ca^{2+}-DEPENDENT PROTEIN KINASE 3 (CPK3) provides a parallel pathway to MITOGEN-ACTIVATED PROTEIN KINASE (MAPK) signalling in *Arabidopsis thaliana* (Mehlmer et al., 2010; Wurzinger et al., 2011). In these cases, just as in the example of SOS1, the importance of Ca^{2+} signalling is recognized, but the components that induce Ca^{2+} signalling in response to the initial stress remain unknown.

Recent research suggests a novel mechanism of osmoperception through salt-dependent protein-nucleic acid interactions (Novak et al., 2011). In the cyanobacterium *Synechocystis* sp. PCC6803 the key enzyme of the glucosyl-glycerol pathway (GgpS) can be non-competitively inhibited by nucleic acids in a sequence- and length-independent manner *via* a salt-dependent electrostatic interaction. The intracellular salt concentration thus could serve as a trigger for GgpS regulation leading to inactivation through nucleic acid binding at low salt concentrations. An increase in salt concentration leads to liberation of GgpS and the accumulation of glucosyl-glycerol facilitates the acclimatization to salt stress. It is unclear if this mechanism is also conserved in plants and other organisms, future research will have to answer this question.

3. Heavy metal sensing

Heavy metals (HMs) have biological importance in plant development and growth, where they play key roles affecting cellular processes such as homeostasis and photosynthesis (Terry, 1980). Many cellular processes are regulated by enzymes whose activities require the presence of heavy metals such as Fe, Zn, Mn, and Cu in the active site or in another position important for the proper enzyme functioning (Hänsch & Mendel, 2009). Thus, some HMs serve as cofactors of different enzymes and are important micronutrients, essential for plant growth. The variety and concentration of HMs depend on different sources. Natural sources include volcanic activity and continental dust. Alternatively HM can originate from human activities. Increased availability (concentration) of HMs strongly affects plant growth and development, resulting mostly in the inhibition of plant growth and toxicity symptoms. Toxic effect of HMs can be subdivided into production of ROS and oxidative stress, inhibition of enzyme activities *via* displacement of essential cofactors, and blocking of protein or metabolite functional groups, for example, by binding to thiol, carboxyl, and histidyl groups (Sharma & Dietz, 2009). The identification and characterization of heavy metal transporters in plants is still ongoing and the mechanisms underlying HM perception through "sensor proteins" are poorly understood. However evidence suggests the existence of HM sensors in plants to perceive the availability of HMs in the environment.

Among HMs iron is of a special importance for plants due to its requirement for photosynthesis and most of the cellular redox-dependent processes. Many studies have been focusing on the iron deficiency response (Briat et al., 2007; Walker & Connolly, 2008; Wintz et al., 2003), less is known about signalling pathways induced by iron excess. Excess of iron is harmful for cells due to its high reactivity with oxygen which leads to ROS production. Accumulation of ferritins, iron storage proteins, in plastids is one of the early responses to iron excess in plants (Lobreaux et al., 1992; Petit et al., 2001). Ferritin gene expression is regulated by other environmental factors including drought, cold, and light intensity (Briat et al., 2010). Recent studies have shown that ferritins participate in defence against oxidative stress by buffering and sequestering free iron (Arosio et al., 2009; Ravet et al., 2009). Detailed gene expression analyses have allowed the identification of HM stress-specific transcriptional profiles and suggested the existence of HM sensors. Transcriptional analysis of plant responses to different HMs suggested the presence of a Zn^{2+} sensor in the cell. Interestingly, this Zn^{2+} sensor might have an important role in response to Cd^{2+} through competitive binding of Cd^{2+} or Zn^{2+}. However, this Zn^{2+} sensor is not yet identified. Following Cd^{2+} exposure expression of AtZIP9, a putative Zn^{2+}/Cd^{2+} transporter, increased in *Arabidopsis thaliana* (Weber et al., 2006). AtZIP9 is also known as a marker of Zn^{2+} depletion (Wintz et al., 2003). Although Zn^{2+} depletion is also rapidly observed under Cd^{2+} treatment, Cd^{2+}-induced gene expression is suggested to be rather a consequence of the competitive binding of Cd^{2+} to the Zn^{2+} sensor in response to Cd^{2+} exposure. Recently, HEAVY METAL ATPase 4 (AtHMA4), a heavy metal pump in *Arabidopsis thaliana,* has been proposed to have a dual role as Zn^{2+} and Cd^{2+} sensor and as a regulator of Zn^{2+} and Cu^{2+} export (Baekgaard et al., 2010). AtHMA4 has an extended C-terminal domain which exhibits high affinity to Zn^{2+} and Cd^{2+}. Heterologous expression of the C-terminal domain, containing 13 cysteine pairs, in the Zn^{2+}-sensitive *zrc1 cot1* yeast strain confers zinc and cadmium tolerance and decreased the accumulation of Zn^{2+} and Cd^{2+}. Another Zn^{2+}/H^{+} antiporter, METAL TOLERANCE PROTEIN 1 (AtMTP1), localizes to the vacuolar membrane and is involved in Zn^{2+} homeostasis. It drives Zn^{2+} detoxification and accumulation in the leaves. Interestingly, a histidine-rich loop might serve as a Zn^{2+} buffering pocket and Zn^{2+} sensor thus being crucial in maintaining proper levels of cytoplasmic Zn^{2+} (Kawachi et al., 2008). It is clear that metal transporters perform the task of metal sensing to regulate import or export of metal ions. How that subsequently translates into a downstream response as reflected in transcriptional adaptation however, is still unclear.

Detoxification of HMs occurs in the cytosol where high affinity targets of HM, so called chelators, bind and sequester HM in order to detoxify the cell. A variety of molecules are able to chelate HM, including both low molecular weight molecules such as organic acids, amino acids, peptides, as well as proteins, such as metallothioneins or phytochelatins (PC). After chelation, HM are transported and further sequestered in the vacuole (Hong-Bo et al., 2010). Indeed, vacuolar compartmentalization is considered as the major tolerance mechanism allowing HM accumulation in plants, mainly in the vacuoles of root cells. However, it has been demonstrated for some HM hyperaccumulator species, that shoots can contain higher HM levels than roots (Hong-Bo et al., 2010). It is possible that chelators act as HM sensors and that the modification of these HM sensors/scavengers might be recognized by other proteins to decipher the information. On the other hand, indirect HM sensing *via* ROS signalling is well documented. After Cd^{2+} transport into the plant cell, glutathione (GSH) could act as an initial ligand to form GS_2-Cd^{2+} complexes. Cd^{2+} can also interact with GSH-derived peptides such as phytochelatins (PC) (Cobbett & Goldsbrough, 2002; Grill et al., 1985). The GS_2-Cd^{2+} and PC-Cd^{2+} complexes formation lead to GSH depletion which will induce oxidative stress through redox imbalance (De Vos et al., 1992; Ortega-Villasante et al., 2005; Schützendubel & Polle, 2002). *Arabidopsis* mutants deficient in GSH such as *cad2* are very sensitive to Cd^{2+} and other HMs (Howden et al., 1995; Hugouvieux et al., 2009). To date, PC synthesis is suggested to be at least one of the specific responses to heavy metal accumulation, since it has been described only after metal excess challenges. Indeed, although the PC synthase gene is constitutively expressed, the protein activity depends on metal ions and/or metal-GS complexes (Maier et al., 2003; Vatamaniuk et al., 2000). PC generation can thus be considered as a direct sensor of the presence of metal in plant tissues. In addition to PCs other potential metal binding targets might be used as markers of the metal perception if their activity and or function are altered after reacting with metal. However, their specificity to heavy metal stress should be assessed. Moreover, reverse genetic will help to understand the role of the specifically induced genes in response to heavy metal during metal stress signalling.

4. Sensing light stress

Similarly to salts and some heavy metals, which are important nutrients required for plant life, plants require light in order to thrive. However, excessive exposure to high light intensities can cause considerable damage to plants. Plants utilize light as their primary source of energy converting light to usable chemical energy through photosynthesis. Light is an essential prerequisite for Chlorophyll (Chl) biosynthesis and chloroplast development; events that do not take place in darkness. Early light perception involves three classes of wavelength-specific photoreceptors, phytochromes (PHYs), cryptochromes (CRYs) and phototropines (PHOTs).

PHYs, sensing red and far-red light, are synthesized in darkness in an inactive Pr form and localize to the cytosol. Upon light perception, PHYs are converted into the biologically active Pfr form and translocate to the nucleus to initiate signalling pathways through direct interaction with PHYTOCHROME INTERACTING FACTORS (PIFs), a subfamily of basic helix-loop-helix (bHLH) transcription factors (Castillon et al., 2007; Leivar & Quail, 2011). *Arabidopsis* contains 5 genes encoding PHY apoproteins, PHYA-E, with PHYA and PHYB playing the main role in the photomorphogenetic response. Isolation of a quintuple mutant lacking all five *PHY* genes has shown that the PHYs are the major but not the sole sensors of red-light in *Arabidopsis* since the mutant is able to respond to red light and accumulate chlorophyll (Strasser et al., 2010). The quintuple mutant has several developmental defects; some of these defects can be bypassed by blue light exposure indicating functional CRYs in the absence of PHYs. In contrast to the red light receptor complement in *Arabidopsis thaliana,* PHYs are the only red light photoreceptors in rice, since *phyA phyB phyC* triple mutant lack chlorophyll biosynthesis and changes in gene expression in response to red light (Takano et al., 2009). Inactive CRY1 and CRY2 are localized to the nucleus where they interact with

CONSTITUTIVELY PHOTOMORPHOGENIC 1 (COP1), an E3 ubiquitin ligase, a key regulator of light signalling. Blue light induces activation and translocation of CRY1 to the cytoplasm, whereas CRY2 remains in the nucleus.

The proteins acting downstream of PHYs and CRYs are mainly transcription factors regulating photosynthesis-related genes necessary for chloroplast biogenesis (Vandenbussche et al., 2007; Waters & Langdale, 2009). PIFs are transcriptional inhibitors, and their degradation upon phytochrome activation promotes the expression of photomorphogenesis-related genes. PIF1 regulates the expression of *PORC*, encoding Pchlide oxidoreductase, an enzyme in chlorophyll biosynthesis pathway (Huq et al., 2004). PIF3 is a negative regulator of genes *HEMA1* and *GENOMES UNCOUPLED 5 (GUN5)*, encoding enzymes of Chl biosynthesis, and also genes encoding for components of Photosystem I (PSI) (Shin et al., 2009). Consequently, *pif1* and *pif3* mutants accumulate high amounts of the Chl intermediates when grown in darkness which leads to photobleaching after transfer to light. Quadruple mutant *pif1 pif3 pif4 pif5* (*pifQ*) is constitutively photomorphogenic and the gene expression profiles of the dark-grown *pifQ* mutant and the red-light-grown wild type *Arabidopsis* are similar (Shin et al., 2009). LONG HYPOCOTYLS 5 (HY5) is a bHLH transcription factor, acting downstream of light signalling cascades induced by activation of photoreceptors. HY5 positively regulates a number of photosynthesis-related genes (Andronis et al., 2008; Lee et al., 2007a). Activity of HY5 is mediated by protein phosphorylation and degradation, where COP1 plays an important role. CRY1 and HY5 have been shown to regulate many highlight responsive genes, including EARLY LIGHT INDUCIBLE PROTEINs (ELIPs) and ASCORBATE PEROXIDASE 2 (APX2), in young seedlings (Kleine et al., 2007). The *cry1* mutant has enhanced sensitivity to high irradiance which is demonstrated by photobleaching and increased photoinhibition of the photosystem II (PSII) under high light (Kleine et al., 2007).

Belonging to another class of blue light photoreceptors, PHOT1 and PHOT2 have a minor role in blue light dependent photomorphogenesis compared to CRYs. However, PHOT1 and PHOT2 play a key role in blue light-dependent chloroplast movement, which serves as a rapid response to different light regimes (Celaya & Liscum, 2005; Demarsy & Fankhauser, 2009). Knockout of both PHOTs is sufficient to eliminate all chloroplast movements (Sakai et al., 2001). Analysis of single *phot* mutants demonstrated the exclusive role of PHOT2 in high light response and redundant role of both PHOT proteins in low light response (Ohgishi et al., 2004). PHOTs have also partially redundant functions in the regulation of phototropism and stomatal opening (Kinoshita et al., 2001). PHOT proteins are plasma membrane localized protein kinases, containing two light oxygen voltage (LOV) domains, essential for their function. PHOT1 and PHOT2 can autophosphorylate (Inoue et al., 2008; Matsuoka & Tokutomi, 2005). This autophosphorylation mediates *in vivo* interaction of PHOT with 14-3-3 proteins (Sullivan et al., 2009). In addition, PHOT interacts directly with PHYTOCHROME KINASE SUBSTRATE family proteins (de Carbonnel et al., 2010).

Unlike light in the visible spectrum, even moderate levels of UV-B irradiation cause oxidative stress. Studies of UV-B light signalling suggested its perception by a specific receptor, different from those discussed above (Ulm & Nagy, 2005). UV RESPONSE LOCUS 8 (UVR8) was recently shown to be a direct sensor of UV-B in *Arabidopsis* (Rizzini et al., 2011). The signalling pathway leading to transcriptional response includes also COP1 and HY5 (Brown & Jenkins, 2008; Oravecz et al., 2006). In the absence of UV-B UVR8 localizes to the nucleus and to the cytosol (Kaiserli & Jenkins, 2007). Upon UV-B irradiation UVR8 translocates to the nucleus where it interacts directly with COP1 in a UV-B dependent manner which, in turn, affects the interaction between COP1 and HY5 (Favory et al., 2009; Kaiserli & Jenkins, 2007).

Mature chloroplasts, in turn, work as environmental sensors regulating stress response and dynamic acclimation of the photosynthetic mechanism to environmental fluctuations. Environmental factors such as light, temperature and nutrient availability exert a strong effect on the function of plant chloroplasts. Changes in light intensity lead to rearrangement of the light harvesting antenna complexes and changes in composition of the two photosystems. Low light requires optimizing of light harvesting whereas high light leads to increased energy dissipation and photoinhibition. Mechanisms of regulation are referred to as nonphotochemical quenching (NPQ) (Bailey & Grossman, 2008; Holt et al., 2004). The *Arabidopsis npq4* mutant, lacking the minor subunit S of PSII (PSBS), is deficient in dissipation of excess energy and is more sensitive for photoinhibition during high light. Overexpression of *PsbS* leads to enhanced resistance to high light stress (Li et al., 2002). Moreover, the transcript and protein levels of *PsbS* are elevated upon exposure of *Arabidopsis* to high light (Giacomelli et al., 2006; Kimura et al., 2003). Screens for *npq* mutants in *Chlamydomonas* identified stress-related chlorophyll a/b binding proteins 2 and 3 as proteins required for proper energy dissipation (Bonente et al., 2011; Peers et al., 2009).

Changes in light quality induce imbalanced excitation of the two photosystems which have distinct absorption spectra. The mechanisms of acclimation include reversible phosphorylation and migration of a part of the light-harvesting antenna complexes between photosystems to balance the energy distribution (so-called state transitions). The process is controlled by the redox status of the plastoquinone pool (PQ). The genetic screen for mutants with impaired state transitions in *Chlamydomonas* resulted in identification of Stt7 protein kinase responsible for phosphorylation of LHCII (Depège et al., 2003). Two orthologs of Stt7 with different substrate specificity exist in *Arabidopsis*: STN7 protein kinase phosphorylates Light Harvesting Complex II (LHCII), and STATE TRANSITION 8 (STN8) protein kinase is involved in phosphorylation of PSII core proteins (Bonardi et al., 2005; Vainonen et al., 2005). STN7, activated by the reduced PQ pool, is required for state transitions (Bonardi et al., 2005) and supposed to be involved in retrograde signalling *via* a yet unknown mechanism (Pesaresi et al., 2010; Pfannschmidt, 2010). Oxidation of the PQ pool inactivates the STN7 kinase which leads to dephosphorylation of light-harvesting proteins by THYLAKOID-ASSOCIATED PHOSPHATASE 38 (TAP38/PPH1) protein phosphatase identified in a reverse genetics screens (Pribil et al., 2010; Shapiguzov et al., 2010). Thus, the pair STN7/TAP38(PPH1) is a key regulator of state transitions and short-term acclimation in *Arabidopsis*. The lack of STN8 protein kinase in *stn8* and *stn7 stn8* single and double mutants results in ineffective repair of PSII and higher susceptibility to photoinhibition under high light stress (Tikkanen et al., 2008).

The photosynthetic reactions in chloroplasts are a continuous source of ROS: several different ROS are produced inside plastids as by-products of photosynthesis. Superoxide ($O_2^{\cdot-}$) is converted to hydrogen peroxide (H_2O_2) by a chloroplast-associated superoxide dismutase (SOD). Singlet oxygen ($^1O_2$) is produced at PSII by energy transfer from excited chlorophyll molecules to oxygen. Imbalanced electron flow due to high light or environmental stress favour production of $^1O_2$ inside plastids. Production of ROS in chloroplasts leads to dramatic changes in nuclear gene expression including genes encoding for photosystem subunits (Bechtold et al., 2008; Fey et al., 2005). The extensive study of the molecular mechanisms of signalling suggests several partially overlapping and redundant pathways (Nott et al., 2006).

The *Arabidopsis fluorescent in blue light* (*flu*) mutant (Meskauskiene et al., 2001) has played a key role as a tool in dissecting signalling pathways specific to singlet oxygen generation within plastids and it has been used to identify genes specifically regulated by singlet oxygen release within plastids (op den Camp et al., 2003). FLU is a nuclear-encoded plastid protein and a negative regulator of Chl biosynthesis. Dark-grown *flu* seedlings overaccumulate protochlorophyllide (Pchlide), an intermediate in the Chl biosynthetic pathway. Pchlide acts as a photosensitizer and produces singlet oxygen upon illumination. Transfer of *flu* seedlings from dark to light leads to their rapid bleaching and death. Light is required for the induction of early responsive genes and the onset of programmed cell death (PCD) in *flu* seedlings after the dark-light shift. CRY1 was shown to mediate the transcriptional response, thus, linking ROS signalling to photomorphogenesis (Danon et al., 2006). Under continuous light the *flu* mutant grows as wild type; however, placing mature *flu* plants in darkness followed by transfer to light, results in growth inhibition.

Extensive mutant screens have been made with mutagenized *flu* seeds to identify second-site mutants that do not show bleaching of seedlings or growth inhibition of mature plants after the dark-to-light shift (Lee et al., 2007b; Wagner et al., 2004). Another approach has utilised a transgenic line of the *flu* mutant with a $^1O_2$-responsive reporter gene (Baruah et al., 2009a; Baruah et al., 2009b). The majority of mapped mutations targets genes encoding for chloroplast proteins. EXECUTER 1 (EX1), identified in a suppressor screen, is required for induction of a larger part of $^1O_2$-responsive genes. The double mutant *ex1 flu* accumulates the same level of the chlorophyll intermediate in darkness, but grows similarly to wild type and does not bleach after a transfer to light. This result suggests that in *ex1 flu* the signalling pathway from $^1O_2$ in chloroplasts to the nucleus is impaired. The closely related EXECUTER 2 works as a modulator of EX1 activity. Both nuclear-encoded proteins are localized to the thylakoid membranes inside chloroplasts; the molecular mechanism of their action is so far unknown.

Another group of second site mutants, *soldat* mutants, suppress the PCD of *flu* seedlings after the dark-light shift. Both characterized mutants (*soldat8 flu, soldat10 flu*) accumulate the same level of Pchlide and exhibit almost the same transcriptional response as the parental *flu* plants but their seedlings remain green after illumination. Identification of impaired genes resulted in SIGMA6 factor of plastid RNA polymerase in *soldat8* and mTERF-related plastid protein in *soldat10*. The only non-plastid regulator identified so far in the mutant screens is PLEIOTROPHIC RESPONSE LOCUS 1 (PRL1) protein, localized to the nucleus and known to mediate multiple stress responses, including energy deprivation (Baruah et al., 2009a). The mutant was isolated by constitutive expression of a $^1O_2$-responsive gene. Gene expression data showed that PRL1 regulates only a subset of $^1O_2$-responsive genes and, in addition, it regulates other genes. The *prl1* mutant has enhanced resistance to combined (cold/high light) stress than wild type (Baruah et al., 2009a). In summary, using the *flu* mutant and genetic approaches, it has been shown that $^1O_2$ signalling from plastids is mediated by several signal transduction pathway and is linked to a variety of environmental stresses.

5. ROS are a common factor of abiotic stresses

A common factor between most stresses is the active production of reactive oxygen species (ROS) (Jaspers & Kangasjärvi, 2010). Also salt acclimation and osmotic stress result in the increased production of ROS (Miller et al., 2010). It has become clear that ROS are not only toxic to cells but serve important functions as signalling molecules. However, we do not currently have a clear understanding of how ROS signalling works and how ROS signals are perceived. The regulation of ROS production occurs in different subcellular locations including the chloroplasts, peroxisomes, mitochondria and in the apoplast. Important components of the apoplastic oxidative burst are the NADPH oxidases respiratory burst oxidase homologues (AtRBOH). AtRBOHD and AtRBOHF have important functions during the defence against pathogens but also in the response to abiotic stress (Miller et al., 2009). It has been proposed that the ROS produced by the NADPH oxidases in the extracellular space serve mostly as signalling molecules rather than being directly cytotoxic (Jaspers & Kangasjärvi, 2010). While the importance of ROS in signalling during stress adaptation and development in plants has been recognized, their perception has remained largely elusive. Several regulators of transcription have been found to be redox regulated in plants, most notably NONEXPRESSER OF PR GENES 1 (NPR1) and the transcription factor TGA2 (see below) (Després et al., 2003; Fan & Dong, 2002; Fobert & Després, 2005). In addition, methionine oxidation of proteins might be another way to regulate the enzymatic activity of proteins directly through redox control (Hardin et al., 2009). These proteins however rely on intracellular ROS for redox regulation. The perception of apoplastic ROS is so far not well understood. Extracellular ROS might be perceived in different ways. Either proteins might be redox modified leading to ROS sensing, or other components of cell wall or membranes might be oxidized and those in turn sensed by receptors, leading to signalling into the cytosol (Hancock et al., 2006; Jaspers & Kangasjärvi, 2010). In particular cysteine-rich proteins have been proposed as candidates for ROS perception in the apoplast (Reddie & Carroll, 2008). Candidates include receptor-like kinases (RLKs) and small apoplastic proteins and peptides. The members of the Cysteine-rich receptor-like kinase (CRKs) subfamily of the receptor-like kinases (RLKs) contain two copies of a conserved cysteine motif C-8X-C-2X-C which has been proposed to serve as a suitable target for redox regulation (Acharya et al., 2007; Chen et al., 2003; Chen et al., 2004; Czernic et al., 1999). *CRKs* have been found to be transcriptionally regulated by oxidative stress, pathogen attack and plant hormones (Chen et al., 2004; Czernic et al., 1999; Wrzaczek et al., 2010). Plants ectopically overexpressing CRKs have altered responses to pathogens and show spontaneous induction of cell death (Acharya et al., 2007; Chen et al., 2003; Chen et al., 2004). The small extracellular GRIM REAPER (GRI) protein is another candidate for ROS perception (Wrzaczek et al., 2009a; Wrzaczek et al., 2009b). A peptide derived from GRI can induce cell death in plants in the presence of superoxide. Interestingly cell death induced by GRI-peptide is dependent on superoxide produced by AtRBOHD but not AtRBOHF which could point to different timing of ROS production by different NADPH oxidases and different functions. However, while evidence could support a role for CRKs and GRI in ROS perception, this hypothesis has yet to be experimentally tested.

The intracellular redox status is sensed also by transcription factors. Interestingly, in yeast and bacteria redox-sensitive transcription factors are involved in the transcriptional regulation of genes encoding for antioxidant enzymes (Georgiou, 2002). However, only few studies have reported the presence of such sensors in plants. During plant immunity, NPR1 plays a critical role in the activation of defence related gene expression, such as *PATHOGENESIS-RELATED 1* (*PR1*). The monomerization of NPR1 occurred upon redox state changes, allowing the translocation of the protein to the nucleus, thereby enhanced the transcription of defence genes after interaction with TGAs (TGA2). The conformation changes of NPR1 have been demonstrated to result from thioredoxin activity and S-nitrosylation (Tada et al., 2008). Recently, the use of cytosolic *ascorbate peroxidase1* (*apx1*) mutants, *Catalase* (*Cat*) antisense lines and pharmacological approach demonstrated that cytosolic H_2O_2 accumulation prevents the translocation of NPR1 protein to the nucleus (Peleg-Grossman et al., 2010), suggesting the important role of NPR1 in the sensing of redox state changes which occur in the cell. Based on the literature in mammals and yeast, other transcription factors, such as heat shock transcription factors have been suggested to act as direct H_2O_2 sensors (Miller & Mittler, 2006).

Another indirect mechanism through which ROS are perceived is *via* the oxidatively modified molecules that are produced by the action of ROS. In principle any oxidatively damaged biomolecule has the potential to act as a signal. In particular the lipids represent a rich source of potential ROS dependant signals. Oxidative membrane damage, especially the process of membrane lipid peroxidation chain reactions are a common and very well documented consequence of oxidative stress. Oxidized lipids are generally themselves highly reactive and toxic; however, many have demonstrated biological activity, especially at low concentrations (Mueller & Berger, 2009). Reactive oxylipins are thiol reactive and able to modify thiol groups on a large number of proteins. Accordingly, similar sets of genes are regulated in transcriptomics experiments by reactive oxylipins as are by other thiol active compounds such as ROS and reactive nitrogen species (RNS) (Mueller & Berger, 2009). Oxylipins represent good example of how oxidized lipids can act as signal molecules. 12-oxo-phytodienoic acid (12OPDA) is a cyclopentone oxylipin intermediate on the jasmonic acid biosynthesis pathway. Interestingly, 12OPDA can be produced both enzymatically and non-enzymatically *via* a free radical catalyzed peroxidation pathway. 12OPDA induces physiological responses similar to those of jasmonates including upregulation of stress and general detoxification pathways.

6. Antioxidant systems

As described above, ROS are actively produced and used as signalling molecules by cells in response to most abiotic stresses. In addition, ROS are continuously produced as by-products during different metabolic pathways in plants. Due to the highly reactive nature of ROS their production and detoxification needs to be strictly controlled. ROS production occurs in virtually all cell compartments; but most notably, in different organelles including mitochondria, chloroplasts and peroxisomes, due to their highly oxidizing metabolic activity, and also in the apoplast.

Plants have developed various antioxidant systems which are specifically localized to different subcellular loci and can be induced upon a stimulus. Diffusion of ROS is typically limited due to their high reactivity. This property requires ROS scavenging and detoxification to take place at or close to the location of ROS production. The ubiquitous presence of the antioxidant system is critical in order to prevent and survive oxidative stress. However, due to its presence and flexibility, the antioxidant systems could also present an efficient means to sense ROS through the antioxidant status of the cell.

Oxidative stress is defined as a disturbance in the equilibrium status of oxidants and antioxidants. Most of the antioxidant components act as free radical scavengers. By binding and inactivating the free radicals, antioxidants protect against oxidative stress. The tight cooperation between enzymatic and non-enzymatic antioxidant systems provides to the cell an elaborated and efficient system to regulate ROS levels (De Gara et al., 2010; Foyer & Noctor, 2009; Miller et al., 2010; Shao et al., 2008).

The non-enzymatic antioxidant system includes ascorbic acid (AsA), GSH and α-tocopherol. AsA and GSH, the most abundant soluble antioxidants in plants, play a key role in plant defence against oxidative stress (Foyer & Noctor, 2011). These antioxidants are present predominantly in the reduced form in the majority of subcellular compartments. In response to stress, GSH can be oxidized into its disulphide form, GSSG, which can accumulate to high levels in plant cells. Glutathione redox status (GSSG/GSH ratio) reflects the level of oxidative stress, therefore GSH is frequently considered to be a suitable oxidative stress marker (Foyer & Noctor, 2009; Foyer & Noctor, 2011). GSH is also able to react with nitrous acid to form S-nitrosothiols called S-nitrosoglutathione (GSNO), which constitute a nitric oxide (NO) reservoir and are emerging to play key role in NO signalling pathways in plants (Besson-Bard et al., 2008; Del Rio, 2011; Lindermayr et al., 2005). In addition, glutathionylation of proteins can modify their activity (Palmieri et al., 2010). GSH and AsA are thought to function in a coordinated manner to regulate redox homeostasis plants during development and environmental responses (Foyer & Noctor, 2005; Foyer & Noctor, 2009; Foyer & Noctor, 2011). α-tocopherol, an abundant vitamin E compound, is a lipid soluble antioxidant found in chloroplasts where it counteracts lipid peroxidation through scavenging of lipid peroxyl radicals and detoxifies singlet oxygen and hydroxyl radicals (Munné-Bosch, 2005).

The major elements of the enzymatic antioxidant system are summarized in the Table 1. SOD, APX and CAT are three main enzymes present ubiquitously permitting the tightly control of ROS levels by scavenging directly ROS and converting them into less reactive and less harmful species. They can be considered as intracellular ROS sensors due to their direct interaction with ROS. Another group of enzymes, monodehydroascorbate reductase (MDAR), dehydroascorbate reductase (DHAR) and glutathione reductase (GR), is involved in the reduction of oxidized AsA or GSH, thus, balancing the redox status of the cell (Asada, 1999; Foyer & Noctor, 2011). Up-regulation of the enzymes involved in the antioxidant system both at the transcript and the protein levels in response to ROS accumulation has been shown for a variety of abiotic stresses (Gill & Tuteja, 2010).

The low molecular-weight antioxidants and the enzymatic antioxidant system constitute a complex and diversified antioxidant system which participates efficiently in all the different organelles of plant cells to maintain redox state. The antioxidant system is one of the first barriers against ROS overproduction during abiotic stresses. The ratios between the oxidized and the reduced forms of antioxidants have been suggested to play a key role in the oxidative stress response and the control of the antioxidant systems (Karpinski et al., 1997). Other studies have underlined the importance of low ascorbate levels and variations of GSH pools during pathogen response, cell death and dormancy (Barth et al., 2004; Gomez et al., 2004; Kranner et al., 2002; Kranner et al., 2006; Mou et al., 2003; Parisy et al., 2007; Pastori et al., 2003)

7. Beyond signal perception and ROS detoxification

So far this review has addressed the perception mechanisms of abiotic stress. However, after a given stress has been perceived the various signals need to be integrated and transmitted into an appropriate response for adaptation. Protein kinases are important components of the signal transduction cascade that links abiotic stress perception to signal integration and transcriptional reprogramming of the cell (de la Fuente van Bentem et al., 2008; Nakagami et al., 2010; Sugiyama et al., 2008). MAPKs and phosphatases alongside with CPKs are prominent participants in the abiotic stress signalling network (Ichimura et al., 2000; Jonak et al., 2004; Mehlmer et al., 2010; Miller et al., 2008; Mizoguchi et al., 1997; Takahashi et al., 2011; Wurzinger et al., 2011; Yuasa et al., 2001).

Enzyme	Reaction catalyzed	Subcellular localization	References
Superoxide dismutase (SOD)	$O_2^{\cdot -} + O_2^{\cdot -} + 2H^+ = 2H_2O_2 + O_2$	cytosol (Cu/Zn-SOD) chloroplast (Cu/Zn-SOD, Fe-SOD) mitochondria and peroxisome (Mn-SOD)	(Alscher et al., 2002; Del Rio, 2011)
Catalase (CAT)	$H_2O_2 = H_2O + \frac{1}{2} O_2$	peroxisome	(Mhamdi et al., 2010; Willekens et al., 1997)
Ascorbate peroxidase (APX)	H_2O_2 + AsA = H_2O + DHA	chloroplast, cytosol, mitochondria	(Santos et al., 1996)
Monodehydroascorbate reductase (MDHAR)	MDHA + NAD(P)H = AA + $NAD(P)^+$	chloroplast, mitochondria, cytosol	(Asada, 1999)
Dehydroascorbate reductase (DHAR)	DHA + 2GSH = AsA + GSSG	chloroplast, cytosol	(Urano et al., 2000)
Glutathione reductase (GR)	GSSG + NAD(P)H = 2GSH + $NAD(P)^+$	cytosol,peroxisomes (AtGR1) chloroplast, mitochondria (AtGR2)	(Kaur & Hu, 2009)

Table 1. Major enzymes involved in ROS detoxification in plants.

Also forward genetic screens have yielded significant insight into the metabolism and signalling in response to abiotic stress and in ROS signalling. Paraquat (methyl viologen) and ozone (O_3) are two of the most commonly used stress inducers in these screens. Paraquat leads to ROS production in the chloroplast as it unlinks the electron transport chain in PSI and is used as a model to address the role and metabolism of chloroplastic ROS. A large-scale screen has identified several new proteins of mostly unknown function involved in the response to oxidative stress (Luhua et al., 2008). O_3 breaks down in the leaf apoplastic space to form secondary ROS that both mimic and induce an apoplastic oxidative burst similar to that of other stresses, most notably the pathogen response. Therefore O_3 is used as a model to investigate the signalling processes associated with apoplastic ROS. Many mutants isolated in screens for O_3-sensitivity are related to antioxidants (Conklin et al., 1996; Conklin & Last, 1995), stomatal biology (Vahisalu et al., 2008) and stress hormones (Ahlfors et al., 2008; Mahalingam et al., 2003; Vahala et al., 2003a; Vahala et al., 2003b), which are covered elsewhere in this chapter or have been previously reviewed (Overmyer et al., 2003). Some mutants have helped to define novel ROS signalling pathways; such as the *rcd1* mutant. The *rcd1* mutant was originally identified in a screen for O_3-sensitivity (Overmyer et al., 2000). The *RCD1* gene is also highly induced during high light stress

(Bechtold et al., 2008) providing a link to light induced chloroplastic ROS signalling. Intriguingly *rcd1* shows enhanced tolerance to paraquat (Ahlfors et al., 2004; Overmyer et al., 2000). RCD1 also shows a subtle salt-sensitive phenotype and is able to interact with the Na^+/H^+ antiporter SOS1 in yeast 2-hybrid (Katiyar-Agarwal et al., 2006). SOS1 is crucial for salt stress tolerance and *sos1* mutants are hypersensitive to salt stress, thus the interaction with RCD1 provides one more link between salt stress and ROS signalling. RCD1 is able to interact with several transcription factors in yeast 2-hybrid analysis (Jaspers et al., 2009; Jaspers et al., 2010b). This together with physiological data makes RCD1 a promising candidate as a central integration node for ROS signalling and indicates that highly divergent perception mechanisms, such as those described here for highlight, salt, dehydration (stomatal), and herbicide induced stresses, all converge into a common pathway (Jaspers et al., 2009; Jaspers et al., 2010a; Jaspers & Kangasjärvi, 2010).

8. Conclusions

The signalling pathways leading to an appropriate and coordinated response to abiotic stress have been the target of intensive research in the past decades. Global warming and intensive farming make it necessary to understand the underlying mechanisms and modify plant stress tolerance through selective breeding or genetic modification. However, with the exception of light perception, relatively little is known about the mechanisms of abiotic stress sensing. Recent research has identified several potential candidates as sensors for salt and heavy metals but a final proof of their function as true receptors is still needed. Similarly to biotic interactions, the production of ROS in abiotic stress has been well established in the last years. Increasing amount of evidence suggests that the generation of specific types of ROS in defined subcellular compartments is an important component of the stress response. Rather than being just cytotoxic by-products of biochemical processes ROS are likely to play central roles as regulators of stress adaptation, PCD, and even plant development (Apel & Hirt, 2004; Jaspers & Kangasjärvi, 2010; Miller et al., 2008). While ROS are biochemically simple molecules the intricate specificity observed during ROS signalling is still insufficiently understood (Gadjev et al., 2006). Some components of ROS perception inside the cell through modification of transcription factors and monitoring of the oxidative potential by means of ROS scavengers have been identified. However these do not explain the high specificity of transcriptional reprogramming induced by production of different ROS in specific subcellular locations.

9. Acknowledgements

We thank Dr. Jorma Vahala and Dr. Hannes Kollist for critical comments on the manuscript. Our work is supported by the Academy of Finland Centre of Excellence Programme and Post-doctoral research grant to Adrien Gauthier. Kirk Overmyer is supported by the University of Helsinki.

10. References

Acharya, B. R., Raina, S., Maqbool, S. B., Jagadeeswaran, G., Mosher, S. L., Appel, H. M., Schultz, J. C., Klessig, D. F., and Raina, R. (2007) Overexpression of CRK13, an

Arabidopsis cysteine-rich receptor-like kinase, results in enhanced resistance to *Pseudomonas syringae*. *Plant Journal* Vol. 50, No. 3, pp. 488-499

Ahlfors, R., Brosché, M., Kollist, H., and Kangasjärvi, J. (2008) Nitric oxide modulates ozone-induced cell death, hormone biosynthesis and gene expression in *Arabidopsis thaliana*. *Plant Journal* Vol. 58, No. 1, pp. 1-12

Ahlfors, R., Lång, S., Overmyer, K., Jaspers, P., Brosché, M., Taurianinen, A., Kollist, H., Tuominen, H., Belles-Boix, E., Piippo, M., Inzé, M., Palva, E. T., and Kangasjärvi, J. (2004) Arabidopsis radical-induced cell death1 belongs to the WWE protein-protein interaction domain protein family and modulates abscisic acid, ethylene, and methyl jasmonate responses. *Plant Cell* Vol. 16, No. 7, pp. 1925-1937

Alscher, R. G., Erturk, N., and Heath, L. S. (2002) Role of superoxide dismutases (SODs) in controlling oxidative stress in plants. *Journal of Experimental Botany* Vol. 53, No. 372, pp. 1331-1341

Andronis, C., Barak, S., Knowles, S. M., Sugano, S., and Tobin, E. M. (2008) The clock protein CCA1 and the bZIP transcription factor HY5 physically interact to regulate gene expression in *Arabidopsis*. *Molecular Plant* Vol. 1, No. 1, pp. 58-67

Apel, K. and Hirt, H. (2004) Reactive oxygen species: metabolism, oxidative stress, and signal transduction. *Annual Review of Plant Biology* Vol. 55, pp. 373-399

Arosio, P., Ingrassia, R., and Cavadini, P. (2009) Ferritins: a family of molecules for iron storage, antioxidation and more. *Biochimica et Biophysica Acta* Vol. 1790, No. 7, pp. 589-599

Asada, K. (1999) The Water-Water cycle in chloroplasts: Scavenging of Active Oxygens and Dissipation of Excess Photons. *Annual Review of Plant Physiology and Plant Molecular Biology* Vol. 50, pp. 601-639

Baekgaard, L., Mikkelsen, M. D., Sorensen, D. M., Hegelund, J. N., Persson, D. P., Mills, R. F., Yang, Z., Husted, S., Andersen, J. P., Buch-Pedersen, M. J., Schjoerring, J. K., Williams, L. E., and Palmgren, M. G. (2010) A combined zinc/cadmium sensor and zinc/cadmium export regulator in a heavy metal pump. *Journal of Biological Chemistry* Vol. 285, No. 41, pp. 31243-31252

Bailey, S. and Grossman, A. (2008) Photoprotection in cyanobacteria: regulation of light harvesting. *Photochemistry and Photobiology* Vol. 84, No. 6, pp. 1410-1420

Barth, C., Moeder, W., Klessig, D. F., and Conklin, P. L. (2004) The timing of senescence and response to pathogens is altered in the ascorbate-deficient Arabidopsis mutant *vitamin c-1*. *Plant Physiology* Vol. 134, No. 4, pp. 1784-1792

Baruah, A., Simkova, K., Apel, K., and Laloi, C. (2009a) Arabidopsis mutants reveal multiple singlet oxygen signaling pathways involved in stress response and development. *Plant Physiology* Vol. 70, No. 5, pp. 547-563

Baruah, A., Simkova, K., Hincha, D. K., Apel, K., and Laloi, C. (2009b) Modulation of $^1O_2$-mediated retrograde signaling by the PLEIOTROPIC RESPONSE LOCUS 1 (PRL1) protein, a central integrator of stress and energy signaling. *Plant Journal* Vol. 60, No. 1, pp. 22-32

Bechtold, U., Richard, O., Zamboni, A., Gapper, C., Geisler, M., Pogson, B., Karpinski, S., and Mullineaux, P. M. (2008) Impact of chloroplastic- and extracellular-sourced

ROS on high light-responsive gene expression in *Arabidopsis*. *Journal of Experimental Botany* Vol. 59, No. 2, pp. 121-133

Besson-Bard, A., Pugin, A., and Wendehenne, D. (2008) New insights into nitric oxide signaling in plants. *Annual Review of Plant Biology* Vol. 59, pp. 21-39

Bonardi, V., Pesaresi, P., Becker, T., Schleiff, E., Wagner, R., Pfannschmidt, T., Jahns, P., and Leister, D. (2005) Photosystem II core phosphorylation and photosynthetic acclimation require two different protein kinases. *Nature* Vol. 437, No. 7062, pp. 1179-1182

Bonente, G., Ballottari, M., Truong, T. B., Morosinotto, T., Ahn, T. K., Fleming, G. R., Niyogi, K. K., and Bassi, R. (2011) Analysis of LhcSR3, a protein essential for feedback de-excitation in the green alga *Chlamydomonas reinhardtii*. *PLoS Biology* Vol. 9, No. 1, e1000577

Briat, J. F., Curie, C., and Gaymard, F. (2007) Iron utilization and metabolism in plants. *Current Opinions in Plant Biology* Vol. 10, No. 3, pp. 276-282

Briat, J. F., Ravet, K., Arnaud, N., Duc, C., Boucherez, J., Touraine, B., Cellier, F., and Gaymard, F. (2010) New insights into ferritin synthesis and function highlight a link between iron homeostasis and oxidative stress in plants. *Annals of Botany* Vol. 105, No. 5, pp. 811-822

Brown, B. A. and Jenkins, G. I. (2008) UV-B signaling pathways with different fluence-rate response profiles are distinguished in mature Arabidopsis leaf tissue by requirement for UVR8, HY5, and HYH. *Plant Physiology* Vol. 146, No. 2, pp. 576-588

Castillon, A., Shen, H., and Huq, E. (2007) Phytochrome Interacting Factors: central players in phytochrome-mediated light signaling networks. *Trends in Plant Science*. Vol. 12, No. 11, pp. 514-521

Celaya, R. B. and Liscum, E. (2005) Phototropins and associated signaling: providing the power of movement in higher plants. *Photochemistry and Photobiology* Vol. 81, No. 1, pp. 73-80

Chefdor, F., Bénédetti, H., Depierreux, C., Delmotte, F., Morabito, D., and Carpin, S. (2006) Osmotic stress sensing in *Populus*: components identification of a phosphorelay system. *FEBS Letters* Vol. 580, No. 1, pp. 77-81

Chen, K., Du, L., and Chen, Z. (2003) Sensitization of defense responses and activation of programmed cell death by a pathogen-induced receptor-like protein kinase in *Arabidopsis*. *Plant Molecular Biology* Vol. 53, No. 1-2, pp. 61-74

Chen, K., Fan, B., Du, L., and Chen, Z. (2004) Activation of hypersensitive cell death by pathogen-induced receptor-like protein kinases from *Arabidopsis*. *Plant Molecular Biology* Vol. 56, No. 2, pp. 271-283

Cobbett, C. and Goldsbrough, P. (2002) Phytochelatins and metallothioneins: roles in heavy metal detoxification and homeostasis. *Annual Review of Plant Biology* Vol. 53, pp. 159-182

Conklin, P. L. and Last, R. L. (1995) Differential accumulation of antioxidant mRNAs in *Arabidopsis thaliana* exposed to ozone. *Plant Physiology* Vol. 109, No. 1, pp. 203-212

Conklin, P. L., Williams, E. H., and Last, R. L. (1996) Environmental stress sensitivity of an ascorbic acid-deficient Arabidopsis mutant. *Proceedings of the National Academy of Sciences of the United States of America* Vol. 93, No. 18, pp. 9970-9974

Culham, D. E., Henderson, J., Crane, R. A., and Wood, J. M. (2003a) Osmosensor ProP of *Escherichia coli* responds to the concentration, chemistry, and molecular size of osmolytes in the proteoliposome lumen. *Biochemistry* Vol. 42, No. 2, pp. 410-420

Culham, D. E., Hillar, A., Henderson, J., Ly, A., Vernikovska, Y. I., Racher, K. I., Boggs, J. M., and Wood, J. M. (2003b) Creation of a fully functional cysteine-less variant of osmosensor and proton-osmoprotectant symporter ProP from *Escherichia coli* and its application to assess the transporter's membrane orientation. *Biochemistry* Vol. 42, No. 40, pp. 11815-11823

Czernic, P., Visser, B., Sun, W., Savouré, A., Deslandes, L., Marco, Y., Van Montagu, M., and Verbruggen, N. (1999) Characterization of an Arabidopsis thaliana receptor-like protein kinase gene activated by oxidative stress and pathogen attack. *Plant Journal* Vol. 18, No. 3, pp. 321-327

Danon, A., Coll, N. S., and Apel, K. (2006) Cryptochrome-1-dependent execution of programmed cell death induced by singlet oxygen in *Arabidopsis thaliana*. *Proceedings of the National Academy of Sciences of the United States of America* Vol. 103, No. 45, pp. 17036-17041

de Carbonnel, M., Davis, P., Roelfsema, M. R., Inoue, S., Schepens, I., Lariguet, P., Geisler, M., Shimazaki, K., Hangarter, R., and Fankhauser, C. (2010) The Arabidopsis PHYTOCHROME KINASE SUBSTRATE2 protein is a phototropin signaling element that regulates leaf flattening and leaf positioning. *Plant Physiology* Vol. 152, No. 3, pp. 1391-1405

De Gara, L., Locato, V., Dipierro, S., and de Pinto, M. C. (2010) Redox homeostasis in plants. The challenge of living with endogenous oxygen production. *Respiratory Physiology & Neurobiology* Vol. 173 Suppl, S13-S19

de la Fuente van Bentem, S., Anrather, D., Dohnal, I., Roitinger, E., Csaszar, E., Joore, J., Buijnink, J., Carreri, A., Forzani, C., Lorkovic, Z. J., Barta, A., Lecourieux, D., Verhounig, A., Jonak, C., and Hirt, H. (2008) Site-specific phosphorylation profiling of *Arabidopsis* proteins by mass spectrometry and peptide chip analysis. *Journal of Proteome Research* Vol. 7, No. 6, pp. 2458-2470

De Vos, C. H., Vonk, M. J., Vooijs, R., and Schat, H. (1992) Glutathione depletion due to copper-Induced phytochelatin synthesis causes oxidative stress in *Silene cucubalus*. *Plant Physiology* Vol. 98, No. 3, pp. 853-858

Del Rio, L. A. (2011) Peroxisomes as a cellular source of reactive nitrogen species signal molecules. *Archives of Biochemistry and Biophysics* Vol. 506, No. 1, pp. 1-11

Demarsy, E. and Fankhauser, C. (2009) Higher plants use LOV to perceive blue light. *Current Opinions in Plant Biology* Vol. 12, No. 1, pp. 69-74

Depège, N., Bellafiore, S., and Rochaix, J. D. (2003) Role of chloroplast protein kinase Stt7 in LHCII phosphorylation and state transition in *Chlamydomonas*. *Science* Vol. 299, No. 5612, pp. 1572-1575

Després, C., Chubak, C., Rochon, A., Clark, R., Bethune, T., Desveaux, D., and Fobert, P. R. (2003) The Arabidopsis NPR1 disease resistance protein is a novel cofactor that confers redox regulation of DNA binding activity to the basic domain/leucine zipper transcription factor TGA1. *Plant Cell* Vol. 15, No. 9, pp. 2181-2191

Epstein, W. (1992) Kdp, a bacterial P-type ATPase whose expression and activity are regulated by turgor pressure. *Acta Physiologica Scandinavica Supplements* Vol. 607, pp. 193-199

Fan, W. and Dong, X. (2002) In vivo interaction between NPR1 and transcription factor TGA2 leads to salicylic acid-mediated gene activation in Arabidopsis. *Plant Cell* Vol. 14, No. 6, pp. 1377-1389

Favory, J. J., Stec, A., Gruber, H., Rizzini, L., Oravecz, A., Funk, M., Albert, A., Cloix, C., Jenkins, G. I., Oakeley, E. J., Seidlitz, H. K., Nagy, F., and Ulm, R. (2009) Interaction of COP1 and UVR8 regulates UV-B-induced photomorphogenesis and stress acclimation in Arabidopsis. *EMBO Journal* Vol. 28, No. 5, pp. 591-601

Fey, V., Wagner, R., Bräutigam, K., Wirtz, M., Hell, R., Dietzmann, A., Leister, D., Oelmüller, R., and Pfannschmidt, T. (2005) Retrograde plastid redox signals in the expression of nuclear genes for chloroplast proteins of *Arabidopsis thaliana*. *Journal of Biological Chemistry* Vol. 280, No. 7, pp. 5318-5328

Fobert, P. R. and Després, C. (2005) Redox control of systemic acquired resistance. *Current Opinions in Plant Biology* Vol. 8, No. 4, pp. 378-382

Foyer, C. H., Bloom, A. J., Queval, G., and Noctor, G. (2009) Photorespiratory metabolism: genes, mutants, energetics, and redox signaling. *Annual Review of Plant Biology* Vol. 60, pp. 455-484

Foyer, C. H. and Noctor, G. (2005) Redox homeostasis and antioxidant signaling: a metabolic interface between stress perception and physiological responses. *Plant Cell* Vol. 17, No. 7, pp. 1866-1875

Foyer, C. H. and Noctor, G. (2009) Redox regulation in photosynthetic organisms: signaling, acclimation, and practical implications. *Antioxidants & Redox Signaling* Vol. 11, No. 4, pp. 861-905

Foyer, C. H. and Noctor, G. (2011) Ascorbate and glutathione: the heart of the redox hub. *Plant Physiology* Vol. 155, No. 1, pp. 2-18

Gadjev, I., Vanderauwera, S., Gechev, T. S., Laloi, C., Minkov, I. N., Shulaev, V., Apel, K., Inzé, D., Mittler, R., and Van Breusegem, F. (2006) Transcriptomic footprints disclose specificity of reactive oxygen species signaling in Arabidopsis. *Plant Physiology* Vol. 141, No. 2, pp. 436-445

Georgiou, G. (2002) How to flip the (redox) switch. *Cell* Vol. 111, No. 5, pp. 607-610

Giacomelli, L., Rudella, A., and van Wijk, K. J. (2006) High light response of the thylakoid proteome in Arabidopsis wild type and the ascorbate-deficient mutant *vtc2-2*. A comparative proteomics study. *Plant Physiology* Vol. 141, No. 2, pp. 685-701

Gill, S. S. and Tuteja, N. (2010) Reactive oxygen species and antioxidant machinery in abiotic stress tolerance in crop plants. *Plant Physiology and Biochemistry* Vol. 48, No. 12, pp. 909-930

Gomez, L. D., Noctor, G., Knight, M. R., and Foyer, C. H. (2004) Regulation of calcium signalling and gene expression by glutathione. *Journal of Experimental Botany* Vol. 55, No. 404, pp. 1851-1859

Grill, E., Winnacker, E. L., and Zenk, M. H. (1985) Phytochelatins: the principal heavy-metal complexing peptides of higher plants. *Science* Vol. 230, No. 4726, pp. 674-676

Hamann, K., Zimmann, P., and Altendorf, K. (2008) Reduction of turgor is not the stimulus for the sensor kinase KdpD of *Escherichia coli*. *Journal of Bacteriology* Vol. 190, No. 7, pp. 2360-2367

Hancock, J., Desikan, R., Harrison, J., Bright, J., Hooley, R., and Neill, S. (2006) Doing the unexpected: proteins involved in hydrogen peroxide perception. *Journal of Experimental Botany* Vol. 57, No. 8, pp. 1711-1718

Hänsch, R. and Mendel, R. R. (2009) Physiological functions of mineral micronutrients (Cu, Zn, Mn, Fe, Ni, Mo, B, Cl). *Current Opinions in Plant Biology* Vol. 12, No. 3, pp. 259-266

Hardin, S. C., Larue, C. T., Oh, M. H., Jain, V., and Huber, S. C. (2009) Coupling oxidative signals to protein phosphorylation via methionine oxidation in *Arabidopsis*. *Biochemical Journal* Vol. 422, No. 2, pp. 305-312

Heermann, R. and Jung, K. (2010) The complexity of the 'simple' two-component system KdpD/KdpE in *Escherichia coli*, *FEMS Microbiology Letters* Vol. 304, No. 2, pp. 97-106

Heermann, R., Lippert, M. L., and Jung, K. (2009) Domain swapping reveals that the N-terminal domain of the sensor kinase KdpD in *Escherichia coli* is important for signaling. *BMC Microbiology* Vol. 9, 133

Hirayama, T. and Shinozaki, K. (2010) Research on plant abiotic stress responses in the post-genome era: past, present and future. *Plant Journal* Vol. 61, No. 6, pp. 1041-1052

Holt, N. E., Fleming, G. R., and Niyogi, K. K. (2004) Toward an understanding of the mechanism of nonphotochemical quenching in green plants. *Biochemistry* Vol. 43, No. 26, pp. 8281-8289

Hong-Bo, S., Li-Ye, C., Cheng-Jiang, R., Hua, L., Dong-Gang, G., and Wei-Xiang, L. (2010) Understanding molecular mechanisms for improving phytoremediation of heavy metal-contaminated soils. *Critical Reviews in Biotechnology* Vol. 30, No. 1, pp. 23-30

Horie, T., Hauser, F., and Schroeder, J. I. (2009) HKT transporter-mediated salinity resistance mechanisms in *Arabidopsis* and monocot crop plants. *Trends in Plant Science* Vol. 14, No. 12, pp. 660-668

Howden, R., Andersen, C. R., Goldsbrough, P. B., and Cobbett, C. S. (1995) A cadmium-sensitive, glutathione-deficient mutant of Arabidopsis thaliana. *Plant Physiology* Vol. 107, No. 4, pp. 1067-1073

Hugouvieux, V., Dutilleul, C., Jourdain, A., Reynaud, F., Lopez, V., and Bourguignon, J. (2009) Arabidopsis putative selenium-binding protein1 expression is tightly linked to cellular sulfur demand and can reduce sensitivity to stresses requiring glutathione for tolerance. *Plant Physiology* Vol. 151, No. 2, pp. 768-781

Huq, E., Al-Sady, B., Hudson, M., Kim, C., Apel, K., and Quail, P. H. (2004) Phytochrome-interacting factor 1 is a critical bHLH regulator of chlorophyll biosynthesis. *Science* Vol. 305, No. 5692, pp. 1937-1941

Ichimura, K., Mizoguchi, T., Yoshida, R., Yuasa, T., and Shinozaki, K. (2000) Various abiotic stresses rapidly activate *Arabidopsis* MAP kinases ATMPK4 and ATMPK6. *Plant Journal* Vol. 24, No. 5, pp. 655-665

Inoue, S., Kinoshita, T., Matsumoto, M., Nakayama, K. I., Doi, M., and Shimazaki, K. (2008) Blue light-induced autophosphorylation of phototropin is a primary step for signaling. *Proceedings of the National Academy of Sciences of the United States of America* Vol. 105, No. 14, pp. 5626-5631

Jaspers, P., Blomster, T., Brosché, M., Salojärvi, J., Ahlfors, R., Vainonen, J. P., Reddy, R. A., Immink, R., Angenent, G., Turck, F., Overmyer, K., and Kangasjärvi, J. (2009) Unequally redundant RCD1 and SRO1 mediate stress and developmental responses and interact with transcription factors. *Plant Journal* Vol. 60, No. 2, pp. 268-279

Jaspers, P., Brosché, M., Overmyer, K., and Kangasjärvi, J. (2010a) The transcription factor interacting protein RCD1 contains a novel conserved domain. *Plant Signaling & Behavior* Vol. 5, No. 1, pp. 78-80

Jaspers, P. and Kangasjärvi, J. (2010) Reactive oxygen species in abiotic stress signaling. *Physiologia Plantarum* Vol. 138, No. 4, pp. 405-413

Jaspers, P., Overmyer, K., Wrzaczek, M., Vainonen, J. P., Blomster, T., Salojärvi, J., Reddy, R. A., and Kangasjärvi, J. (2010b) The RST and PARP-like domain containing SRO protein family: analysis of protein structure, function and conservation in land plants. *BMC Genomics* Vol. 11, 170

Jonak, C., Nakagami, H., and Hirt, H. (2004) Heavy metal stress. Activation of distinct mitogen-activated protein kinase pathways by copper and cadmium. *Plant Physiology* Vol. 136, No. 2, pp. 3276-3283

Kaiserli, E. and Jenkins, G. I. (2007) UV-B promotes rapid nuclear translocation of the *Arabidopsis* UV-B specific signaling component UVR8 and activates its function in the nucleus. *Plant Cell* Vol. 19, No. 8, pp. 2662-2673

Karpinski, S., Escobar, C., Karpinska, B., Creissen, G., and Mullineaux, P. M. (1997) Photosynthetic electron transport regulates the expression of cytosolic ascorbate peroxidase genes in Arabidopsis during excess light stress. *Plant Cell* Vol. 9, No. 4, pp. 627-640

Katiyar-Agarwal, S., Zhu, J., Kim, K., Agarwal, M., Fu, X., Huang, A., and Zhu, J. K. (2006) The plasma membrane Na+/H+ antiporter SOS1 interacts with RCD1 and functions in oxidative stress tolerance in *Arabidopsis*. *Proceedings of the National Academy of Sciences of the United States of America* Vol. 103, No. 49, pp. 18816-18821

Kaur, N. and Hu, J. (2009) Dynamics of peroxisome abundance: a tale of division and proliferation. *Current Opinions in Plant Biology* Vol. 12, No. 6, pp. 781-788

Kawachi, M., Kobae, Y., Mimura, T., and Maeshima, M. (2008) Deletion of a histidine-rich loop of AtMTP1, a vacuolar Zn^{2+}/H^{+} antiporter of *Arabidopsis thaliana*, stimulates the transport activity. *Journal of Biological Chemistry* Vol. 283, No. 13, pp. 8374-8383

Kimura, M., Yamamoto, Y. Y., Seki, M., Sakurai, T., Sato, M., Abe, T., Yoshida, S., Manabe, K., Shinozaki, K., and Matsui, M. (2003) Identification of *Arabidopsis* genes regulated by high light-stress using cDNA microarray. *Photochemistry and Photobiology* Vol. 77, No. 2, pp. 226-233

Kinoshita, T., Doi, M., Suetsugu, N., Kagawa, T., Wada, M., and Shimazaki, K. (2001) Phot1 and phot2 mediate blue light regulation of stomatal opening. *Nature* Vol. 414, No. 6864, pp. 656-660

Kleine, T., Kindgren, P., Benedict, C., Hendrickson, L., and Strand, A. (2007) Genome-wide gene expression analysis reveals a critical role for CRYPTOCHROME1 in the response of Arabidopsis to high irradiance. *Plant Physiology* Vol. 144, No. 3, pp. 1391-1406

Kranner, I., Beckett, R. P., Wornik, S., Zorn, M., and Pfeifhofer, H. W. (2002) Revival of a resurrection plant correlates with its antioxidant status. *Plant Journal* Vol. 31, No. 1, pp. 13-24

Kranner, I., Birtic, S., Anderson, K. M., and Pritchard, H. W. (2006) Glutathione half-cell reduction potential: a universal stress marker and modulator of programmed cell death? *Free Radical Biology & Medicine* Vol. 40, No. 12, pp. 2155-2165

Laimins, L. A., Rhoads, D. B., and Epstein, W. (1981) Osmotic control of kdp operon expression in *Escherichia coli*. *Proceedings of the National Academy of Sciences of the United States of America* Vol. 78, No. 1, pp. 464-468

Lee, J., He, K., Stolc, V., Lee, H., Figueroa, P., Gao, Y., Tongprasit, W., Zhao, H., Lee, I., and Deng, X. W. (2007a) Analysis of transcription factor HY5 genomic binding sites revealed its hierarchical role in light regulation of development. *Plant Cell* Vol. 19, No. 3, pp. 731-749

Lee, K. P., Kim, C., Landgraf, F., and Apel, K. (2007b) EXECUTER1- and EXECUTER2-dependent transfer of stress-related signals from the plastid to the nucleus of *Arabidopsis thaliana*. *Proceedings of the National Academy of Sciences of the United States of America* Vol. 104, No. 24, pp. 10270-10275

Leivar, P. and Quail, P. H. (2011) PIFs: pivotal components in a cellular signaling hub. *Trends in Plant Science* Vol. 16, No. 1, pp. 19-28

Li, X. P., Müller-Moule, P., Gilmore, A. M., and Niyogi, K. K. (2002) PsbS-dependent enhancement of feedback de-excitation protects photosystem II from photoinhibition. *Proceedings of the National Academy of Sciences of the United States of America* Vol. 99, No. 23, pp. 15222-15227

Lindermayr, C., Saalbach, G., and Durner, J. (2005) Proteomic identification of S-nitrosylated proteins in Arabidopsis. *Plant Physiology* Vol. 137, No. 3, pp. 921-930

Lobreaux, S., Massenet, O., and Briat, J. F. (1992) Iron induces ferritin synthesis in maize plantlets. *Plant Molecular Biology* Vol. 19, No. 4, pp. 563-575

Luhua, S., Ciftci-Yilmaz, S., Harper, J., Cushman, J., and Mittler, R. (2008) Enhanced tolerance to oxidative stress in transgenic Arabidopsis plants expressing proteins of unknown function. *Plant Physiology* Vol. 148, No. 1, pp. 280-292

Maeda, S., Sugita, C., Sugita, M., and Omata, T. (2006) A new class of signal transducer in His-Asp phosphorelay systems. *Journal of Biological Chemistry* Vol. 281, No. 49, pp. 37868-37876

Mahalingam, R., Gomez-Buitrago, A., Eckardt, N., Shah, N., Guevara-Garcia, A., Day, P., Raina, R., and Fedoroff, N. V. (2003) Characterizing the stress/defense transcriptome of *Arabidopsis*. *Genome Biology* Vol. 4, No. 3, R20

Maier, T., Yu, C., Küllertz, G., and Clemens, S. (2003) Localization and functional characterization of metal-binding sites in phytochelatin synthases. *Planta* Vol. 218, No. 2, pp. 300-308

Mäser, P., Eckelman, B., Vaidyanathan, R., Horie, T., Fairbairn, D. J., Kubo, M., Yamagami, M., Yamaguchi, K., Nishimura, M., Uozumi, N., Robertson, W., Sussman, M. R., and Schroeder, J. I. (2002a) Altered shoot/root Na+ distribution and bifurcating salt sensitivity in *Arabidopsis* by genetic disruption of the Na+ transporter AtHKT1. *FEBS Letters* Vol. 531, No. 2, pp. 157-161

Mäser, P., Hosoo, Y., Goshima, S., Horie, T., Eckelman, B., Yamada, K., Yoshida, K., Bakker, E. P., Shinmyo, A., Oiki, S., Schroeder, J. I., and Uozumi, N. (2002b) Glycine residues in potassium channel-like selectivity filters determine potassium selectivity in four-loop-per-subunit HKT transporters from plants. *Proceedings of the National Academy of Sciences of the United States of America* Vol. 99, No. 9, pp. 6428-6433

Matsuoka, D. and Tokutomi, S. (2005) Blue light-regulated molecular switch of Ser/Thr kinase in phototropin. *Proceedings of the National Academy of Sciences of the United States of America* Vol. 102, No. 37, pp. 13337-13342

Mehlmer, N., Wurzinger, B., Stael, S., Hofmann-Rodrigues, D., Csaszar, E., Pfister, B., Bayer, R., and Teige, M. (2010) The Ca^{2+}-dependent protein kinase CPK3 is required for MAPK-independent salt-stress acclimation in Arabidopsis. *Plant Journal* Vol. 63, No. 3, pp. 484-498

Meskauskiene, R., Nater, M., Goslings, D., Kessler, F., op den Camp, R., and Apel, K. (2001) FLU: a negative regulator of chlorophyll biosynthesis in *Arabidopsis thaliana*. *Proc. Natl. Acad. Sci. U. S. A* Vol. 98, No. 22, pp. 12826-12831

Mhamdi, A., Queval, G., Chaouch, S., Vanderauwera, S., Van Breusegem, F., and Noctor, G. (2010) Catalase function in plants: a focus on *Arabidopsis* mutants as stress-mimic models. *Journal of Experimental Botany* Vol. 61, No. 15, pp. 4197-4220

Miller, G. and Mittler, R. (2006) Could heat shock transcription factors function as hydrogen peroxide sensors in plants? *Annals of Botany* Vol. 98, No. 2, pp. 279-288

Miller, G., Schlauch, K., Tam, R., Cortes, D., Torres, M. A., Shulaev, V., Dangl, J. L., and Mittler, R. (2009) The plant NADPH oxidase RBOHD mediates rapid systemic signaling in response to diverse stimuli. *Science Signaling* Vol. 2, No. 84, ra45

Miller, G., Shulaev, V., and Mittler, R. (2008) Reactive oxygen signaling and abiotic stress. *Physiologia Plantarum* Vol. 133, No. 3, pp. 481-489

Miller, G., Suzuki, N., Ciftci-Yilmaz, S., and Mittler, R. (2010) Reactive oxygen species homeostasis and signalling during drought and salinity stresses. *Plant Cell Environment* Vol. 33, No. 4, pp. 453-467

Mizoguchi, T., Ichimura, K., and Shinozaki, K. (1997) Environmental stress response in plants: the role of mitogen-activated protein kinases. *Trends in Biotechnology* Vol. 15, No. 1, pp. 15-19

Møller, I. M., Jensen, P. E., and Hansson, A. (2007) Oxidative modifications to cellular components in plants. *Annual Review of Plant Biology* Vol. 58, pp. 459-481

Mou, Z., Fan, W., and Dong, X. (2003) Inducers of plant systemic acquired resistance regulate NPR1 function through redox changes. *Cell* Vol. 113, No. 7, pp. 935-944

Mueller, M. J. and Berger, S. (2009) Reactive electrophilic oxylipins: pattern recognition and signalling. *Phytochemistry* Vol. 70, No. 13-14, pp. 1511-1521

Munné-Bosch, S. (2005) The role of alpha-tocopherol in plant stress tolerance. *Journal of Plant Physiology* Vol. 162, No. 7, pp. 743-748

Munns, R. and Tester, M. (2008) Mechanisms of salinity tolerance. *Annual Review of Plant Biology* Vol. 59, pp. 651-681

Nakagami, H., Sugiyama, N., Mochida, K., Daudi, A., Yoshida, Y., Toyoda, T., Tomita, M., Ishihama, Y., and Shirasu, K. (2010) Large-scale comparative phosphoproteomics identifies conserved phosphorylation sites in plants. *Plant Physiology* Vol. 153, No. 3, pp. 1161-1174

Nott, A., Jung, H. S., Koussevitzky, S., and Chory, J. (2006) Plastid-to-nucleus retrograde signaling. *Annual Review of Plant Biology* Vol. 57, pp. 739-759

Novak, J. F., Stirnberg, M., Roenneke, B., and Marin, K. (2011) A novel mechanism of osmosensing, a salt-dependent protein-nucleic acid interaction in the cyanobacterium *Synechocystis* species PCC 6803. *Journal of Biological Chemistry* Vol. 286, No. 5, pp. 3235-3241

Ohgishi, M., Saji, K., Okada, K., and Sakai, T. (2004) Functional analysis of each blue light receptor, cry1, cry2, phot1, and phot2, by using combinatorial multiple mutants in *Arabidopsis*. *Proceedings of the National Academy of Sciences of the United States of America* Vol. 101, No. 8, pp. 2223-2228

op den Camp, R. G., Przybyla, D., Ochsenbein, C., Laloi, C., Kim, C., Danon, A., Wagner, D., Hideg, E., Göbel, C., Feussner, I., Nater, M., and Apel, K. (2003) Rapid induction of distinct stress responses after the release of singlet oxygen in Arabidopsis. *Plant Cell* Vol. 15, No. 10, pp. 2320-2332

Oravecz, A., Baumann, A., Máté, Z., Brzezinska, A., Molinier, J., Oakeley, E. J., Adám, E., Schäfer, E., Nagy, F., and Ulm, R. (2006) CONSTITUTIVELY PHOTOMORPHOGENIC1 is required for the UV-B response in *Arabidopsis*. *Plant Cell* Vol. 18, No. 8, pp. 1975-1990

Ortega-Villasante, C., Rellan-Alvarez, R., Del Campo, F. F., Carpena-Ruiz, R. O., and Hernandez, L. E. (2005) Cellular damage induced by cadmium and mercury in Medicago sativa. *Journal of Experimental Botany* Vol. 56, No. 418, pp. 2239-2251

Overmyer, K., Brosché, M., and Kangasjärvi, J. (2003) Reactive oxygen species and hormonal control of cell death. *Trends in Plant Science* Vol. 8, No. 7, pp. 335-342. ISSN 1360-1385

Overmyer, K., Tuominen, H., Kettunen, R., Betz, C., Langebartels, C., Sandermann, H., and Kangasjärvi, J. (2000) Ozone-sensitive Arabidopsis *rcd1* mutant reveals opposite roles for ethylene and jasmonate signaling pathways in regulating superoxide-dependent cell death. *Plant Cell* Vol. 12, No. 10, pp. 1849-1862. ISSN 1040-4651

Paithoonrangsarid, K., Shoumskaya, M. A., Kanesaki, Y., Satoh, S., Tabata, S., Los, D. A., Zinchenko, V. V., Hayashi, H., Tanticharoen, M., Suzuki, I., and Murata, N. (2004) Five histidine kinases perceive osmotic stress and regulate distinct sets of genes in *Synechocystis*. *Journal of Biological Chemistry* Vol. 279, No. 51, pp. 53078-53086

Palmieri, M. C., Lindermayr, C., Bauwe, H., Steinhauser, C., and Durner, J. (2010) Regulation of plant glycine decarboxylase by S-nitrosylation and glutathionylation. *Plant Physiology* Vol. 152, No. 3, pp. 1514-1528

Parisy, V., Poinssot, B., Owsianowski, L., Buchala, A., Glazebrook, J., and Mauch, F. (2007) Identification of *PAD2* as a γ-glutamylcysteine synthetase highlights the importance of glutathione in disease resistance of Arabidopsis. *Plant Journal* Vol. 49, No. 1, pp. 159-172

Pastori, G. M., Kiddle, G., Antoniw, J., Bernard, S., Veljovic-Jovanovic, S., Verrier, P. J., Noctor, G., and Foyer, C. H. (2003) Leaf vitamin C contents modulate plant defense transcripts and regulate genes that control development through hormone signaling. *Plant Cell* Vol. 15, No. 4, pp. 939-951

Peers, G., Truong, T. B., Ostendorf, E., Busch, A., Elrad, D., Grossman, A. R., Hippler, M., and Niyogi, K. K. (2009) An ancient light-harvesting protein is critical for the regulation of algal photosynthesis. *Nature* Vol. 462, No. 7272, pp. 518-521

Peleg-Grossman, S., Melamed-Book, N., Cohen, G., and Levine, A. (2010) Cytoplasmic H_2O_2 prevents translocation of NPR1 to the nucleus and inhibits the induction of PR genes in arabidopsis. *Plant Signaling & Behavior* Vol. 5, No. 11, pp. 1401-1406

Pesaresi, P., Hertle, A., Pribil, M., Schneider, A., Kleine, T., and Leister, D. (2010) Optimizing photosynthesis under fluctuating light: The role of the Arabidopsis STN7 kinase. *Plant Signaling & Behavior* Vol. 5, No. 1, pp. 21-25

Petit, J. M., Briat, J. F., and Lobreaux, S. (2001) Structure and differential expression of the four members of the *Arabidopsis thaliana* ferritin gene family. *Biochemical Journal* Vol. 359, No. Pt 3, pp. 575-582

Pfannschmidt, T. (2010) Plastidial retrograde signalling--a true "plastid factor" or just metabolite signatures? *Trends in Plant Science* Vol. 15, No. 8, pp. 427-435

Pribil, M., Pesaresi, P., Hertle, A., Barbato, R., and Leister, D. (2010) Role of plastid protein phosphatase TAP38 in LHCII dephosphorylation and thylakoid electron flow. *PLoS Biology* Vol. 8, No. 1, e1000288

Qiu, Q. S., Guo, Y., Quintero, F. J., Pardo, J. M., Schumaker, K. S., and Zhu, J. K. (2004) Regulation of vacuolar Na^+/H^+ exchange in *Arabidopsis thaliana* by the salt-overly-sensitive (SOS) pathway. *Journal of Biological Chemistry* Vol. 279, No. 1, pp. 207-215

Quintero, F. J., Martinez-Atienza, J., Villalta, I., Jiang, X., Kim, W. Y., Ali, Z., Fujii, H., Mendoza, I., Yun, D. J., Zhu, J. K., and Pardo, J. M. (2011) Activation of the plasma membrane Na/H antiporter Salt-Overly-Sensitive 1 (SOS1) by phosphorylation of an auto-inhibitory C-terminal domain. *Proceedings of the National Academy of Sciences of the United States of America* Vol. 108, No. 6, pp. 2611-2616

Ravet, K., Touraine, B., Boucherez, J., Briat, J. F., Gaymard, F., and Cellier, F. (2009) Ferritins control interaction between iron homeostasis and oxidative stress in Arabidopsis. *Plant Journal* Vol. 57, No. 3, pp. 400-412

Reddie, K. G. and Carroll, K. S. (2008) Expanding the functional diversity of proteins through cysteine oxidation. *Current Opinions in Chemical Biology* Vol. 12, No. 6, pp. 746-754

Rizzini, L., Favory, J. J., Cloix, C., Faggionato, D., O'Hara, A., Kaiserli, E., Baumeister, R., Schäfer, E., Nagy, F., Jenkins, G. I., and Ulm, R. (2011) Perception of UV-B by the *Arabidopsis* UVR8 protein. *Science* Vol. 332, No. 6025, pp. 103-106

Rübenhagen, R., Rönsch, H., Jung, H., Krämer, R., and Morbach, S. (2000) Osmosensor and osmoregulator properties of the betaine carrier BetP from *Corynebacterium glutamicum* in proteoliposomes. *Journal of Biological Chemistry* Vol. 275, No. 2, pp. 735-741

Sakai, T., Kagawa, T., Kasahara, M., Swartz, T. E., Christie, J. M., Briggs, W. R., Wada, M., and Okada, K. (2001) *Arabidopsis* nph1 and npl1: Blue light receptors that mediate both phototropism and chloroplast relocation. *Proceedings of the National Academy of Sciences of the United States of America* Vol. 98, No. 12, pp. 6969-6974

Santos, M., Gousseau, H., Lister, C., Foyer, C., Creissen, G., and Mullineaux, P. (1996) Cytosolic ascorbate peroxidase from *Arabidopsis thaliana* L. is encoded by a small multigene family. *Planta* Vol. 198, No. 1, pp. 64-69

Schachtman, D. P. and Schroeder, J. I. (1994) Structure and transport mechanism of a high-affinity potassium uptake transporter from higher plants. *Nature* Vol. 370, No. 6491, pp. 655-658

Schützendubel, A. and Polle, A. (2002) Plant responses to abiotic stresses: heavy metal-induced oxidative stress and protection by mycorrhization. *Journal of Experimental Botany* Vol. 53, No. 372, pp. 1351-1365

Shao, H. B., Chu, L. Y., Shao, M. A., Jaleel, C. A., and Mi, H. M. (2008) Higher plant antioxidants and redox signaling under environmental stresses. *CR Biology* Vol. 331, No. 6, pp. 433-441

Shapiguzov, A., Ingelsson, B., Samol, I., Andres, C., Kessler, F., Rochaix, J. D., Vener, A. V., and Goldschmidt-Clermont, M. (2010) The PPH1 phosphatase is specifically involved in LHCII dephosphorylation and state transitions in Arabidopsis. *Proceedings of the National Academy of Sciences of the United States of America* Vol. 107, No. 10, pp. 4782-4787

Sharma, S. S. and Dietz, K. J. (2009) The relationship between metal toxicity and cellular redox imbalance. *Trends in Plant Science* Vol. 14, No. 1, pp. 43-50

Shi, H., Ishitani, M., Kim, C., and Zhu, J. K. (2000) The *Arabidopsis thaliana* salt tolerance gene *SOS1* encodes a putative Na^+/H^+ antiporter. *Proceedings of the National Academy of Sciences of the United States of America* Vol. 97, No. 12, pp. 6896-6901

Shi, H., Quintero, F. J., Pardo, J. M., and Zhu, J. K. (2002) The putative plasma membrane Na^+/H^+ antiporter SOS1 controls long-distance Na^+ transport in plants. *Plant Cell* Vol. 14, No. 2, pp. 465-477

Shin, J., Kim, K., Kang, H., Zulfugarov, I. S., Bae, G., Lee, C. H., Lee, D., and Choi, G. (2009) Phytochromes promote seedling light responses by inhibiting four negatively-acting phytochrome-interacting factors. *Proceedings of the National Academy of Sciences of the United States of America* Vol. 106, No. 18, pp. 7660-7665

Shoumskaya, M. A., Paithoonrangsarid, K., Kanesaki, Y., Los, D. A., Zinchenko, V. V., Tanticharoen, M., Suzuki, I., and Murata, N. (2005) Identical Hik-Rre systems are involved in perception and transduction of salt signals and hyperosmotic signals but regulate the expression of individual genes to different extents in *Synechocystis*. *Journal of Biological Chemistry* Vol. 280, No. 22, pp. 21531-21538

Strasser, B., Sánchez-Lamas, M., Yanovsky, M. J., Casal, J. J., and Cerdan, P. D. (2010) *Arabidopsis thaliana* life without phytochromes. *Proceedings of the National Academy of Sciences of the United States of America* Vol. 107, No. 10, pp. 4776-4781

Sugiyama, N., Nakagami, H., Mochida, K., Daudi, A., Tomita, M., Shirasu, K., and Ishihama, Y. (2008) Large-scale phosphorylation mapping reveals the extent of tyrosine phosphorylation in *Arabidopsis*. *Molecular Systems Biology* Vol. 4, 193

Sullivan, S., Thomson, C. E., Kaiserli, E., and Christie, J. M. (2009) Interaction specificity of *Arabidopsis* 14-3-3 proteins with phototropin receptor kinases. *FEBS Letters* Vol. 583, No. 13, pp. 2187-2193

Tada, Y., Spoel, S. H., Pajerowska-Mukhtar, K., Mou, Z., Song, J., Wang, C., Zuo, J., and Dong, X. (2008) Plant immunity requires conformational changes of NPR1 via S-nitrosylation and thioredoxins. *Science* Vol. 321, No. 5891, pp. 952-956

Takahashi, F., Mizoguchi, T., Yoshida, R., Ichimura, K., and Shinozaki, K. (2011) Calmodulin-dependent activation of MAP kinase for ROS homeostasis in *Arabidopsis*. *Molecular Cell* Vol. 41, No. 6, pp. 649-660

Takano, M., Inagaki, N., Xie, X., Kiyota, S., Baba-Kasai, A., Tanabata, T., and Shinomura, T. (2009) Phytochromes are the sole photoreceptors for perceiving red/far-red light in rice. *Proceedings of the National Academy of Sciences of the United States of America* Vol. 106, No. 34, pp. 14705-14710

Terry, N. (1980) Limiting factors in photosynthesis: I. Use of iron stress to control photochemical capacity *in vivo*. *Plant Physiology* Vol. 65, No. 1, pp. 114-120

Tikkanen, M., Nurmi, M., Kangasjärvi, S., and Aro, E. M. (2008) Core protein phosphorylation facilitates the repair of photodamaged photosystem II at high light. *Biochimica et Biophysica Acta* Vol. 1777, No. 11, pp. 1432-1437

Tran, L. S., Urao, T., Qin, F., Maruyama, K., Kakimoto, T., Shinozaki, K., and Yamaguchi-Shinozaki, K. (2007) Functional analysis of AHK1/ATHK1 and cytokinin receptor histidine kinases in response to abscisic acid, drought, and salt stress in *Arabidopsis*. *Proceedings of the National Academy of Sciences of the United States of America* Vol. 104, No. 51, pp. 20623-20628

Ulm, R. and Nagy, F. (2005) Signalling and gene regulation in response to ultraviolet light. *Current Opinions in Plant Biology* Vol. 8, No. 5, pp. 477-482

Urano, J., Nakagawa, T., Maki, Y., Masumura, T., Tanaka, K., Murata, N., and Ushimaru, T. (2000) Molecular cloning and characterization of a rice dehydroascorbate reductase. *FEBS Letters* Vol. 466, No. 1, pp. 107-111

Urao, T., Yakubov, B., Satoh, R., Yamaguchi-Shinozaki, K., Seki, M., Hirayama, T., and Shinozaki, K. (1999) A transmembrane hybrid-type histidine kinase in *Arabidopsis* functions as an osmosensor. *Plant Cell* Vol. 11, No. 9, pp. 1743-1754

Urao, T., Yamaguchi-Shinozaki, K., and Shinozaki, K. (2000) Two-component systems in plant signal transduction. *Trends in Plant Science* Vol. 5, No. 2, pp. 67-74

Urao, T., Yamaguchi-Shinozaki, K., and Shinozaki, K. (2001) Plant histidine kinases: an emerging picture of two-component signal transduction in hormone and environmental responses. *Science Signaling* Vol. 2001, No. 109, re18

Vaahtera, L. and Brosché, M. (2011) More than the sum of its parts - How to achieve a specific transcriptional response to abiotic stress. *Plant Science* Vol. 180, No. 3, pp. 421-430

Vahala, J., Keinänen, M., Schützendübel, A., Polle, A., and Kangasjärvi, J. (2003a) Differential effects of elevated ozone on two hybrid aspen genotypes predisposed to chronic ozone fumigation. Role of ethylene and salicylic acid. *Plant Physiology* Vol. 132, No. 1, pp. 196-205

Vahala, J., Ruonala, R., Keinänen, M., Tuominen, H., and Kangasjärvi, J. (2003b) Ethylene insensitivity modulates ozone-induced cell death in birch. *Plant Physiology* Vol. 132, No. 1, pp. 185-195

Vahisalu, T., Kollist, H., Wang, Y. F., Nishimura, N., Chan, W. Y., Valerio, G., Lamminmäki, A., Brosché, M., Moldau, H., Desikan, R., Schroeder, J. I., and Kangasjärvi, J. (2008) SLAC1 is required for plant guard cell S-type anion channel function in stomatal signalling. *Nature* Vol. 452, No. 7186, pp. 487-491

Vainonen, J. P., Hansson, M., and Vener, A. V. (2005) STN8 protein kinase in *Arabidopsis thaliana* is specific in phosphorylation of photosystem II core proteins. *Journal of Biological Chemistry* Vol. 280, No. 39, pp. 33679-33686

van der Heide, T., Stuart, M. C., and Poolman, B. (2001) On the osmotic signal and osmosensing mechanism of an ABC transport system for glycine betaine. *EMBO Journal* Vol. 20, No. 24, pp. 7022-7032

Vandenbussche, F., Habricot, Y., Condiff, A. S., Maldiney, R., Van Der Straeten, D., and Ahmad, M. (2007) HY5 is a point of convergence between cryptochrome and cytokinin signalling pathways in *Arabidopsis thaliana*. *Plant Journal* Vol. 49, No. 3, pp. 428-441

Vatamaniuk, O. K., Mari, S., Lu, Y. P., and Rea, P. A. (2000) Mechanism of heavy metal ion activation of phytochelatin (PC) synthase: blocked thiols are sufficient for PC synthase-catalyzed transpeptidation of glutathione and related thiol peptides. *Journal of Biological Chemistry* Vol. 275, No. 40, pp. 31451-31459

Wagner, D., Przybyla, D., op den Camp, R., Kim, C., Landgraf, F., Lee, K. P., Würsch, M., Laloi, C., Nater, M., Hideg, E., and Apel, K. (2004) The genetic basis of singlet oxygen-induced stress responses of *Arabidopsis thaliana*. *Science* Vol. 306, No. 5699, pp. 1183-1185

Walker, E. L. and Connolly, E. L. (2008) Time to pump iron: iron-deficiency-signaling mechanisms of higher plants. *Current Opinions in Plant Biology* Vol. 11, No. 5, pp. 530-535

Waters, M. T. and Langdale, J. A. (2009) The making of a chloroplast. *EMBO Journal* Vol. 28, No. 19, pp. 2861-2873

Weber, M., Trampczynska, A., and Clemens, S. (2006) Comparative transcriptome analysis of toxic metal responses in *Arabidopsis thaliana* and the Cd^{2+}-hypertolerant facultative metallophyte Arabidopsis halleri. *Plant Cell and Environment* Vol. 29, No. 5, pp. 950-963

Willekens, H., Chamnongpol, S., Davey, M., Schraudner, M., Langebartels, C., Van Montagu, M., Inzé, D., and Van Camp, W. (1997) Catalase is a sink for H_2O_2 and is indispensable for stress defence in C3 plants. *EMBO Journal* Vol. 16, No. 16, pp. 4806-4816

Wintz, H., Fox, T., Wu, Y. Y., Feng, V., Chen, W., Chang, H. S., Zhu, T., and Vulpe, C. (2003) Expression profiles of *Arabidopsis thaliana* in mineral deficiencies reveal novel transporters involved in metal homeostasis. *Journal of Biological Chemistry* Vol. 278, No. 48, pp. 47644-47653

Wohlbach, D. J., Quirino, B. F., and Sussman, M. R. (2008) Analysis of the *Arabidopsis* histidine kinase ATHK1 reveals a connection between vegetative osmotic stress sensing and seed maturation. *Plant Cell* Vol. 20, No. 4, pp. 1101-1117

Wood, J. M., Bremer, E., Csonka, L. N., Kraemer, R., Poolman, B., van der Heide, T., and Smith, L. T. (2001) Osmosensing and osmoregulatory compatible solute accumulation by bacteria. *Comparative Biochemistry and Physiology Part A: Molecular & Integrative Physiology* Vol. 130, No. 3, pp. 437-460

Wrzaczek, M., Brosché, M., and Kangasjärvi, J. (2009a) Scorched earth strategy: Grim Reaper saves the plant. *Plant Signaling & Behavior* Vol. 4, No. 7, pp. 631-633

Wrzaczek, M., Brosché, M., Kollist, H., and Kangasjärvi, J. (2009b) *Arabidopsis* GRI is involved in the regulation of cell death induced by extracellular ROS. *Proceedings of the National Academy of Sciences of the United States of America* Vol. 106, No. 13, pp. 5412-5417

Wrzaczek, M., Brosché, M., Salojärvi, J., Kangasjärvi, S., Idänheimo, N., Mersmann, S., Robatzek, S., Karpinski, S., Karpinska, B., and Kangasjärvi, J. (2010) Transcriptional regulation of the CRK/DUF26 group of receptor-like protein kinases by ozone and plant hormones in Arabidopsis. *BMC Plant Biology* Vol. 10, 95

Wurzinger, B., Mair, A., Pfister, B., and Teige, M. (2011) Cross-talk of calcium-dependent protein kinase and MAP kinase signalling. *Plant Signaling & Behavior* Vol. 6, No. 1, pp. 8-12

Yokoi, S., Quintero, F. J., Cubero, B., Ruiz, M. T., Bressan, R. A., Hasegawa, P. M., and Pardo, J. M. (2002) Differential expression and function of *Arabidopsis thaliana* NHX Na^+/H^+ antiporters in the salt stress response. *Plant Journal* Vol. 30, No. 5, pp. 529-539

Yuasa, T., Ichimura, K., Mizoguchi, T., and Shinozaki, K. (2001) Oxidative stress activates ATMPK6, an Arabidopsis homologue of MAP kinase. *Plant and Cell Physiology* Vol. 42, No. 9, pp. 1012-1016

Zhu, J. K. (2002) Salt and drought stress signal transduction in plants. *Annual Review of Plant Biology* Vol. 53, pp. 247-273

3

Plant Organelles-to-Nucleus Retrograde Signaling

Nadezhda Yurina and Margarita Odintsova
A.N. Bach Institute of Biochemistry, Russian Academy of Sciences
Russia

1. Introduction

Plant cells contain two types of energy producing organelles: chloroplasts, which convert solar into chemical energy, and mitochondria, which convert stored energy into ATP. Cell organelles comprise thousands of various proteins; most of them are encoded by nuclear genes; only few genes, which encode mainly the components of the system of gene expression per se and of the respiratory (in mitochondria) or photosynthetic (in chloroplasts) chains, are localized in the genomes of organelles. Thus, the maintenance of cell organelle functional activity during cell growth and development depends predominantly on the nuclear genome encoding most organelle proteins and their own genomes encoding the limited but substantial number of proteins. In this connection, organelle ribosomes, photosystems, complexes of mitochondrial respiratory chain are mosaic in their origin; they are multiprotein complexes comprising subunits encoded by the nucleus and organelles. Therefore, special mechanisms are required for coordination of gene expression in various cell compartments. Since the nuclear genome plays a key role in the biogenesis and functioning of cell organelles, the main recent attention was paid to the analysis of so called anterograde regulation, which controls the flow of information from the nucleus and cytoplasm to organelles. Along with anterograde regulation, retrograde signaling occurs in the cells, when signals emitted by organelles control gene expression in the nucleus (Pesaresi et al. 2007; Yurina & Odintsova, 2007). In this review, we summarize the recent understanding of chloroplast-to-nuclear and mitochondria-to-nuclear retrograde regulation in plant, which involves multiple potential signaling pathways in relation to abiotic stress.

2. Plastid-generated signals and their role in nuclear gene expression

The very first evidence for the existence of plastid-generated signals that control the expression of nuclear genes encoding the chloroplast proteins was obtained about 30 years ago in the studies of *albostrians* and *Saskatoon* mutants of barley (*Hordeum vulgare* L., cv. Haisa) (Emanuel et al., 2004). The subsequent studies of mutant plants with impaired carotenoid biosynthesis (manifested as photobleached plastids) demonstrated that the absence of functionally active chloroplasts considerably diminished the expression of several photosynthetic genes residing in the nucleus (Oelmüller et al., 1986). Similar results were obtained when carotenoid biosynthesis in the seedlings was reduced by norflurazon, an inhibitor of phytoene desaturase, the enzyme participating in carotenoid biosynthesis (Yurina et al., 2006). In addition to photobleaching, the expression of nuclear genes was inhibited when chloroplast development was blocked by inhibiting the expression of plastid genes with tagetitoxin or nalidixic acid (Gray et al., 1995). Numerous signaling pathways were found to function in the cell and coordinate the expression of nuclear genes depending on chloroplast requirements (Beck, 2005).

Protein synthesis in plastids was found to correlate with the expression of nuclear genes coding for plastid proteins. The contents of several nuclear-encoded proteins transferred to plastids, e.g., the Calvin cycle enzymes, and also of proteins functionally related to plastids, such as nitrate reductase, and to peroxisomes, such as glycolate oxidase, catalase, and hydroxypyruvate reductase, were reportedly lower in the leaves of *albostrians* barley with deficient chloroplast ribosomes than in the wild-type plants. The activity of nuclear genes encoding the chloroplast-localized enzymes in the seedlings of white mustard (*Sinapis alba* L.) declined in the presence of chloramphenicol, an inhibitor of translation in chloroplasts, whereas the activity of phytochrome-induced cytoplasmic enzymes, such as chalcone synthase, was not affected and in some cases even increased (Oelmüller et al., 1986). In pea (*Pisum sativum* L.) plants, chloramphenicol inhibited the red- and blue light-induced expression of the gene encoding early light-induced protein (ELIP) (Adamska, 1995).

Treating plants with specific inhibitors of translation in chloroplasts, such as lincomycin, erythromycin, and streptomycin, also down-regulated the expression of nuclear genes encoding the photosynthetic proteins. The inhibitors of protein synthesis were shown to affect the expression of nuclear genes only at the early stages of seedling development, during the initial two–three days of germination, and the authors presumed that the products of protein synthesis at the early steps of germination may act as plastid signals (Gray, 1995). As a whole, the evidence already amassed presumes that protein synthesis in plastids generates some signal forerunning the expression of several nuclear genes. These nuclear genes encode the plastid components and also the proteins localized in other cell compartments. However, the data are missing as to whether the inhibition of protein synthesis in the plastids would decrease the expression of any nuclear genes. The current experimental evidence lists among the probable sources of chloroplast signals the reactive oxygen species (ROS), the changes in the redox state of the components of photosynthetic electron transport chain (ETC) of the stromal components of plastids, and also the metabolites produced in the course of photosynthesis (Piippo et al., 2006).

In plants, ROS are continuously synthesized as byproducts of numerous metabolic pathways in various cell compartments. ROS include oxygen ions, free radicals, and inorganic and organic peroxides. The contents of ROS are dramatically elevated under stress conditions, such as high illuminance, low temperature, etc. ROS accumulation provokes oxidative stress and damages cell structures. ROS are neutralized by enzymes, such as superoxide dismutase and catalase, and by antioxidant systems. The singlet oxygen $^1O_2$ produced by PS II and the superoxide anion generated by PS I are the major forms of ROS produced by chloroplasts under the stress conditions of high illuminance; the superoxide anion is rapidly dismutated into hydrogen peroxide (Apel & Hirt, 2004). Both these ROS were shown to act as plastid signals. The singlet oxygen $^1O_2$ generated in plastids affects the expression of several nuclear genes. Plastids of the *flu* mutant of *Arabidopsis thaliana* amass protochlorophyllide (protoChlide), which generates the singlet oxygen under illumination. In *Arabidopsis*, the singlet oxygen was shown to enhance the expression of 70 genes and inhibit the expression of nine genes. Two arguments are put forward to advance the idea that the singlet oxygen serves as a chloroplast signal. First, it is short-lived, about 200 ns, with the transduction distance (the action at the distance) up to 10 nm. Therefore it is thought to perform the role of a plastid-generated signal, which specifically activates the genetically determined program of cell responses to the stress conditions. The EXECUTER 1 protein recently identified in *Arabidopsis* plants is presumed to recognize or transduce this signal. The N-terminal region of this protein resembles the signal sequences of chloroplast proteins encoded by nuclear genes and essential for the import of these proteins (Wagner et al., 2004). Localization of EXECUTER 1 in plastids would establish an important segment of the signaling pathway linking the plastid-generated singlet oxygen to the expression of nuclear genes.

The singlet oxygen is the major constituent of ROS generated by PS II in plants that are deficient in carotenoids due to mutations or following the treatment with herbicides, which inhibit carotenoid synthesis, such as norflurazon. In carotenoid-deficient plants, the plastid structure is considerably damaged, and the expression of several nuclear genes, mostly encoding the photosynthetic proteins, is inhibited. However, the expression profiles of the nuclear genes inhibited by carotenoid deficiency differs from those activated or inhibited by the singlet oxygen generated by illumination in the protoChlide -accumulating *flu* mutants. Therefore the singlet oxygen generated by PS II in the absence of carotenoids most probably activates the signaling pathway different from the pathway activated by the singlet oxygen produced from protoChlide or generated by PS II in the presence of 3-(3,4- dichlorophenyl)-1,1-dimethylurea (DCMU) (Wagner et al., 2004). The idea of alternative signaling pathway was substantiated by the analysis of *gun* mutants with lower sensitivity of gene expression than in the wild-type plants (Beck, 2005).

Hydrogen peroxide is another form of ROS generated by chloroplasts. It accumulates when plants are moved from temperate to strong light, and the superoxide radicals generated by PS I are transformed into hydrogen peroxide. The latter is reduced to water by chloroplast ascorbate peroxidase (APX). While the singlet oxygen, due to its high reaction capacity, mostly remains in the chloroplasts, the excess hydrogen peroxide easily diffuses across the chloroplast envelope into the cytoplasm. Treating plants with H_2O_2 was shown to induce the expression of nuclear genes related to plant responses to stress factors; thus, the expression of *cAPX* , the gene for cytoplasmic APX, is enhanced by excess light. The crucial role of hydrogen peroxide in *cAPX* expression was demonstrated in the experiments with *Arabidopsis* leaves infiltrated with catalase before excess light treatment: while the induction of *APX2* (*Arabidopsis* cAPX) was inhibited, the expression of the superoxide dismutase gene was not affected (Chang et al., 2004).

Recent studies of transgenic tobacco plants expressing the genes for catalase or thylakoid-type APX demonstrated that *APX* expression involved both hydrogen peroxide and the redox state of plastoquinone pool. Under excess light stress, cell levels of H_2O_2 in the transgenic tobacco plants were considerably lower and the induction of *cAPX* was saturated much earlier than in the wild-type plants. It follows that chloroplast-generated H_2O_2 would represent a redox signal to activate *cAPX* expression in the nucleus. The initial induction observed under photooxidative stress conditions was related to the redox state of plastoquinone pool (see below for more details), whereas the subsequent mRNA accumulation was in line with continuous accumulation of H_2O_2 (Yabuta et al., 2004). The correlation between the *APX2* expression and H_2O_2 generation under the excess light stress was also elucidated by comparing the expression of luciferase gene under the *APX2* promoter to H_2O_2 accumulation demonstrated by cytochemical staining with methylviologen (Chang et al., 2004). This dye was shown to enhance the formation of singlet oxygen and hydrogen peroxide in chloroplasts, and the concomitant *cAPX* expression correlated with the methylviologen-generated H_2O_2 accumulation in the cells.

Both the location and the mechanism of action of plastid-generated H_2O_2 are unknown. Similar to water, hydrogen peroxide is thought to freely diffuse across the biological membranes and in this way directly interact with out-of-plastid systems of signal transduction. Nonetheless it is not clear how cells discern between the plastid-generated H_2O_2 and hydrogen peroxide produced in other cell compartments, e.g., on the plasma membrane when cells are attacked by pathogens. The stress responses to pathogens are known to widely differ from those induced by excess light stress (Beck, 2005).

Thus, we conclude that at least two chloroplast-generated forms of ROS, the singlet oxygen and hydrogen peroxide, participate in the transduction of specific signals from the plastids to the nucleus. Two ROS forms induce different responses at the level of gene expression and probably perform in the different signaling pathways. The mechanism of ROS interaction with the nucleus is also unknown. In this aspect, the experiments with mutants, such as *EXECUTER1* where the singlet oxygen participates in the plastid signal blockade, are very promising (Beck, 2005).

The environmentally induced changes in redox state of the components of photosynthetic ETC act as signals that regulate gene expression in the chloroplasts and partly in the nucleus (Pfannschmidt & Liere, 2005). It means that photosynthesis is a source of information essential for the control over the nuclear gene expression that is not recognized by cytoplasmic photoreceptors, while the chloroplasts themselves serve as sensors for the changes in light quantity and quality and in this way induce the physiological responses of photoacclimation (Beck, 2005).

The participation of the redox state of ETC components in nuclear gene expression was established in several ways. To change the redox state of ETC components, plants grown at low light were transferred at regular intervals into strong light and back. The same goal was attained by the light conditions that primarily excited either PS I or PS II, by changing the growth temperature and the level of carbohydrates and electron acceptors, such as O_2 and CO_2 , and by herbicides DCMU and 2,5-dibromo-3-methyl-6-isopropyl-p-benzoquinone (DBMIB), which specifically blocked electron transport from PS II to the cytochrome b_6/f complex at the site upstream (DCMU) or downstream of plastoquinone pool (DBMIB). These data led to the conclusion that the redox state of plastoquinones was the initial signal that regulated the expression of particular genes (Nott et al., 2006). Recently the role of the redox state of ETC as the source of chloroplast signal(s) transferred into the nuclear compartment has been supported by the analysis of expression profile of numerous nuclear genes in *Arabidopsis*: 2661 out of 3292 genes under analysis encoded chloroplast proteins and only 631, nonchloroplast ones (Richly et al., 2003). In the cases when the plants grown under PS I-inducing light were transferred under PS II-inducing illumination, the changes in mRNA contents were registered for 2133 genes: the changes were positive (enhanced expression) for 1121 genes and negative (diminished expression) for 1012 genes. The in depth study demonstrated that among these genes, 286 are immediately regulated by the redox state of photosynthetic ETC signals: the expression of 86 genes was up-regulated and 200 down-regulated (Fey et al., 2005). Whatever small were the changes in the expression levels of most genes, this evidence seems to prove that the redox state of the ETC components regulate the expression of many nuclear genes. In green algae, the redox state of the plastoquinone pool was shown to affect *Lhcb* expression. In higher plants, however, expression of *Lhcb* has been correlated with the phosphorylation status of LhcII. The thylakoid protein STN7 has been identified as the protein kinase for LHCII (Inaba, 2010). The control exerted by the redox state of plastoquinone pool on the nuclear gene expression in the higher plants is of lesser importance than in the green algae.

Chloroplast redox signals were shown to affect the nuclear gene expression at both transcriptional and posttranscriptional (*Fed-1, PetE*) levels by controlling mRNA stability and binding to polyribosomes (Sherameti et al., 2002). The redox state depends on illumination, and therefore the photosynthetic apparatus can be seen as a photoreceptor recording photon quality and quantity (Beck, 2005).

In addition to the redox state of chloroplast ETC components, the nuclear genes are also controlled by the redox state of the stromal components, such as glutathione and the ferredoxin-thioredoxin system. It follows that in mature leaves of higher plants, the redox state of the components at the donor side of PS I is more important for light-dependent changes in the expression of nuclear genes coding for plastid proteins than the redox state of plastoquinone pool, especially at the early steps of signal transduction. Such conclusion confers with the evidence from the experiments with *Synechocystis* sp. 6803: in the latter case, the expression of few genes was related to the redox state of plastoquinone pool (Piippo et al., 2006). Following short exposure to extreme environmental conditions, the leading role in signal transduction from the chloroplasts to the nucleus would pass on to the stromal redox components, the products of CO_2 fixation, and the ATP/ADP ratio.

Recently many studies were focused on the role of Mg-protoporphyrin IX (MgProtoIX) and its methyl ester (MgProtoIXMe) as the regulators of the nuclear gene expression. The first proof for the participation of chlorophyll precursors in the transduction of plastid signals was obtained in the experiments with *Chlamydomonas reinhardtii* cells. In synchronous cultures of this alga, mRNA of the gene *Lhcb* accumulated already in two hours after the cultures were transferred into the light, mostly due to enhanced transcription. The studies of higher plants support the idea that chlorophyll precursors participated in the transduction of plastid-generated signals. Thujaplicin treatment of etiolated cress seedlings interfered with protochlorophyllide synthesis and resulted in MgProtoIX accumulation and decline in the light-induced synthesis of *Lhcb* mRNA. When treated with 5-aminolevulinic acid (ALA), such seedlings produced only half of light-induced mRNA *Lhcb* as compared to the seedlings treated with water. When treated with amitrole, an inhibitor of carotenoid biosynthesis, the ethiolated barley seedlings accumulated ALA, MgProtoIX, and MgProtoIXMe, with the concomitant decline in light-induced *Lhcb* and *RbcS* expression. In *Arabidopsis*, several nuclear genes encode proteins essential for plastid-to-nucleus signal transduction have been identified (Susek et al., 1993). The *Lhcb* promoter equipped with the proper selection marker was built into the nuclear genome, and seeds were mutagenized with ethyl methanesulfonate. These lines were used to isolate mutants with *Lhcb* expression independent of the chloroplast functional state. While due to chloroplast photobleaching, the expression of reporter genes under the *Lhcb* promoter was low in the wild-type plants grown under continuous illumination in the presence of norflurazon, the mutants exhibited high expression of *Lhcb* on the norflurazon-supplemented medium. Such screening identified five nonallelic loci with diminished capacity for plastid-to-nucleus signaling;

these mutants were called "genomes uncoupled", or *gun*-mutants (Susek et al., 1993). Four *gun*-mutants have been already characterized at the molecular level: several impaired enzymes, such as heme oxygenases, phytochromobiline synthases, and H subunits of Mg-chelatase or MgProtoIX-binding protein, were localized in the plastids and shown to participate in the porphyrin biosynthesis. The mutations at these genes decreased MgProtoIX accumulation. Damaging other enzymes in the biosynthetic pathway leading to MgProtoIX also diminished the plastid control of *Lhcb* expression. The plastid signal transduction is apparently affected by plant developmental stage: while the expression of *Lhcb, HEMA1,* and *Elip* was inhibited in norflurazon-treated seedlings of *Arabidopsis* and barley, similar treatment of adult plants downregulated only *Lhcb* expression (Pogulskaya et al., 2006). A considerable progress has been recently reported in characterizing the sequences of the target genes for plastid signals, such as MgProtoIX and others. The study of light-responsive promoters of the genes controlled by plastid signals and mostly encoding the components of photosynthetic apparatus demonstrated that light and plastid signals affected the nuclear gene expression at one and the same *cis*-elements. In these experiments (Kusnetsov et al., 1996), the plastid signal controlling the nuclear gene expression was activated with norflurazon, which promoted photooxidation of the thylakoid membrane in the light.

Based on the above-mentioned data, it was suggested that in higher plants a plastid-to-nucleus signal is induced by MgProtoIX and/or MgProtoIXMe (Nott et al., 2006; Pogulskaya et al., 2006; Strand et al., 2003; Yurina et al., 2006).

Subsequent studies showed that accumulation of MgProtoIX is not always accompanied by inhibition of *Lhcb* expression (Mochizuki et al., 2008; Mochizuki et al., 2010; Moulin et al., 2008;). Determination of ProtoIX, MgProtoIX and MgProtoIXMe concentrations in *Arabidopsis* seedlings grown in the presence of norflurazon showed that norflurazon inhibits significantly the formation of the intermediates of tetrapyrrole biosynthesis, the inhibition was more pronounced in older seedlings. The expression levels of some other genes, such as *Lhcb1, RbcS, HEMA, BAM3* (encodes beta-amylase) and *CA1* (encodes carbonic anhydrase) were also reduced. It was also shown that the *Arabidopsis* mutants *cs* and *ch42* with impaired Mg-chelatase subunit I (ChlI) did not display *gun* phenotype, although the production of MgProtoIX in these mutants was considerably decreased (Mochizuki et al., 2001).

The herbicide 2,2′-dipyridyl supposed to induce accumulation of the MgProtoIXMe inhibited *gun2* and *gun5* phenotypes (Mochizuki et al., 2008). In norflurazon-treated seedlings, the *Lhcb* expresssion was decreased, however no accumulation of porphyrins was observed (Gadjieva et al., 2005).

Contradictory data concerning the signaling role of MgProtoIX in the repression of nuclear plastid protein genes may be associated with problems of accurate quantitative determination of tetrapyrrole biosynthesis intermediates. A detailed study of the role of MgProtoIX (and other chlorophyll biosynthesis intermediates) as signaling molecules during retrograde regulation revealed the absence of correlation between MgProtoIX accumulation and expression of nuclear plastid protein genes. It was shown that in norflurazon-treated plants at different growth conditions, nuclear gene expression was inhibited, however, the accumulation of MgProtoIX or other intermediates of chlorophyll biosynthesis did not occur (Moulin et al., 2008). Conversely, elevation of the MgProtoIX endogenous level by the addition of the tetrapyrrole precursor ALA caused an induction not repression of the nuclear photosynthetic genes. Chemical or genetic modification of the tetrapyrrole levels and light conditions revealed no correlation between the intermediates of tetrapyrrole biosynthesis (including MgProtoIX) and *Lhcb* expression (Mochizuki et al., 2010). *Gun* mutations had no effect on heme accumulation (Voigt et al., 2010). The determination of tetrapyrrole content in *Arabidopsis* plants under light stress conditions by a conventional procedure revealed no significant changes in MgProtoIX concentration in norflurazon-treated plants as compared to control plants (our data). A certain decrease in MgProtoIX concentration was even recorded (unpublished data). Thus, it remains unclear whether MgProtoIX serves as a signal during retrograde regulation in higher plants.

Based on conflicting experimental data it was suggested that not MgProtoIX molecules but their derivatives, degradation products or ROS, such as singlet oxygen or hydrogen peroxide, may induce retrograde signaling cascades (Mochizuki et al., 2010; Moulin et al., 2008). There is experimental evidence that expression of nuclear plastid protein genes *Lhcb* and *RbcS* is controlled by hydrogen peroxide and singlet oxygen (La Rocca et al., 2001). Based on these data it was hypothesized that a simple signaling cascade mechanism connecting MgProtoIX accumulation with inhibition of nuclear plastid protein gene expression is unlikely (Mochizuki et al., 2008; Moulin et al., 2008).

The discrepant results may partially arise from different conditions of norflurazon treatment of seedlings (Zhang et al., 2011). It cannot be excluded that MgProtoIX-generated signals are short-living and function in particular cellular compartments and thus are difficult to detect. Transcriptome analysis of the green unicellular algae *C. reinhardtii* grown in the presence of MgProtoIX or heme showed that tetrapyrrole biosynthesis intermediates transiently but considerably changed expression of approx. 1000 genes. They include a limited number of photosynthesis-related genes, the genes for the tricarbonic acid cycle enzymes, the genes for heme-binding proteins, several stress-responsive genes, as well as genes involved in protein folding and degradation. Both tetrapyrroles act as secondary messengers in adaptive response of the whole cell, not only cellular organelles (Voss et al., 2011).

Another important process that may generate a plastid signal is protein import into plastids (Inaba, 2010). It was shown that in a mutant with a deficiency of the main receptor for the imported proteins Toc159, expression of nuclear plastid protein genes is inhibited. This abnormality in protein import into the plastids serves as a plastid signal (Kakizaki & Inaba, 2010). By this mechanism retrograde signals regulate expression of nuclear plastid protein genes in accordance with the requirements of these organelles and provide efficient assembly of multisubunit complexes encoded both by nuclear and chloroplast genes. It should be noted that the signaling molecules triggering retrograde regulation have not been identified so far (Pfannschmidt, 2010).

2.1 Components of the retrograde signaling pathways

The GUN1 protein plays an important role in retrograde signal transduction. It was shown that the *gun1* mutation is not associated with tetrapyrrole biosynthesis. This follows from the differences between the *gun1* and *gun2-gun5* genes (Cottage et al., 2010). Treatment of plastids with lincomycin, an inhibitor of protein synthesis, represses expression of nuclear photosynthetic genes in wild-type *Arabidopsis* seedlings and *gun2-gun5* mutants, but not in *gun1* mutants (Koussevitzky et al., 2007). The *gun1* gene codes for a protein of plastid nucleoids, which contains 10 copies of a pentatricopeptide repeat (PPR) and the so-called SMR-domain (a minor MutS-associated domain) close to the C-terminus of the polypeptide chain (Cottage et al., 2010; Koussevitzky et al., 2007). The SMR-domain binds DNA, while the PPR-motif participates in RNA processing and is found in many mitochondrial and plastid proteins. Since lincomycin does not inhibit expression of nuclear photosynthesis-related genes in *gun1* mutants, it was suggested that GUN1 may be involved in signal transduction induced by impaired expression of plastid genes (Armbruster et al., 2011). GUN1-dependent signal transduction may also be involved in coordination of nuclear photosynthetic gene expression with the efficiency of protein import into the plastids (Kakizaki & Inaba, 2010). It was demonstrated that GUN1 mediates signals induced by tetrapyrrole biosynthesis intermediates and redox state of the electron transport chain. Since *gun1* mutants display abnormal reaction to high-intensity light, it was hypothesized that GUN1 integrates signals induced by norflurazon, lyncomycin and high-intensity light (Koussevitzky et al., 2007). A comparative study of anthocyan and *Lhcb1* transcript accumulation in *gun1-1* mutants and wild-type *Arabidopsis* seedlings indicated that the deficiency of functional GUN1 impairs early development of seedlings and alters the sensitivity of plants to sucrose and abscisic acid (see below). However, the mode of GUN1 action is presently unknown (Cottage et al., 2010).

In the retrograde signaling pathway downstream of GUN1 operate two nuclear transcription factors: AP2 (Apetala 2)-like transcription factor ABI4 and GLK1. The *Arabidopsis abi4* mutant has a weak *gun* phenotype. The *abi4* gene is highly expressed in seeds in contrast to seedlings. The *abi4* expression is induced by glucose and probably other sugars. ABI4 is a negative regulator of the *Lhcb* expression. In response to plastid signals ABI4 competitively binds to G-box of the *cis*-element and inhibits *Lhcb* expression (Inaba, 2010). The transcription factors of the Apetala 2 (AP2)- type act as repressors of transcription in the presence of abscisic acid, ethylene and jasmonic acid (Koussevitzky et al., 2007). By a still unknown mechanism GUN1 activates ABI4. It was shown that just this mechanism is used by retrograde signals induced by tetrapyrrole biosynthesis intermediates, plastid gene expression and redox state of the electron transport chain, but differs from the mechanisms utilized by signals associated with abnormal protein import into the plastids (Inaba, 2010). In this case, the transcription factor GLK1 operates. In contrast to ABI4, GLK1 is a positive regulator of *Lhcb* expression, which coordinates expression of nuclear photosynthetic genes (Kamikaze & Inaba, 2010). Under stressful conditions, such as norflurazon treatment and impaired protein import, GLK1 expression is considerably decreased that leads to inhibition of photosynthesis-related genes. For GLK1 repression, GUN1 is necessary.

It was shown that many nuclear photosynthetic genes controlled by plastid signals contain the ACGT sequence in their promoter regions that also serves as the major (core) element involved in cell response to abscisic acid. This points to the involvement of components of ABA-signaling cascade (or abscisic acid levels) in retrograde signaling (Jung & Chory, 2010; Koussevitzky et al., 2007). Analysis of the ABA-deficient (*aba*) and ABA-insensitive (*abi*) *Arabidopsis* mutants showed that in regulation of nuclear gene expression and plant adaptation to stressful conditions, the components of the ABA cascade are tightly associated with the Mg-Proto retrograde signaling pathway. The interaction between the plastid and abscisic acid signals is mediated by ABI4, which is a 'master switch' that controls expression of a large number of genes in response to diverse signals. The ABI4 transcription factor is not only a component of the plastid retrograde signaling pathway but is also involved in mitochondrial retrograde signaling, thus it serves as a convergence point of these signaling pathways (Cottage et al., 2010).

It was shown that the H subunit of Mg-chelatase (ChlH) participates in signal transduction from the plastids to the nucleus and in ABA-dependent signaling cascades. The ChlH protein is a receptor of ABA-induced signals. In *Arabidopsis* it is a positive regulator of signaling induced by Mg-Proto and chlorophyll biosynthesis intermediates. Plants with enhanced expression of this protein are supersensitive to abscisic acid, while ChlH-deficient mutants possess an ABA-insensitive phenotype. It was assumed that the plastid proteins EX1 and EX2 (Inaba, 2010) and protein kinases, in particular the thylakoid membrane protein STN7, which is a Lhcb protein kinase (Inaba, 2010; Pfannschmidt, 2010), are also involved in retrograde signaling. However these proteins are located in the plastids. The cytosolic mediators of the plastid signals have not been identified. The nuclear transcription factor Whirly1 (Why1), which was originally described as a telomere-binding protein, is also supposed to be involved in retrograde signaling.

Abundant experimental data indicate that plastid retrograde signaling cascades interact with light signaling. Light activates expression of a large number of photosynthesis-related nuclear genes, such as *Cab* encoding chlorophyll *a/b* - binding proteins, *RbcS* (encoding the small subunit of Rubisco) and *Pc* coding for plastocyanin. Retrograde signals may act synergistically with the light signals or transform them from inducers into repressors of photosynthetic gene expression (Larkin et al., 2008; Osipenkova et al., 2010). It is generally accepted that the plastid retrograde signals are endogenous regulators of the light signaling and that integration of the light and plastid signals helps plants to overcome chloroplast dysfunction during organelle biogenesis under unfavorable light conditions (Larkin et al., 2008). Both types of signals act on the same *cis*-elements in the promoters of the nuclear plastid protein genes (Kusnetsov et al., 1996). The ABI4 participates both in retrograde and light signaling. This protein binds to the promoter region of the gene regulated by a retrograde signal through the conserved CCAC motif, which is a core element necessary for binding. In such a way ABI4 inhibits G-box mediated light-inducible expression of photosynthesis-related genes when the chloroplast development is inhibited (Koussevitzky et al., 2007).

Earlier a strong correlation between the light and sucrose signaling pathways was revealed (Dijkwel et al., 1997). It was shown that the sucrose signaling pathways may interact (at least in some tissues) with other retrograde signaling pathways and thus modulate the response of nuclear genes to retrograde signals. In the study of *sun6* (sucrose-uncoupled6) *Arabidopsis* mutants (allelic *abi4*-mutants), a correlation between redox signals and sucrose-regulated gene expression was established (Oswald et al., 2001). It was also shown that the SUN6 protein is involved in sucrose repression of phytochrome A-dependent signaling pathways (Dijkwel et al., 1997). These data indicate interactions of sucrose, light and plastid signaling pathways.

The data presented indicate that plant cells possess a complex network of signaling pathways that function independently or interact with each other due to common intermediates or final cascade components. Tetrapyrroles play a key role in this network, their metabolism regulates the plant cell functions.

The transfer of numerous genes from cell organelles into the nucleus in the course of cell evolution (Odintsova & Yurina, 2006) led to gene redistribution in cell compartments and the coordinated expression of nuclear and organellar genes. The expression of nuclear genes in plastids and mitochondria must be harmonized with the functional state of these organelles, and such coordination was provided by evolution of the so-called retrograde (organelle-to-nucleus) control over the expression of nucleus-coded organellar genes. The pathways of such signal transduction have not been sufficiently clarified: we do not know how the plastid or mitochondrial signal crosses the organelle membranes, what signal molecules stay in the cytosol and nucleus, and whether their action upon the regulatory proteins of the nucleus is direct or mediated by the induction of other signals. In most cases the plastid signals are related to the redox state of organelles, e.g., the ETC components and redox-active stromal compounds, such as thioredoxin and glutathione, or with the chlorophyll biosynthesis. ROS, the side products of photosynthesis and mitochondrial respiration, also can participate as signal molecules in the organelle-to-nucleus signal transduction. ROS are formed by both types of organelles, and therefore can provide for the coordinated expression of nuclear genes in the plastids and mitochondria at the transcriptional level. The analysis of the nucleus-encoded chloroplast transcriptome demonstrated that the transcription of the nuclear genes coding for plastid proteins was controlled by several types of plastid signals comprising the complicated signal network within plant cells (Leister, 2005). The attempts to investigate this network using the mutant plants are problematical because alternative control mechanisms may function in the mutants impaired in one and the same photosynthetic complex (Beck, 2005).

3. Mitochondrial retrograde regulation in plants

Most data about mitochondrial retrograde regulation (MRR) in plants were obtained on material with disturbed functioning of mitochondria. Expression of some nuclear genes is activated in response to disturbances in mtETC, tricarbonic acid cycle, and also mtDNA (Rhoads & Subbaiah, 2007). Plant mitochondria dysfunction induced by mutations results frequently in male sterility, embryo lethal phenotype, or chlorosis in plants, which could not complete their life cycle (Newton et al., 2004). Plant mitochondria dysfunction can be induced by biotic and abiotic stresses. A bulk of information was obtained demonstrating a great contribution of mitochondria in general plant response to stress.

Mitochondrial retrograde regulation in plants is usually studied on the systems with disturbed mitochondrial functions induced by mutations, chemical agents, or biotic and abiotic stresses. Dysfunction of mitochondria leads to changes in the nuclear gene expression. Plant response to mitochondrial function disturbance is an expression induction of genes encoding proteins involved in the restoration of mitochondria functioning, such as alternative oxidase (AO) or alternative NADPH-dehydrogenases, and genes encoding antioxidant enzymes, such as glutathione transferases, catalases, ascorbate peroxidases, and superoxide dismutases, normalizing ROS level (Rhoads & Subbaiah, 2007). At present, most studied plant response to mitochondria dysfunction is the changes in expression of the nuclear gene encoding AO (Mackenzie & McIntosh, 1999; Rhoads & Vanlerberghe, 2004). It is believed that the two pathways of signal transduction to the nucleus operate at AO expression: one with the involvement of ROS and another with the involvement of organic acids (Gray et al., 2004).

Cytoplasmic male sterility (CMS), i.e., plant failure to produce viable pollen, is induced by signals from mitochondria and is one of the characteristic examples of MRR in plants. Male sterility is often results from mutation-induced mitochondria dysfunction. Changes in mtDNA or mitochondrial gene expression induce changes in nuclear gene expression, result in the modification of stamen phenotype, and finally in the suppression of pollen formation. CMS is a commonly occurring phenomenon observed in more than 150 plant species. It is met frequently in hybrid lines obtained due to intra- or inter-specific crosses, i.e., in alloplasmic lines where the nucleus of one species functions in the cytoplasm of another species.

The fact that CMS is determined by the interaction between the nucleus and mitochondria was proved by maternal inheritance of male sterility phenotype and by suppression of male sterility with nuclear genes of fertile restoration (Rf genes). These genes (except a single gene) encode proteins belonging to the family of PPR proteins (proteins containing pentatricopeptide repeats). These proteins in *Arabidopsis* are supposed to be localized in mitochondria or plastids. Some of them participate in the processing of various RNAs in organelles (Carlsson et al., 2008).

It should be noted that not only retrograde signaling pathways but also other mechanisms could result in CMS, i.e., diverse mechanisms could determine male sterility. However, the target genes of retrograde signaling pathways and functions of genes for fertility restoration are evidently common for many CMS systems. Since several Rf genes could suppress the phenotype of male sterility and restore male fertility in various species, we can suppose that changes in transcription, translation, or RNA processing in mitochondria are represent the common mechanism of restoration. The homeotic genes, i.e., genes, which mutations result in changes of floral organs, are most probable targets for the retrograde signaling pathways directed from CMS-inducing mitochondria to the nucleus. However, the signal for CMS origin is unknown (Carlsson et al., 2008; Kubo & Newton, 2008).

Plant mitochondria respond to abiotic and biotic stresses. They can serve stress sensors and initiate responses (or be involved in responses) to particular stress types. Thus, MRR is involved in the plant cell response to oxygen stress. It is known that, under definite conditions (flooding for example), oxygen is a limiting factor for plant growth and survival. Under normal growth and development, plant also could experience hypoxia because oxygen diffusion in tissues with closely packed cells and small intercellular spaces is hindered. It is also known that even plants acclimated to land conditions manifest a great tolerance to oxygen deficit, which indicates the presence in them of highly sensitive system for oxygen sensing and well developed responses to oxygen stress (Bailey-Serres & Chang, 2005). Adaptation to a rapidly changing oxygen level includes rapid changes in gene expression (potentially induced by preceding rapid changes in the level of ROS, redox state, and/or energetic status) permitting a correction of the metabolism (for example, suppression of carbon entry in TCA cycle with parallel increase in the carbon entry into glycolytic and other enzymic pathways) (Rhoads & Subbaiah, 2007). Reactions leading to programmed cell death are also one of the plant responses to anaerobiosis (Subbaiah & Sachs, 2003).

The genes controlled by oxygen could be roughly divided into aerobic genes (transcribed at atmospheric oxygen level), hypoxic genes (induced at lowered oxygen level), anoxic genes (induced in the absence of oxygen), and hyperoxic genes (transcribed under oxidative conditions). Certainly, such a division is rather conditional (Rhoads & Subbaiah, 2007).

Since oxygen is primarily absorbed by cell mitochondria, changes in the level of available oxygen should be primarily sensed by these organelles, resulting in MRR of nuclear genes. The levels of transcripts and proteins (TCA enzymes or mtETC components) were shown to be reduced at oxygen removal and rapidly restored after oxygen level restoration (Branco-

Price et al., 2005). In experiments with maize cell culture, it was shown that, in the absence of oxygen, immediate and reversible increase in the Ca^{2+} level in the cytosol ($[Ca^{2+}]_{cyt}$) occurs. Changes in the Ca^{2+} level are observed around mitochondria and disappeared by the blocker of mitochondrial calcium channel ruthenium red. This dye suppresses induction of genes sensitive to anoxia, alcohol dehydrogenase 1 (*adh1*) and sucrose synthase (sh1) in maize seedlings and cell culture (Subbaiah et al., 1998). Maize seedlings treated with ruthenium red are especially sensitive to anoxic stress. Ca^{2+} addition neutralizes ruthenium red effects, confirming the supposition that ion changes in maize evidently initiated by mitochondria are the signals for responses to anoxia. Calcium ions were shown to be an important component of the signal transduction pathway related to hypoxia in *Arabidopsis*, rice, and barley, which indicates the conserved nature of this pathway in plants (Rhoads & Subbaiah, 2007). The conclusion that mitochondria are a source of $[Ca^{2+}]_{cyt}$-signal initiating nuclear gene activation is supported by dynamics of mitochondrial Ca^{2+} ($[Ca^{2+}]_{mt}$) in response to anoxia. $[Ca^{2+}]_{mt}$ release was shown to occur in maize cells immediately after ceasing oxygen influx.

Heat stress is known to reduce crop yield. One of the ways to decline heat shock effects is the induction of genes encoding heat shock proteins (HSPs), including low-molecular HSPs (sHSPs). Trasgenic plants expressing HSPs are a convenient model for studying mitochondrial responses to heat shock. Thus, transgenic *Arabidopsis* plants were obtained expressing transgene encoding maize mitochondrial sHSP, ZmHSP22 (Rhoads et al., 2007). This protein is weakly constitutively synthesized in etiolated maize seedlings and induced by heat shock. In transgenic line, this transgene under the control of cauliflower mosaic virus 35S promoter was actively expressed constitutively. Using confocal immunofluorescent microscopy and the analysis of isolated mitochondria, it was shown that ZmHSP22 penetrates *Arabidopsis* mitochondria and processed there with the formation of mature protein. In transgenic plants subjected to heat stress, expression of several nuclear genes encoding endogenous mitochondrial sHSP of *Arabidopsis* (AtHSP23.6) and also HSPs localized in chloroplasts (AtHSP25.3 and AtHSP70-6) and cytosol (AtHSP17.4 and AtHSP70-1) was changed. In wild-type plants, AtHSP23.6 is weakly expressed but induced by heat stress. These data allow a supposition that heat-induced MRR could affect HSP expression.

The common component of various abiotic stresses is an oxidative stress induced by ROS generation. Oxidative stress could induce mitochondria dysfunction in plants and result in inactivation of definite hormonal signals (Rhoads & Subbaiah, 2007). Proteomic approach for identification of mitochondrial components sensitive to oxidative damage showed that they are enzymes of the TCA cycle, components of mtETC and oxidative phosphorylation (Sweetlove & Foyer, 2004). In the leaves, glycine decarboxylase, the enzyme of the photorespiratory pathway, turned out to be especially prone to oxidative damage (Taylor et al., 2002).

3.1 Components of signal transduction pathways related to plant mitochondrial retrograde regulation

The common feature of abiotic and biotic stresses in plants is an increased level of ROS in the cells, which leads to changes in expression of nuclear genes (Vranova et al., 2002). However, the precise role of ROS in plant responses to stresses is unknown. A specific ROS property is that they are damaging compounds and simultaneously signal molecules during plant responses to stresses. However, at different stresses, ROS elevated level induces different changes in gene expression; ROS can interact with other stress-factors, such as calcium, hormones, or changes in the cell redox state (Gadiev et al., 2006). mtROS are the part of the total ROS pool generated in the cells in response to stress. The contribution of mitochondria and other cell compartments into the ROS level is undetermined; changes in nuclear gene expression induced by each of these compartments are unknown as well.

It was shown that inhibition of the cytochrome respiratory pathway results in the increase in the mtROS content (especially hydrogen peroxide), which evidently induces MRR signaling and AO gene expression, which leads to the reduction in mtROS generation. However, this does not mean that mtROS are required for all MRR pathways functioning in the plants. Monofluoroacetate is a mighty inducer of AtAOX1a gene expression (Zarkovic et al., 2005), but it does not induce a strong enhancement in ROS generation in the cell (Rhoads & Subbaiah, 2004). Although the role of mtROS and MRR at inhibition of plant mtETC by antimycin seems to be proven, other signaling components of mitochondrial regulation downstream ROS are not identified.

It was shown that cell redox state and signals emitted by the photosynthetic ETC are involved in the control of nuclear and chloroplast gene expression. Changes in the redox state have also a great significance for plant responses to stresses (Foyer & Noctor, 2005). Mitochondria can change the cell redox state, in particular due to ROS generation and functioning of the glutathione–ascorbate cycle. It is supposed that in plants glutathione and ascorbate are the components of redox signaling, which induce expression of defense genes (Dutilleul et al., 2003; Foyer & Noctor, 2005).

Glutathione is also supposed to be a component of the signal transduction pathway related to cold stress (Kocsy et al., 2001). Until now, only few components of the MRR signaling pathways in plants are identified. Among data available, especially interesting ones are that calcium is involved in MRR related to hypoxia; indeed, Ca^{2+} ions participate in signal transduction in mitochondrial retrograde signaling pathways most frequently met in higher eukaryotes (Butow & Avadhani, 2004). It is supposed that Ca^{2+} is involved as a signal not only in plant MRR under hypoxia but also in other stress types. All these data argue for an important role of Ca^{2+} transport for plant responses to stresses. It is unknown how a signal from the increase in the $[Ca^{2+}]_{cyt}$ determined by mitochondria is amplified, transmitted to the nucleus, and results in changes in nuclear gene expression. It is believed that Ca^{2+}-binding proteins participate in this process, viz., calcium-dependent protein kinases, transcription factors, and regulatory proteins (such as 14-3-3 proteins). It is not excluded that Ca^{2+} signal per se is transferred to the nucleus. Some data are obtained that changes in the Ca^{2+} content in the nucleus affect directly gene expression.

Plant genomes (at least sequenced genomes of *Arabidopsis* and rice) contain more genes encoding transcription factors than sequenced genomes of other organisms; in particular, they comprise unique gene families characteristic of only plants (Chen et al., 2002). It is supposed that these factors could be additional factors tuning plant responses to stresses. Some of them are evidently involved in MRR, although no transcription factors specific for MRR are identified until now. It might be that such factors as WRKY, bZIP, and Dof are involved in plant MRR because they participate in responses to biotic and abiotic stresses, including oxidative stress (Chen et al., 2002).

The elucidation of molecular mechanisms of plant MRR is only at its start. Mitochondria responses to stresses, which are the part of total plant responses, could determine the fate of plant cells, resulting in the restoration of their vital activity or in their death. Thus, induction of AO and alternative NADPH-dehydrogenases in tobacco cell suspension at suppression of cytochrome respiratory pathway prevent cell death. In the cells lacking AO, ROS accumulation induced by mtETC inhibition evidently shifts metabolism toward programmed cell death (Robson & Vanlerberghe, 2002). A capability of the induction of AO expression could also affect the independent mitochondrial pathways of programmed cell death.

Several retrograde mitochondrial signaling pathways could function in the plant cells (Zarkovic et al., 2005), and they could initiate specific changes in nuclear gene expression in response to specific disturbances in mitochondria functioning.

Retrograde mitochondrial signaling pathways interact with each other. In addition, they interact with retrograde chloroplast signaling pathways (Pesaresi et al., 2006, 2007) and other signal transduction pathways in the plant cell including sugars (Pesaresi et al., 2007), hormones (Kwak et al., 2006), enzymes (Subbaiah et al., 2006), etc.

The interaction between retrograde signaling from mitochondria and plastids is of a great interest because these signals control expression of nuclear genes encoding organelle components in dependence on organelle functional state. Chloroplast and mitochondrial metabolisms are known to be connected. Photosynthesis provides substrates for mitochondrial respiration and, in its turn, depends on some compounds synthesized by mitochondria. In darkness, mitochondria are the major source of ATP for cell processes, including those in chloroplasts. In addition in darkness, ATP supports the proton gradient across the thylakoid membrane, thus protecting chloroplasts against photoinhibition after the start of illumination. In the light, mitochondria provide chloroplasts with carbon-containing compounds, produced in the TCA cycle, for NH_4^+ assimilation, whereas ATP supports diverse biosynthetic reactions, including the restoration of photosystem II functions (Pesaresi et al., 2006).

Data concerning signal molecules and components of inter-organelle signaling pathways are few. The molecular analysis of these pathways supposes their multiple interactions. It is believed that NO, ascorbate, and ROS could fulfill the role of signals in the mitochondria and chloroplast interactions. However, the ways of signal transduction from organelles to the nucleus and the incorporation of these signals into the general system of expression regulation are not known so far (Pesaresi et al., 2006).

4. Conclusion

Retrograde regulation of nuclear gene expression by organelles is best studied in budding yeast *S. cerevisiae* and higher plant chloroplasts. In chloroplasts, retrograde signals are mostly related to the redox state of organelles or to chlorophyll biosynthesis. Intermediates of tetrapyrrole biosynthesis and the products of organellar protein synthesis are evidently major retrograde signals (Pesaresi et al., 2006, 2007). Changes in the nuclear gene expression dependent on plastid retrograde signaling pathways include a multilevel control of transcription and the involvement of the ABI4 transcription factor (Pesaresi et al., 2007). ROS, harmful by-products of photosynthesis and mitochondrial respiration, could also serve signal molecules transmitting a signal from organelles to the nucleus. Since ROS are generated in both types of organelles, they could provide for coordinated expression of nuclear genes of plastid and mitochondria at the level of transcription. Most identified regulons are shown to contain nuclear genes of chloroplast and mitochondria, and this allows coordinated regulation of activities of these organelles (Rhoads & Subbaiah, 2007). Despite the fact that investigations of plant organelles-to-nucleus retrograde signaling have been the subject of intensive research for several decades, the available data are fragmentary. The molecules, which induce the signal, are still unknown, as well as the mechanisms of signal transduction and the components of the signaling cascade involved. Not much is known on the specificity and cross-talk between different signaling pathways in plants. Usage of classic and novel methodological approaches help identify signal molecules involved in mitochondrial retrograde regulation in plants. Further research exploiting the traditional methods of biochemistry, direct and reverse genetics, proteomics and metabolomics will help to identify the signaling molecules triggering retrograde cascades, to unravel the mechanisms of signal transduction.

5. Acknowledgment

This study was supported by the Molecular and Cell Biology program of the Presidium of RAS.

6. References

Adamska, I. (1995). Regulation of Early Light-Inducible Protein Gene Expression by Blue and Red Light in Etiolated Seedlings Involves Nuclear and Plastid Factors. *Plant Physiol.*, Vol.107, No.4, (April 1999), pp. 1167-1175, ISSN 1532-2548

Apel, K. & Hirt, H. (2004). Reactive Oxygen Species: Metabolism, Oxidative Stress, and Signal Transduction. *Annu. Rev. Plant Biol.*, Vol. 55., pp. 373-399, ISSN 1543-5008

Armbruster, U.; Pesaresi, P.; Pribil, M.; Hertle, A. & Leister, D. (2011). Update on Chloroplast Research: New Tools, New Topics, and New Trends. *Mol. Plant.*, Vol.4, No.1, (January, 2011), pp. 1-16, ISSN 1752-9867

Bailey-Serres, J. & Chang, R. (2005). Sensing and Signaling in Response to Oxygen Deprivation in Plants and Other Organisms. *Ann. Bot.*, Vol.96, No.4, (September, 2005), pp. 507-518, ISSN 0305-7364

Beck, Ch.F. (2005). Signaling Pathways from the Chloroplast to the Nucleus. *Planta*, Vol. 222, No.5, (November 2005), pp. 743-756, ISSN 0032-0935

Branco-Price, C.; Kawaguchi, R.; Ferreira, R.B. & Bailey-Serres, G. (2005). Genome-Wide Analysis of Transcript Abundance and Translation in Arabidopsis Seedlings Subjected to Oxygen Deprivation. *Ann. Bot.*, Vol. 96, No.4, (September, 2005), pp. 647-660, ISSN 0305-7364

Butow, R.A. & Avadhani, N.G. (2004). Mitochondrial Signaling, the Retrograde Response. *Mol. Cell.*, Vol.14, No.1, (April, 2004), pp. 1-15, ISSN 1097-2765

Carlsson, J.; Leino, M.; Sohlberg, J.; Sundstrom, J.F. & Glimelius, K. (2008). Mitochondrial Regulation of Flower Development. *Mitochondrion*, Vol. 8, No.1, (January, 2008), pp. 74-86, ISSN 1567-7249

Chang, C.C.; Ball, L.; Fryer, M.J.; Baker, N.R.; Karpinski, S. & Mullineaux, P.M. (2004). Induction of Ascorbate Peroxidase 2 Expression in Wounded Arabidopsis Leaves Does Not Involve Known Wound-Signalling Pathways but Is Associated with Changes in Photosynthesis. *Plant J.*, Vol. 38, No.3, (May, 2004), pp. 499-511, ISSN 0960-7412

Chen, W.; Provart, N.J.; Glazebrook, J.; Katagiri, F.; Chang, H.S.; Eulgem, T.; Mauch, F.; Luan, S.; Zou, G.; Whitham, S.A.; Budworth, P.R.; Tao, Y.; Xie, Z.; Chen, X.; Lam, S.; Kreps, J.A.; Harper, J.F.; Si-Ammour, A.; Mauch-Mani, B.; Heinlein, M.; Kobayashi, K.; Hohn, T.; Dangl, J.L.; Wang, X. & Zhu, T. (2002). Expression Profile Matrix Arabidopsis Transcription Factor Genes Suggests Their Putative Functions in Response to Environmental Stresses. *Plant Cell*, Vol.14, No.3, (March, 2002), pp. 559-574, ISSN 1040-4651

Cottage, A.; Mott, E.K.; Kempster, J.A. & Gray, J.C. (2010). The Arabidopsis Plastid-Signalling Mutant gun1 (genomes uncoupled1) Shows Altered Sensitivity to Sucrose and Abscisic Acid and Alterations in Early Seedling Development. *J. Exp. Bot.*, Vol.61, No.13, (July, 2010), pp. 3773-3786, ISSN 0022-0957

Dijkwel, P.P.; Huijser, C.; Weisbeek, P.J.; Chua, N.H. & Smeekens, S.C. (1997). Sucrose Control of Phytochrome A Signaling in Arabidopsis. *Plant Cell*, Vol.9, No.4, (April, 1997), pp. 583-595, ISSN 1040-4651

Dutilleul, C.; Garmier, M.; Noctor, G.;., Mathieu,C.; Chetrit, P.; Foyer, C.H. & De Paepe, R. (2003). Leaf Mitochondria Modulate Whole Cell Redox Homeostasis, Set Antioxidant Capacity, and Determine Stress Resistance Through Altered Signaling and Diurnal Regulation. *Plant Cell*, Vol.5, No.5, (May, 2003), pp. 1212-1226, ISSN 1040-4651

Emanuel, C.; Weihe, A.; Graner, A.; Hess, W.R. & Börner, T. (2004). Chloroplast Development Affects Expression of Phage-Type RNA Polymerases in Barley Leaves. *Plant J.*, Vol.38, No.3, (May, 2004), pp. 460-472, ISSN 0960-7412

Fey, V.; Wagner, R.; Bräutigam, K.; Wirtz, M.; Hell, R.; Dietzmann, A.; Leister, D.; Oelmüller, R. & Pfannschmidt, T. (2005). Retrograde Plastid Redox Signals in the Expression of Nuclear Genes for Chloroplast Proteins of Arabidopsis thaliana. *J. Biol. Chem.*, Vol. 80, No.7 , (November, 2004), pp. 5318-5328, ISSN 0021-9258

Foyer, C.N. & Noctor, G. (2005). Redox Homeostasis and Antioxidant Interface between Stress Perception and Physiological Responses. *Plant Cell*, Vol. 17, No.7, (July, 2005), p. 1866-1875, ISSN 1040-4651.

Gadiev, I.; Vanderauwera, S.; Gechev, T.S.; Laloi, C.; Minkov, I.N.; Shulaev, V.; Apel, K.; Inze, D.; Mittler, R. & Van Breusegem, F. (2006). Transcriptomic Footprints Disclose Specificity of Reactive Oxygen Species Signaling in Arabidopsis. *Plant Physiol.*, Vol.141, No.2 , (January, 2006), pp. 436-445, ISSN 1532-2548

Gadjieva, R.; Axelsson, E.; Olsson, U. &, Hansson, M. (2005). Analysis of gun Phenotype in Barley Magnesium Chelatase and Mg-protoporphyrin IX Monomethyl Ester Cyclase Mutants. *Plant Physiol. Biochem.*, Vol.43, No.10-11, (October, 2005), pp. 901–908, ISSN 0981- 9428

Gray, G.R.; Maxwell, D.P.; Villarimo, A.R. & McIntosh, L. (2004). Mitochondria / Nuclear Signaling of Alternative Oxidase Expression Occurs through Distinct Pathways Involving Organic Acids and Reactive Oxygen Species. *Plant Cell Rep.*, Vol.23, No.7, (August, 2004), pp. 497-503, ISSN 0721-7714

Gray, J.C.; Sornarajah, R.; Zabron, A.A.; Duckett,C.M. & Khan, M.S. (1995). Chloroplast Control of Nuclear Gene Expression, In: *Photosynthesis: From Light to Biosphere*, Mathis P. (Ed.), pp. 543-550, Kluwer Academic Publishers, ISBN 978-0-7923-3862-8, Dordrecht, Netherlands

Inaba, T. (2010). Bilateral Communication Between Plastid and the Nucleus: Plastid Protein Import and Plastid-to-nucleus Retrograde Signaling. *Biosci. Biotechnol. Biochem.* Vol.74, No.3 , (March 2010), pp. 471-476, ISSN 0916-8451

Jung, H.-S. & Chory, J. (2010). Signaling between Chloroplasts and the Nucleus: Can a Systems Biology Approach Bring Clarity to a Complex and Highly Regulated Pathway? *Plant Physiol.* Vol.152, No.2, (November, 2009), pp. 453–459, ISSN 1532-2548.

Kakizaki, T. & Inaba, T. (2010). New Insights into the Retrograde Signaling Pathway between the Plastids and the Nucleus. *Plant Signal Behav.*, Vol. 5, No.2, (February, 2010), pp. 196-199, ISSN 1559-2316

Kocsy, G.; Galiba, G. & Brunold, C. (2001). Role of Glutathione in Plants. *Physiol. Plant.*, Vol. 113, No.2, (October, 2001), pp. 158-164, ISSN 0032-0889

Koussevitzky, S.; Nott, A.; Mockler, T.C.; Hong, F.; Sachetto-Martins, G.; Surpin, M.; Lim, J.; Mittler, R. & Chory, J. (2007). Signals from Chloroplasts Converge to Regulate Nuclear Gene Expression. *Science,* Vol.316, No.5825, (May, 2007), pp. 715–719, ISSN 0036-8075

Kubo, T. & Newton, K.J. (2008). Angiosperm Mitochondrial Genomes and Mutations. *Mitochondrion*, Vol. 8, No.1, (Novermber, 2007), pp. 5-14, ISSN 1567-7249

Kusnetsov, V.; Bolle, C.; Lubberstedt, T.; Sopory, S.; Herrmann, R.G. & Oelmüller, R. (1996). Evidence That the Plastid Signal and Light Operate via the Same Cis-Acting Elements in the Promoters of Nuclear Genes for Plastid Proteins. *Mol. Gen. Genet.*, Vol.252, No.6, (October, 1996), pp. 631-639.

Kwak, J.M.; Nguyen, V. & Schroeder, J.I. (2006). The Role of Reactive Oxygen Species in Hormonal Responses. *Plant Physiol.*,Vol.141, No.2, (June, 2006), pp. 323-329, ISSN 1532-2548

La Rocca, N.; Rascio, N.; Oster, U. & Rüdiger, W. (2001). Amitrole Treatment of Etiolated Barley Seedlings Leads to Deregulation of Tetrapyrrole Synthesis and to Reduced Expression of *Lhc* and *RbcS* Genes. *Planta*, Vol.213, No.1, (May, 2001), pp. 101–108, ISSN 0032-0935

Larkin, R.M. & Ruckle, M.E. (2008). Integration of Light and Plastid Signals. *Curr. Opin. Plant Biol.*, Vol.11, No.6, (October, 2008), pp. 593–599. ISSN 1369- 5266

Leister, D. (2005). Genomics-Based Dissection of the Cross-Talk of Chloroplasts with the Nucleus and Mitochondria in Arabidopsis. *Gene,* Vol. 354, No.12, (October, 2005), pp. 110-116, ISSN 0378-1119

Mackenzie, S. & McIntosh, L. (1999). Higher Plant Mitochondria. *Plant Cell,* Vol.11, No.4, (April, 1999), pp. 571-586, ISSN 1040-4651

Mochizuki, N.; Brusslan, J.A.; Larkin, R.; Nagatani, A. & Chory, J. (2001). Arabidopsis Genomes Uncoupled (GUN 5) Mutant Reveals the Involvement of Mg-chelatase H Subunit in Plastid-to-nucleus Signal Transduction. *Proc. Natl. Acad. Sci. USA,* Vol.98, No.4, (February, 2001), pp. 2053–2058, ISSN 1091-6490

Mochizuki, N.; Tanaka, R.; Grimm, B.; Masuda, T.; Moulin, M.; Smith, A.G.; Tanaka, A. & Terry, M.J. (2010) The Cell Biology of Tetrapyrroles: a Life and Death Struggle. *Trends Plant Sci.*, Vol. 15, No.9, (July, 2010), pp. 488–498, ISSN 1360-1385

Mochizuki, N.; Tanaka, R.; Tanaka, A.; Masuda, T. & Nagatani, A. (2008). The Steady-state Level of Mg protoporphyrin IX is not a Determinant of Plastid-to-nucleus Signaling in Arabidopsis. *Proc. Natl. Acad. Sci. USA*, Vol.105, No.39, (September, 2008), pp. 15184–15189, ISSN 1091-6490

Moulin, M.; McCormac, A.C.; Terry, M.Y. & Smith, A.G. (2008). Tetrapyrrole Profiling in Arabidopsis Seedlings Reveals that Retrograde Plastid Nuclear Signaling is not Due to Mg-protoporphyrin IX Accumulation. *Proc. Natl. Acad. Sci. U S A*, Vol. 105, No.39, (September, 2008), pp. 15178–15183, ISSN 1091-6490

Newton, K.J.; Gabay-Laughnan, S. & DePaepe, R. (2004). Mitochondrial Mutations in Plants, In: *Advances in Photosynthesis and Respiration*. Day, D.A., Millar, A.H., Whelan J. (Eds.), pp. 121-142, Kluwer, ISBN 978-90-481-3882-1, Dordrecht. Netherlands

Nott, A.; Jung, H.-S.; Koussevitzky, S. & Chory, J. (2006). Plastid-to-Nucleus Retrograde Signaling. *Annu. Rev. Plant. Biol.*, Vol. 57, pp. 739-759, ISSN 1040-2519

Odintsova, M.S. & Yurina, N.P. (2006) Chloroplast Genomics of Land Plants and Algae, In: *Biotechnological Applications of Photosynthetic Proteins: Biochips, Biosensors and Biodevices*, Giardi M.T., Piletska E.V., (Eds.), pp. 57-72, Landes Bioscience/Eurekah. Com., ISBN 0-387-33009-7, Georgetown (Texas), USA

Oelmüller, R.; Levitan, I.; Bergfeld, R.; Rajasekhar, V.K. & Mohr, H. (1986). Expression of Nuclear Genes as Affected by Treatments Acting on the Plastids. *Planta*, Vol. 168, No.4, (November, 1986), pp. 482-492, ISSN 0032-0935

Osipenkova, O.V.; Odintsova, M.S. &Yurina, N.P. (2010). The Influence of Light, Hormonal, and Carbohydrate Signal Systems on ELIP genes expression in gun_mutants Arabidopsis thaliana. *Applied Biochemistry and Microbiology*, Vol. 46, No. 3, pp. 331–338. ISSN 0003_6838

Oswald, O.; Martin, T.; Dominy, P.J. & Graham, I.A. (2001). Plastid Redox State and Sugars: Interactive Regulators of Nuclear-encoded Photosynthetic Gene Expression. *Proc. Natl. Acad. Sci. U S A*, Vol. 98, No.4, (January, 2001), pp. 2047-2052, ISSN 1091-6490

Pesaresi, P.; Masiero, S.; Eubel, H.; Braun, H-P.; Bhushan, S.; Glaser, E.; Salamini, F. & Leister, D. (2006). Nuclear Photosynthetic Gene Expression Is Synergistically Modulated by Rates of Protein Synthesis in Chloroplasts and Mitochondria. *The Plant Cell*. Vol.128, No.4, (March, 2006), pp. 970-991, ISSN 1040-4651

Pesaresi, P.; Schneider, A.; Kleine, T. & Leister, D. (2007) Interorganellar Communication. *Curr. Opinion in Plant Biology*, Vol.10, No.6, (August, 2007), pp. 600-606, ISSN 1369-5266

Pfannschmidt, T. & Liere, K. (2005). Redox Regulation and Modification of Proteins Controlling Chloroplast Gene Expression. *Antioxidants Redox Signal*, Vol.7, No.5-6, (May-June, 2005), pp. 607-618, ISSN 1523-0864

Pfannschmidt, T. (2010). Plastidial Retrograde Signalling – a True "Plastid Factor" or Just Metabolite Signatures? *Trends Plant Sci.*, Vol. 15, No.8, (January, 2010), pp. 427–435, ISSN 1360-1385

Piippo, M.; Allahverdiyeva, Y.; Paakkarinen, V.; Suoranta, U.-M.; Battchikova, N. & Aro, E.-M. (2006). Chloroplast-Mediated Regulation of Nuclear Genes in Arabidopsis thaliana in the Absence of Light Stress. *Physiol. Genomics*. Vol. 25. pp. 142-152, ISSN 1094-8341

Pogulskaya, E.N.; Yurina, N.P. & Karapetyan, N.V. (2006) Tetrapyrrole Involvement in Expression Regulation of a Nuclear Gene of Low-Molecular-Weight Plastid Protein ELIP. *Applied Biochemistry and Microbiology*, Vol. 42, No. 3, pp. 320–325, ISSN 0003-6838

Rhoads, D.M. & Subbaiah, C.C. (2007). Mitochondrial Retrograde Regulation in Plants. *Mitochondrion*. Vol.7, No.3, (May, 2007), pp. 177-194, ISSN 1567-7249

Rhoads, D.M. & Vanlerberghe, G.C. (2004). Mitochondria-Nucleus Interactions, Evidence for Mitochondrial Retrograde Communication in Plant Cells, In: *Advances in Photosynthesis and Respiration*. Day, D.A., Millar, A.H., Whelan J. (Eds.), Vol. 17, pp. 83-106, Kluwer Academic Publishers, ISBN 978-90-481-3882-1, Dordrecht. Netherlands

Rhoads, D.M.; White, S.J.; Zhou, Y.; Muralidharan, M. & Elthon, T.E. (2005). Altered Gene Expression in Plants with Constitutive Expression of a Mitochondrial Small Heat Shock Protein Suggests the Involvement of Retgrograde Regulation in the Heat Stress Response. *Physiol. Plant.*, Vol.123, No.4, (March, 2005), pp. 435-444, ISSN 0031-9317

Richly, E.; Dietzmann, A.; Biehl, A.; Kurth, J.; Laloi, C.; Apel, K.; Salamini, F. & Leister, D. (2003). Covariations in the Nuclear Chloroplast Transcriptome Reveal a Regulatory Master-Switch. *EMBO Rep.*, Vol.4, No.5, (May, 2003), pp. 491-498, ISSN 1469-221X

Robson, A.C. & Vanlerberghe, G.C. (2002). Transgenic Plant Cells Lacking Mitochondrial Alternative Oxidase Have Increased Susceptibility to Mitochondria-Dependent and -Independent Pathways of Programmed Cell Death. *Plant Physiol*, Vol.129, No.4, (August, 2002), pp. 1908-1920, ISSN 1532-2548

Sherameti, I.; Nakamura, M.; Yamamoto, Y.Y.; Pfannschmidt, T.; Obokata J. & Oelmüller R. (2002). Polyribosome Loading of Spinach mRNAs for Photosystem I Subunits Is Controlled by Photosynthetic Electron Transport. A Crucial Cis Element in the Spinach PsaD Gene Is Located in the 5'-Untranslated Region. *Plant J.*, Vol. 32, No.5, (December, 2002), pp. 631-639, ISSN 0960-7412

Strand, A.; Asami, T.; Alonso, J.; Ecker, J.R. & Chory, J. (2003). Chloroplast to Nucleus Communication Triggered by Accumulation of Mg-protoporphyrin IX. *Nature*, Vol. 421, No.6918, (January, 2003), pp. 79–83, ISSN 0028-0836

Subbaiah, C.C. &, Sachs, M.M. (2003). Molecular and Cellular Adaptations of Maize to Flooding Stress. *Ann. Bot.*, Vol. 91, No.2, (January, 2002), pp. 119-127, ISSN 0003-3847

Subbaiah, C.C.; Bush, D.S. & Sachs, M.M. (1998). Mitochondrial Contribution to the Anoxic Ca^{2+} Signal in Maize Suspension-Cultured Cells. *Plant Physiol.*, Vol.118, No.3, (November, 1998), pp. 759-771, ISSN 1532-2548

Subbaiah, C.C.; Palaniappan, A.; Duncan, K.; Rhoads, D.M. & Huber, S.C. (2006). Mitochondrial Localization and Putative Signaling Function of Sucrose Synthase in Maize. *J. Biol. Chem.*, Vol. 81, No.23, pp. 15625-15635, ISSN 0021-9258

Susek, R.E.; Ausubel, F.M. & Chory, J. (1993). Signal Transduction Mutants of Arabidopsis Uncouple Nuclear CAB and RBCS Gene Expression from Chloroplast Development. *Cell*, Vol.74, No.5, (September, 1993), pp. 787-799, ISSN 0092-8674

Sweetlove, L.J. & Foyer, C.H. (2004). Roles for Reactive Oxygen Species and Antioxidants in Plant Mitochondria Plant Mitochondria: From Genome to Function, In: *Advances in Photosynthesis and Respiration*. Day, D.A., Millar, A.H., Whelan J. (Eds.), Vol. 17, pp. 307-320, Kluwer, ISBN 978-90-481-3882-1, Dordrecht. Netherlands

Taylor, N.L.; Day, D.A. & Millar, A.H. (2002). Environmental Stress Causes Oxidative Damage to Plant Mitochondria Leading to Inhibition of Glycine Decarboxylase. *J. Biol. Chem.*, Vol.277, No.45, (November, 2002), pp. 42663-42668, ISSN 0021-9258

Voigt, C.; Oster, U.; Börnke, F.; Jahns, P.; Dietz, K.J.; Leister, D. & Kleine, T. (2010). In-depth Analysis of the Distinctive Effects of Norflurazon Implies that Tetrapyrrole Biosynthesis, Organellar Gene Expression and ABA Cooperate in the GUN-type of Plastid Signalling. *Physiol. Plant.* Vol.138, No.4, (December, 2009), pp. 503-519, ISSN 0031-9317

Voss, B.; Meinecke, L.; Kurz, T.; Al-Babili, S.; Beck, C.F. & Hess, W.R. (2011). Hemin and Mg-protoporphyrin IX Induce Global Changes in Gene Expression in *Chlamydomonas reinhardtii. Plant Physiol.*, Vol.155, No.2, (December, 2010), pp. 892-905, ISSN 1532-2548

Vranova, E.; Inze, D. & Van Breusegem, F. (2002). Signal Transduction during Oxidative Stress. *J. Exp. Bot.*, Vol.53, No.372, (May, 2002), pp. 1227-1236, ISSN 0022-0957

Wagner, D.; Przybyla, D.; Camp, R.; Kim, C.; Landgraf, F.; Lee, K.P.; Würsch, M.; Laloi, C.; Nater, M.; Hideg, E. & Apel, K. (2004). The Genetic Basis of Singlet Oxygen-Induced Stress Responses of Arabidopsis thaliana. *Science,* Vol.306, No.5699, (November, 2004), pp. 1183-1185, ISSN 0036-8075

Yabuta, Y.; Maruta, T.; Yoshimura, K.; Ishikawa, T. & Shigeoka, S. (2004). Two Distinct Redox Signaling Pathways for Cytosolic APX Induction under Photooxidative Stress . *Plant Cell Physiol.*, Vol. 45, No.11, (Novemer, 2004), pp. 1586-1594, ISSN 0032-0781

Yurina, N.P. & Odintsova, M.S., (2007) Plant Signaling Systems. Plastid_Generated Signals and Their Role in Nuclear Gene Expression. *Russ. J. Plant Physiol.*, Vol.54, No.4, pp. 427-438, ISSN 1021-4437

Yurina, N.P.; Pogulskaya, E.N. & Karapetyan, N.V.. (2006) Effect of Photodestruction of Plastids from Norflurazon_Treated Barley Seedlings on Expression of Nuclear Genes Encoding Chloroplast Stress Proteins. *Biochemistry (Moscow)*, Vol.71, No.4, pp. 430-436, ISSN 0006_2979

Zarkovic, J.; Anderson, S.L. & Rhoads, D.M. (2005). A Reporter Gene System Used to Study Developmental Expression of Alternative Oxidase and Isolate Mitochondrial Retrograde Regulation Mutants in Arabidopsis. Plant Mol. Biol., Vol.57, No.6, pp. 871-888, ISSN 1573-5028

Zhang, Z.W.; Yuan, S.; Feng, H.; Xu, F.; Cheng, J.; Shang, J.; Zhang, D.W. & Lin, H.H. (2011) Transient Accumulation of Mg-protoporphyrin IX Regulates Expression of PhANGs - New Evidence for the Signaling Role of Tetrapyrroles in Mature *Arabidopsis* plants. J. Plant Physiol., Vol.168, No.7, (January, 2011), pp. 714-721, ISSN 1532-2548

Part 2

Nucleic Acids, Proteins and Enzymes

4

Post-Translational Modifications of Nuclear Proteins in the Response of Plant Cells to Abiotic Stresses

Jennifer Dahan[1,2,3], Emmanuel Koen[1,2], Agnès Dutartre[1,2], Olivier Lamotte[1,2] and Stéphane Bourque[1,2,4]

[1]*Université de Bourgogne, UMR Plante-Microbe-Environnement*
Dijon cedex
[2]*UMR 1088 INRA - 5184 CNRS*
Dijon cédex
[3]*Institut Jean-Pierre Bourgin, INRA Centre de Versailles-Grignon*
Versailles Cedex
France

1. Introduction

For a long time, in plant cells as in animal cells, the nucleus was only considered as the organelle in which fundamental mechanisms such as replication and transcription occurred. While strong efforts were deployed in order to identify important families of transcription factors such as MYB, WRKY or TGA families (Dubos *et al.*, 2010; Rushton *et al.*, 2010), a few attention was devoted to our lack of knowledge about their regulation in regard to the physiological conditions of the plant cells. Whereas the major importance of post-translational modification of proteins is well established for several decades regarding cytosolic proteins, the last years have been characterized by the discovery that the plant cell nucleus also contains all the enzymes necessary to assume these fundamental reactions in terms of signal transduction. For example, Mitogen-Activated Protein Kinases (MAPK) are well known protein kinases (PKs) involved in response to both biotic and abiotic stresses (for review see Dahan *et al.*, 2009). These MAPKs play a crucial role in the regulation of specific gene expression by phosphorylating particular transcription factors. However, while they are well described in the cytosol, only recently researchers focused on their presence and involvement in the nucleus of plant cells challenged by abiotic stresses (for example, Ahlfors *et al.*, 2004). Unfortunately, these authors like the other ones involved in plant cell nucleus studies did not try to identify the targets of these MAPK. This example highlights the fact that our knowledge of the incidence of protein posttranslational modifications regarding the cellular activities is still poorly rudimental, and particularly in the field of abiotic stress responses.

Amongst the targets of these post-translational modifications, histones will be a piece of choice, being one of the favourite substrates for acetylation or methylation for example. Histones are small basic protein associated with DNA to form the chromatin. Chromatin contains histone octamere composed by two copies of each histone H2A, H2B, H3 and H4 to form the histone core. Histones composing the histone core are extremely conserved in the different kingdoms. For example, histone H3 differs only by two amino acids in its sequence in rat and pea. The histone core is rolled by approximately 146 bps of DNA to form a nucleosome, the repeating unit of chromatin which provided a first level of DNA compaction. Association of nucleosome with other protein like histone H1 creates a structure named 30 nm chromatin fiber (Kornberg and Lorch, 1999; Hayes and Hansen, 2001). This fiber can be unfold by several protein grouped under the term chromatin remodeling complex to generate a 11 nm fiber, the template for transcription. The 11 nm fiber can also be repressive to transcription process but different types of protein complexes can affect the chromatin to modulate DNA accessibility to transcriptional machinery and at the end gene transcription. Histone tails and globular domains are subjected to a variety of posttranslational modifications such as acetylation, methylation, phosphorylation, ubiquitination, sumoylation, ADP ribosylation, deimination and proline isomerization. These covalent modifications of histones are important in chromatin dynamics (Kouzarides, 2007). It has been proposed that all the histone postranslationnal modifications constitute a code, "the histone code" associating to all the possible combinations of modification a particular state allowing biological process such as transcription of gene. This code is written by particular proteins: "the writers", interpreted by other proteins: "the readers" and erased by a last class of protein, "the erasers" (Figure 1). Readers could also be able to modify histones or to recruit other protein actin on chromatin (Strahl and Allis, 2000).

The goal of this book chapter is to summarize our current knowledge of the molecular actors and their regulations that lead to posttranslational modifications of nuclear proteins, and *in fine* to the regulation of specific target gene expression. For this purpose, a large number of nuclear enzymes that are involved in (de)phosphorylation, (de)acetylation, and (de)methylation of nuclear proteins, but also in nuclear protein degradation pathway associated with sumoylation and ubiquitination, and in changes of the redox state of the nuclear proteins will be presented and their roles illustrated by various but non exhaustive cell responses to abiotic stresses.

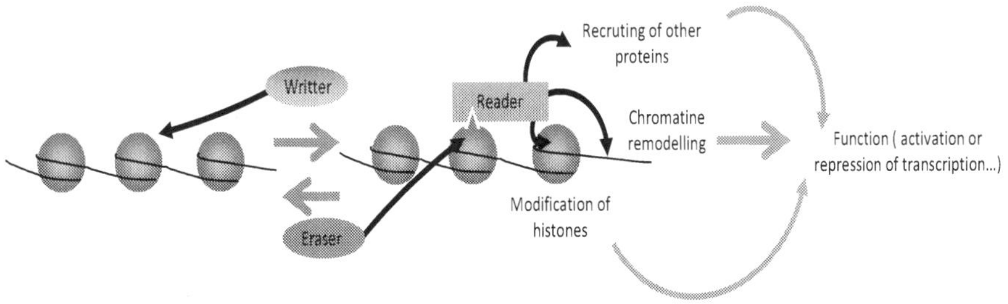

Fig. 1. Schematic representation of histone code. Writers can recognize chromatin area and add marks on histone (methylation, acetylation…). Theses labeling is dynamic because marks can be remove by erasers. They can also be recognized par readers, another group of proteins able to act directly or indirectly on chromatin structure. The final consequence can be a modification of the gene transcription in concerned loci.

2. Nuclear protein phosphorylation / dephosphorylation

Phosphorylation is probably the most prominent and major posttranslational modification in living organisms. This reversible modification consists on the covalent binding of a phosphate on aminoacids hydroxyl residues. In eukaryotes, phosphorylation occurs mainly on Ser, Thr and Tyr residues. This modification depends on the action of two types of enzymes with antagonistic activities: (i) PKs which phosphorylates and (ii) protein phosphatases (PPs) which remove phosphoryl groups from target proteins. The binding of the phosphoryl group alters the functional properties of target proteins in terms of activity, subcellular localization, protein-protein interactions or stability (Cohen, 2000). This versatility has made phosphorylation one of the major means for modulation of cellular activities, as phosphorylation status can be finely tuned as the result of the balance between PKs and PPs activities at a given time point on specific substrates. These enzymes are thus involved at every steps of signal transduction, from plasma membrane to final effectors. Although numerous studies like the ones regarding brassinosteroids, have investigated PKs, PPs and their targets in the plasma membrane and the cytoplasm (Li, 2005), only few ones has focused on those localized, temporarily or permanently, in the nuclear compartment, despite their involvement in critical processes for cell surviving (Dahan *et al.*, 2009). A recent phosphoproteomic analysis of the nuclear proteins from *Arabidopsis thaliana* showed that the identified phosporylated proteins cover a wide range of nuclear activities, indicating an equal importance of phosporylation in the modulation of nuclear activities as for the cytoplasmic ones (Jones *et al.*, 2009).

In plants, PKs constitute a superfamily of proteins, and according to their importance in cellular homeostasis, it was shown that around 4 % of *A. thaliana* genome encodes putative PKs (Champion *et al.*, 2004). According to their substrate specificities, PKs and PPs can be classified into three families: (i) Ser/Thr PKs/PPs, (ii) Tyr PKs/PPs and (iii) dual-specificity (Ser/Thr-Tyr) PKs/PPs. In plants, no functional Tyr PK has been described to date, although predicted (Miranda-Saavedra and Barton, 2007). Consistent with this, the relative abundance of phosphor-Ser, phosphor-Thr and phosphor-Tyr has been estimated in *A. thaliana* to 85 %, 11 % and 4 % respectively (Sugiyama *et al.*, 2008). Although Tyr (de)phosphorylation is thus accomplished through dual-specificity PKs/PPs, to date, the vast majority of studies have been focused on Ser/Thr PKs/PPs in plant.

More than half of the putative PKs from the *A. thaliana* genome fall in the clade of the so-called Receptor-Like protein Kinases (RLKs), which are transmembrane proteins probably acting as receptors of environmental stimuli (Tör *et al.*, 2009). The other part of plant PKs, which is of interest in this article, constitutes the clade of the "soluble" ones, and concerns all PKs found in the cytoplasm and the nucleus. This clade can be divided into numerous families and subfamilies, based on sequence similarities of the kinase domain and features of flanking sequences determining regulation properties (Hanks and Hunter, 1995; Champion *et al.*, 2004; Miranda-Saavedra and Barton, 2007); readers are invited to report to these works for more details). Although for a large part studied in the cytoplasm or at the plasma membrane level, members from almost all families have been found in the nuclear compartment (for a detailed review, see Dahan *et al.*, 2010). The majority of the available data on these nuclear PKs comes from subcellular distribution studies. Indeed for most of them their nuclear localization has been investigated using chimerical PKs fused to a fluorescent reporter, or immunolocalization studies based on specific antibodies. The activity of the PKs in the nucleus has only been assessed in rare cases, impairing our understanding of their roles and functionality.

While the core catalytic domain of PKs derived from a single ancestor (Hanks and Hunter, 1995), PPs origins are more diversified (Moorhead *et al.*, 2009). In plants, around 160 sequences coding for putative catalytic subunit of PPs have been retrieved (Kerk *et al.*, 2002). They can be divided in three main families, based on catalytic domain sequences and enzymatic features: (i) the PPP (phophoprotein phosphatase) family, (ii) the PPM (metallodependent protein phosphatase) family, and (iii) the PTP (phosphotyrosine phosphatase) including the DSP (dual-specificity phosphatase). Readers are invited to refer to other reviews that detailed the features of these families (Kerk *et al.*, 2002; Luan, 2003; Moorhead *et al.*, 2009). As for plant PKs, PP functionalities in the nucleus have been poorly studied.

The role of several nuclear PKs in response to different abiotic stresses has been investigated in different model plants. To date, and despite the great diversity of PKs, only a few families have been shown to be involved in these signaling pathways at a nuclear level. Among them, the MAPK family is certainly the most studied family of PKs in the nucleus. MAPK is the last component of a cascade of three PKs, which are sequentially activated (Widmann *et al.*, 1999). The perception of an environmental stimulus drives the activation of the first PK of this transduction module, namely MAPKKK (MAPK kinase kinase). MAPKKK in turn activates by phosphorylation on a specific motif a MAPKK (MAPK kinase), which phosphorylates MAPK, the final effector, on conserved residues. This chain of phosphorylation is thought to take place in the cytoplasm, and the MAPK is then translocated upon activation by phosphorylation into the nucleus, where it modulates gene expression by acting on transcription factors. However the presence of MAPKK and MAPKKK in the nucleus was already reported (Dahan *et al.*, 2009).

Figure 2 illustrates, although partially, the known nuclear PKs that were shown to be involved in response to various abiotic stresses. Even though MAPK was involved in the response to many abiotic stresses on the basis of their activation in the cytosol, few of them have been demonstrated to localize and to act in the nuclear compartment. In fact in the context of abiotic stresses only one study showed the translocation in *A. thaliana* of two MAPKs, AtMPK3 and AtMPK6, following ozone exposure and using immunolocalization of the native protein (Ahlfors *et al.*, 2004). However contradictory results were obtained when these two MAPKs were expressed fused to GFP, and showed a constitutive nuclear and cytoplasmic localization (Yoo *et al.*, 2008). In addition to ozone exposure, these two MAPKs are furthermore activated in response to several abiotic stresses, such as cold, salt, drought, wounding and touch for AtMPK6 (Ichimura *et al.*, 2000) and osmotic stress for AMPK3 (Droillard *et al.*, 2002). The orthologs of AtMPK3 and AtMPK6 in tobacco, respectively WIPK (Wounding-Induced Protein Kinase) and SIPK (Salicylic acid-Induced Protein Kinase), are also activated in response to several environmental cues, such as high salinity and osmotic stresses, wouding and ozone exposure (Zhang and Klessig, 1998; Mikolajczyk *et al.*, 2000; Samuel and Ellis, 2002). They were shown to be nuclear, independently of any stimuli when fused to GFP (Menke *et al.*, 2005; Yap *et al.*, 2005), and the activity of SIPK could be retrieved in nuclear extracts of osmotically stressed tobacco cells (Dahan *et al.*, 2009), demonstrating its functionality in the nuclear compartment in response to environmental stresses.. A few substrates of MAPKs have been identified, and so their precise roles in response to abiotic stresses remain elusive. For the most part, their known targets are transcription factors. For example, SIPK was shown to phosphorylate *in vitro* the transcription factor WRKY1, involved in the expression of defense-related genes (Menke *et al.*, 2005). This suggests that

MAPKs, in response to different abiotic stresses, could be involved in the reprogramming of transcription patterns to allow the plant to cope with the perceived stresses. Given that, the overexpression or knock-out of several MAPKs impairs the outcome of the response to some abiotic stresses. The extinction of *Atmpk6* expression using RNAi and a loss-of-function mutation of *Atmpk3* both generate increased sensitivity to ozone leading to cell death, implying that these MAPK play an important role in the management of ozone stress by *A. thaliana* (Ahlfors *et al.*, 2004). Another *A. thaliana* MAPK, AtMPK4, has been shown to reside permanently in the nucleus along with the cytoplasm (Andreasson *et al.*, 2005; Kosetsu *et al.*, 2010). Although more studied for its potential role in cytokinesis (Kosetsu *et al.*, 2010) and its involvement in plant defense reactions (Andreasson *et al.*, 2005; Gao *et al.*, 2008), its activity is observed upon application of stresses such as cold, salt and osmotic stresses (Droillard *et al.*, 2004; Teige *et al.*, 2004). One substrate for this MAPK has been described in vitro: MKS1, which is a protein interacting with two WRKY transcription factors (Andreasson *et al.*, 2005). The phosphorylation of MKS1 by AtMPK4 is thought to result in the release of the WRKY33, which could then play its role as a transcriptional regulator (Qiu *et al.*, 2008). However, this interaction takes place in the context of defense reaction, and no data is available regarding functionality of AtMPK4 in response to abiotic stresses.

Stress	Kinase/phosphatase	HDAC/HAT
Cold	MPK6 (Ichimura *et al.*, 2000) MPK4 (Droillard *et al.*, 2004; Teige *et al.*, 2004)	AtGCN5/HAG1 (Servet *et al.*, 2010) HDA18, HDA19 (Alinsug *et al.*, 2009) SIR2 (Bond *et al.*, 2009)
Heat		HDA7 (Alinsug *et al.*, 2009)
Salt	MPK6 (Ichimura *et al.*, 2000) SnRK2 (Halford and Hey, 2009) MPK4 (Droillard *et al.*, 2004; Teige *et al.*, 2004)	HDA6 (Chen *et al.*, 2010) HDA19 (Chen and Wu, 2010) HDA2, HDA14 (Alinsug *et al.*, 2009) HD2C (Sridha and Wu, 2006)
Osmotic	MPK3 (Droillard *et al.*, 2002) MPK4 (Droillard *et al.*, 2004; Teige *et al.*, 2004) SnRK2 (Halford and Hey, 2009)	
Drought	MPK6 (Ichimura *et al.*, 2000) SnRK2 (Halford and Hey, 2009)	
Light		HDA19 (Tian *et al.*, 2003) HDA7 (Alinsug *et al.*, 2009) GCN5 (Benhamed *et al.*, 2006)
Ozone	MPK3, MPK6 (Ahlfors *et al.*, 2004)	
Wounding	MPK6 (Ichimura *et al.*, 2000)	HDA19 (Zhou *et al.*, 2005)

Fig. 2. An overview of nuclear enzymes involved in phosphorylation/dephosphorylation or acetylation/deacetylation processes in response to various abiotic stresses.

Not only MAPKs are the nuclear crossroads of abiotic stresses. Data that are piling up to put PKs and PPs at the centre of the regulation of the adaptative reponses of plant cells to osmotic, saline and hydric stresses, in link with ABA (abscisic acid), point out the role of other nuclear PK. In particular SnRK2 (for Snf1-Related protein Kinase) family is constituted of PKs initially characterized as ABA signalling factors. They can be categorized into three subclasses, and subsequent studies on rice and *A. thaliana* PKs demonstrated that all of the subclasses could be activated in response to osmotic stress, with subclasses II and III being also responsive to ABA (Halford and Hey, 2009). ABA is a phytohormone essential for the establishment of adaptative responses to drought and salinity stresses. Its accumulation leads to expression of ABA-responsive genes helping cells to cope with hydric and osmotic variations (Hubbard *et al.*, 2010). It was recently shown that at the heart of its signalling pathway in *A. thaliana* is a core complex composed of three subclass III SnRK2 called SnRK2D, E and I, and their cognate PPs PP2C. Indeed SnRK2s are rendered active upon phosphorylation on two critical residues in their activation loop (Belin *et al.*, 2006; Burza *et al.*, 2006; Boudsocq *et al.*, 2007). Recently, several PPs from the PP2C family have been shown to dephosphorylate these residues, thus inactivating the PKs. These partners have been colocalized in the nuclear compartment, where they interact in permanence (Umezawa *et al.*, 2009). Interestingly, the same SnRK2s were also shown to associate in the nucleus with AREB1, a transcription factor whose expression occurs during drought (Fujii *et al.*, 2009). A triple mutant plant impaired in the expression of SnRK2D, E and I exhibited a strongly reduced tolerance to drought, comforting the involvement of these PKs in response to water shortage. Accordingly, drastic changes was observed in response to ABA, high salinity stress and drought at the transcriptional level, with down-regulation of ABA responsive-genes as compared to wild type plants (Fujita *et al.*, 2009). Upon perception, water and osmotic stresses lead to production of ABA which in turn activates the transduction module composed of the PP2Cs and the SnRK2s. The signaling pathway is then achieved by phosphorylation of specific transcription factors, controlling a set of stress and ABA-responsive genes. Likewise, other SnRK2s have been characterized in several plant species that can phosphorylate ABA-responsive transcription factors. In wheat, the SnRK2 PKABA1 has been shown to interact with and phosphorylate *in vitro* TaABF, an ortholog of AREB1 (Johnson *et al.*, 2002). Furthermore, three rice SnRK2s were found to phosphorylate and activate another ortholog of AREB1, TRAB1 (Kobayashi *et al.*, 2005). However in the last two cases, the subcellular partitioning of the PK was not investigated. Recent data suggest that the transduction module composed of the SnRK2 and PP2C could be directly activated in the nucleus, meaning that the ABA receptor is expected to translocate after binding to its ligand.

Calcium is a well recognized second messenger, and as such its variations in the cytosol specifically induced by many stimuli promote specific signal transduction through decoding by a panel of Ca^{2+} sensor molecules (Kudla *et al.*, 2010). It is now well established that apart from the cytosol, nucleus is able to generate its own Ca^{2+} signatures following diverse environmental cues (Mazars *et al.*, 2009). Accordingly, Ca^{2+}-binding proteins have been localized in the nucleus, where they can decode Ca^{2+} variations into an appropriate response, like modulation of the expression of specific set of genes (Xiong *et al.*, 2006). Accordingly, several PKs and PPs potentially regulated by Ca^{2+} have been localized in the nucleus, where they are thought to decode nuclear Ca^{2+} variations into modifications of phosphorylation state of specific target proteins. The concerned PKs are mainly CDPKs

(Ca^{2+}-Dependent Protein Kinases) and CIPKs (Calcineurin B-like[CBL]-Interacting Protein Kinases). CDPKs are specifics to plant and exhibit in their sequence a CaM-like domain with three to four EF-hand Ca^{2+} binding motifs, rendering them directly dependent on Ca^{2+} binding for their activities (Klimecka and Muszynska, 2007). A study on the subcellular targeting performed for 9 of the 34 putative CDPK in *A. thaliana* showed that two of them, AtCPK3 and AtCPK4 are constitutively cytosolic and nuclear (Dammann *et al.*, 2003). AtCPK3 has been shown to be activated at the plasma membrane level following a salt stress, and in accordance to this putative involvement knock-out *Atcpk3* mutant plant exhibited altered resistance to salt stress, whereas it was improved in *Atcpk3* overexpressor mutant line (Mehlmer *et al.*, 2010). Intriguingly, no transcriptional changes in the expression of traditionally salt-stress related genes were observed, raising the question of AtCPK3 nuclear role (Mehlmer *et al.*, 2010). Apart salt stress, another CDPK, AtCPK32, has been found to be involved in ABA signaling, through interaction and phosphorylation with an ABA-responsive transcription factors, ABF4. Interestingly these two proteins are colocalized in the nucleus (Choi *et al.*, 2005). Whereas most of the CDPK studied show a constitutive partitioning between different sublocation and the nucleus, two CDPK from different species are suggested to translocate from the plasma membrane to the nucleus upon stresses. McCPK1 from *Mesembryanthemum crystallinum* is transcriptionnaly induced following salt- or water deficit stresses and when fused to GFP, translocate from the plasma membrane to the nuclear compartment (Chehab *et al.*, 2004). However the mechanisms and the functions of this phenomenon are not explained to date. Even if CDPKs are thought to have their activities regulated by Ca^{2+}concentrations in the cytosol (Harmon, 2003), recent data suggest that they could also be part of phosphorylation cascade, as for NtCDPK2 and NtCDPK3, complicating conventional signaling schemes (Witte *et al.*, 2010). Despite recent advances in the understanding of the functionality of Ca^{2+} variations, the ion still constitute a missing when it comes down to evaluate its part in the activation of specific signaling pathways.

3. Nuclear protein acetylation / deacetylation

The first descriptions of a nuclear protein modification by the chemical binding of an acetyl group on the amine function of lysine residues was observed on histone proteins (Allfrey *et al.*, 1964), explaining why enzymes responsible for the acetylation and deacetylation processes were named histone acetyltransferases (HATs; Gallwitz, 1971) and histone deacetylases (HDACs; Inoue and Fujimoto, 1969), respectively. However, based on their ability to act on proteins unrelated to histone such as transcription factors or coregulators of gene transcription, HATs and HDACs are more generally considered as lysine acetyl transferases and lysine deacetylases (Chen and Tian, 2007). All these proteins were reported to be located, exclusively or not, into the nucleus of plant cell.

3.1 Histone deacetylases (HDACs)

Concerning HDACs, plant genomes contain members that belong to two different families that are common to eukaryotes (Pandey *et al.*, 2002): the RPD3 (Reduced Potassium Dependency Protein 3) - HDA1 (Histone Deacetylase 1) superfamily, and SIR2 (Silent Information Regulator 2) family. The third family, termed HD2 (type-2 Histone Deacetylase), is specific to plant cells (Fu *et al.*, 2007).

The first member of the RPD3-HDA1 superfamily was identified in yeast in a complementation screen of a mutant for a high affinity potassium transporter (Vidal *et al.*, 1990). It was latter shown that RPD3 encodes the catalytic subunit of the HDB (Histone Deacetylase B) complex activity (Rundlett *et al.*, 1996). The first plant RPD3-HDAC member, ZmRpd3, was identified in maize thanks to its ability to functionally complement a yeast *rpd3* null mutant (Rossi *et al.*, 1998). Among the eleven *RPD3* genes identified in the maize genome, three of them (*ZmHDA101, ZmHDA102* and *ZmHDA108*) exhibit the same expression pattern, suggesting a possible functional redundancy among this gene family (Varotto *et al.*, 2003). In term of specificity, ZmRPD3s, like its counterpart in yeast, seem to specifically remove the acetyl tail of lysines of histone H4 (H4K5, H4K12) prior to its incorporation in chromatin (Lechner *et al.*, 2000). Homologs of ZmRPD3 were also identified in *A. thaliana* and classified in three different clades : Class I for the RPD3 group, Class II for the HDA1-like group, and Class IV for the AtHDA2 group (Alinsug *et al.*, 2009). Whereas 18 RPD3-like genes were identified from the genome analysis data, only *AtHD1* (also called *AtHDA19* or *RPD3A*) and *AtHDA6* were largely characterized. *AtHD1*, that is constitutively expressed, encodes a protein that localises both in the cytoplasm and in the nucleus, but is predominantly accumulated in the euchromatic region and excluded from the nucleolus (Varotto *et al.*, 2003; Fong *et al.*, 2006). AtHD1 exhibits a histone deacetylase activity that removes the acetyl group of histone H3 (H3K9) and H4 (H4K5, H4K8, H4K12, H4K16; Tian *et al.*, 2003). AtHDA6, although it is the close homolog of AtHD1, exhibits major differences in term of expression and specificity. AtHDA6 is largely accumulated in the nucleoli (Probst *et al.*, 2004) and is a broad-specificity HDAC that removes the acetyl group of various lysines of histone H3 and H4 (H3K14, H4K5, H4K12; Earley *et al.*, 2006). In agreement with these differences in their expression profiles, AtHD1 and AtHDA6 are involved in different signalling pathways: AtHD1 regulates several developmental processes such as early senescence, floral organ identity or late flowering (Wu *et al.*, 2000; Tian and Chen, 2001; Tian *et al.*, 2005), by controlling the expression of a set of genes involved in protein synthesis, ionic homeostasis or plant hormonal regulation (Tian *et al.*, 2005). AtHDA6, due to its localisation in the nucleolus, is involved in the inhibition of the NORs (nucleolus organisation region; Probst *et al.*, 2004; Earley *et al.*, 2006), and evidences supporting roles in silencing transgenes and transposons were also reported (Murfett *et al.*, 2001; Aufsatz *et al.*, 2002; Lippman *et al.*, 2003; May *et al.*, 2005). In a structural point of view, a huge study was conducted with the RPD3-like HDACs of *A. thaliana* (Alinsug *et al.*, 2009). All members are characterized by a variable catalytical domain characterized by a pocket that can either bind a Zn^{2+} cation necessary to ensure the deacetylation reaction, or different inhibitors like sodium butyrate or trichostatine A (Finnin *et al.*, 1999). Different members of this family like HDA6, HDA7 or HDA19 contain a NLS and / or NES, confirming the nuclear localisation of these enzymes, but also that a cytosolic / nuclear shuttling could be of importance for their mode of action.

Several studies, mainly based on the use of mutants in *RPD3-HDA1 A. thaliana* genes, have reported that they play major roles in various biotic and abiotic stress responses (Figure 2; Alinsug *et al.*, 2009). Amongst the 18 HDAC *A. thaliana* genes belonging to the RPD3-HDA1 family, most of them appear to be partly or largely involved in response to various environmental stresses. For example, in the Class I RP3-HDAC, AtHDA19 that is the best characterized member of this family, is highly expressed in germinating and imbibed seeds and also strongly accumulated in response to cold stress (Alinsug *et al.*,

2009), wounding (Zhou *et al.*, 2005) or light response (Tian *et al.*, 2003; Tian *et al.*, 2005). In fact, Tian *et al.* (2005) showed that approximately 7 % of the Arabidopsis genome is up- or down-regulated in *athda19* mutants. For example, Tian *et al.* (2003) showed that *hda19* mutant lines were affected in their flowering process under long day conditions, exhibiting the opposite phenotype of mutants altered in the expression of the histone acetyltraferase GCN5. Interstingly, the double mutant *gcn5:hda19* is characterized by a normal phenotype in response to light, suggesting that both HDA19 and GCN5 could target the same genes. This was shown by Benhamed *et al.* (2006): a reduced histone acetylation was monitored in the promoter of *CAB2*, *RBCS-1A* and *IAA3* genes in the *gcn5* mutants, whereas an increase histone acetylation was observed in the promoter of the same genes in the *hda19* mutants. Furthermore AtHDA19 expression is strongly induced in response to wounding and the stress related hormons JA and ethylene, suggesting that it could be involved in many abiotic stresses leading to the production of both compounds (Zhou *et al.*, 2005). AtHDA19 was also reported to be involved in response to abiotic through its interaction with the transcription corepressor LEUNIG by repressing gene transcription (Gonzalez *et al.*, 2007). In the same way, overexpression of AtHDA19 in *Brassica napus* demonstrated that it was involved in response to cold stress by interacting with bnKCP1, a novel protein containing a putative kinase-inducible domain (Gao *et al.*, 2003).

AtHDA6, another RPD3-type histone deacetylase in *A. thaliana*, is involved in response to ABA and salt stress. Chen *et al.* (2010) showed that an AtHDA6 mutant, *axe1-5*, as well as *HDA6* RNA-interfering plants are hypersensitive to salt and ABA treatment, due to the down-regulation of various abiotic stress responsive genes like *ABI1*, *ABI2*, *RD29A* or *RD29B*. This hypersensitivity to salt stress is correlated with changes in H3 acetylation pattern. Without ABA treatment, an increase of H3 acetylation was shown both in the promoter and in exons of these genes in the *axe1-5* mutant compared to Col0. In response to ABA, an increase of H3 acetylation was only monitored in Col0 but not in the *axe1-5* mutant, indicating that HDA6 is required for the induction of acetylation by ABA and salt treatment. A similar phenotype was obtained with the *A. thaliana HDA19* T-DNA insertion mutant, *hda19-1* (Chen and Wu, 2010). On the contrary, other RPD3 HDAC like AtHDA7 are not differentially expressed in regard to developmental stages but are induced in response to both biotic (*Pseudomonas syringae*) and abiotic stimuli such light intensity or heat stress, according to microarray data analyses (Alinsug *et al.*, 2009).

Little information is available regarding the properties of HDA1-like HDACs in plants. The study of one member of this family in maize, ZmHDA1, reported that it is expressed in an inactive form of high molecular weight (84-kDa) that needs to be processed in an active form of 48-kDa by a proteolytic cleavage of the C-terminal end to regulate gene transcription (Pipal *et al.*, 2003). According to the data of the sequenced genomes, *A. thaliana* contains five members of the HDA1 group also termed class II. Members of this Class II are also involved in response to various environmental stresses: while heat stress seems to upregulate most of the Class II HDAs, NaCl treatment only stimulates AtHDA14 and AtHDA2 expression (Alinsug *et al.*, 2009). Furthermore, AtHDA18 expression is induced in response to cold treatment. Demetriou *et al.* (2009) reported that members of this HDAC family in Barley are also regulated by JA in response to abiotic stresses: whereas *HvHDAC1-I-1* is slightly reduced after 6 hr of treatment, other genes like *HvHDAC1-II-1* or *HvHDAC1-IV-1* are strongly accumulated after 6 hr.

The SIR2-like HDACs are characterised by their NAD-dependent histone deacetylase activity. They are homolog to the yeast SIR2 HDAC that is involved in longevity (Imai *et al.*, 2000). Although the genome of all sequenced plants contains several Sir2-like genes, their functions are yet to be determined. A recent study in *A. thaliana* showed that AtSIR2 regulates several defence related genes involved in the synthesis of SA, implying potentially SIR2 members in plant defence reactions (Wang *et al.*, 2010). To our knowledge, only one study recently reported the involvement of SIR2 in response to low temperature in the context of vernalization (Bond *et al.*, 2009). SIR2 class of HDAC would repress the expression of the MADS box transcription factor *FLC* (for Flowering Locus C, a central gene in the vernalization process) by altering the acetylation pattern of histones H3 and H4.

The last HDAC family, termed HD2-like, is specific to plants and do not share any sequence similarities with other HDAC proteins (Pandey *et al.*, 2002). Three genes HD2 HDAC genes /or homologues are coded in maize genome, four in *A. thaliana*, two in rice and in barley. All HD2 display similarities with the FKBP family peptidyl-propyl cis-trans isomerase (Aravind and Koonin, 1998; Dangl *et al.*, 2001) and display the same architecture: a conserved EFWG motif in the N-terminal region that comprises the catalytical domain, a central acidic domain involved both in enzymatic activity and its regulation, and in the C-terminal part a NLS and a Zn^{2+} finger that may be involved in protein/protein or protein/DNA interactions. Several members of this family like ZmHD2 in maize and AtHD2A in *A. thaliana* were shown to be localized in the nucleolus and to deacetylate the lysine 9 of histone H3, suggesting that they should be involved in the regulation of ribosomal RNA expression (Lusser *et al.*, 1997; Earley *et al.*, 2006). However their mode of action is still largely unknown; they seem to be mainly involved in plant defence reactions (Bourque *et al.*, 2011) and in seed development (Wu *et al.*, 2000). In *A. thaliana*, Sridha and Wu (2006) showed that overexpression of *AtHD2C* enhances tolerance to salt stress and drought compare to wild-type plants by affecting the expression of several abscisic acid-responsive genes. However, we do not know whether or not it affects the acetylation of these genes.

3.2 Histone acetyltransferases (HATs)

The action of the HDAC proteins is reversed by the one of the HATs that catalyse the transfer of an acetyl group from acetyl~SCoA to the amine function of a lysine residue, particularly in histone proteins (Servet *et al.*, 2010). Like HDACs, HATs are divided into several groups based on primary homology with yeast and mammalian HATs: GNAT (for GCN5-related N-acetyltransferase), MYST (for MOZ, Ybf2/Sas2 and Tip60), p300/CBP (for CREB-binding protein) and TAF1 (for TATA-binding protein (TBP)-associated factor 1) groups. The GNAT group is usually subdivided into three subfamilies termed GCN5 (for General Control Nonderepressive protein 5), ELP3 (for transcriptional ELongator complex Protein 3) and HAT1. In a structural point of view, HATs are usually constituted by two fundamental domains: the catalytical acetyltransferase domain, and the bromodomain. The bromodomain is composed of 110 aminoacids that can bind to acetylated lysine residues (Owen *et al.*, 2000). Bromodomains are generally found in proteins that regulate chromatin structure and gene expression, such as HATs and the ATPase component of certain nucleosomes-remodeling complexes. The mode of recognition of acetyl-lysine by the bromodomain is similar to the one of acetyl~SCoA by HATs, since the bromodomain is the only domain known to interact with acetylated lysine containing peptides. Among the different HATs expressed in *A. thaliana*, AtGCN5/HAG1, which belongs to the GNAT HAT family, is one of the best characterized in terms of function and is involved both in abiotic and developmental responses (see Servet *et al.*, 2010 for a recent review). Regarding its specificity, the major targets of this HAT are histone H3 (H3K9, H3K14, H3K18 and H3K27) and histone H4 (H4K5, H4K8, H4K12 and H4K16; Zhang *et al.*, 2007). The bromodomain of AtGCN5, that was shown to bind to acetylated histone lysine motifs, is probably not required for its binding to most of its targets (Benhamed *et al.*, 2008). Among its targets, in addition to histones H3 and H4, AtGCN5 also interacts and acetylates other proteins such as AtADA2 that could regulate HAT activity (Servet *et al.*, 2010). Furthermore, AtGCN5 activity could be regulated by phosphorylation/dephosphorylation, since it interacts with protein phosphatase 2C *in vitro* and mutation in PP2C gene increases H3K14 acetylation, one of the targets of AtGCN5. The other HAT proteins are less characterized in terms of target and mode of action.

Works in the group of D.X Zhou have shown that AtGN5 is required for light-regulated gene expression by promoting the acetylation of the promoter of target genes (Benhamed *et al.*, 2006; Benhamed *et al.*, 2008; Servet *et al.*, 2010). Several studies reported that abiotic stresses involve the association/regulation of HAT activities with various proteinaceous partners in the regulation of specific genes leading to the establishment of the biological response. For example, the GCN5 HAT is known to physically interact with the transcriptional coactivator Ada2 (Stockinger *et al.*, 2001), this one enhancing the ability of GCN5 to acetylate histones *in vitro* and enabling GCN5 to acetylate nucleosomal histones (Mao *et al.*, 2006). Hark *et al.* (2009) showed that mutants of one of both *ADA2* genes, *ADA2b*, display hypersensitive phenotype to salt stress and altered responses to low temperature stress, a phenotype close to the one of AtGCN5 mutants. A recent study showed that ADA2b and GCN5 interact with a third partner, the coactivator SGF29a, to enhance the acetylation in the promoter region of target genes like COR6.6, RAB18, and RD29b (Kaldis *et al.*, 2011). In the same vein, Gao *et al.* (2007) showed that AtGCN5 HAT activity is required in response to cold and stress treatment by its physical interaction with the transcription factor AtEML. The authors suggest that AtEML would co-ordinates the expression of target stress regulated genes through involvement in recruiting AtGCN5 to their promoters.

4. Protein methylation

Protein methylation involves transfer of a methyl group from S-adenosylmethionine (the universal methyl donor in cells) to acceptor groups on substrate proteins (Aletta *et al.*, 1998). It commonly occurs on carboxyl groups of glutamate, leucine, and isoprenylated cysteine, or on the side-chain nitrogen atoms of lysine, arginine, and histidine residues (Clarke, 1993). In eukaryotes nuclei, one of the best known examples of protein methylation is probably histone methylation. Like acetylation, histone methylation is considerate as an important process regulating the chromatine dynamics and function (Strahl and Allis, 2000; Jenuwein and Allis, 2001; Zhang and Reinberg, 2001; Kouzarides, 2002). Histone methylation can occur at different residues and on distinct sites. For one residue, different numbers of methyl groups can be added. Even if some methylation has been characterized in the globular domain of H3 in animals (Feng *et al.*, 2002), methylation seems to concern most of the time the N-terminal part of histones (named histone tail) (Bannister and Kouzarides, 2005). Histones methylation has been described mainly on histones H3 and H4 and occurs on both arginine and lysine residues.

4.1 Histone lysine methylations

Lysine methylation consists on addition of one or more methylgroups to the ε-amino group of lysine residues, resulting in mono-, di-, or trimethylated lysine. Unlike acetylation, methylation of these residues does not change the charge but progressively increases the bulk and hydrophobicity. This may impact intra- or intermolecular hydrogen-bond interactions of the amino group or create new sites recognized by reader proteins that bind preferentially to the methylated domain (Lee *et al.,* 2005). In eukaryotes, histone lysine methylation occurs on histone H3 at lysines 4, 9, 14, 27, 36, and 79 and on histone H4 at lysines 20 and 59 (Strahl and Allis, 2000; Berger, 2002; Zhang *et al.*, 2002; Zhang *et al.*, 2003). A recent study provide evidences that methylation can also occurs on Lysine 37 of histone H2B *in vivo* (Gardner *et al.*, 2011).
In *A. thaliana,* the best known histone lysine methylation occurs at Lys4 (K4), Lys9 (K9), Lys27 (K27), and Lys36 (K36) of histone H3 while methylation on lysine 20 of H4 has only been observed with immunostaining. Presence of methylation on H3 K79 which is highly conserved in non-plant systems has not been reported in plant. Methylations of lysine are catalyzed by HKMTs (histone lysine methyltransferases) that almost all share a SET [Su(var), Enhancer of zeste, trithorax] domain, a conserved motif containing approximately 130 amino acids which was originally identified in Drosophila. This domain is responsible for catalysis and binding of cofactor S-adenosyl-lmethionine (Lee *et al.*, 2005). Plant genome encodes number of protein with set domain; for example, *A. thaliana* genome encodes 41 while grapes encode 32 and maize 37 (http://www.chromdb.org). Function of their homology with their Drosophila homologues E(Z), TRX, ASH1 and SU(VAR)3-9, they are assigned to four groups (Jenuwein *et al.,* 1998). In *A. thaliana,* certain families of HKMTs seem to catalyse methylation on one particular site. It is the case for HKMTs homologues of SU(VAR)3-9 that are acting on H3K9 and for HKMTs relative to E(Z) that methylate H3K27. E(z) proteins are associated with other members in PRC2 (polycomb-group repressive complex 2) to performed H3K27 methylation. Members of other families seem to be able to act at different sites. It is the case of TRX homologue where different members can contribute to methylation of H3K4 or H3K27. Also ASH1 (for Absent, Small, or Homeotic discs 1) family seems to act on both H3K36 and H3K4. Nevertheless the link between this protein and histone methylation can be direct or indirect since methyl transferase activity has been not tested for all members of these different families (for review Liu *et al.,* 2010). Methylated lysine can be then recognized by protein reader leading in a direct or indirect way to a particular function like gene silencing (Schotta *et al.,* 2002; Jackson *et al.,* 2004). These proteins comported domain that can recognized methylated lysine such as chromolike domains of the royal superfamily (including chromodomain, tudor domain, malignant brain tumor (MBT), PWWP, and plant Agenet module), or a plant homeodomain finger (PHD) or the WD40 repeat (Taverna *et al.,* 2007). At this time only fews readers are known in plants and the way by which they translate the histone marks to direct downstream functions it is not fully understood.

4.2 Stoechiometrie and localization of histone lysine methylation in plant

In plants like in animals the three degrees of methylation (mono, di et trimethylation) on histone H3 at lysines 4, 9, 14, 27, 36 are found but proportions between the different degrees change. For example in *A. thaliana,* higher levels of H3K4 di-methylation (H3K4me2) and lower H3K9me2 and H3K9me3 levels have been detected compare to animals (Jackson *et al.,*

2004; Guo *et al.*, 2006). Histones methylation profile has mainly be determined in *A. thaliana* thanks to Immunostaining of nuclei, chromatin immunoprecipitation (ChIP) , ChIP coupled with high-resolution microarray analysis (ChIP-chip) and mass spectrometry in combination with high-performance liquid chromatography (HPLC) separation methods (Liu *et al.*, 2010). H3K4me1/2/3 are highly enriched in euchromatin in *A. thaliana* and about 90 % of annotated genes carry one or more of the H3K4 methylation marks, suggesting an important role of this histone modification in the control of gene expression (van Dijk *et al.*, 2010). H3K9 methylation is critical for maintenance of transcriptional gene silencing and genome stability (Vaillant and Paszkowski, 2007). H3K9me1 and H3K9me2 are predominant and enriched in heterochromatin (Johnson *et al.*, 2004) while H3K9me3 is enriched in euchromatin (Mathieu *et al.*, 2005; Turck *et al.*, 2007). H3K9me2 is particularly present in transposons and repeated sequences according to the repressing transposon activities that have been attribute to this histone mark (Lippman *et al.*, 2004; Bernatavichute *et al.*, 2008). H3K27me1 and H3K27me2 are enriched in heterochromatin (Mathieu *et al.*, 2005; Fuchs *et al.*, 2006) while H3K27me3 is localized in euchromatin (Turck *et al.*, 2007). H3K27me3 is found in about 4400 genes and is often localized upstream of promoters and of 5'UTR suggesting like for H3K4, an important role of this methylation in controle of gene expression in plant (Turck *et al.*, 2007; Zhang *et al.*, 2007). H3K36me1/3 are enriched in euchromatin while H3K36me2 is present in both euchromatin and heterochromatin (Lin *et al.*, 2008).

4.3 Histones arginine methylation

Arginine methylation can impact histone and more generally protein properties. In fact, arginine is a positively charged amino acid that has five potential hydrogen bond donors positioned for favorable interactions with biological hydrogen bond acceptors (Bedford and Clarke, 2009). These biochemical properties give to this amino acid a crucial role for protein structure and interaction with other molecules. In proteins interacting with DNA, arginine residues are the most frequent hydrogen bond donors to backbone phosphate groups and to thymine, adenine, and guanine bases (Luscombe *et al.*, 2001). Protein arginine methylation results in the addition of one or two methyl groups to the guanidino nitrogen atoms of arginine (Gary and Clarke, 1998). Each addition of a methyl group to an arginine residue changes its form and removes a potential hydrogen bond donor (Bedford and Clarke, 2009). Since no protein able to interact with histone arginine methylation are known, it is possible that the only fact of methylate arginine is to impact interaction between histones and DNA, leading to an impact of chromatin structure and finally of gene expression.

In Eukaryotes, the best known arginine methylation concerned Arg2 (R2), Arg8 (R8), Arg17 (R17), Arg26 (R26) of histone H3, and Arg3 (R3) of histone H4. In *A. thaliana*, at this time only methylation on H4R3 and H3R17 have been detect *in vivo*. Protein arginine methylation is catalyzed by a family of protein named arginine methyltransferases (PRMTs). PRMTs are classified into four classes depending of the final methylated product of reaction that they can catalyze. Type I and type II enzymes are among others, involved in histone methylation and are one of the best characterized (Bedford and Richard, 2005). Both of them catalyze first addition of single methyl group on the terminal nitrogen atom of Arginine to form Mono methylated Arginine (MMA). Then, Type I PRMTs form asymmetric di-methylated Arg meaning that two methyl groups are added on the same nitrogen atom of the guanidine leading to an ω-NG,NG-di-methyl arginine, while type II PRMTs performed symmetric dimethylation of Arg meaning that the two methyl groups are on two different nitrogen atome leading to an ω-NG,N_G-d i-methyl arginine (Mitchell *et al.*, 1992). *A.thaliana* genome codes for nine PRMTs (Niu *et al.*, 2007). AtPRMT4a and AtPRMT4b, are homologs of human CARM1 and can performed asymmetrically H3R17me2a *in vivo*. H4R3 can be di-methylated symmetrically by AtPRMT5/SKB1 (Pei *et al.*, 2007; Wang *et al.*, 2007; Schmitz *et al.*, 2008) and asymmetrically by AtPRMT1a, AtPRMT1b (Yan *et al.*, 2007) and AtPRMT10 (Niu *et al.*, 2007).

4.4 Histones demethylation

Histone methylation has been considered as irreversible until the discovery in 2004 of LSD1 (lysine-specific demethylase 1) in animals (Shi *et al.*, 2004). This discovery proved that histone methylation is a dynamic process regulated by HMTs (the writers) but also histones demethylases (the erasers). Two types of demethylases exist with distinct mechanisms to remove lysine methylation: Flavin adenine dinucleotide (FAD)-dependent KDM1/LSD act by amine oxidation and need the cofactor Flavin adenine dinucleotide. JmjC domain-containing proteins act by hydroxylation and most of them use Fe(II) and α-ketoglutarate (Shi *et al.*, 2004; Tsukada *et al.*, 2006). These two classes of enzymes act on different substrates: Flavin adenine dinucleotide (FAD)-dependent KDM1/LSD1 are able to demethylate mono and dimethylation while JmjC domain-containing proteins demethylase act one mono-di and tri methylated lysines (Klose and Yi, 2007). In *A. thaliana*, potential histone demethylases have been predicted based on conservation of cofactor-binding amino acids (Lu *et al.*, 2008). There are four KDM1/LSD1 homologs in *A. thaliana* and one of them, LDL1, has been shown to demethylate di- and mono-methylated H3K4 (Spedaletti *et al.*, 2008). *A. thaliana* genome also contains 21 JmjC domain-containing proteins (JMJs). These JMJs are grouped into five subfamilies according to sequence similarities. The biochemical properties and biological functions of histone demethylases are emerging and how these enzymes work, are recruited to their target loci, and play roles are still largely unknown (Liu *et al.*, 2010).

In animals, H3 and H4 arginines can be desiminate by PADI4 (peptidyl arginine deiminase 4). Deimination could be antagonist of arginine methylation since citrulline prevents arginine from being methylated (Cuthbert *et al.*, 2004). Moreover it could be a way to remove methylation of arginine. In fact, monomethylated arginine could be converted to citrullin by PADI4 (Wang *et al.*, 2004). In addition a family of amine oxidases may be able to demethylate arginine residues using a similar mechanism as they demethylate lysine residues (Bedford and Richard, 2005). In plant, arginine demethylation process is still unknown.

4.5 Histone methylation in responses to abiotic stresses

Abiotic stresses modulate expression of different genes. As it has been explained before, gene regulation can be due to chromatin remodeling involving histone modification. In plants, modification of histone methylation occurs during stress responses even if the mode of action and consequence of these changes are still not well understand. These modifications could control stress relative genes. For example, hundreds of stress-responsive genes are targets for H3K27me3 in *A. thaliana* (Zhang *et al.*, 2007).

During a cold stress, methylation of H3K27me3 gradually decreases at the loci of two cold-responsive genes, COR15A and AtGolS3 while expressions of genes increase. Even if Tri-methylation of histone H3 Lys27 (H3K27me3) is generally considered as a negative marker of transcription (Zhang *et al.*, 2007), the link between reduction of H3K27me3 and increase of transcription seems to be not obvious in this case : first COR15A and ATGOLS3 are not targets for LHP1 (Zhang *et al.*, 2007), an *A. thaliana* protein that binds H3K27me3 *in vitro* (Zhang *et al.*, 2007; Exner *et al.*, 2009) and that has been shown to be required for silencing of genes with H3K27me3 (Mylne *et al.*, 2006; Sung *et al.*, 2006; Zhang *et al.*, 2007; Exner *et al.*, 2009). Also, when cold-exposed plants are returned to normal growth conditions, transcription of COR15A and ATGOLS3 was repressed to the initial level before cold exposure while decrease in H3K27me3 is still maintained. Also, this decrease does not enhance the induction of transcription when plants are returned to cold temperatures (Kwon *et al.*, 2009). According to these results, it has been proposed that H3K27me3 could act as a memory marker for recent transcriptional activity in *A. thaliana*. In this case, previous exposure of plants to certain environmental stresses may negatively affect the level of H3K27me3 and lower the chance of stress-responsive genes being silenced.

Dehydration stress also induces variations on histone methylation. In fact, histone H3 modifications at the coding regions of four dehydration stress responsive genes, RD29A, RD29B, RD20, and an AP2 transcription factor have been reported during a drought stress in *A. thaliana* (Kim *et al.*, 2008), characterized by an enrichment of H3K4me3 and H3K9ac (a positive marker of gene activation) at these four loci. Another study presents the whole-genome distribution patterns of histone H3 H3K4me1, H3K4me2, and H3K4me3 and its modification after a drought stress in *A. thaliana*. While H3K4me1 and H3K4me2 levels changed modestly during dehydration stress, drastic changes in the H3K4me3 levels are observed. These changes are correlated with modification in level of transcription of responding genes: a large increase of H3K4me3 level was found on nucleosomes of the genes which had a high expression and a large decrease in H3K4me3 levels has been reported in highly down-regulated genes (van Dijk *et al.*, 2010). By analyzing the H3K4me3 distribution profiles on nucleosomes of stress-induced genes, this study provided specific chromatin pattern associated with many genes involved in dehydration stress response and confirm the putative role of H3K4me3 in transcription activation.

The *A. thaliana* SKB1 protein is a type II Arg methyltransferase homologue to PRMT5 in mammals that catalyzes Arg symmetric dimethylation H4R3sme2. SKB1 is associated to chromatin region of FLC promoting flowering by suppressing its expression through H4R3sme2 (Wang *et al.*, 2007; Schmitz *et al.*, 2008). SKB1 is also associated in chromatin of other genes involving among others in stress responses like HAB1, who is really important in ABA and salt stress (Saez *et al.*, 2004; Saez *et al.*, 2006) where it represses transcription through H4R3sme2. Salt stress and ABA treatment (that is accumulated under a salt stress) lead to dissociation of SKB1 from chromatin leading to a reduced level of H4R3sme2 and to an higher expression of genes in ABA and salt response like HAB1, suggesting a direct mechanism by which salt and ABA impact gene transcription (Figure 3). This theory is supporting by the fact that SKB1 invalidation leads to a decrease of H4R3sme2 levels, an increase of HAB1 and some other stress-responsive gene expression and a bigger susceptibility to salt stress (Zhang *et al.*, 2011). During a salt stress, SKB1 not only leaves chromatin of some loci but also methylates U6 small nuclear ribonucleoprotein (snRNP)-specific Sm-like protein LSM4. SKB1 invalidation leads to splicing defects in hundreds of genes that are involved in many biological processes, including the abiotic stress responses. Furthermore, *lsm4* mutant, similarly to *Skb1*, is hypersensitive to salt and shows similar splicing defects in some genes (Zhang *et al.*, 2011). In conclusion, SKB1 plays a dual role in salt response by altering the methylation status of H4R3sme2 and LSM4.

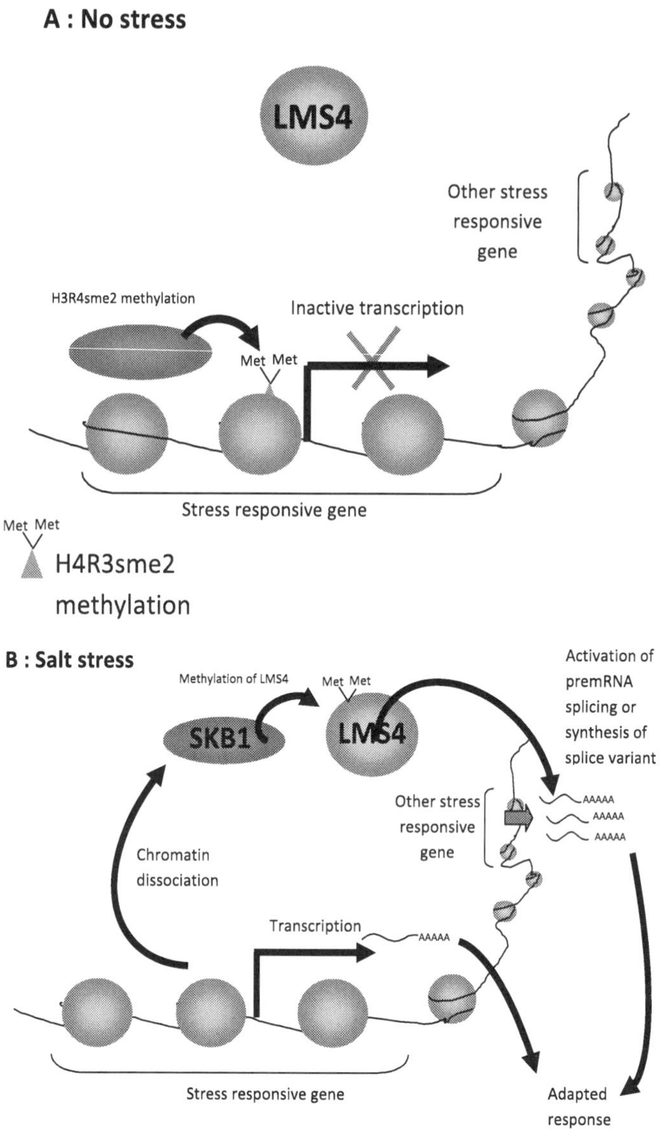

Fig. 3. Involvement of methylation in salt stress responses. A : Without stress, SKB1 a PRMT is associated to the chromatin of some stress related genes where it methylates histone on H4R3sme2. These methylations are repressive mark leading to transcription inactivation of these genes. B: During a salt stress, there is a chromatin dissociation of SKB1. This dissociation has a dual role in stress related gene control to insure an adapted response: Firstly SKB1 do not maintains H3R4sme2 in the loci where it was associated leading to an

higher expression of many stress relative genes. Secondly, SKB1 will methylated another protein LMS4 connected to splicing. This last methylation leads to an activation of premRNA splicing and /or synthesis of splice variant of some other genes connected to stress responses.

5. Redox modification of nuclear proteins

5.1 ROS and RNS in the nucleus

Reactive oxygen species (ROS) are chemically reactive species of oxygen formed by successive one-electron reduction of molecular oxygen. It includes superoxide anion ($O_2^{\cdot -}$), hydrogen peroxide (H_2O_2), perhydroxyl radical ($HO_2\cdot$) and the highly reactive hydroxyl radical ($OH\cdot$) (Apel and Hirt, 2004). ROS is produced in different organelles such as mitochondria, chloroplast, peroxizome, endoplasmic reticulum and in the cytosol (Gill and Tuteja, 2010) but to our knowledge, only one study has described ROS accumulation in the nucleus, in response to an elicitor of plant defense (Ashtamker *et al.*, 2007). However, recent data indicate that plant cell nucleus possesses antioxidant redox system to control ROS homeostasis (Pulido *et al.*, 2009a; Pulido *et al.*, 2009b). Furthermore, ROS production in other cellular compartments results in changes in gene expression, indicating that ROS production can influence gene transcription in the nucleus. Another reactive radical, Nitric Oxide (NO), has been known for years as a signaling molecule in animal and plant cells (Besson-Bard *et al.*, 2008). Reactive Nitrogen Species (RNS) contains all the NO derived molecules and their chemistry have been documented elsewhere (Stamler *et al.*, 1992). For instance, NO can be oxidized or reduced to NO^+ or NO^- or can react with $O_2^{\bullet -}$ to form the strong oxidant peroxynitrite ($ONOO^-$). Through its different chemical forms, NO can react with a great variety of molecules including proteins, lipids, metals, molecular oxygen or nucleic acids. Evidences for NO production in the nucleus of various plant cell types (stomata guard cells, stomata subsidiary cells, epidermal cells) have been documented in some studies in response to different stimuli such as heat stress, green light, osmotic stress, plant defense elicitors (Foissner *et al.*, 2000; Gould *et al.*, 2003; Vitecek *et al.*, 2008). As observed for ROS, NO can have an impact on gene expression (Palmieri *et al.*, 2008).

5.2 Redox-based post-translational modification in plant cell nuclei

ROS and RNS could affect gene expression through different mechanisms. These reactive compounds can activate components of signaling pathways controlling gene expression or can directly affect the DNA binding activity of several kinds of transcription factors. Heat shock transcription factors (HSF) bind to a consensus sequence found in the promoter of many stress-responsive genes. HSF transcription factors have been thought to be ROS sensors as reviewed by (Miller and Mittler, 2006). Other transcription factors are regulated in a redox dependent manner through a dithiol/disulfide exchange and some examples are detailed bellow.

The activity of transcription factors from the R2R3 MYB family have been shown to be involved in abiotic stress responses including salt stress, drought stress (Jung *et al.*, 2008). Using the typical R2R3 MYB protein P1 from *Zea mays*, Heine *et al.* (2004) have shown that DNA binding of this transcription factor is redox-dependent. More precisely, the two cysteines residues 49 and 53 are necessary for DNA binding and under non-reducing conditions, they formed a disulfide bridge that prevents DNA binding. Another R2R3 MYB plant transcription factor of *A. thaliana*, AtMYB2, has been shown to be redox-modulated (Serpa *et al.*, 2007). AtMYB2 has been suggested to play a role in response to stresses that induce NO production such as ABA or salt treatment. Contrary to P1, cystein 49 is absent in AtMYB2 (Heine *et al.*, 2004) and the S-nitrosylation of Cys 53 by NO donors blocks its binding to a specific DNA sequence (Serpa *et al.*, 2007). In this case, it indicates that the S-nitrosylation of AtMYB2 could be a mechanism to turn off the activity of this protein. However, the role of the redox regulation of R2R3 MYB transcription family protein needs to be addressed *in vivo* in response to abiotic stresses.

The DNA binding of RAP2.4a (At1g36060), a AP2/DREB-type transcription factors, to the promoter of 2-Cys peroxiredoxin-A gene (2CPA) is redox-regulated by dithiol/disulfide transition of regularoty cysteinyl residues (Wormuth *et al.*, 2007; Shaikhali *et al.*, 2008). Reducing conditions lead to the monomerisation of RAP2.4a whereas oxidizing conditions dimerize or oligomerize RAP2.4a proteins through the formation of disulfide bridges which increased binding affinity of the protein to DNA. Loss-of-function of *RAP2.4a* affects the adaptation of plants to changes in environmental conditions such as naturally fluctuating light conditions. Furthermore, RAP2.4a transcription factor regulates the expression of genes known to be induced by ROS and involved in abiotic stress tolerance such as ZAT 10 (Mittler *et al.*, 2006). A closed homologue, RAP2.4b (At1g78080) has also been involved in stress response. Mutations in *RAP2.4b* cause altered expression of light and drought-responsive genes and defects in developmental processes or drought tolerance. However, its redox control as observed with RAP2.4a has not been investigated.

In the last decades, studies in the animal field have indicated that the main mode of action of NO is based on the post-translational modification of proteins. Firstly, S-nitrosylation consists of the oxidation by NO of reduced sulfidryl groups of cysteyls residues of proteins thus forming a nitrosothiol (SNO). Secondly, tyrosine nitration is based on the addition of a NO_2 group on a tyrosyl residue of a protein. And thirdly, NO can bind covalently transition metal of metalloproteins. This last process is called metal nitrosation. These NO-based post-translational modifications of proteins finally affect the activity of the modified proteins. While thousand proteins have been identified in animals to be modified by NO, only a few have been characterized in plants (Astier *et al.*, 2011; Seth and Stamler, 2011). Among them, some are linked with the nucleus function. The best example on the role of NO in the modification of proteins associated with nuclear functions is the transcription coactivator NPR1 (for Non-expressor of Pathogenesis Related-1) in *A. thaliana*. Although data indicate that NPR1 is important for plant resistance to abiotic stresses (Quilis *et al.*, 2008; Yasuda *et al.*, 2008; Rao *et al.*, 2002), its mode of action and its regulation have been discovered in plants infected by avirulent pathogens or treated by elicitors of plant defense reactions. In untreated cells, NPR1 forms a homo multimeric complex that is sequestered in the cytosol and stabilized by intermolecular disulphide bonds (Mou *et al.*, 2003). In this case, NPR1 monomers still exist. To prevent target gene activation in the absence of inducing stimulus, they are translocated to the nucleus at low rate and finally targeted to the proteasome (Spoel *et al.*, 2010). A role of NO in this process has been recently discovered. It was reported that S-nitrosylation of NPR1 by S-nitrosoglutathione (GSNO) at cysteine-156 facilitates its oligomerization and blocks its nuclear translocation (Tada *et al.*, 2008). After the perception of stimuli such as pathogen infection, redox changes dependent on the production of salicylic acid occur. Multimeric NPR1 is reduced to monomers by thioredoxins and NPR1 monomers are then translocated to the nucleus where their interaction with transcription factors allowed the regulation of the expression of several genes (Despres *et al.*, 2003; Tada *et al.*, 2008; Fan and Dong, 2002; Boyle *et al.*, 2009). Particularly, NPR1 activates the binding of the b-ZIP transcription factor TGA1 to the activation sequence-1 (as-1) element of the promoter region of defense genes. Disulfide bridge formation involving cysteyl residues 260 and 266 precludes the interaction of TGA1 with NPR1 and then prevents binding of TGA1 to its target DNA element (Despres *et al.*, 2003). Recent mass spectrometry analysis indicated that these cysteyl residues of TGA1 are S-nitrosylated and S-glutathionylated after GSNO treatment (Lindermayr *et al.*, 2010). Furthermore, GSNO enhanced the DNA binding activity of TGA1 to its DNA element in the presence of NPR1. It can be the consequence of conformational changes of both proteins which allow a more effective TGA1-NPR1 interaction and finally a more effective DNA-binding of TGA1. Interestingly, a GSNO-induced nuclear translocation of NPR1 has also been observed by these authors. It could be due to SA-mediated redox-changes since NO induced SA production (Durner *et al.*, 1998; Huang *et al.*, 2004). All these results indicate that NO has a major regulatory role of NPR1 functions by controlling its translocation from the cytosol to the nucleus and by affecting the NPR1/TGA1 complex and downstream dependent responses. Other proteins might be involved in this process. The small oxidoreductases glutaredoxins that mediate redox regulation of proteins through the reduction of disulphide bridges or the glutathionylation of cysteyl residues (Dalle-Donne *et al.*, 2009; Rouhier *et al.*, 2010) has been recently shown to interact with transcription factors from the TGA family in the nucleus of plant cells (Ndamukong *et al.*, 2007; Li *et al.*, 2009). However, the role of GRX in TGA function has not been described yet.

Another protein is redox-modified by NO. In a recent paper, Wawer *et al.* (2010) have shown that *Nicotiana tabacum* GAPDH (glyceraldehyde-3-phosphate dehydrogenase) localized in the nucleus and in the cytosol of plant cells. GAPDH was transiently S-nitrosylated after the salt treatment of *N. tabacum* cell suspensions but the role of this NO-dependent modification is not known.

6. Ubiquitin and ubuquitin-like post-ranslationnal modifications

Protein post-translational modifications by ubiquitin and ubiquitin-like proteins are essential for a plethora of cell functions in eukaryotic cells and are involved in development processes but also in the responses to biotic and abiotic stresses (Fu *et al.*, 2010; Miura and Hasegawa, 2010; Trujillo and Shirasu, 2010). Ubiquitination is a post-translational modification of proteins corresponding to the reversible attachment of the 76-amino acid protein ubiquitin to target proteins through a well-characterized process (Fu *et al.*, 2010). This three-step enzymatic cascade is catalyzed by ubiquitin-activating enzyme (E1), ubiquitin-conjugating enzyme (E2) and ubiquitin ligase enzyme (E3). A major function of protein ubiquitination is to address proteins to the proteasome for their degradation. However, it can also control intracellular localization of proteins, transcription of DNA and cell cycle. Thus, ubiquitination regulates a great number of cellular processes. There are evidences that the ubiquitination system is involved in regulating signaling pathways controlling plant adaptation to stresses and especially abiotic stresses. Some proteins belonging to the ubiquitination enzymatic machinery and involved in abiotic stress tolerance have been shown to localized in the nucleus of plant cell. AtHOS1, a E3-ubiquitin ligase that control cold tolerance in *A. thaliana* have been shown to localized in the nucleus after cold stress (Lee *et al.*, 2001, see below). Furthermore, AtDRIP1 and AtDRIP2 (DREB2A-interacting protein) are two E3-ubiquitin ligase (Qin *et al.*, 2008). AtDRIP1 is localized in the nucleus of *A. thaliana* cells where it interact with AtDREB2A (Dehydratation-responsive element binding protein 2A), a transcription factor that is important for drought tolerance and controls the expression of water deficit-inducible gene expression. As proposed by (Qin *et al.*, 2008), the ubiquitination of AtDREB2A may occur in a no-stress condition leading to its degradation by the proteasome. However, during drought stress, AtDREB2A is activated by an unknown mechanism. It is possible that the ubiquitination and degradation of AtDREB2A is blocked leading to the accumulation of effective AtDREB2A protein that in turn activate the expression of drought-responsive genes. This result highlights the importance of ubiquitination process in controlling gene transcription in the nucleus.

In addition to ubiquitin, post-translational modification of proteins by ubiquitin-like proteins (Ubls) such as SUMO (small ubiquitin-like modifier), RUB (related to ubiquitin), NEDD8 (neural precursor cell expressed, developmentally down-regulated), ATG8 and ATG12 (autophagy 8 and 12) have been shown to be functional in plants and to control essential cellular processes as observed in other eukaryotes (reviewed by (Miura and Hasegawa, 2010). Among them, modification of proteins by SUMO (SUMOylation) has been one of the most described Ubl-post-translational modifications in plants. SUMO proteins are synthesized as precursors that need to be cleaved by specific proteases to expose a glycyl residue necessary for their ligation to target proteins. The addition of SUMO proteins to a lysyl residue of the target protein is catalyzed in a three-step enzymatic reaction similar to ubiquitinylation (Kurepa *et al.*, 2003).

In *A. thaliana*, genetic analysis of SUMOylation process indicates that this post-translational modification of proteins is essential for plant development. Thus, mutation in the gene coding AtSCE1 (Arabidopis Sumo-conjugating Enzyme1) or AtSAE2 (Arabidopsis Sumo-activating Enzyme 2) proteins results in embryo lethality (Saracco *et al.*, 2007). The same phenotype was observed in plants in which the genes coding both AtSUMO1 and AtSUMO2 proteins were mutated (Saracco *et al.*, 2007). Number of proteins modified by SUMO have been shown to increase *in planta* after abiotic stresses including heat, oxidative stress, ethanol, phosphate starvation, salt and cold stress (Kurepa *et al.*, 2003; Miura *et al.*, 2005; Yoo *et al.*, 2006; Miura *et al.*, 2007; Conti *et al.*, 2008; Miller *et al.*, 2010). Furthermore, mutants defective in proteins of the SUMO-conjugating pathway such as AtSIZ1 (arabidopsis SUMO E3 ligase) are impaired in stress responses such as thermotolerance and have been shown to exhibit exaggerated phosphate starvation responses (Miura *et al.*, 2005; Miura *et al.*, 2007) indicating that SUMOylation of proteins is an essential process mediating stress acclimation. As observed in other eukaryotic cells, SUMOylation enzymes (*e.g.* AtSCEA and SUMO1/2) localized in the nucleus in Arabidopsis (Lois *et al.*, 2003) indicating a role of SUMOylation in controlling many aspects of nuclear function. In rice (*Oryza sativa*), SUMO-conjugating enzyme (OsSCE1) has been shown to localize in the nucleus. Yeast-two hybrid experiments indicated that OsSCE1 interacts with the heat-inducible pyrophosphatase (PPIase) OsFKBP20, a class of proteins that assist molecular chaperones in reactions associated with protein folding and protein transport across membrane. It was proposed that OsSCE1 and OsFKBP20 proteins mediate in concert the stress response of rice plants (Nigam *et al.*, 2008). In Arabidopsis, heat stress dramatically increased the pool of SUMO conjugates which were mainly detected in the nucleus (Saracco *et al.*, 2007). In a recent proteomic analyses, Miller *et al.* (2010) have identified 357 SUMOylated proteins in Arabidopsis. Many of them are nuclear proteins that participate in a wide range of processes related to nuclear function, such as chromatin modification, DNA maintenance/repair or gene transcription. Furthermore, nuclear SUMOylated proteins identified in non-stressed plants were enriched in oxidative and/or heat-stressed plants. Among the nuclear SUMO conjugates, some transcription factors have been identified (Miura and Hasegawa, 2010). For instance the SUMOylation of AtICE1, a MYC transcription factor involved in cold stress responses in Arabidopsis has been characterized (see below). Protein SUMOylation is a reversible process and de-SUMOylation is catalyzed by SUMO proteases a class of enzymes that is also involved in the generation of mature SUMO proteins. Recently, the Arabidopsis mutants *overly tolerant to salt 1* (*ots1*) and *2* (*ots2*) has been shown to be mutated in two SUMO proteases OTS1 and OTS2 that act redundantly to regulate salt stress response (Conti *et al.*, 2008). Both proteins are localized in the nucleus indicating that SUMOylation/deSUMOylation of nuclear proteins is likely to control essential processes required for salt stress responses.

It has been shown recently that both ubiquitination and SUMOylation can affect the same protein. During cold stress, Arabidopsis cold responsive genes are induced transiently indicating that their expression is finely regulated. Among the transcription factors that controls cold-gene expression and cold stress tolerance, AtICE1 is a MYC transcription factor that is constitutively expressed. It controls the cold induction of other transcription factors such as AtCBF3 which in turn drive the expression of cold responsive genes. On the contrary, AtMYB15 is a repressor of such genes (Agarwal *et al.*, 2006). ICE1 is post translationnally modified by both ubiquitynation and SUMOylation processes and these processes are thought to provide a fine-tuning for the expression of cold-responsive genes. More precisely, the nuclear-localized SUMO E3 ligase AtSIZ1 (Cheong *et al.*, 2009) mediate the SUMOylation of AtICE1. SUMOylation of AtICE1 is thought to stabilize or activate the protein, leading to the expression of genes required for low temperature tolerance, such as AtCBF3 (Miura *et al.*, 2007). Furthermore, the RING-E3-ubiquitin ligase HOS1 (for *high expression of osmotically responsive gene*) have been shown to relocalize to the nucleus after a cold treatment (Lee *et al.*, 2001) where it interacts with ICE1. HOS1 mediates the polyubiquitination of ICE1 targeting this transcription factor for degradation by the proteasome (Dong *et al.*, 2006). This leads to the repression of cold responsive genes such as CBF3 by the transcription factor AtMYB15. A model for the opposite role of ubiquitination and SUMOylation in the control of cold-responsive genes during cold episode has been proposed (Miura *et al.*, 2007; Miura and Hasegawa, 2010).

7. Conclusion

As we illustrate in this review using studies reporting the involvement of nuclear posttranslational modifications in response to numerous environmental changes, almost all known major abiotic stresses induce one or more nuclear protein modifications to regulate the expression of specific target genes.

After reading this chapter, three main conclusions emerge. Firstly, histone modifications and associated chromatin and gene expression changes appear to be a critical point necessary for establishment of an appropriate biological response. Understanding the regulation of histone modifications, competitions that can occur between different posttranslational modifications (the famous "histone code") and interpretation of these

modifications by readers proteins is being initiated and should greatly highlight our comprehension of gene expression regulation in response to stresses in the coming years.
Secondly, we should keep in mind that protein modifications do not occur only on histones, but also concern many others proteins such as transcription regulators. However the number of non-histone proteins targeted by the nuclear posttranslational modification machinery remains sparse probably due to the technical difficulties for identification and purification of these very low abundant proteins in the nuclear compartment.
Thirdly, in almost all cases, the chemical and / or physical signals leading to nuclear machinery through the nuclear envelop are still unknown. Here too our knowledge about nuclear pore channels functioning in plant is emerging, but recent discoveries in this field should extend our understanding of how nuclear protein posttranslational modifications are controlled and lead the plant in an appropriate response to stress.

8. Acknowledgements

We would like to thank the Conseil Régional de Bourgogne for its financial support. Jennifer Dahan was supported by a grant from the Ministère de l'Enseignement Supérieur et de la Recherche. Agnès Dutartre is supported by a grant from the Université de Bourgogne and the Conseil Régional de Bourgogne, Emmanuel Koen by the Ministère de l'Agriculture et de la Pêche and Olivier Lamotte by the Centre National de la Recherche Scientifique and by the Conseil Régional de Bourgogne (FABER grant n° 20109201AAO047S05280).

9. References

Agarwal M, Hao Y, Kapoor A, Dong CH, Fujii H, Zheng X and Zhu JK (2006) A R2R3 type MYB transcription factor is involved in the cold regulation of CBF genes and in acquired freezing tolerance. Journal of Biological Chemistry 281: 37636-37645

Ahlfors R, Macioszek V, Rudd J, Brosché M, Schlichting R, Scheel D and Kangasjärvi J (2004) Stress hormone-independent activation and nuclear translocation of mitogen-activated protein kinases in *Arabidopsis thaliana* during ozone exposure. The Plant Journal 40: 512-522

Aletta JM, Cimato TR and Ettinger MJ (1998) Protein methylation: a signal event in post-translational modification. Trends in Biochemical Sciences 23: 89-91

Alinsug M, Yu C-W and Wu K (2009) Phylogenetic analysis, subcellular localization, and expression patterns of RPD3/HDA1 family histone deacetylases in plants. BMC Plant Biology 9: 37

Allfrey VG, Faulkner R and A.E. M (1964) Acetylation and methylation of histones and their possible role in the regulation of RNA synthesis. Proceedings of the National Academy of Sciences of the United States of America 51: 786-794

Andreasson E, Jenkins T, Brodersen P, Thorgrimsen S, Petersen NHT, Zhu S, Qiu J-L, Micheelsen P, Rocher A, Petersen M, Newman M-A, Bjorn Nielsen H, Hirt H, Somssich I, Mattsson O and Mundy J (2005) The MAP kinase substrate MKS1 is a regulator of plant defense responses. Embo Journal 24: 2579-2589

Apel K and Hirt H (2004) Reactive oxygen species: metabolism, oxidative stress, and signal transduction. Annual Review of Plant Biology 55: 373 - 379

Aravind L and Koonin EV (1998) Second family of histone deacetylases. Science 280: 1167

Ashtamker C, Kiss V, Sagi M, Davydov O and Fluhr R (2007) Diverse subcellular locations of cryptogein-induced reactive oxygen species production in tobacco Bright Yellow-2 cells. Plant Physiology 143: 1817-1826

Astier J, Rasul S, Koen E, Manzoor H, Besson-Bard A, Lamotte O, Jeandroz S, Durner J, Lindermayr C and Wendehenne D (2011) S-nitrosylation: An emerging post-translational protein modification in plants. Plant Science In Press, Corrected Proof

Aufsatz W, Mette MF, van der Winden J, Matzke M and Matzke AJM (2002) HDA6, a putative histone deacetylase needed to enhance DNA methylation induced by double-stranded RNA. Embo Journal 21: 6832-6841

Bannister AJ and Kouzarides T (2005) Reversing histone methylation. Nature 436: 1103-1106

Bedford MT and Clarke SG (2009) Protein arginine methylation in mammals: who, what, and why. Molecular Cell 33: 1-13

Bedford MT and Richard S (2005) Arginine methylation: An emerging regulator of protein function. Molecular Cell 18: 263-272

Belin C, de Franco P-O, Bourbousse C, Chaignepain S, Schmitter J-M, Vavasseur A, Giraudat J, Barbier-Brygoo H and Thomine S (2006) Identification of features regulating OST1 kinase activity and OST1 function in guard cells. Plant Physiology 141: 1316-1327

Benhamed M, Martin-Magniette ML, Taconnat L, Bitton F, Servet C, De Clercq R, De Meyer B, Buysschaert C, Rombauts S, Villarroel R, Aubourg S, BeynonnJ, Bhalerao RP, Coupland G, Gruissem W, Menke FLH, Weisshaar B, Renou JP, Zhou DX and Hilson P (2008) Genome-scale *Arabidopsis* promoter array indetifies targets of the histone acetylreanferase. Plant Journal 56: 493-504

Benhamed M, Bertrand C, Servet C and Zhou DX (2006) Arabidopsis GCN5, HD1, and TAF1/HAF2 interact to regulate histone acetylation required for light-responsive gene expression. Plant Cell 18: 2893 - 2903

Berger SL (2002) Histone modifications in transcriptional regulation. Current Opinion in Genetics & Development 12: 142-148

Bernatavichute YV, Zhang XY, Cokus S, Pellegrini M and Jacobsen SE (2008) Genome-Wide Association of Histone H3 Lysine Nine Methylation with CHG DNA Methylation in Arabidopsis thaliana. Plos One 3: e3156

Besson-Bard A, Pugin A and Wendehenne D (2008) New insights into nitric oxide signaling in plants. Annual Review of Plant Biology 59: 21-39

Bond DM, Dennis ES, Pogson BJ and Finnegan EJ (2009) Histone acetylation, VERNALIZATION INSENSITIVE 3, FLOWERING LOCUS C, and the vernalization response. Molecular Plant 2: 724-737

Boudsocq M, Droillard M-J, Barbier-Brygoo H and Laurière C (2007) Different phosphorylation mechanisms are involved in the activation of sucrose non-fermenting 1 related protein kinases 2 by osmotic stresses and abscisic acid. Plant Molecular Biology 63: 491-503

Bourque S, Dutartre AH, V., Dahan J, Jeandroz S, Pichereaux C, Rossignol M and Wendehenne D (2011) Type-2 histone deacetylases as new regulators of elicitor-1 induced cell death in plants. New Phytol.: In press

Boyle P, Le Su E, Rochon A, Shearer HL, Murmu J, Chu JY, Fobert PR and Despres C (2009) The BTB/POZ domain of the Arabidopsis disease resistance protein NPR1 interacts with the repression domain of TGA2 to negate its function. Plant Cell 21: 3700-3713

Burza AM, Pekala I, Sikora J, Siedlecki P, Malagocki P, Bucholc M, Koper L, Zielenkiewicz P, Dadlez M and Dobrowolska G (2006) *Nicotiana tabacum* osmotic stress-activated kinase is regulated by phosphorylation on Ser-154 and Ser-158 in the kinase activation loop. Journal of Biological Chemistry 281: 34299-34311

Champion A, Kreis M, Mockaitis K, Picaud A and Henry Y (2004) *Arabidopsis* kinome: after the casting. Functional & Integrative Genomics 4: 163-187

Chehab EW, Patharkar OR, Hegeman AD, Taybi T and Cushman JC (2004) Autophosphorylation and subcellular localization dynamics of a salt- and water deficit-induced Calcium-Dependent Protein Kinase from ice plant. Plant Physiology 135: 1430-1446

Chen L-T, Luo M, Wang Y-Y and Wu K (2010) Involvement of Arabidopsis histone deacetylase HDA6 in ABA and salt stress response. Journal of Experimental Botany 61: 3345-3353

Chen L-T and Wu K (2010) Role of histone deacetylases HDA6 and HDA19 in ABA and abiotic stress response. Plant Signaling Behaviour 5: 1318-1320

Chen ZJ and Tian L (2007) Roles of dynamic and reversible histone acetylation in plant development and polyploidy. Biochimica et Biophysica Acta (BBA) - Gene Structure and Expression 1769: 295-307

Cheong MS, Park HC, Hong MJ, Lee J, Choi W, Jin JB, Bohnert HJ, Lee SY, Bressan RA and Yun DJ (2009) Specific domain structures control abscisic acid-, salicylic acid-, and stress-mediated SIZ1 phenotypes. Plant Physiology 151: 1930-1942

Choi H-i, Park H-J, Park JH, Kim S, Im M-Y, Seo H-H, Kim Y-W, Hwang I and Kim SY (2005) *Arabidopsis* Calcium-Dependent Protein Kinase AtCPK32 interacts with ABF4, a transcriptional regulator of abscisic acid-responsive gene expression, and modulates its activity. Plant Physiology 139: 1750-1761

Clarke S (1993) Protein methylation. Current Opinion in Cell Biology 5: 977-983

Cohen P (2000) The regulation of protein function by multisite phosphorylation - a 25 year update. Trends in Biochemical Sciences 25: 596-601

Conti L, Price G, O'Donnell E, Schwessinger B, Dominy P and Sadanandom A (2008) Small ubiquitin-like modifier proteases OVERLY TOLERANT TO SALT1 and -2 regulate salt stress responses in Arabidopsis. Plant Cell 20: 2894-2908

Cuthbert GL, Daujat S, Snowden AW, Erdjument-Bromage H, Hagiwara T, Yamada M, Schneider R, Gregory PD, Tempst P, Bannister AJ and Kouzarides T (2004) Histone deimination antagonizes arginine methylation. Cell 118: 545-553

Dahan J, Pichereaux C, Rossignol M, Blanc S, Wendehenne D, Pugin A and Bourque S (2009) Activation of a nuclear-localized SIPK in tobacco cells challenged by cryptogein, an elicitor of plant defence reactions. Biochemical Journal 418: 191-200

Dahan J, Wendehenne D, Ranjeva R, Pugin A and Bourque S (2010) Plant protein kinases: still enigmatic components in plant cell signalling. New Phytologist 185: 355-368

Dalle-Donne I, Rossi R, Colombo G, Giustarini D and Milzani A (2009) Protein S-glutathionylation: a regulatory device from bacteria to humans. Trends in Biochemical Sciences 34: 85-96

Dammann C, Ichida A, Hong B, Romanowsky SM, Hrabak EM, Harmon AC, Pickard BG and Harper JF (2003) Subcellular targeting of nine Calcium-Dependent Protein Kinase isoforms from *Arabidopsis*. Plant Physiology 132: 1840-1848

Dangl M, Brosch G, Haas H, Loidl P and Lusser A (2001) Comparative analysis of HD2 type histone deacetylases in higher plants. Planta 213: 280-285

Demetriou K, Kapazoglou A, Bladenopoulos K and Tsaftaris A (2009) Epigenetic chromatin modifiers in barley: II. Characterization and expression analysis of the *HDA1* family of barley histone deacetylases during development and in response to jasmonic acid. Plant Molecular Biology Reporter 28: 9-21

Despres C, Chubak C, Rochon A, Clark R, Bethune T, Desveaux D and Fobert PR (2003) The Arabidopsis NPR1 disease resistance protein is a novel cofactor that confers redox regulation of DNA binding activity to the basic domain/leucine zipper transcription factor TGA1. Plant Cell 15: 2181-2191

Dong CH, Agarwal M, Zhang Y, Xie Q and Zhu JK (2006) The negative regulator of plant cold responses, HOS1, is a RING E3 ligase that mediates the ubiquitination and degradation of ICE1. Proceeding of the National Academy of Science of the United States of America 103: 8281-8286

Droillard M-J, Boudsocq M, Barbier-Brygoo H and Laurière C (2002) Different protein kinase families are activated by osmotic stresses in *Arabidopsis thaliana* cell suspensions: Involvement of the MAP kinases AtMPK3 and AtMPK6. FEBS Letters 527: 43-50

Droillard M-J, Boudsocq M, Barbier-Brygoo H and Laurière C (2004) Involvement of MPK4 in osmotic stress response pathways in cell suspensions and plantlets of *Arabidopsis thaliana*: activation by hypoosmolarity and negative role in hyperosmolarity tolerance. FEBS Letters 574: 42-48

Dubos C, Stracke R, Grotewold E, Weisshaar B, Martin C and Lepiniec L (2010) MYB transcription factors in Arabidopsis. Trends in Plant Science 15: 573-581

Durner J, Wendehenne D and Klessig DF (1998) Defense gene induction in tobacco by nitric oxide, cyclic GMP, and cyclic ADP-ribose. Proceedings of the National Academy of Sciences of the United States of America 95: 10328-10333

Earley K, Lawrence RJ, Pontes O, Reuther R, Enciso AJ, Silva M, Neves N, Gross M, Viegas W and Pikaard CS (2006) Erasure of histone acetylation by Arabidopsis HDA6 mediates large-scale gene silencing in nucleolar dominance. Genes & Development 20: 1283-1293

Exner V, Aichinger E, Shu H, Wildhaber T, Alfarano P, Caflisch A, Gruissem W, Kohler C and Hennig L (2009) The chromodomain of LIKE HETEROCHROMATIN PROTEIN 1 is essential for H3K27me3 binding and function during Arabidopsis development. Plos One 4: e5335

Fan W and Dong X (2002) *In vivo* interaction between NPR1 and transcription factor TGA2 leads to salicylic acid-mediated gene activation in *Arabidopsis*. Plant Cell 14: 1377-1389

Feng Q, Wang HB, Ng HH, Erdjument-Bromage H, Tempst P, Struhl K and Zhang Y (2002) Methylation of H3-lysine 79 is mediated by a new family of HMTases without a SET domain. Current Biology 12: 1052-1058

Finnin MS, Donigian JR, Cohen A, Richon VM, Rifkind RA, Marks PA, Breslow R and Pavletich NP (1999) Structures of a histone deacetylase homologue bound to the TSA and SAHA inhibitors. Nature 401: 188-193

Foissner I, Wendehenne D, Langebartels C and Durner J (2000) *In vivo* imaging of an elicitor-induced nitric oxide burst in tobacco. Plant Journal 23: 817-824

Fong PM, Tian L and Chen ZJ (2006) *Arabidopsis thaliana* histone deacetylase 1 (AtHD1) is localized in euchromatic regions and demonstrates histone deacetylase activity *in vitro*. Cell Research 16: 479-488

Fu H, Lin YL and Fatimababy AS (2010) Proteasomal recognition of ubiquitylated substrates. Trends in Plant Science 15: 375-386

Fu W, Wu K and Duan J (2007) Sequence and expression analysis of histone deacetylases in rice. Biochemical and Biophysical Research Communications 356: 843-850

Fuchs J, Demidov D, Houben A and Schubert I (2006) Chromosomal histone modification patterns - from conservation to diversity. Trends in Plant Science 11: 199-208

Fujii H, Chinnusamy V, Rodrigues A, Rubio S, Antoni R, Park S-Y, Cutler SR, Sheen J, Rodriguez PL and Zhu J-K (2009) *In vitro* reconstitution of an abscisic acid signalling pathway. Nature 462: 660-664

Fujita Y, Nakashima K, Yoshida T, Katagiri T, Kidokoro S, Kanamori N, Umezawa T, Fujita M, Maruyama K, Ishiyama K, Kobayashi M, Nakasone S, Yamada K, Ito T, Shinozaki K and Yamaguchi-Shinozaki K (2009) Three SnRK2 protein kinases are the main positive regulators of abscisic acid signaling in response to water stress in Arabidopsis. Plant and Cell Physiology 50: 2123-2132

Gallwitz D (1971) Orgab specificity of histone acetyltranferases. FEBS Letters 13: 306-308

Gao M-J, Hegedus D, Sharpe A, Robinson S, Lydiate D and Hannoufa A (2007) Isolation and characterization of a GCN5-interacting protein from *Arabidopsis thaliana*. Planta 225: 1367-1379

Gao M-J, Schäfer UA, Parkin IAP, Hegedus DD, Lydiate DJ and Hannoufa A (2003) A novel protein from *Brassica napus* has a putative KID domain and responds to low temperature. Plant Journal 33: 1073-1086

Gao M, Liu J, Bi D, Zhang Z, Cheng F, Chen S and Zhang Y (2008) MEKK1, MKK1/MKK2 and MPK4 function together in a mitogen-activated protein kinase cascade to regulate innate immunity in plants. Cell Research 18: 1190-1198

Gardner KE, Zhou L, Parra MA, Chen X and Strahl BD (2011) Identification of Lysine 37 of Histone H2B as a Novel Site of Methylation. Plos One 13: e16244

Gary JD and Clarke S (1998) RNA and protein interactions modulated by protein arginine methylation. Progress in Nucleic Acid Research and Molecular Biology, Vol 61 61: 65-131

Gill SS and Tuteja N (2010) Reactive oxygen species and antioxidant machinery in abiotic stress tolerance in crop plants. Plant Physiology and Biochemistry 48: 909-930

Gonzalez D, Bowen AJ, Carroll TS and Conlan RS (2007) The transcription corepressor LEUNIG interacts with the histone deacetylase HDA19 and mediator components MED14 (SWP) and CDK8 (HEN3) to repress transcription. Molecular and Cell Biology 27: 5306-5315

Gould K, Lamotte O, Klinguer A, Pugin A and Wendehenne D (2003) Nitric oxide production in tobacco leaf cells: a generalized stress response? Plant Cell Environment 16: 1851-1862

Guo L, Yin BL, Zhou JL, Li XY and Deng XW (2006) Development of rabbit monoclonal and polyclonal antibodies for detection of site-specific histone modifications and their application in analyzing overall modification levels. Cell Research 16: 519-527

Halford NG and Hey SJ (2009) Snf1-related protein kinases (SnRKs) act within an intricate network that links metabolic and stress signalling in plants. Biochemical Journal 419: 247-259

Hanks SK and Hunter T (1995) Protein kinases 6. The eukaryotic protein kinase superfamily: kinase (catalytic) domain structure and classification. The FASEB Journal 9: 576-596

Hark AT, Vlachonasios KE, Pavangadkar KA, Rao S, Gordon H, Adamakis I-D, Kaldis A, Thomashow MF and Triezenberg SJ (2009) Two Arabidopsis orthologs of the transcriptional coactivator ADA2 have distinct biological functions. Biochimica et Biophysica Acta (BBA) - Gene Regulatory Mechanisms 1789: 117-124

Harmon AC (2003) Calcium-regulated protein kinases of plants. Gravitatiobal and Space Biology Bulletin 16: 83-90

Hayes JJ and Hansen JC (2001) Nucleosomes and the chromatin fiber. Current Opinion in Genetics & Development 11: 124-129

Heine GF, Hernandez JM and Grotewold E (2004) Two cysteines in plant R2R3 MYB domains participate in REDOX-dependent DNA binding. Journal of Biological Chemistry 279: 37878-37885

Huang X, Stettmaier K, Michel C, Hutzler P, Mueller M and Durner J (2004) Nitric oxide is induced by wounding and influences jasmonic acid signaling in *Arabidopsis thaliana*. Planta 218: 938-946

Hubbard KE, Nishimura N, Hitomi K, Getzoff ED and Schroeder JI (2010) Early abscisic acid signal transduction mechanisms: newly discovered components and newly emerging questions. Genes & Development 24: 1695-1708

Ichimura K, Mizoguchi T, Yoshida R, Yuasa T and Shinozaki K (2000) Various abiotic stresses rapidly activate Arabidopsis MAP kinases ATMPK4 and ATMPK6. Plant Journal 24: 655-665

Imai S, Armstrong CM, Kaeberlein M and Guarente L (2000) Transcriptional silencing and longevity protein Sir2 is an NAD-dependent histone deacetylase. Nature: 795-800

Inoue A and Fujimoto D (1969) Enzymatic deacetylation of histone. Biochemical and Biophysical Research Communications 36: 146-150

Jackson JP, Johnson L, Jasencakova Z, Zhang X, PerezBurgos L, Singh PB, Cheng XD, Schubert I, Jenuwein T and Jacobsen SE (2004) Dimethylation of histone H3 lysine 9 is a critical mark for DNA methylation and gene silencing in *Arabidopsis thaliana*. Chromosoma 112: 308-315

Jenuwein T and Allis CD (2001) Translating the histone code. Science 293: 1074-1080

Jenuwein T, Laible G, Dorn R and Reuter G (1998) SET domain proteins modulate chromatin domains in eu- and heterochromatin. Cellular and Molecular Life Sciences 54: 80-93

Johnson L, Mollah S, Garcia BA, Muratore TL, Shabanowitz J, Hunt DF and Jacobsen SE (2004) Mass spectrometry analysis of Arabidopsis histone H3 reveals distinct combinations of post-translational modifications. Nucleic Acids Research 32: 6511-6518

Johnson LM, Cao XF and Jacobsen SE (2002) Interplay between two epigenetic marks: DNA methylation and histone H3 lysine 9 methylation. Current Biology 12: 1360-1367

Jones AME, MacLean D, Studholme DJ, Serna-Sanz A, Andreasson E, Rathjen JP and Peck SC (2009) Phosphoproteomic analysis of nuclei-enriched fractions from *Arabidopsis thaliana*. Journal of Proteomics 72: 439-451

Jung C, Seo JS, Han SW, Koo YJ, Kim CH, Song SI, Nahm BH, Choi YD and Cheong J-J (2008) Overexpression of AtMYB44 enhances stomatal closure to confer abiotic stress tolerance in transgenic Arabidopsis. Plant Physiology 146: 623-635

Kaldis A, Tsementzi D, Tanriverdi O and Vlachonasios K (2011) *Arabidopsis thaliana* transcriptional co-activators ADA2b and SGF29a are implicated in salt stress responses. Planta 233: 749-762

Kerk D, Bulgrien J, Smith DW, Barsam B, Veretnik S and Gribskov M (2002) The complement of protein phosphatase catalytic subunits encoded in the genome of Arabidopsis. Plant Physiology 129: 908-925

Kim JM, To TK, Ishida J, Morosawa T, Kawashima M, Matsui A, Toyoda T, Kimura H, Shinozaki K and Seki M (2008) Alterations of lysine modifications on the histone H3 N-Tail under drought stress conditions in *Arabidopsis thaliana*. Plant and Cell Physiology 49: 1580-1588

Klimecka M and Muszynska G (2007) Structure and functions of plant calcium-dependent protein kinases. Acta Biochimica Polonica 54: 219-233

Klose RJ and Yi Z (2007) Regulation of histone methylation by demethylimination and demethylation. Nature Reviews Molecular Cell Biology 8: 307-318

Kobayashi Y, Murata M, Minami H, Yamamoto S, Kagaya Y, Hobo T, Yamamoto A and Hattori T (2005) Abscisic acid-activated SNRK2 protein kinases function in the gene-regulation pathway of ABA signal transduction by phosphorylating ABA response element-binding factors. Plant Journal 44: 939-949

Kornberg RD and Lorch YL (1999) Twenty-five years of the nucleosome, fundamental particle of the eukaryote chromosome. Cell 98: 285-294

Kosetsu K, Matsunaga S, Nakagami H, Colcombet J, Sasabe M, Soyano T, Takahashi Y, Hirt H and Machida Y (2010) The MAP kinase MPK4 is required for cytokinesis in *Arabidopsis thaliana*. Plant Cell 22: 3778-3790

Kouzarides T (2002) Histone methylation in transcriptional control. Current Opinion in Genetics & Development 12: 198-209

Kouzarides T (2007) Chromatin modifications and their function. Cell 128: 693 - 705

Kudla J, Batistic O and Hashimoto K (2010) Calcium signals: the lead currency of plant information processing. Plant Cell 22: 541-563

Kurepa J, Walker JM, Smalle J, Gosink MM, Davis SJ, Durham TL, Sung DY and Vierstra RD (2003) The small ubiquitin-like modifier (SUMO) protein modification system in Arabidopsis. Accumulation of SUMO1 and -2 conjugates is increased by stress. Journal of Biological Chemistry 278: 6862-6872

Kwon CS, Lee D, Choi G and Chung WI (2009) Histone occupancy-dependent and -independent removal of H3K27 trimethylation at cold-responsive genes in Arabidopsis. Plant Journal 60: 112-121

Lechner T, Lusser A, Pipal A, Brosch G, Loidl A, Goralik-Schramel M, Sendra R, Wegener S, Walton JD and Loidl P (2000) RPD3-type histone deacetylases in maize embryos. Biochemistry 39: 1683-1692

Lee DY, Teyssier C, Strahl BD and Stallcup MR (2005) Role of protein methylation in regulation of transcription. Endocrine Reviews 26: 147-170

Lee H, Xiong L, Gong Z, Ishitani M, Stevenson B and Zhu JK (2001) The Arabidopsis HOS1 gene negatively regulates cold signal transduction and encodes a RING finger protein that displays cold-regulated nucleo--cytoplasmic partitioning. Genes and Development 15: 912-924

Li J (2005) Brassinosteroid signaling: from receptor kinases to transcription factors. Current Opinion in Plant Biology 8: 526-531

Li S, Lauri A, Ziemann M, Busch A, Bhave M and Zachgo S (2009) Nuclear activity of ROXY1, a glutaredoxin interacting with TGA factors, is required for petal development in *Arabidopsis thaliana*. Plant Cell 21: 429-441

Lin C-H, Li B, Swanson S, Zhang Y, Florens L, Washburn MP, Abmayr SM and Workman JL (2008) Heterochromatin protein 1a stimulates histone H3 lysine 36 demethylation by the Drosophila KDM4A demethylase. Molecular cell 32: 696-706

Lindermayr C, Sell S, Müller B, Leister D and Durner J (2010) Redox regulation of the NPR1-TGA1 system of *Arabidopsis thaliana* by nitric oxide. Plant Cell 22: 2894-2907

Lippman Z, Gendrel AV, Black M, Vaughn MW, Dedhia N, McCombie WR, Lavine K, Mittal V, May B, Kasschau KD, Carrington JC, Doerge RW, Colot V and Martienssen R (2004) Role of transposable elements in heterochromatin and epigenetic control. Nature 430: 471-476

Lippman Z, May B, Yordan C, Singer T and Martienssen R (2003) Distinct mechanisms determine transposon inheritance and methylation via small interfering RNA and mistone modification. PLoS Biol 1: e67

Liu CY, Lu FL, Cui X and Cao XF (2010) Histone Methylation in Higher Plants. Annual Review of Plant Biology, 61: 395-420

Lois LM, Lima CD and Chua NH (2003) Small ubiquitin-like modifier modulates abscisic acid signaling in Arabidopsis. Plant Cell 15: 1347-1359

Lu FL, Li GL, Cui X, Liu CY, Wang XJ and Cao XF (2008) Comparative analysis of JmjC domain-containing proteins reveals the potential histone demethylases in Arabidopsis and rice. Journal of Integrative Plant Biology 50: 886-896

Luan S (2003) Protein phosphatases in plants. Annual Review of Plant Biology 54: 63-92

Luscombe NM, Laskowski RA and Thornton JM (2001) Amino acid-base interactions: a three-dimensional analysis of protein-DNA interactions at an atomic level. Nucleic Acids Research 29: 2860-2874

Lusser A, Brosch G, loidl A, Haas H and Loidl P (1997) Identification of histone deacetylase HD2 as an acidic nucleolar phosphoprotein. Science 277: 88-91

Mao Y, Pavangadkar KA, Thomashow MF and Triezenberg SJ (2006) Physical and functional interactions of Arabidopsis ADA2 transcriptional coactivator proteins with the acetyltransferase GCN5 and with the cold-induced transcription factor CBF1. Biochimica et Biophysica Acta (BBA) - Gene Structure and Expression 1759: 69-79

Mathieu O, Probst AV and Paszkowski J (2005) Distinct regulation of histone H3 methylation at lysines 27 and 9 by CpG methylation in Arabidopsis. Embo Journal 24: 2783-2791

May BP, Lippman ZB, Fang Y, Spector DL and Martienssen RA (2005) Differential regulation of strand-specific transcripts from *Arabidopsis* centromeric satellite repeats. PLoS Genet 1: e79

Mazars C, Bourque S, Mithöfer A, Pugin A and Ranjeva R (2009) Calcium homeostasis in plant cell nuclei. New Phytologist 181: 261-274

Mehlmer N, Wurzinger B, Stael S, Hofmann-Rodrigues D, Csaszar E, Pfister B, Bayer R and Teige M (2010) The Ca^{2+}-dependent protein kinase CPK3 is required for MAPK-independent salt-stress acclimation in *Arabidopsis*. Plant Journal 63: 484-498

Menke FLH, Kang H-G, Chen Z, Park JM, Kumar D and Klessig DF (2005) Tobacco transcription factor WRKY1 is phosphorylated by the MAP kinase SIPK and mediates HR-like cell death in tobacco. Molecular Plant-Microbe Interactions 18: 1027-1034

Mikolajczyk M, Awotunde OS, Muszynska G, Klessig DF and Dobrowolska G (2000) Osmotic stress induces rapid activation of a salicylic acid–induced protein kinase and a homolog of protein kinase ASK1 in tobacco cells. Plant Cell 12: 165-178

Miller GAD and Mittler RON (2006) Could heat shock transcription factors function as hydrogen peroxide sensors in plants? Annals of Botany 98: 279-288

Miller MJ, Barrett-Wilt GA, Hua Z and Vierstra RD (2010) Proteomic analyses identify a diverse array of nuclear processes affected by small ubiquitin-like modifier conjugation in Arabidopsis. Proceeding of the National Academy of Science of the United States of America 107: 16512-16517

Miranda-Saavedra D and Barton GJ (2007) Classification and functional annotation of eukaryotic protein kinases. Proteins: Structure, Function, and Bioinformatics 68: 893-914

Mitchell JBO, Thornton JM, Singh J and Price SL (1992) Towards an understanding of the arginine aspartate interaction. Journal of Molecular Biology 226: 251-262

Mittler R, Kim Y, Song LH, Coutu J, Coutu A, Ciftci-Yilmaz S, Lee H, Stevenson B and Zhu JK (2006) Gain- and loss-of-function mutations in Zat10 enhance the tolerance of plants to abiotic stress. Febs Letters 580: 6537 - 6542

Miura K and Hasegawa PM (2010) Sumoylation and other ubiquitin-like post-translational modifications in plants. Trends in Cellular Biology 20: 223-232

Miura K, Jin JB, Lee J, Yoo CY, Stirm V, Miura T, Ashworth EN, Bressan RA, Yun DJ and Hasegawa PM (2007) SIZ1-mediated sumoylation of ICE1 controls CBF3/DREB1A expression and freezing tolerance in Arabidopsis. Plant Cell 19: 1403-1414

Miura K, Rus A, Sharkhuu A, Yokoi S, Karthikeyan AS, Raghothama KG, Baek D, Koo YD, Jin JB, Bressan RA, Yun DJ and Hasegawa PM (2005) The Arabidopsis SUMO E3 ligase SIZ1 controls phosphate deficiency responses. Proceeding of the National Academy of Science of the United States of America 102: 7760-7765

Moorhead GBG, De wever V, Templeton G and Kerk D (2009) Evolution of protein phosphatases in plants and animals. Biochemical Journal 417: 401-409

Mou Z, Fan W and Dong X (2003) Inducers of plant systemic acquired resistance regulate NPR1 function through redox changes. Cell 113: 935-944

Murfett J, Wang X-J, Hagen G and Guilfoyle TJ (2001) Identification of Arabidopsis histone deacetylase hda6 mutants that affect transgene expression. Plant Cell 13: 1047-1062

Mylne JS, Barrett L, Tessadori F, Mesnage S, Johnson L, Bernatavichute YV, Jacobsen SE, Fransz P and Dean C (2006) LHP1, the Arabidopsis homologue of HETEROCHROMATIN PROTEIN1, is required for epigenetic silencing of FLC. Proceedings of the National Academy of Sciences of the United States of America 103: 5012-5017

Ndamukong I, Abdallat AA, Thurow C, Fode B, Zander M, Weigel R and Gatz C (2007) SA-inducible Arabidopsis glutaredoxin interacts with TGA factors and suppresses JA-responsive PDF1.2 transcription. Plant Journal 50: 128-139

Nigam N, Singh A, Sahi C, Chandramouli A and Grover A (2008) SUMO-conjugating enzyme (Sce) and FK506-binding protein (FKBP) encoding rice (Oryza sativa L.) genes: genome-wide analysis, expression studies and evidence for their involvement in abiotic stress response. Molecular Genetics and Genomics 279: 371-383

Niu LF, Lu FL, Pei YX, Liu CY and Cao XF (2007) Regulation of flowering time by the protein arginine methyltransferase AtPRMT10. EMBO Reports 8: 1190-1195

Owen DJ, Ornaghi P, Yang J-C, Lowe N, Evans PR, Ballario P, Neuhaus D, Filetici P and Travers AA (2000) The structural basis for the recognition of acetylated histone H4 by the bromodomain of histone acetyltransferase Gcn5p. EMBO Journal 19: 6141-6149

Palmieri MC, Sell S, Huang X, Scherf M, Werner T, Durner J and Lindermayr C (2008) Nitric oxide-responsive genes and promoters in Arabidopsis thaliana: a bioinformatics approach. Journal of Experimental Botany 59: 177-186

Pandey R, Müller A, Napoli CA, Selinger DA, Pikaard CS, Richards EJ, Bender J, Mount DW and Jorgensen RA (2002) Analysis of histone acetyltransferase and histone deacetylase families of *Arabidopsis thaliana* suggests functional diversification of chromatin modification among multicellular eukaryotes. Nucleic Acids Research 30: 5036-5055

Pei YX, Niu LF, Lu FL, Liu CY, Zhai JX, Kong XF and Cao XF (2007) Mutations in the type II protein arginine methyltransferase AtPRMT5 result in pleiotropic developmental defects in Arabidopsis(1[C][OA]). Plant Physiology 144: 1913-1923

Pipal A, Goralik-Schramel M, Lusser A, Lanzanova C, Sarg B, Loidl A, Lindner H, Rossi V and Loidl P (2003) Regulation and processing of maize histone deacetylase Hda1 by limited proteolysis. Plant Cell 15: 1904-1917

Probst AV, Fagard M, Proux F, Mourrain P, Boutet S, Earley K, Lawrence RJ, Pikaard CS, Murfett J, Furner I, Vaucheret H and Scheid OM (2004) Arabidopsis distone deacetylase HDA6 is required for maintenance of transcriptional gene silencing and determines nuclear organization of rDNA repeats. Plant Cell 16: 1021-1034

Pulido P, Cazalis R and Cejudo FJ (2009a) An antioxidant redox system in the nucleus of wheat seed cells suffering oxidative stress. Plant Journal 57: 132-145

Pulido P, Dominquez F and Cejudo FJ (2009b) A hydrogen peroxide detoxification system in the nucleus of wheat seed cells: protection or signaling role? Plant Signaling Behavior 4: 23-25

Qin F, Sakuma Y, Tran LS, Maruyama K, Kidokoro S, Fujita Y, Fujita M, Umezawa T, Sawano Y, Miyazono K, Tanokura M, Shinozaki K and Yamaguchi-Shinozaki K (2008) Arabidopsis DREB2A-interacting proteins function as RING E3 ligases and negatively regulate plant drought stress-responsive gene expression. Plant Cell 20: 1693-1707

Qiu J-L, Fiil BK, Petersen K, Nielsen HB, Botanga CJ, Thorgrimsen S, Palma K, Suarez-Rodriguez MC, Sandbech-Clausen S, Lichota J, Brodersen P, Grasser KD, Mattsson O, Glazebrook J, Mundy J and Petersen M (2008) *Arabidopsis* MAP kinase 4 regulates gene expression through transcription factor release in the nucleus. EMBO Journal 27: 2214-2221

Quilis J, Penas G, Messeguer J, Brugidou C and Segundo BS (2008) The Arabidopsis AtNPR1 inversely modulates defense responses against fungal, bacterial, or viral pathogens while conferring hypersensitivity to abiotic stresses in transgenic rice. Molecular Plant-Microbe Interactions 21: 1215-1231

Rao MV, Lee H-i and Davis KR (2002) Ozone-induced ethylene production is dependent on salicylic acid, and both salicylic acid and ethylene act in concert to regulate ozone-induced cell death. Plant Journal 32: 447-456

Rossi V, Hartings H and Motto M (1998) Identification and caracterisation of an RDP3 homologue from maize (*Zea mays* L.) that is able to complement an rpd3 null mutant of *Saccharomyces cerevisiaeI*. Molecular and General Genetics 258: 288-296

Rouhier N, Couturier J, Johnson MK and Jacquot J-P (2010) Glutaredoxins: roles in iron homeostasis. Trends in Biochemical Sciences 35: 43-52

Rundlett SE, Carmen AA, Kobayashi R, Bavykin S, Turner BM and Grunstein M (1996) HDA1 and RPD3 are members of distinct yeast histone deacetylase complexes that regulate silencing andâ€‰transcription. Proceedings of the National Academy of Sciences of the United States of America 93: 14503-14508

Rushton PJ, Somssich IE, Ringler P and Shen QJ (2010) WRKY transcription factors. Trends in Plant Science 15: 247-258

Saez A, Apostolova N, Gonzalez-Guzman M, Gonzalez-Garcia MP, Nicolas C, Lorenzo O and Rodriguez PL (2004) Gain-of-function and loss-of-function phenotypes of the protein phosphatase 2C HAB1 reveal its role as a negative regulator of abscisic acid signalling. Plant Journal 37: 354-369

Saez A, Robert N, Maktabi MH, Schroeder JI, Serrano R and Rodriguez PL (2006) Enhancement of abscisic acid sensitivity and reduction of water consumption in Arabidopsis by combined inactivation of the protein phosphatases type 2C ABI1 and HAB1. Plant Physiology 141: 1389-1399

Samuel MA and Ellis BE (2002) Double jeopardy: both overexpression and suppression of a redox-activated plant mitogen-activated protein kinase render tobacco plants ozone sensitive. Plant Cell 14: 2059-2069

Saracco SA, Miller MJ, Kurepa J and Vierstra RD (2007) Genetic analysis of SUMOylation in Arabidopsis: conjugation of SUMO1 and SUMO2 to nuclear proteins is essential. Plant Physiology 145: 119-134

Schmitz RJ, Sung S and Amasino RM (2008) Histone arginine methylation is required for vernalization-induced epigenetic silencing of FLC in winter-annual *Arabidopsis thaliana*. Proceedings of the National Academy of Sciences of the United States of America 105: 411-416

Schotta G, Ebert A, Krauss V, Fischer A, Hoffmann J, Rea S, Jenuwein T, Dorn R and Reuter G (2002) Central role of Drosophila SU(VAR)3-9 in histone H3-K9 methylation and heterochromatic gene silencing. EMBO Journal 21: 1121-1131

Serpa V, Vernal J, Lamattina L, Grotewold E, Cassia R and Terenzi H (2007) Inhibition of AtMYB2 DNA-binding by nitric oxide involves cysteine S-nitrosylation. Biochemical and Biophysical Research Communications 361: 1048-1053

Servet C, Conde E, Silva N and Zhou DX (2010) Histone acetyltransferase AtGCN5/HAG1 is a versatile regulator of developmentat and inducible gene expression in *Arabidopsis*. Molecular Plant 3: 670-677

Seth D and Stamler JS (2011) The SNO-proteome: causation and classifications. Current Opinion in Chemical Bioogyl 15: 129-136

Shaikhali J, Heiber I, Seidel T, Stroher E, Hiltscher H, Birkmann S, Dietz K-J and Baier M (2008) The redox-sensitive transcription factor Rap2.4a controls nuclear expression of 2-Cys peroxiredoxin A and other chloroplast antioxidant enzymes. BMC Plant Biology 8: 48

Shi YJ, Lan F, Matson C, Mulligan P, Whetstine JR, Cole PA, Casero RA and Shi Y (2004) Histone demethylation mediated by the nuclear arnine oxidase homolog LSD1. Cell 119: 941-953

Spedaletti V, Polticelli F, Capodaglio V, Schinina ME, Stano P, Federico R and Tavladoraki P (2008) Characterization of a lysine-specific histone demethylase from *Arabidopsis thaliana*. Biochemistry 47: 4936-4947

Spoel SH, Tada Y and Loake GJ (2010) Post-translational protein modification as a tool for transcription reprogramming. New Phytologist 186: 333-339

Sridha S and Wu K (2006) Identification of AtHD2C as a novel regulator of abscisic acid responses in Arabidopsis. Plant Journal 46: 124-133

Stamler JS, Singel DJ and Loscalzo J (1992) Biochemistry of nitric oxide and its redox-activated forms. Science 258: 1898-1902

Stockinger EJ, Mao Y, Regier MK, Triezenberg SJ and Thomashow MF (2001) Transcriptional adaptor and histone acetyltransferase proteins in Arabidopsis and their interactions with CBF1, a transcriptional activator involved in cold-regulated gene expression. Nucleic Acids Research 29: 1524-1533

Strahl BD and Allis CD (2000) The language of covalent histone modifications. Nature 403: 41-45

Sugiyama N, Nakagami H, Mochida K, Daudi A, Tomita M, Shirasu K and Ishihama Y (2008) Large-scale phosphorylation mapping reveals the extent of tyrosine phosphorylation in Arabidopsis. Molecular System Biology 4

Sung SB, He YH, Eshoo TW, Tamada Y, Johnson L, Nakahigashi K, Goto K, Jacobsen SE and Amasino RM (2006) Epigenetic maintenance of the vernalized state in *Arabidopsis thaliana* requires LIKE HETEROCHROMATIN PROTEIN 1. Nature Genetics 38: 706-710

Tada Y, Spoel SH, Pajerowska-Mukhtar K, Mou Z, Song J, Wang C, Zuo J and Dong X (2008) Plant immunity requires conformational charges of NPR1 via S-nitrosylation and thioredoxins. Science 321: 952-956

Taverna SD, Li H, Ruthenburg AJ, Allis CD and Patel DJ (2007) How chromatin-binding modules interpret histone modifications: lessons from professional pocket pickers. Nature Structiral & Molecular Biology 14: 1025-1040

Teige M, Scheikl E, Eugelm T, Doczi R, Ichimura K, Shinozaki K, Dangl J and Hirt H (2004) The MKK2 pathway mediates cold and salt stress signaling in *Arabidopsis*. Molecular Cell 15: 141-152

Tian L and Chen ZJ (2001) Blocking histone deacetylation in Arabidopsis induces pleiotropic effects on plant gene regulation and development. Proceedings of the National Academy of Sciences of the United States of America 98: 200-205

Tian L, Fong MP, Wang JJ, Wei NE, Jiang H, Doerge RW and Chen ZJ (2005) Reversible histone acetylation and deacetylation mediate genome-wide, promoter-dependent and locus-specific changes in gene expression during plant development. Genetics 169: 337-345

Tian L, Wang J, Fong MP, Chen M, Cao H, Gelvin SB and Chen ZJ (2003) Genetic control of developmental changes induced by disruption of *Arabidopsis* histone deacetylase 1 (AtHD1) expression. Genetics 165: 399-409

Tör M, Lotze MT and Holton N (2009) Receptor-mediated signalling in plants: molecular patterns and programmes. Journal of Experimental Botany 60: 3645-3654

Trujillo M and Shirasu K (2010) Ubiquitination in plant immunity. Current Opinion in Plant Biology 13: 402-408

Tsukada Y, Fang J, Erdjument-Bromage H, Warren ME, Borchers CH, Tempst P and Zhang Y (2006) Histone demethylation by a family of JmjC domain-containing proteins. Nature 439: 811-816

Turck F, Roudier F, Farrona S, Martin-Magniette ML, Guillaume E, Buisine N, Gagnot S, Martienssen RA, Coupland G and Colot V (2007) Arabidopsis TFL2/LHP1 specifically associates with genes marked by trimethylation of histone H3 lysine 27. Plos Genetics 3: 855-866

Umezawa T, Sugiyama N, Mizoguchi M, Hayashi S, Myouga F, Yamaguchi-Shinozaki K, Ishihama Y, Hirayama T and Shinozaki K (2009) Type 2C protein phosphatases directly regulate abscisic acid-activated protein kinases in Arabidopsis. Proceeding of the National Academy of Science of the United States of America 106: 17588-17593

Vaillant I and Paszkowski J (2007) Role of histone and DNA methylation in gene regulation. Current Opinion in Plant Biology 10: 528-533

van Dijk K, Ding Y, Malkaram S, Riethoven JJM, Liu R, Yang JY, Laczko P, Chen H, Xia YN, Ladunga I, Avramova Z and Fromm M (2010) Dynamic changes in genome-wide histone H3 lysine 4 methylation patterns in response to dehydration stress in Arabidopsis thaliana. BMC Plant Biology 10

Varotto S, Locatelli S, Canova S, Pipal A, Motto M and Rossi V (2003) Expression profile and cellular localization of maize Rpd3-type histone deacetylases during plant development. Plant Physiology 133: 606-617

Vidal M, Buckley AM, Hilger F and Gaber RF (1990) Direct selection for mutants with increased K^+ transport in *Saccharomyces cerevisiae*. Genetics 125: 313-320

Vitecek J, Reinohl V and Jones RL (2008) Measuring NO production by plant tissues and suspension cultured cells. Molecular Plant 1: 270-284

Wang C, Gao F, Wu J, Dai J, Wei C and Li Y (2010) Arabidopsis putative deacetylase AtSRT2 regulates basal defense by supressing PAD4, EDS5 and SID2 expression. Plant Cell Physiology 51: 1291-1299

Wang X, Zhang Y, Ma QB, Zhang ZL, Xue YB, Bao SL and Chong K (2007) SKB1-mediated symmetric dimethylation of histone H4R3 controls flowering time in Arabidopsis. EMBO Journal 26: 1934-1941

Wang Y, Wysocka J, Sayegh J, Lee YH, Perlin JR, Leonelli L, Sonbuchner LS, McDonald CH, Cook RG, Dou Y, Roeder RG, Clarke S, Stallcup MR, Allis CD and Coonrod SA (2004) Human PAD4 regulates histone arginine methylation levels via demethylimination. Science 306: 279-283

Wawer I, Bucholc M, Astier J, Anielska-Mazur A, Dahan J, Kulik A, Wyslouch-Cieszynska A, Zareba-Koziol M, Krzywinska E, Dadlez M, Dobrowolska G and Wendehenne D (2010) Regulation of *Nicotiana tabacum* osmotic stress-activated protein kinase and its cellular partner GAPDH by nitric oxide in response to salinity. Biochemical Journal 429: 73-83

Widmann C, Gibson S, Jarpe MB and Johnson GL (1999) Mitogen-Activated Protein Kinase: conservation of a three-kinase module from yeast to human. Physiological Reviews 79: 143-180

Witte C-P, Keinath N, Dubiella U, Demoulière R, Seal A and Romeis T (2010) Tobacco Calcium-dependent Protein Kinases are differentially phosphorylated *in vivo* as part of a kinase cascade that regulates stress response. Journal of Biological Chemistry 285: 9740-9748

Wormuth D, Heiber I, Shaikali J, Kandlbinder A, Baier M and Dietz K-J (2007) Redox regulation and antioxidative defence in Arabidopsis leaves viewed from a systems biology perspective. Journal of Biotechnology 129: 229-248

Wu K, Malik K, Tian L, Brown D and Miki B (2000) Functional analysis of a RPD3 histone deacetylase homologue in *Arabidopsis thaliana*. Plant Molecular Biology 44: 167-176

Wu K, Tian L, Zhou C, Brown D and Miki B (2000) Repression of gene expression by *Arabidopsis* HD2 hitone deacetylases. Plant Journal 34: 241-247

Xiong T-C, Bourque S, Lecourieux D, Amelot N, Grat S, Brière C, Mazars C, Pugin A and Ranjeva R (2006) Calcium signaling in plant cell organelles delimited by a double membrane. Biochimica et Biophysica Acta (BBA) - Molecular Cell Research 1763: 1209-1215

Yan D, Zhang Y, Niu L, Yuan Y and Cao X (2007) Identification and characterization of two closely related histone H4 arginine 3 methyltransferases in *Arabidopsis thaliana*. Biochemical Journal 408: 113-121

Yap Y-K, Kodama Y, Waller F, Chung KM, Ueda H, Nakamura K, Oldsen M, Yoda H, Yamaguchi Y and Sano H (2005) Activation of a novel transcription factor through phosphorylation by WIPK, a wound-induced mitogen-activated protein kinase in tobacco plants. Plant Physiology 139: 127-137

Yasuda M, Ishikawa A, Jikumaru Y, Seki M, Umezawa T, Asami T, Maruyama-Nakashita A, Kudo T, Shinozaki K, Yoshida S and Nakashita H (2008) Antagonistic interaction between systemic acquired resistance and the abscisic acid-mediated abiotic stress response in Arabidopsis. Plant Cell 20: 1678-1692

Yoo CY, Miura K, Jin JB, Lee J, Park HC, Salt DE, Yun DJ, Bressan RA and Hasegawa PM (2006) SIZ1 small ubiquitin-like modifier E3 ligase facilitates basal thermotolerance in Arabidopsis independent of salicylic acid. Plant Physiology 142: 1548-1558

Yoo S-D, Cho Y-H, Tena G, Xiong Y and Sheen J (2008) Dual control of nuclear EIN3 by bifurcate MAPK cascades in C_2H_4 signalling. Nature 451: 789-795

Zhang K, Sridhar VV, Zhu J, Kapoor A and Zhu JK (2007) Distinctive core histone post-translational modification patterns in *Arabisopsis thaliana*. PLoS One 2: e1210

Zhang KL, Tang H, Huang L, Blankenship JW, Jones PR, Xiang F, Yau PM and Burlingame AL (2002) Identification of acetylation and methylation sites of histone H3 from chicken erythrocytes by high-accuracy matrix-assisted laser desorption ionization-time-of-flight, matrix-assisted laser desorption ionization-postsource decay, and nanoelectrospray ionization tandem mass spectrometry. Analytical Biochemistry 306: 259-269

Zhang LW, Eugeni EE, Parthun MR and Freitas MA (2003) Identification of novel histone post-translational modifications by peptide mass fingerprinting. Chromosoma 112: 77-86

Zhang S and Klessig DF (1998) The tobacco wounding-activated mitogen-activated protein kinase is encoded by SIPK. Proceedings of the National Academy of Sciences 95: 7225-7230

Zhang X, Germann S, Blus BJ, Khorasanizadeh S, Gaudin V and Jacobsen SE (2007) The Arabidopsis LHP1 protein colocalizes with histone H3 Lys27 trimethylation. Nature Structural & Molecular Biology 14: 869-871

Zhang XY, Clarenz O, Cokus S, Bernatavichute YV, Pellegrini M, Goodrich J and Jacobsen SE (2007) Whole-genome analysis of histone H3 lysine 27 trimethylation in Arabidopsis. Plos Biology 5: 1026-1035

Zhang Y and Reinberg D (2001) Transcription regulation by histone methylation: interplay between different covalent modifications of the core histone tails. Genes & Development 15: 2343-2360

Zhang ZL, Zhang SP, Zhang Y, Wang X, Li D, Li QL, Yue MH, Li Q, Zhang YE, Xu YY, Xue YBA, Chong K and Bao SL (2011) Arabidopsis floral initiator SKB1 confers high salt tolerance by regulating transcription and pre-mRNA splicing through altering histone H4R3 and small nuclear ribonucleoprotein LSM4 methylation. Plant Cell 23: 396-411

Zhou C, Zhang L, Duan J, Miki B and Wu K (2005) HISTONE DEACETYLASE 19 is involved in jasmonic acid and ethylene signalling of pathogen response in *Arabidopsis*. Plant Cell 17: 1196-1204

5

Facing the Environment: Small RNAs and the Regulation of Gene Expression Under Abiotic Stress in Plants

Inês Trindade[1], Dulce Santos[2], Tamas Dalmay[3] and Pedro Fevereiro[1]

[1]Laboratório de Biotecnologia de Células Vegetais
Instituto de Tecnologia Química e Biológica
Universidade Nova de Lisboa
[2]Centro de Veterinária e Zootecnia, Instituto de Investigação Científica e Tropical
Universidade Técnica de Lisboa
[3]School of Biological Sciences
University of East Anglia
[1,2]Portugal
[3]United Kingdom

1. Introduction

Plant growth and development is highly dependent on a variety of environmental conditions such as temperature, light, water availability and soil conditions that strongly affect productivity worldwide. Over the past years, several reports have raised awareness for the fact that extreme weather conditions are predicted to become more frequent in a near future, which is likely to have a strong impact in crop production. For instance, it is estimated that by 2030, global water demand solely for agriculture may have increased by more than 30% as a consequence of foreseen climate changes (Foresight, 2011). Therefore, there is an urgency to understand how plants behave when facing adverse conditions, in order to reduce productivity losses in sub-optimal environments.

Plants have developed different strategies to cope with abiotic stress conditions. Upon environmental stimuli, changes in metabolism, signal transduction pathways and gene expression can be detected (Shinozaki & Yamaguchi-Shinozaki, 2007). Post-transcriptional regulatory mechanisms, as well as epigenetic and post-translational modifications, like ubiquitination and sumoylation, seem to play an important role in the regulation of gene expression in stress conditions (reviewed in Hirayama & Shinozaki, 2010).

In this chapter we will provide evidences of the involvement of small RNAs (sRNAs) in the regulation of gene expression as response to abiotic stress. The role of microRNAs (miRNAs) and other sRNAs under water deficit, high salinity, cold and oxidative stress, among others, as well as their relation to hormone signalling in plants, will be reviewed. Moreover, we will show that key enzymes involved in sRNA synthesis pathways also seem to be regulated upon environmental stimuli, affecting the expression of most sRNAs and consequently of several genes.

2. Plant small RNAs

Evidence for the existence of RNA-mediated silencing mechanisms in plants first appeared in the late 1990's, when short antisense RNA molecules were isolated from tomato plants where post-transcriptional gene silencing (PTGS) had been detected (Hamilton & Baulcombe, 1999). Since then, the knowledge on sRNAs has broadened and these molecules have been identified as important players in a wide variety of processes in plants. Back in 2002, when the first set of plant microRNAs (miRNAs) was cloned (Reinhart et al., 2002), there were 218 entries in the public miRNA database (miRBase; Griffiths-Jones, 2004) whereas nowadays more than 15 000 entries can be found.

The biosynthesis of sRNAs is triggered by the presence of double-stranded RNA (dsRNA) molecules that are processed into small RNA duplexes by RNase III Dicer-like (DCL) proteins (reviewed by Bartel, 2004). S-adenosyl methionine-dependent methyltransferase Hua Enhancer 1 (HEN1) methylates these molecules at the 3' end, in a sequence-independent manner, protecting them from uridylation and degradation (Park et al., 2002; Li et al., 2005; Yu et al., 2005). There are 4 loci in arabidopsis and rice (*Oryza sativa*) that encode for DCL proteins (Liu et al., 2005). Although some functional redundancy has been observed, each DCL seems to have a specific role in sRNA biosynthesis pathways (Bouche et al., 2006; Henderson et al., 2006).

Following methylation, sRNAs are incorporated into ARGONAUTE (AGO) proteins, in order to act upon targets in a sequence-specific manner (Llave et al., 2002b; Rhoades et al., 2002; Vazquez et al., 2004). Plants have multiple loci encoding for AGO proteins: 10 are known in arabidopsis and at least 18 in rice (Mallory & Vaucheret, 2010). AGO4, AGO6 and AGO9 are usually involved in silencing at the transcriptional level, by association with 24-nt sRNAs, whereas AGO1 and AGO7 participate in post-transcriptional gene silencing and are usually loaded with 21 and 22-nt sRNAs (reviewed in Mallory & Vaucheret, 2010). Recently, a role for AGO2 in responses to virus infection has also been described (Harvey et al., 2011). With few exceptions, the main characteristics that influence the loading of AGO proteins in plants are both the size of the sRNAs and the nature of the nucleotide at the 5' end (Mi et al., 2008; Mallory & Vaucheret, 2010).

Depending mainly on their origin, sRNAs can be roughly divided into two distinct classes: miRNAs (Fig.1A) and small interfering RNAs (siRNAs, Fig.1B and 1C) (Hamilton et al., 2002; Rajagopalan et al., 2006).

2.1 MicroRNAs

miRNAs are usually transcribed from intergenic regions (Llave et al., 2002a; Reinhart et al., 2002) by Polymerase II (Pol II) into 5' capped and 3' polyadenylated primary transcripts, in a way similar to protein-coding genes (Xie et al., 2005). These precursors (named pri-miRNAs) have regions of self-complementarity and can therefore fold back into stem-loop structures (Park et al., 2002; Meyers et al., 2008), yielding the dsRNA region, needed to trigger RNA silencing mechanisms. The hairpin precursors, called pre-miRNAs, are released from the pri-miRNAs by DCL1 in the nucleus. DCL1 then processes the pre-miRNAs into smaller duplexes bearing the mature miRNA and the complementary strand (miRNA*) (Reinhart et al., 2002; Henderson et al., 2006). In plants, both HYPONASTIC LEAVES1 (HYL1) and SERRATE (SE) proteins seem to interact with DCL1 and are also required for this process (Han et al., 2004; Lobbes et al., 2006).

At this point in the biosynthesis pathway, miRNAs can either stay in the nucleus or be transported to the cytoplasm, a process that seems to be dependent on HASTY protein, an exportin homolog (Bartel, 2004; Park et al., 2005). Either way, mature miRNAs are loaded into AGO proteins, and act upon highly or perfectly complementary targets by promoting their cleavage or repressing translation (Llave et al., 2002b; Rhoades et al., 2002; Chen, 2004). Most miRNAs are 21-nt long and are generally loaded into AGO1 that has both a small RNA-binding PAZ domain and a catalytic PIWI domain, and mediates miRNA-guided cleavage of complementary target transcripts (Vaucheret et al., 2004; Baumberger & Baulcombe, 2005). Opposite to what was known for animals, in 2005 Baumberger and Baulcombe reported that immunoprecipitation assays suggested that AGO1 alone could be enough for RNA induced silencing mechanisms in plants. Recently, however, Hsp90, a chaperone involved in protein folding, was co-purified with AGO1 (Iki et al., 2010), suggesting these two proteins could be part of a plant RNA silencing complex. In the same work, Hsp90 was found to be required for the loading of sRNAs into AGO1, presumably by inducing conformational changes in this protein (Iki et al., 2010).

There are also other miRNAs that are processed into 22-nt molecules, a process that is thought to be related to asymmetry in the hairpin precursor that is cleaved into a 22-nt mature miRNA and a 21-nt miRNA* (Cuperus et al., 2010). Recently, these molecules that are also loaded into AGO1 were found to trigger the production of secondary sRNAs from the 3' product of miRNA-directed cleavage (Chen et al., 2010; Cuperus et al., 2010).

2.2 Small interfering RNAs

A second and less studied group of sRNAs is known to exist in plants. These molecules, generally called siRNAs, are cleaved by DCL proteins from long dsRNA molecules that do not form hairpins (Carrington & Ambros, 2003). These are produced by either the transcription of inverted repeats or by the action of RNA-dependent RNA polymerases (RDRs) that convert single-stranded RNA transcripts into dsRNA (Dalmay et al., 2000).

The best studied siRNAs in plants fall into two distinct categories: RDR2-dependent (or heterochromatin associated) siRNAs (Fig.1B) and *trans*-acting siRNAs (ta-siRNAs, Fig.1C) (Ramachandran & Chen, 2008).

The first sub-group comprises siRNAs that are transcribed by RNA polymerase IV (Pol IV) and further processed to dsRNA by RDR2 (Herr et al., 2005). These sRNAs are mostly cleaved into 24-nt duplexes by DCL3 (Gasciolli et al., 2005) and loaded into AGO4 to guide target silencing, usually by promoting DNA or histone methylation (Zilberman et al., 2003; Herr et al., 2005).

The other sub-group, gathers the small RNAs that act upon transcripts that are very different from the genes that originated them (Vazquez et al., 2004). The biogenesis of these ta-siRNAs requires miRNA-AGO1-mediated cleavage of target transcripts, called *TAS* genes (Allen et al., 2005; Allen & Howell, 2010), and seems to be related to the production of 22-nt miRNAs (Chen et al., 2010; Cuperus et al., 2010). The RNA polymerase RDR6, together with SUPRESSOR OF GENE SILENCING 3 (SGS3) protein, processes the cleavage products (Vazquez et al., 2004; Montgomery et al., 2008b), yielding long dsRNA molecules that are subjected to DCL4-mediated phased cleavage of 21-nt dsRNA duplexes (Gasciolli et al., 2005; Yoshikawa et al., 2005). One of the strains from each of these duplexes is then loaded into an AGO protein, according to the criteria previously mentioned, in order to promote the silencing of target transcripts (Mi et al., 2008). Four *TAS* genes have been identified in arabidopsis: *TAS1*, *TAS2* (both targeted by miR173) (Montgomery et al., 2008b), *TAS3* (cleaved by miR390) (Allen et al., 2005) and *TAS4* (targeted by miR828) (Rajagopalan et al., 2006). Interestingly, *TAS3* seems to undergo a specific pathway that is conserved in several plant species and different from what is observed for other *TAS* genes. This pathway depends on miR390, a 21-nt long conserved miRNA, and requires AGO7 to be bound to the miRNA in order to cleave *TAS3* transcripts (Adenot et al., 2006; Montgomery et al., 2008a).

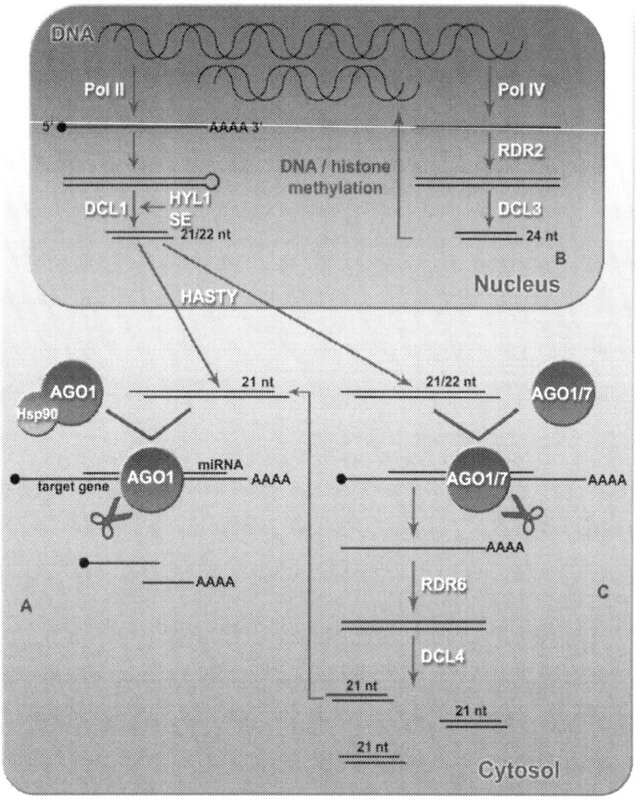

Fig. 1. Biosynthesis and action of the main classes of plant small RNAs. miRNAs are synthesized from stem loop precursors transcribed by Pol II, like protein-coding genes. Most 21-nt miRNAs are loaded into AGO1 and act upon target mRNAs in a sequence dependent manner, by either promoting their cleavage (**A**) or repressing translation (not represented). In the nucleus, siRNAs are synthesized from long dsRNA precursors, transcribed by Pol IV. These small RNAs are mainly loaded into AGO3 and are known to be involved in DNA and histone methylation (**B**). 22-nt long miRNAs are also loaded into AGO1. However, together with miR390, which is 21-nt long and is loaded into AGO7, these miRNAs trigger the biosynthesis of ta-siRNAs by DCL4. The newly formed siRNAs will be loaded into AGO1 and, like miRNAs, promote target cleavage (**C**).

3. Small RNAs and abiotic stress

A large portion of already identified miRNAs are present in several plant species (Reinhart et al., 2002; Bonnet et al., 2004; Wang et al., 2004). Most of these conserved miRNAs are known to have key roles in plant development, targeting a variety of transcription factors (TFs) related to pattern formation and organ differentiation (Llave et al., 2002b; Rhoades et al., 2002; Carrington & Ambros, 2003; Todesco et al., 2010). Lately, however, with high-throughput sequencing techniques becoming more and more accessible, several miRNAs (and other sRNAs) have been identified that seem to be family or species-specific and can be involved in many other processes in plants (Bonnet et al., 2004; Moxon et al., 2008; Szittya et al., 2008; Schreiber et al., 2011; Song et al., 2011). In fact, as the knowledge on small RNAs has broadened and new targets have been identified, it became apparent that, apart from their role in plant development, sRNAs are also involved in other processes in plants (Jones-Rhoades & Bartel, 2004; Sunkar & Zhu, 2004).

Since the early 2000's, several reports have associated sRNAs to abiotic stress responses, showing that post-transcriptional regulation of gene expression plays an important role in these phenomena (reviewed by Covarrubias & Reyes, 2010). The fact that plant miRNAs target transcripts in a sequence-specific manner, allowed Jones-Rhoades and Bartel (2004) to computationally predict targets for a conserved group of these riboregulators. In this work, the authors found that the expression of miR395 depends on sulfate availability, and that some other miRNAs can target transcripts that are potentially involved in responses to abiotic stress (Jones-Rhoades & Bartel, 2004). By then, Sunkar and Zhu (2004), analysed two week-old arabidopsis seedlings, and identified a variety of conserved miRNAs that were differentially expressed upon cold, dehydration, NaCl and ABA treatments. The authors found that, for instance, miR393, miR397b and miR402 were up-regulated in all the implemented conditions. On the other hand, miR389a, which was later found to be related to ta-siRNAs (Allen & Howell, 2010), was down-regulated under all stresses (Sunkar & Zhu, 2004).

Recently, Amor and co-workers (2009) analysed a group of arabidopsis long non-protein coding RNAs (npcRNAs), and found that 22, out of 76, showed differential expression upon water deficit, phosphate starvation and NaCl treatment. Interestingly, these npcRNAs are thought to regulate gene expression in either this long form, or by being precursors for siRNAs and miRNAs (Amor et al., 2009). Moreover, changes in DNA and histone methylation, processes that can be regulated by siRNAs, are known to occur in plants subjected to adverse conditions (Chinnusamy & Zhu, 2009). In fact, it was recently found that in arabidopsis stress-induced transgenerational global genome methylation, and consequent induced stress tolerance, is impaired in *dcl2 dcl3* mutants (Boyko et al., 2010). Together, these observations show that, besides miRNAs, other sRNAs can also be involved in responses to environmental stimuli in plants.

In the following sections we will summarize some of the most important results obtained in this field.

3.1 Osmotic stress: Water deprivation, salinity and abscisic acid

Water deprivation and salt stress alter the osmotic balance inside the cells, and considerable crosstalk is known to exist between responses to both conditions (Table 1) (Shinozaki & Yamaguchi-Shinozaki, 2007). Stomata closure, reduction in cell growth and photosynthesis and higher respiration rates are all general phenomena that can be observed in plants subjected to osmotic stress (Shinozaki & Yamaguchi-Shinozaki, 2007). Moreover, changes are also registered at the cellular and molecular levels, and several proteins and osmolytes are known to accumulate inside the cells to overcome osmotic changes (Shinozaki & Yamaguchi-Shinozaki, 2007).

In 2007, Zhao and co-workers analyzed the expression of a set of rice miRNAs from seedlings subjected to dehydration stress, induced by PEG treatment, and found that miR169g, miR393 and miR397b were up-regulated in these conditions. Recently, both miR169 and miR393 were also found to be induced by different concentrations of PEG in tobacco (Frazier et al., 2011). However, miR169 was shown to be down-regulated in adult rice plants after water withholding for several days (Zhou et al., 2010), suggesting that sRNA regulation depends on several factors, like the developmental stage and how a certain stress is imposed (Jia et al., 2009b). On the other hand, Zhao and co-workers (2007) showed that, among the miR169 family members, only miR169g seemed to be induced by PEG. In fact, it has been reported that different members of the same miRNA family are often regulated in different ways upon a certain stimuli (Lu et al., 2008). Therefore, it may also be possible, that in both works, different miRNAs were analysed that have distinct expression profiles under similar conditions.

Members of the miR169 family, have also been shown to be induced under high salinity conditions in both arabidopsis and rice (Zhao et al., 2009), and miR393 was also induced after treatment with ABA and NaCl in common bean (*Phaseolus vulgaris*; Arenas-Huertero et al., 2009). These observations suggest that these miRNAs can be involved in general responses to osmotic stress.

The elements of the miR169 family are known to target CCAAT-box Binding Factors (Rhoades et al., 2002), some of which have been shown to be involved in water deficit tolerance in maize (Nelson et al., 2007). However, these TFs were identified in maize as positive regulators of these responses, and their over-expression was shown to improve tolerance to water deprivation and increased yield (Nelson et al., 2007). Therefore, a reduction in the expression of miR169 should be expected in these conditions, and in most cases the opposite is observed. Further studies are necessary to unveil the relation between miR169 and CCAAT-box Binding Factors and their role in responses to osmotic stress.

MiR393, on the other hand, is known to target *TRANSPORT INHIBITOR RESPONSE1* (*TIR1*) transcripts, which are involved in auxin signalling pathways (Jones-Rhoades & Bartel, 2004). MiR393-mediated down-regulation of *TIR1* under stress conditions could result in reduced hypocotyl elongation and lateral root growth, as observed in *tir1* mutants (Ruegger et al., 1998). Strikingly, miR393 was also shown to be induced upon *Pseudomonas syringae* infection and repress bacterial growth, thus establishing a relation between responses to both biotic and abiotic stress conditions (Navarro et al., 2006).

In barrel medic (*Medicago truncatula)*, we have shown that miR408 is highly up-regulated in both shoots and roots of adult plants subjected to water deprivation for one week (Fig. 2) (Trindade et al., 2010). This up-regulation was also recently observed in barley (*Hordeum vulgare*) leaves after dehydration treatment for 4h and 8h (Kantar et al., 2010). MiR408 is known to target plantacyanin-like transcripts in several species, hence it has been related to the control of copper homeostasis in plants (Abdel-Ghany & Pilon, 2008; Trindade et al., 2010). Therefore, it seems that there may be a relation between the reduction of water uptake and copper deficiency, although further studies are needed to confirm this hypothesis.

Despite the intensive crosstalk observed in responses to water deficit and salt stress, not all miRNAs seem to be involved in pathways that are common to both. For instance, different concentrations of NaCl down-regulated the expression of miR159 and miR167, and induced the expression of miR172 and miR396 in tobacco (*Nicotiana tabacum*) plants (Frazier et al., 2011). On the other hand, these miRNAs were induced after treatment with different concentrations of PEG, in plants grown in the exact same conditions (Frazier et al., 2011). Interestingly, when compared to control plants, most miRNAs exhibited higher differences upon treatment with lower PEG concentrations, than with higher concentrations (Frazier et al., 2011). This observation suggests that miRNAs could be involved mainly in early responses to abiotic stress conditions, like reported for some TFs (Zhu, 2002).

Several water deficit and high salinity-responsive genes are also differentially expressed upon treatments with exogenous abscisic acid (ABA; Table 1) (Zhu, 2002; Nakashima et al., 2009). In fact, it is widely known that responses to osmotic stress comprise both ABA-dependent and ABA-independent pathways (Bohnert & Sheveleva, 1998; Nakashima et al., 2009). For instance, the expression of miR159 was shown to be induced by ABA and drought in arabidopsis, regulating abiotic stress responses in germinating seedlings (Reyes & Chua, 2007).

Moreover, dehydration-responsive elements (DREs) and abscisic acid responsive elements (ABREs), known to be present in the promoter region of water deficit and high salinity-inducible genes (Nakashima et al., 2009), have been identified in the region upstream of several rice *MIR169* genes (Zhao et al., 2007; Zhao et al., 2009). This observation suggests that, besides regulating a large number of TFs, miRNAs can themselves be regulated at the

miRNA	Regulation	Target transcripts	References
miR159	↑ water deficit ↑ ABA ↓ NaCl	*MYB101, MYB33*	Reyes & Chua (2007); Frazier et al. (2011)
miR165/166	↑ cold	*PHABULOSA*, homeobox genes	Zhou et al. (2008); Lv et al. (2010)
miR167	↑ water deficit ↓ NaCl		Frazier et al. (2011)
miR168	↑ water deficit ↑ NaCl	*ARGONAUTE 1*	Liu et al. (2008)
miR169	↓ water withholding ↑ PEG treatment ↑ NaCl	CCAAT-box Binding Factors	Zhao et al. (2007); Zhao et al. (2009); Zhou et al. (2010); Frazier et al. (2011)

miRNA	Regulation	Target transcripts	References
miR171	↑ water deficit ↑ NaCl	*SCARECROW*-like	Liu et al. (2008)
miR172	↑ water deficit ↑ NaCl ↑ cold	*APETALA2*	Frazier et al. (2011)
miR393	↑ water deficit ↑ ABA ↑ NaCl ↑ cold	*TRANSPORT INHIBITOR RESPONSE1*	Sunkar & Zhu (2004); Zhao et al. (2007); Arenas-Huertero et al. (2009); Frazier et al. (2011);
miR396	↑ water deficit ↑ NaCl ↑ cold	*GROWTH REGULATING FACTOR*	Liu et al. (2008); Frazier et al. (2011)
miR397b	↑ water deficit ↑ ABA ↑ NaCl ↑ cold	Laccases	Sunkar & Zhu (2004); Zhao et al. (2007);
miR398	↑ water deficit ↑ ABA ↑ NaCl	*COPPER SUPEROXIDE DISMUTASE, CYTHOCHROME c OXIDASE SUBUNIT V*	Jia et al. (2009a); Trindade et al. (2010)
miR402	↑ water deficit ↑ ABA ↑ NaCl ↑ cold	*ARGONAUTE 2*	Sunkar & Zhu (2004);
miR408	↑ water deficit	*PLANTACYANIN*	Trindade et al. (2010); Kantar et al. (2010)
miR2118	↑ water deficit ↑ ABA	unknown	Arenas-Huertero et al. (2009)

↑ indicates up-regulation; ↓ indicates down-regulation.

Table 1. Conserved miRNAs involved in responses to osmotic and cold stress.

transcriptional level. Interestingly, the over-expression of miR169c in tomato was reported to improve drought tolerance through a reduction in stomata aperture, a process that is controlled by ABA, and consequent reduction of transpiration and leaf water losses (Chinnusamy et al., 2008; Zhang et al., 2011b).

Non-conserved miRNAs have also been reported to be involved in responses to osmotic stress (Table 2). In common bean, the legume-specific miR1514a and miR2119, and the conserved miR2118 were up-regulated upon water deficit and ABA treatments (Arenas-Huertero et al., 2009). Interestingly, miR1514a was predicted to target transcripts for Mn-superoxide dismutase (SOD) and a cysteine protease, while miR2119 was predicted to cleave Aldehyde Dehydrogenase 1 and zinc finger protein mRNAs, among others, (Arenas-Huertero et al., 2009), supporting the hypothesis of their involvement in responses to different types of stresses. Recently, the grass-specific miR1432 and miR1867 (predicted to target phenylalanine tRNA synthetase and a protein from the DUF1242 superfamily, respectively) were also shown to be induced by dehydration in roots and shoots of wild emmer wheat (*Triticum turgidum* ssp. *dicoccoides*), respectively (Kantar et al., 2011).

Besides miRNAs, other sRNAs have been shown to have a role in osmotic stress responses. For instance, Borsani and co-workers (2005) identified in arabidopsis a new group of plant siRNAs, known as natural antisense siRNAs (nat-siRNAs), that is formed from overlapping genes (Borsani et al., 2005). Their biogenesis, requires DCL2, RDR6, SGS3 and Pol IV to generate a 24-nt nat-siRNA from D1-pyroline-5-carboxylate dehydrogenase (P5CDH) cleavage, that will set the phase for subsequent cleavage by DCL1 of 21-nt nat-siRNAs from the same transcript (Borsani et al., 2005). In the same work, the authors found that the expression of SRO5 (a transcript of unknown function) is required for the production of nat-siRNAs and that it is induced upon salt treatment. This mechanism represents a unique regulation of salt stress responses, because cleavage of P5CDH leads to accumulation of proline (improving salt tolerance) and reactive oxygen species (ROS), an effect that is counteracted by SRO5 (Borsani et al., 2005). Moreover, the authors mention that they have identified a variety of other siRNAs synthesized from overlapping genes that were only identified under certain abiotic stress conditions, implying that these molecules could have a more generalized role in the responses to environmental constrains.

3.2 Cold stress

Cold stress generally leads to a reduction in metabolic rates and water availability, due to chilling-induced inhibition of absorption (Chinnusamy et al., 2007). Therefore, it is not surprising that crosstalk between water deficit, salt stress and cold has been reported (Table 1) (Bohnert & Sheveleva, 1998).

As for other stress conditions, post-transcriptional regulation of gene expression also plays an important role in response to low temperatures (Chinnusamy et al., 2007), and alternative-splicing in particular is known to have an important role under cold stress conditions (Iida et al., 2004).

In 2008, miR168, miR171 and miR396 were shown to be induced by drought, cold and salt stress in arabidopsis seedlings (Liu et al., 2008), suggesting that miRNAs can be involved in the pathways common to all these stimuli. Following this, Zhou and co-workers (2008) identified four arabidopsis MIRNA genes that are inducible by cold stress, using a computational approach based on transcriptome and promoter analysis data, coupled with experimental validation. Northern blot analysis revealed that miR165/miR166, miR169 and miR172 were also up-regulated upon cold treatment (Zhou et al., 2008). Some elements of the miR166 family were also found to be up-regulated in similar conditions in rice, while miR168, miR169 and miR171 showed opposite expression profiles (Lv et al., 2010). These observations show once again that miRNA expression upon abiotic stress can be complex and depend on a variety of parameters. Interestingly, most of these conserved cold-regulated miRNAs are known to target TFs with known roles in plant development (Jones-Rhoades & Bartel, 2004), suggesting that miRNA-mediated responses to this kind of stress could be mainly at the structural level.

miRNA	Regulation	Predicted target transcript	Other species	References
pvu-miR1514a	↑ water deficit ↑ ABA	*MnSOD*, *CYSTEINE PROTEASE*	soybean	Arenas-Huertero et al. (2009)
pvu-miR2119	↑ water deficit ↑ ABA	*ALDEHYDE DEHYDROGENASE1*, Zinc finger domain	soybean; barrel medic	Arenas-Huertero et al. (2009)
ttu-miR1432	↑ water deficit ↑ ABA	*PHENYLALANINE tRNA SYNTHETASE*	rice, maize, sorghum	Kantar et al. (2011)
ttu-miR1867	↑ water deficit ↑ ABA	*DUF1242*	rice	Kantar et al. (2011)
osa-miR1320	↓ cold stress	Clathrin assembly, *FUCOSYLTRANSFERASE7*, *REMORIN*	------	Lv et al. (2010)
osa-miR1435	↑ late cold stress	B3 DNA Binding domain, *UDP GLUCOSYLTRANSFERASE*	sorghum	Lv et al. (2010)
osa-miR1876	↓ cold stress	*HISTONE DEACETYLASE6*	------	Lv et al. (2010)
osa-miR1884	↓ cold stress	*ATPase*, *URICASE*, ABA stress-ripening, TMV response-related	------	Lv et al. (2010)
osa-miR1850	↑ oxidative stress	unknown	------	Li et al. (2010)
ptc-miR475a,b	↓ early cold stress	Pentatricopeptide repeat	------	Lu et al. (2008)
ptc-miR476a	↓ cold stress	Pentatricopeptide repeat	------	Lu et al. (2008)

pvu- common bean; *ttu*- wild emmer wheat; *osa*- rice; *ptc*- black cottonwood. ↑ indicates up-regulation; ↓ indicates down-regulation.

Table 2. Non-conserved miRNAs involved in responses to abiotic stress in plants.

A deeper look into the promoter region of some cold-inducible MIRNA genes, revealed an abundance of motifs that are known to be involved in responses to cold, namely W-box (TTGAC), ABRE-core (ACGTGG/TC) and LTRE-core (A/GCCGAC) (Zhou et al., 2008). This observation supports once again the idea that stress-responsive miRNAs can be regulated at the transcriptional level.

Like previously reported for water deficit and salt stress, non-conserved miRNAs are also involved in cold stress responses in plants (Table 2). For instance, a set of rice miRNAs, namely miR1435, miR1876, miR1320, and miR1884, not previously detected in arabidopsis, were recently found to be differentially expressed upon cold treatment (Lv et al., 2010). Moreover, in black cottonwood (*Populus trichocarpa*) the species-specific miR475a,b and miR476a, both predicted to target transcripts coding for pentatricopeptide repeat-containing proteins (PPRs), were shown to be down-regulated in similar conditions (Lu et al., 2008). Interestingly, PPRs were identified in arabidopsis as RNA-binding proteins that seem to be involved in post-transcriptional regulation in mitochondria and chloroplasts (Lurin et al., 2004), suggesting sRNAs can also regulate organelle metabolism.

3.3 Flooding and hypoxia

Flooding can also severely compromise plant growth and productivity, since it is associated to a reduction of oxygen availability to the cells (Licausi, 2010).

The expression of some maize miRNAs was found to be induced after submergence (Zhang et al., 2008b). Strikingly, it seems that different miRNAs are regulated at different time points and in most cases this induction is transient. For instance, miR166, miR167 and miR171 were induced in the early stages, while the expression of miR159 was reduced by this time and further induced after 24h of submergence (Zhang et al., 2008b). Most of these miRNAs were shown to have *cis*-acting elements in their promoter regions that are associated with anaerobic stress, namely AREs (anaerobic responsive elements) and GC-motifs (Zhang et al., 2008b). A member of the miR167 family was also up-regulated in submerged arabidopsis plants, together with elements from the miR156, miR157, miR158, miR161, miR169, miR172, miR391, miR395 and miR775 families (Moldovan et al., 2010). With the exception of miR172, all the other miRNAs exhibited similar expression patterns upon treatment for mitochondrial inhibition, suggesting they are actually hypoxia-inducible miRNAs (Moldovan et al., 2010).

Like previously observed for osmotic stress, other sRNAs can be involved in responses to oxygen deprivation. Recently, Moldovan and co-workers (2010) have shown that several tas-siRNAs derived from the *TAS1*, *TAS2* and *TAS3* loci are also up-regulated in arabidopsis under hypoxia conditions.

3.4 UV radiation and light

According to some recent reports, miRNAs can also be involved in responses to light and radiation stimuli. Zhou and co-workers (2007) gathered data from promoter element analysis and the expression of protein-coding genes predicted to be targeted by miRNAs, and identified *in silico* a group of potentially UV-B inducible arabidopsis miRNAs. Later, using a home-made array, Jia and co-workers (2009) identified miRNAs in black cottonwood that were differentially expressed upon 2h of treatment with the same radiation. As previously predicted for arabidopsis (Zhou et al., 2007), miR156, miR160, miR165/166, miR167 and miR398 were induced in these conditions, whereas the same pattern was not observed for miR159 and miR393 in both species (Jia et al., 2009a). Moreover, miR169, miR395 and miR472 were down-regulated by UV-B radiation, while the opposite was observed for miR168, miR398 and miR408 (Jia et al., 2009a). These results show that, although several miRNAs are conserved among plant species, their regulation upon certain stimuli can actually be different.

Also, in both arabidopsis and common aspen (*Populus tremula*), the majority of the selected miRNAs exhibited stress and light-inducible *cis*-acting elements, like GT-1 site (GGTTAA) and I-boxes (GATAAGA), upstream of their coding genes (Zhou et al., 2007; Jia et al., 2009a). Besides being regulated by UV radiation, some miRNAs also seem to be regulated by the photoperiod. For instance, miR167, miR168, miR171 and miR393 were shown to accumulate during the light period in arabidopsis (Sire et al., 2009). This expression pattern was absent in plants grown under continuous light, showing that these miRNAs were actually induced by light and not by the circadian clock (Sire et al., 2009).

3.5 Oxidative stress

In plants, oxidative stress usually emerges as a secondary effect of other stress conditions and is highly responsible for losses in crop productivity (Bartels & Sunkar, 2005; Sunkar et al., 2006). To overcome the higher production of ROS upon stress stimuli, plants have developed enzymatic and non-enzymatic detoxification strategies (Bohnert & Sheveleva, 1998), and during the last decade, several works have reported the involvement of sRNAs in these mechanisms.

In an attempt to mimic oxidative stress, Li and co-workers (2010) treated rice seedlings with different concentrations of H_2O_2, and found that the conserved miR169, miR397 and miR827 were up-regulated in these conditions, while miR319a and miR528 were down-regulated. Moreover, the authors identified a rice-specific miRNA, miR1850, that was induced by the applied stress (Li et al., 2010), providing extra evidence to the involvement of miRNAs in both general and species-specific mechanisms of abiotic stress response.

The expression of miR398 has been shown to be repressed upon oxidative stress (Sunkar et al., 2006; Jagadeeswaran et al., 2009), and, as expected, it is down-regulated under several adverse conditions, like salt stress, ozone fumigation and even infection with *Pseudomonas syringae* (Jagadeeswaran et al., 2009). Besides, targeting cytochrome c oxidase (Jones-Rhoades & Bartel, 2004; Trindade et al., 2010), this miRNA is known to target both cytosolic and plastidic Cu/Zn superoxide dismutases (CSD1 and CSD2, respectively) (Jones-Rhoades & Bartel, 2004; Sunkar et al., 2006), that are responsible for the conversion of O_2^- to H_2O_2 (Kliebenstein et al., 1998). Therefore the reduction in the levels of miR398, leads to an increase in CSD transcripts, thus protecting the cells against oxidative damage (Sunkar et al., 2006).

However, the expression of miR398 is not always decreased under abiotic stress conditions. In adult barrel medic plants subjected to water deficit for one week, this miRNA was highly up-regulated in both shoots and roots (Fig. 2) (Trindade et al., 2010), a result that was also recently observed in both PEG-treated tobacco (Frazier et al., 2011) and dehydrated wild emmer wheat (Kantar et al., 2011). A similar expression profile had previously been observed after sucrose addition in arabidopsis (Dugas & Bartel, 2008) and in common aspen subjected to short time treatments with ABA, NaCl and UV-B radiation (Jia et al., 2009a). Interestingly, it was recently shown that the expression of this miRNA not only depends on the type of stress imposed, but it also seems to oscillate with the degree of stress (Frazier et al., 2011).

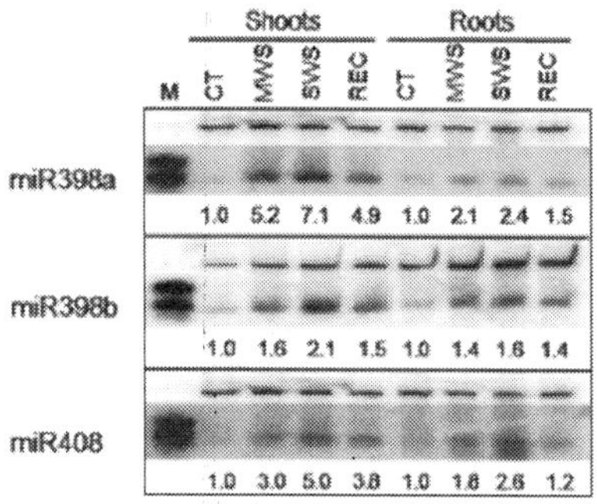

Fig. 2. Northern blot analysis of the expression of miR398a, miR398b and miR408 in both shoots and roots of 8 week-old *Medicago truncatula* plants subjected to different degrees of water deficit: CT- control; MWS - moderate water deficit; SWS- severe water deficit; REC- recovery; M-microRNA size marker from New England Biolabs (the darker bands correspond to 24 and 21 nt). Small nuclear RNA U6 was used as a loading control (Trindade et al., 2010).

Besides having a role in response to oxidative stress, miR398, along with other miRNAs, has been associated to the maintenance of copper homeostasis in plants (Abdel-Ghany & Pilon, 2008). Overall these results seem to suggest that this miRNA can be regulated in opposite ways by at least two distinct pathways in order to protect the cells from damage. One pathway seems to be dependent on the copper status of the plant, while the other seems to be related to oxidative stress. Some authors have hypothesized that the induction of miR398, and consequent degradation of transcripts of copper proteins, could be a way of saving copper for vital proteins like plastocyanin, known to be involved in the electron transport chain of the photosynthetic pathway (Abdel-Ghany & Pilon, 2008). According to Dugas and Bartel (2008), the induction of this miRNA, and further reduction of CSD transcripts, under these conditions could be caused by a reduction in ROS production as a consequence of some photosynthetic impairment. It could also be that different MIR398 genes are being regulated in different ways, like previously proposed for MIR159 (Reyes & Chua, 2007). A deeper analysis on the expression pattern and the promoter regions of each miR398 family member could shed some light on this matter.

3.6 Hormone signalling

Plant hormones are also known to play a role in abiotic stress responses. As mentioned above, many genes that have been shown to be induced by drought, salt stress and cold are also regulated upon exogenous ABA treatments (Zhu, 2002).

Besides ABA, other phytohormones like auxins, cytokinins, gibberellins, among others, can also be directly or indirectly involved in responses to environmental stimuli. Over the last years, several reports have shown that some sRNAs can be involved in these signalling pathways, either by being regulated by these hormones, or by targeting hormone-responsive transcripts (Guo et al., 2005; Reyes & Chua, 2007). In 2005, Sorin and co-workers reported that *ago1* mutants were practically unable to form adventitious roots and that this abnormal development was related to impairment in the auxin metabolism, suggesting an involvement of miRNAs in these processes (Sorin et al., 2005). Previously, it had also been shown that *hyl1* mutants exhibited altered sensitivity to different hormone treatments (Lu & Fedoroff, 2000). Additionally, some abiotic stress-inducible miRNAs exhibit hormone-responsive *cis*-acting elements, like ABRE, gibberellic acid (GA3)-responsive element (GARE), methyl jasmonate-responsive element (CGTCA motif) and salicylic acid-responsive element in their promoter regions (Liu et al., 2009; Lv et al., 2010). Taken together, these data strongly suggest that there is an involvement of plant hormones in sRNA-mediated responses to abiotic stress in plants.

Auxins, for instance, were recently shown to participate in responses to salt and oxidative stresses in arabidopsis through regulation of the redox metabolism, in which TIR1 and AFB2 auxin receptors, known to be targeted by miR393 (Jones-Rhoades & Bartel, 2004), seem to be involved (Iglesias et al., 2010). Also, some light and cold-inducible miRNAs are known to be involved in the regulation of auxin signalling pathways (Zhou et al., 2007; Zhou et al., 2008). In rice, miR167 and miR169 were down-regulated by ABA and miR168 and miR169 were down and up-regulated, respectively, after auxin treatment (Liu et al., 2009). miR167 is known to target ARF6 and ARF8, which are regulators of female and male reproduction (Wu et al., 2006). These observations led to the hypothesis that ABA-induced down-regulation of this miRNA could be related to early reproduction, a phenomenon that is generally observed in plants subjected to stress conditions (Liu & Chen, 2009). Moreover, there seems to be a relation between auxin and ABA signalling pathways, in which miR167 is apparently involved. Also, miR319 was shown to be up-regulated by both GA_3 and ABA, and down-regulated by cytokinins (Liu et al., 2009), suggesting that miRNAs can be involved in processes of crosstalk between different phytohormones and between those and abiotic stress conditions.

Upon certain stress stimuli, plants can respond by altering their growth and development, in order to avoid or reduce stress-inducible damage. Auxins are known to be involved in the regulation of plant development, and could play an important role in these mechanisms (reviewed by Vanneste & Friml, 2009). For instance, the previously mentioned up-regulation of miR393 under stress conditions, and consequent regulation of auxin signalling pathways, has been related to a reduction of plant growth observed in these situations (Sunkar & Zhu, 2004).

Abnormal root growth and development has also been observed in several plant species subjected to osmotic stress (Bartels & Sunkar, 2005). Recently, it was found that lateral root growth in arabidopsis depends on the tight feedback regulation between miR390, involved in the production of *TAS3*-derived ta-siRNAs, and an auxin responsive factor (ARF4), known to be targeted by these siRNAs (Allen et al., 2005; Marin et al., 2010). These results show that siRNAs can be involved in hormone signalling pathways during plant development as well, suggesting they could also regulate hormone-mediated responses to abiotic stress.

As mentioned earlier, miR393 was found to be involved in responses to bacterial infection in arabidopsis by regulating a group of auxin receptors (Navarro et al., 2006). Recently, other miRNAs, known to target hormone-related transcripts, were shown to be differentially expressed upon infection with different bacterial strains (Zhang et al., 2011a). These observations show that the relation between sRNAs and phytohormones can go beyond plant development and abiotic stress tolerance.

4. Regulation of small RNA pathways by abiotic stress

RNA-mediated regulation of gene expression as response to abiotic stress conditions depends, not only on the expression of sRNAs and their targets, but it also seems to rely on the regulation of the proteins involved in their biosynthesis. In 2003, Szittya and co-workers have shown that virus and transgene-induced RNA silencing was dependent on temperature. The authors found that the accumulation of siRNAs, but not of miRNAs, was lower under cold stress and speculated that the activity of some DCL enzymes could be affected in these conditions (Szittya et al., 2003). Recently, in our laboratory we have shown that several *AGO* and *DCL* genes are differentially expressed in barrel medic adult plants subjected to water deficit (Capitão et al., 2011).

One striking feature of miRNA-mediated gene silencing in plants is that it is subjected to feedback self-regulation. Not only is *DCL1*, which is related to the biogenesis of most plant miRNAs, targeted by miR162 (Xie et al., 2003), but *AGO1* is also known to stabilize miR168 and, simultaneously, be targeted by this miRNA (Rhoades et al., 2002; Vaucheret et al., 2004; Vaucheret et al., 2006). Strikingly, both miR162 and miR168 have been reported to be differentially expressed in abiotic stress conditions, like hypoxia (Zhang et al., 2008b), UV-B radiation (Jia et al., 2009a), water deficit (Zhou et al., 2010) and salt stress (Ding et al., 2009). This observation supports the idea that the expression of sRNAs, and therefore of target genes, can be regulated indirectly through the regulation of key enzymes of the biosynthesis pathway.

Interestingly, Moldovan and co-workers (2010), have reported a reduction in 24-nt long sRNAs and an increase in 21-nt long species in arabidopsis plants subjected to hypoxia, when compared to controlled conditions. This observation suggests that other enzymes, besides DCL1 and AGO1 can be regulated by environmental conditions. In fact, *hen1*, *hyl1*, *hsty* and *se* mutants are characterized by hypersensitivity to ABA, leading to a lower tolerance to osmotic stress (Lu & Fedoroff, 2000; Zhang et al., 2008a). Also, the expression of RDR6, known to be involved in the biosynthesis of ta-siRNAs and nat-siRNAs, was recently shown to be induced by ABA (Yang et al., 2008).

Taken together, these results support the notion that altering sRNA biosynthesis pathways can affect the response to environmental conditions in plants. Whether the regulation of these proteins can also be mediated by other sRNAs, like it was found for DCL1 and AGO1, is yet to be investigated.

5. Concluding remarks

Despite the large amount of work published over the last years on sRNA-mediated regulation of gene expression under abiotic stress conditions, a lot of information is still missing on the functionality of most sRNAs. So far, this type of studies has been undertaken for just a few conserved miRNAs, several of them related to nutrient status in plants (reviewed by Sunkar et al., 2007).

In fact, the diversity of sRNAs, the observed redundancy in some biosynthesis pathways and their feedback regulation, unveil a higher complexity than previously thought when RNA silencing mechanisms were first identified in plants. Also, as reviewed in this manuscript, the expression of sRNAs seems to depend on a wide variety of factors, from plant species, to developmental stage, to growth conditions, among others, making it hard to fully understand their role under abiotic stress conditions.

Additionally, some recently identified non-conserved miRNAs seem to have no specific targets, or their experimental validation has proven difficult. In fact, computational tools for target prediction have been reported to yield a high percentage of false positives for these miRNAs (Moxon et al., 2008), which, at some extent, can account for that limitation. Nevertheless, a few of them were already shown to be differentially expressed in abiotic stress conditions, and are probably involved in more specific and less studied response mechanisms. In 2008, Addo-Quaye and co-workers have reported the sequencing of arabidopsis degradome, which is a technique that allows the high-throughput sequencing of mRNAs without the 5'-end cap, including degraded transcripts and targets of sRNA-mediated cleavage. Using this technique that does not rely on *in silico* predictions, the authors were able to confirm several previously identified targets, identify new targets for some miRNAs and ta-siRNAs, and also to identify new *TAS* transcripts (Addo-Quaye et al., 2008). The sequencing of the degradome is now being adopted for other plant species (Addo-Quaye et al., 2009; Pantaleo et al., 2010) and is expected to retrieve interesting new information on the function and regulation of sRNAs in plants subjected to different conditions.

Phytohormones and stress-inducible TFs seem to play an important role in regulating the expression of miRNAs under abiotic stress conditions, in a similar way to what happens with protein-coding genes. Supporting this idea is the common observation that most miRNAs, besides targeting several TFs, also have stress-related *cis*-acting elements in their promoter regions. These data unveil a complex interaction between transcriptional and post-transcriptional regulators of gene expression that seems to be crucial to fine-tune responses to different environmental stimuli.

Additionally, when it comes to abiotic stress tolerance, most studies on sRNA-mediated regulation of gene expression have been performed inside laboratories, in highly controlled systems. However, in the wild, plants are hardly found in these conditions, and are usually subjected to different types of stresses at the same time. Interestingly, according to several studies, the responses to a situation of multiple simultaneous stresses can actually be different from the responses observed when two or three types of stress are induced separately (reviewed by Mittler, 2006). Therefore, some of the results already verified might even not be observed in the field, and care must be taken when analysing the function of sRNAs in these processes.

Over the last years, it has become clear that sRNAs play an important role in abiotic stress responses in plants. Moreover, they have also been related to pathogen infection and symbiosis (Navarro et al., 2006; Branscheid et al., 2010), revealing their importance in biotic stress conditions as well. Nevertheless, although a lot has been said on the biosynthesis and function of sRNAs, at the moment it seems to be just the tip of the iceberg and exciting new developments are expected in the near future.

6. Acknowledgments

The authors would like to thank José Ricardo Salvado for designing Figure 1. Inês Trindade was funded by a PhD fellowship from Fundação para a Ciência e Tecnologia (SFRH/BD/40738/2007).

7. References

Abdel-Ghany, S.E. & Pilon, M. (2008). MicroRNA-mediated systemic down-regulation of copper protein expression in response to low copper availability in Arabidopsis. *Journal of Biological Chemistry* 283: 15932-15945.

Addo-Quaye, C., Eshoo, T.W., Bartel, D.P. & Axtell, M.J. (2008). Endogenous siRNA and miRNA targets identified by sequencing of the Arabidopsis degradome. *Current Biology* 18: 758-762.

Addo-Quaye, C., Snyder, J.A., Park, Y.B., Li, Y.F., Sunkar, R. & Axtell, M.J. (2009). Sliced microRNA targets and precise loop-first processing of MIR319 hairpins revealed by analysis of the Physcomitrella patens degradome. *RNA* 15: 2112-2121.

Adenot, X., Elmayan, T., Lauressergues, D., Boutet, S., Bouche, N., Gasciolli, V. & Vaucheret, H. (2006). DRB4-dependent TAS3 trans-acting siRNAs control leaf morphology through AGO7. *Current Biology* 16: 927-932.

Allen, E. & Howell, M.D. (2010). miRNAs in the biogenesis of trans-acting siRNAs in higher plants. *Seminars in Cell & Developmental Biology* 21: 798-804.

Allen, E., Xie, Z., Gustafson, A.M. & Carrington, J.C. (2005). microRNA-directed phasing during trans-acting siRNA biogenesis in plants. *Cell* 121: 207-221.

Amor, B.B., Wirth, S., Merchan, F., Laporte, P., d'Aubenton-Carafa, Y., Hirsch, J., Maizel, A., Mallory, A., Lucas, A., Deragon, J.M., Vaucheret, H., Thermes, C. & Crespi, M. (2009). Novel long non-protein coding RNAs involved in Arabidopsis differentiation and stress responses. *Genome Research* 19: 57-69.

Arenas-Huertero, C., Perez, B., Rabanal, F., Blanco-Melo, D., De la Rosa, C., Estrada-Navarrete, G., Sanchez, F., Covarrubias, A.A. & Reyes, J.L. (2009). Conserved and novel miRNAs in the legume Phaseolus vulgaris in response to stress. *Plant Molecular Biology* 70: 385-401.

Bartel, D.P. (2004). MicroRNAs: genomics, biogenesis, mechanism, and function. *Cell* 116: 281-297.

Bartels, D. & Sunkar, R. (2005). Drought and Salt Tolerance in Plants. *Critical Reviews in Plant Sciences* 24: 23–58.

Baumberger, N. & Baulcombe, D.C. (2005). Arabidopsis ARGONAUTE1 is an RNA Slicer that selectively recruits microRNAs and short interfering RNAs. *Proceedings of the National Academy of Sciences of the United States of America* 102: 11928-11933.

Bohnert, H.J. & Sheveleva, E. (1998). Plant stress adaptations--making metabolism move. *Current Opinion in plant Biology* 1: 267-274.

Bonnet, E., Wuyts, J., Rouze, P. & Van de Peer, Y. (2004). Detection of 91 potential conserved plant microRNAs in Arabidopsis thaliana and Oryza sativa identifies important target genes. *Proceedings of the National Academy of Sciences of the United States of America* 101: 11511-11516.

Borsani, O., Zhu, J., Verslues, P.E., Sunkar, R. & Zhu, J.K. (2005). Endogenous siRNAs derived from a pair of natural cis-antisense transcripts regulate salt tolerance in Arabidopsis. *Cell* 123: 1279-1291.

Bouche, N., Lauressergues, D., Gasciolli, V. & Vaucheret, H. (2006). An antagonistic function for Arabidopsis DCL2 in development and a new function for DCL4 in generating viral siRNAs. *EMBO Journal* 25: 3347-3356.

Boyko, A., Blevins, T., Yao, Y., Golubov, A., Bilichak, A., Ilnytskyy, Y., Hollunder, J., Meins, F., Jr. & Kovalchuk, I. (2010). Transgenerational adaptation of Arabidopsis to stress requires DNA methylation and the function of Dicer-like proteins. *PLoS One* 5: e9514.

Branscheid, A., Sieh, D., Pant, B.D., May, P., Devers, E.A., Elkrog, A., Schauser, L., Scheible, W.R. & Krajinski, F. (2010). Expression pattern suggests a role of MiR399 in the regulation of the cellular response to local Pi increase during arbuscular mycorrhizal symbiosis. *Molecular Plant-Microbe Interactions* 23: 915-926.

Capitão, C., Paiva, J.A.P., Santos, D.M. & Fevereiro, P. (2011). Small RNA pathways are modulated by water deficit in Medicago truncatula. *BMC Plant Biology*. 11:79

Carrington, J.C. & Ambros, V. (2003). Role of microRNAs in plant and animal development. *Science* 301: 336-338.

Chen, H.M., Chen, L.T., Patel, K., Li, Y.H., Baulcombe, D.C. & Wu, S.H. (2010). 22-Nucleotide RNAs trigger secondary siRNA biogenesis in plants. *Proceedings of the National Academy of Sciences of the United States of America* 107: 15269-15274.

Chen, X. (2004). A microRNA as a translational repressor of APETALA2 in Arabidopsis flower development. *Science* 303: 2022-2025.

Chinnusamy, V., Gong, Z. & Zhu, J.K. (2008). Abscisic acid-mediated epigenetic processes in plant development and stress responses. *Journal of Integrative Plant Biology* 50: 1187-1195.

Chinnusamy, V., Zhu, J. & Zhu, J.K. (2007). Cold stress regulation of gene expression in plants. *Trends in Plant Science* 12: 444-451.

Chinnusamy, V. & Zhu, J.K. (2009). Epigenetic regulation of stress responses in plants. *Current Opinion in plant Biology* 12: 133-139.

Covarrubias, A.A. & Reyes, J.L. (2010). Post-transcriptional gene regulation of salinity and drought responses by plant microRNAs. *Plant, Cell & Environment* 33: 481-489.

Cuperus, J.T., Carbonell, A., Fahlgren, N., Garcia-Ruiz, H., Burke, R.T., Takeda, A., Sullivan, C.M., Gilbert, S.D., Montgomery, T.A. & Carrington, J.C. (2010). Unique functionality of 22-nt miRNAs in triggering RDR6-dependent siRNA biogenesis from target transcripts in Arabidopsis. *Nature Structural & Molecular Biology* 17: 997-1003.

Dalmay, T., Hamilton, A., Rudd, S., Angell, S. & Baulcombe, D.C. (2000). An RNA-dependent RNA polymerase gene in Arabidopsis is required for posttranscriptional gene silencing mediated by a transgene but not by a virus. *Cell* 101: 543-553.

Ding, D., Zhang, L., Wang, H., Liu, Z., Zhang, Z. & Zheng, Y. (2009). Differential expression of miRNAs in response to salt stress in maize roots. *Annals of Botany* 103: 29-38.

Dugas, D.V. & Bartel, B. (2008). Sucrose induction of Arabidopsis miR398 represses two Cu/Zn superoxide dismutases. *Plant Molecular Biology* 67: 403-417.

Foresight (2011). The Future of Food and Farming. Final Project Report. London: The Government Office for Science.

Frazier, T.P., Sun, G., Burklew, C.E. & Zhang, B. (2011). Salt and Drought Stresses Induce the Aberrant Expression of microRNA Genes in Tobacco. *Molecular Biotechnology*.

Gasciolli, V., Mallory, A.C., Bartel, D.P. & Vaucheret, H. (2005). Partially redundant functions of Arabidopsis DICER-like enzymes and a role for DCL4 in producing trans-acting siRNAs. *Current Biology* 15: 1494-1500.

Griffiths-Jones, S. (2004). The microRNA Registry. *Nucleic Acids Res* 32: D109-111.

Guo, H.S., Xie, Q., Fei, J.F. & Chua, N.H. (2005). MicroRNA directs mRNA cleavage of the transcription factor NAC1 to downregulate auxin signals for arabidopsis lateral root development. *Plant Cell* 17: 1376-1386.

Hamilton, A., Voinnet, O., Chappell, L. & Baulcombe, D. (2002). Two classes of short interfering RNA in RNA silencing. *EMBO Journal* 21: 4671-4679.

Hamilton, A.J. & Baulcombe, D.C. (1999). A species of small antisense RNA in posttranscriptional gene silencing in plants. *Science* 286: 950-952.

Han, M.H., Goud, S., Song, L. & Fedoroff, N. (2004). The Arabidopsis double-stranded RNA-binding protein HYL1 plays a role in microRNA-mediated gene regulation. *Proceedings of the National Academy of Sciences of the United States of America* 101: 1093-1098.

Harvey, J.J., Lewsey, M.G., Patel, K., Westwood, J., Heimstadt, S., Carr, J.P. & Baulcombe, D.C. (2011). An Antiviral Defense Role of AGO2 in Plants. *PLoS One* 6: e14639.

Henderson, I.R., Zhang, X., Lu, C., Johnson, L., Meyers, B.C., Green, P.J. & Jacobsen, S.E. (2006). Dissecting Arabidopsis thaliana DICER function in small RNA processing, gene silencing and DNA methylation patterning. *Nature Genetics* 38: 721-725.

Herr, A.J., Jensen, M.B., Dalmay, T. & Baulcombe, D.C. (2005). RNA polymerase IV directs silencing of endogenous DNA. *Science* 308: 118-120.

Hirayama, T. & Shinozaki, K. (2010). Research on plant abiotic stress responses in the post-genome era: past, present and future. *Plant Journal* 61: 1041-1052.

Iglesias, M.J., Terrile, M.C., Bartoli, C.G., D'Ippolito, S. & Casalongue, C.A. (2010). Auxin signaling participates in the adaptative response against oxidative stress and salinity by interacting with redox metabolism in Arabidopsis. *Plant Molecular Biology* 74: 215-222.

Iida, K., Seki, M., Sakurai, T., Satou, M., Akiyama, K., Toyoda, T., Konagaya, A. & Shinozaki, K. (2004). Genome-wide analysis of alternative pre-mRNA splicing in Arabidopsis thaliana based on full-length cDNA sequences. *Nucleic Acids Research* 32: 5096-5103.

Iki, T., Yoshikawa, M., Nishikiori, M., Jaudal, M.C., Matsumoto-Yokoyama, E., Mitsuhara, I., Meshi, T. & Ishikawa, M. (2010). In vitro assembly of plant RNA-induced silencing complexes facilitated by molecular chaperone HSP90. *Molecular Cell* 39: 282-291.

Jagadeeswaran, G., Saini, A. & Sunkar, R. (2009). Biotic and abiotic stress down-regulate miR398 expression in Arabidopsis. *Planta* 229: 1009-1014.

Jia, X., Ren, L., Chen, Q.J., Li, R. & Tang, G. (2009a). UV-B-responsive microRNAs in Populus tremula. *Journal of Plant Physiology* 166: 2046-2057.

Jia, X., Wang, W.X., Ren, L., Chen, Q.J., Mendu, V., Willcut, B., Dinkins, R., Tang, X. & Tang, G. (2009b). Differential and dynamic regulation of miR398 in response to ABA and salt stress in Populus tremula and Arabidopsis thaliana. *Plant Molecular Biology* 71: 51-59.

Jones-Rhoades, M.W. & Bartel, D.P. (2004). Computational identification of plant microRNAs and their targets, including a stress-induced miRNA. *Molecular Cell* 14: 787-799.

Kantar, M., Lucas, S.J. & Budak, H. (2011). miRNA expression patterns of Triticum dicoccoides in response to shock drought stress. *Planta* 233: 471-484.

Kantar, M., Unver, T. & Budak, H. (2010). Regulation of barley miRNAs upon dehydration stress correlated with target gene expression. *Functional & Integrative Genomics* 10: 493-507.

Kliebenstein, D.J., Monde, R.A. & Last, R.L. (1998). Superoxide dismutase in Arabidopsis: an eclectic enzyme family with disparate regulation and protein localization. *Plant Physiology* 118: 637-650.

Li, J., Yang, Z., Yu, B., Liu, J. & Chen, X. (2005). Methylation protects miRNAs and siRNAs from a 3'-end uridylation activity in Arabidopsis. *Current Biology* 15: 1501-1507.

Li, T., Li, H., Zhang, Y.X. & Liu, J.Y. (2010). Identification and analysis of seven H2O2-responsive miRNAs and 32 new miRNAs in the seedlings of rice (Oryza sativa L. ssp. indica). *Nucleic Acids Research.*

Licausi, F. (2010). Regulation of the molecular response to oxygen limitations in plants. *New Phytologist* DOI: 10.1111/j.1469-8137.2010.03562.x.

Liu, B., Li, P., Li, X., Liu, C., Cao, S., Chu, C. & Cao, X. (2005). Loss of function of OsDCL1 affects microRNA accumulation and causes developmental defects in rice. *Plant Physiology* 139: 296-305.

Liu, H.H., Tian, X., Li, Y.J., Wu, C.A. & Zheng, C.C. (2008). Microarray-based analysis of stress-regulated microRNAs in Arabidopsis thaliana. *RNA* 14: 836-843.

Liu, Q. & Chen, Y.Q. (2009). Insights into the mechanism of plant development: interactions of miRNAs pathway with phytohormone response. *Biochemical and Biophysical Research Communications* 384: 1-5.

Liu, Q., Zhang, Y.C., Wang, C.Y., Luo, Y.C., Huang, Q.J., Chen, S.Y., Zhou, H., Qu, L.H. & Chen, Y.Q. (2009). Expression analysis of phytohormone-regulated microRNAs in rice, implying their regulation roles in plant hormone signaling. *FEBS Letters* 583: 723-728.

Llave, C., Kasschau, K.D., Rector, M.A. & Carrington, J.C. (2002a). Endogenous and silencing-associated small RNAs in plants. *Plant Cell* 14: 1605-1619.

Llave, C., Xie, Z., Kasschau, K.D. & Carrington, J.C. (2002b). Cleavage of Scarecrow-like mRNA targets directed by a class of Arabidopsis miRNA. *Science* 297: 2053-2056.

Lobbes, D., Rallapalli, G., Schmidt, D.D., Martin, C. & Clarke, J. (2006). SERRATE: a new player on the plant microRNA scene. *EMBO Reports* 7: 1052-1058.

Lu, C. & Fedoroff, N. (2000). A mutation in the Arabidopsis HYL1 gene encoding a dsRNA binding protein affects responses to abscisic acid, auxin, and cytokinin. *Plant Cell* 12: 2351-2366.

Lu, S., Sun, Y.H. & Chiang, V.L. (2008). Stress-responsive microRNAs in Populus. *Plant Journal* 55: 131-151.

Lurin, C., Andres, C., Aubourg, S., Bellaoui, M., Bitton, F., Bruyere, C., Caboche, M., Debast, C., Gualberto, J., Hoffmann, B., Lecharny, A., Le Ret, M., Martin-Magniette, M.L., Mireau, H., Peeters, N., Renou, J.P., Szurek, B., Taconnat, L. & Small, I. (2004). Genome-wide analysis of Arabidopsis pentatricopeptide repeat proteins reveals their essential role in organelle biogenesis. *Plant Cell* 16: 2089-2103.

Lv, D.K., Bai, X., Li, Y., Ding, X.D., Ge, Y., Cai, H., Ji, W., Wu, N. & Zhu, Y.M. (2010). Profiling of cold-stress-responsive miRNAs in rice by microarrays. *Gene* 459: 39-47.

Mallory, A. & Vaucheret, H. (2010). Form, function, and regulation of ARGONAUTE proteins. *Plant Cell* 22: 3879-3889.

Marin, E., Jouannet, V., Herz, A., Lokerse, A.S., Weijers, D., Vaucheret, H., Nussaume, L., Crespi, M.D. & Maizel, A. (2010). miR390, Arabidopsis TAS3 tasiRNAs, and their AUXIN RESPONSE FACTOR targets define an autoregulatory network quantitatively regulating lateral root growth. *Plant Cell* 22: 1104-1117.

Meyers, B.C., Axtell, M.J., Bartel, B., Bartel, D.P., Baulcombe, D., Bowman, J.L., Cao, X., Carrington, J.C., Chen, X., Green, P.J., Griffiths-Jones, S., Jacobsen, S.E., Mallory, A.C., Martienssen, R.A., Poethig, R.S., Qi, Y., Vaucheret, H., Voinnet, O., Watanabe, Y., Weigel, D. & Zhu, J.K. (2008). Criteria for annotation of plant MicroRNAs. *Plant Cell* 20: 3186-3190.

Mi, S., Cai, T., Hu, Y., Chen, Y., Hodges, E., Ni, F., Wu, L., Li, S., Zhou, H., Long, C., Chen, S., Hannon, G.J. & Qi, Y. (2008). Sorting of small RNAs into Arabidopsis argonaute complexes is directed by the 5' terminal nucleotide. *Cell* 133: 116-127.

Mittler, R. (2006). Abiotic stress, the field environment and stress combination. *Trends in Plant Science* 11: 15-19.

Moldovan, D., Spriggs, A., Yang, J., Pogson, B.J., Dennis, E.S. & Wilson, I.W. (2010). Hypoxia-responsive microRNAs and trans-acting small interfering RNAs in Arabidopsis. *Journal of Experimental Botany* 61: 165-177.

Montgomery, T.A., Howell, M.D., Cuperus, J.T., Li, D., Hansen, J.E., Alexander, A.L., Chapman, E.J., Fahlgren, N., Allen, E. & Carrington, J.C. (2008a). Specificity of ARGONAUTE7-miR390 interaction and dual functionality in TAS3 trans-acting siRNA formation. *Cell* 133: 128-141.

Montgomery, T.A., Yoo, S.J., Fahlgren, N., Gilbert, S.D., Howell, M.D., Sullivan, C.M., Alexander, A., Nguyen, G., Allen, E., Ahn, J.H. & Carrington, J.C. (2008b). AGO1-miR173 complex initiates phased siRNA formation in plants. *Proceedings of the National Academy of Sciences of the United States of America* 105: 20055-20062.

Moxon, S., Jing, R., Szittya, G., Schwach, F., Rusholme Pilcher, R.L., Moulton, V. & Dalmay, T. (2008). Deep sequencing of tomato short RNAs identifies microRNAs targeting genes involved in fruit ripening. *Genome Research* 18: 1602-1609.

Nakashima, K., Ito, Y. & Yamaguchi-Shinozaki, K. (2009). Transcriptional regulatory networks in response to abiotic stresses in Arabidopsis and grasses. *Plant Physiology* 149: 88-95.

Navarro, L., Dunoyer, P., Jay, F., Arnold, B., Dharmasiri, N., Estelle, M., Voinnet, O. & Jones, J.D. (2006). A plant miRNA contributes to antibacterial resistance by repressing auxin signaling. *Science* 312: 436-439.

Nelson, D.E., Repetti, P.P., Adams, T.R., Creelman, R.A., Wu, J., Warner, D.C., Anstrom, D.C., Bensen, R.J., Castiglioni, P.P., Donnarummo, M.G., Hinchey, B.S., Kumimoto, R.W., Maszle, D.R., Canales, R.D., Krolikowski, K.A., Dotson, S.B., Gutterson, N., Ratcliffe, O.J. & Heard, J.E. (2007). Plant nuclear factor Y (NF-Y) B subunits confer drought tolerance and lead to improved corn yields on water-limited acres. *Proceedings of the National Academy of Sciences of the United States of America* 104: 16450-16455.

Pantaleo, V., Szittya, G., Moxon, S., Miozzi, L., Moulton, V., Dalmay, T. & Burgyan, J. (2010). Identification of grapevine microRNAs and their targets using high-throughput sequencing and degradome analysis. *Plant Journal* 62: 960-976.

Park, M.Y., Wu, G., Gonzalez-Sulser, A., Vaucheret, H. & Poethig, R.S. (2005). Nuclear processing and export of microRNAs in Arabidopsis. *Proceedings of the National Academy of Sciences of the United States of America* 102: 3691-3696.

Park, W., Li, J., Song, R., Messing, J. & Chen, X. (2002). CARPEL FACTORY, a Dicer homolog, and HEN1, a novel protein, act in microRNA metabolism in Arabidopsis thaliana. *Current Biology* 12: 1484-1495.

Rajagopalan, R., Vaucheret, H., Trejo, J. & Bartel, D.P. (2006). A diverse and evolutionarily fluid set of microRNAs in Arabidopsis thaliana. *Genes & Development* 20: 3407-3425.

Ramachandran, V. & Chen, X. (2008). Small RNA metabolism in Arabidopsis. *Trends in Plant Science* 13: 368-374.

Reinhart, B.J., Weinstein, E.G., Rhoades, M.W., Bartel, B. & Bartel, D.P. (2002). MicroRNAs in plants. *Genes & Development* 16: 1616-1626.

Reyes, J.L. & Chua, N.H. (2007). ABA induction of miR159 controls transcript levels of two MYB factors during Arabidopsis seed germination. *Plant Journal* 49: 592-606.

Rhoades, M.W., Reinhart, B.J., Lim, L.P., Burge, C.B., Bartel, B. & Bartel, D.P. (2002). Prediction of plant microRNA targets. *Cell* 110: 513-520.

Ruegger, M., Dewey, E., Gray, W.M., Hobbie, L., Turner, J. & Estelle, M. (1998). The TIR1 protein of Arabidopsis functions in auxin response and is related to human SKP2 and yeast grr1p. *Genes & Development* 12: 198-207.

Schreiber, A.W., Shi, B.J., Huang, C.Y., Langridge, P. & Baumann, U. (2011). Discovery of Barley miRNAs through Deep Sequencing of Short Reads. *BMC Genomics* 12: 129.

Shinozaki, K. & Yamaguchi-Shinozaki, K. (2007). Gene networks involved in drought stress response and tolerance. *Journal of Experimental Botany* 58: 221-227.

Sire, C., Moreno, A.B., Garcia-Chapa, M., Lopez-Moya, J.J. & San Segundo, B. (2009). Diurnal oscillation in the accumulation of Arabidopsis microRNAs, miR167, miR168, miR171 and miR398. *FEBS Letters* 583: 1039-1044.

Song, Q.X., Liu, Y.F., Hu, X.Y., Zhang, W.K., Ma, B., Chen, S.Y. & Zhang, J.S. (2011). Identification of miRNAs and their target genes in developing soybean seeds by deep sequencing. *BMC Plant Biology* 11: 5.

Sorin, C., Bussell, J.D., Camus, I., Ljung, K., Kowalczyk, M., Geiss, G., McKhann, H., Garcion, C., Vaucheret, H., Sandberg, G. & Bellini, C. (2005). Auxin and light control of adventitious rooting in Arabidopsis require ARGONAUTE1. *Plant Cell* 17: 1343-1359.

Sunkar, R., Chinnusamy, V., Zhu, J. & Zhu, J.K. (2007). Small RNAs as big players in plant abiotic stress responses and nutrient deprivation. *Trends in Plant Science* 12: 301-309.

Sunkar, R., Kapoor, A. & Zhu, J.K. (2006). Posttranscriptional induction of two Cu/Zn superoxide dismutase genes in Arabidopsis is mediated by downregulation of miR398 and important for oxidative stress tolerance. *Plant Cell* 18: 2051-2065.

Sunkar, R. & Zhu, J.K. (2004). Novel and stress-regulated microRNAs and other small RNAs from Arabidopsis. *Plant Cell* 16: 2001-2019.

Szittya, G., Moxon, S., Santos, D.M., Jing, R., Fevereiro, M.P., Moulton, V. & Dalmay, T. (2008). High-throughput sequencing of Medicago truncatula short RNAs identifies eight new miRNA families. *BMC Genomics* 9: 593.

Szittya, G., Silhavy, D., Molnar, A., Havelda, Z., Lovas, A., Lakatos, L., Banfalvi, Z. & Burgyan, J. (2003). Low temperature inhibits RNA silencing-mediated defence by the control of siRNA generation. *EMBO Journal* 22: 633-640.

Todesco, M., Rubio-Somoza, I., Paz-Ares, J. & Weigel, D. (2010). A collection of target mimics for comprehensive analysis of microRNA function in Arabidopsis thaliana. *PLoS Genetics* 6: e1001031.

Trindade, I., Capitao, C., Dalmay, T., Fevereiro, M.P. & Santos, D.M. (2010). miR398 and miR408 are up-regulated in response to water deficit in Medicago truncatula. *Planta* 231: 705-716.

Vanneste, S. & Friml, J. (2009). Auxin: a trigger for change in plant development. *Cell* 136: 1005-1016.

Vaucheret, H., Mallory, A.C. & Bartel, D.P. (2006). AGO1 homeostasis entails coexpression of MIR168 and AGO1 and preferential stabilization of miR168 by AGO1. *Molecular Cell* 22: 129-136.

Vaucheret, H., Vazquez, F., Crete, P. & Bartel, D.P. (2004). The action of ARGONAUTE1 in the miRNA pathway and its regulation by the miRNA pathway are crucial for plant development. *Genes & Development* 18: 1187-1197.

Vazquez, F., Vaucheret, H., Rajagopalan, R., Lepers, C., Gasciolli, V., Mallory, A.C., Hilbert, J.L., Bartel, D.P. & Crete, P. (2004). Endogenous trans-acting siRNAs regulate the accumulation of Arabidopsis mRNAs. *Molecular Cell* 16: 69-79.

Wang, J.F., Zhou, H., Chen, Y.Q., Luo, Q.J. & Qu, L.H. (2004). Identification of 20 microRNAs from Oryza sativa. *Nucleic Acids Research* 32: 1688-1695.

Wu, M.F., Tian, Q. & Reed, J.W. (2006). Arabidopsis microRNA167 controls patterns of ARF6 and ARF8 expression, and regulates both female and male reproduction. *Development* 133: 4211-4218.

Xie, Z., Allen, E., Fahlgren, N., Calamar, A., Givan, S.A. & Carrington, J.C. (2005). Expression of Arabidopsis MIRNA genes. *Plant Physiology* 138: 2145-2154.

Xie, Z., Kasschau, K.D. & Carrington, J.C. (2003). Negative feedback regulation of Dicer-Like1 in Arabidopsis by microRNA-guided mRNA degradation. *Current Biology* 13: 784-789.

Yang, J.H., Seo, H.H., Han, S.J., Yoon, E.K., Yang, M.S. & Lee, W.S. (2008). Phytohormone abscisic acid control RNA-dependent RNA polymerase 6 gene expression and post-transcriptional gene silencing in rice cells. *Nucleic Acids Research* 36: 1220-1226.

Yoshikawa, M., Peragine, A., Park, M.Y. & Poethig, R.S. (2005). A pathway for the biogenesis of trans-acting siRNAs in Arabidopsis. *Genes & Development* 19: 2164-2175.

Yu, B., Yang, Z., Li, J., Minakhina, S., Yang, M., Padgett, R.W., Steward, R. & Chen, X. (2005). Methylation as a crucial step in plant microRNA biogenesis. *Science* 307: 932-935.

Zhang, J.F., Yuan, L.J., Shao, Y., Du, W., Yan, D.W. & Lu, Y.T. (2008a). The disturbance of small RNA pathways enhanced abscisic acid response and multiple stress responses in Arabidopsis. *Plant, Cell & Environment* 31: 562-574.

Zhang, W., Gao, S., Zhou, X., Chellappan, P., Chen, Z., Zhang, X., Fromuth, N., Coutino, G., Coffey, M. & Jin, H. (2011a). Bacteria-responsive microRNAs regulate plant innate immunity by modulating plant hormone networks. *Plant Molecular Biology* 75: 93-105.

Zhang, X., Zou, Z., Gong, P., Zhang, J., Ziaf, K., Li, H., Xiao, F. & Ye, Z. (2011b). Over-expression of microRNA169 confers enhanced drought tolerance to tomato. *Biotechnology Letters* 33: 403-409.

Zhang, Z., Wei, L., Zou, X., Tao, Y., Liu, Z. & Zheng, Y. (2008b). Submergence-responsive MicroRNAs are potentially involved in the regulation of morphological and metabolic adaptations in maize root cells. *Annals of Botany* 102: 509-519.

Zhao, B., Ge, L., Liang, R., Li, W., Ruan, K., Lin, H. & Jin, Y. (2009). Members of miR-169 family are induced by high salinity and transiently inhibit the NF-YA transcription factor. *BMC Molecular Biology* 10: 29.

Zhao, B., Liang, R., Ge, L., Li, W., Xiao, H., Lin, H., Ruan, K. & Jin, Y. (2007). Identification of drought-induced microRNAs in rice. *Biochemical and Biophysical Research Communications* 354: 585-590.

Zhou, L., Liu, Y., Liu, Z., Kong, D., Duan, M. & Luo, L. (2010). Genome-wide identification and analysis of drought-responsive microRNAs in Oryza sativa. *Journal of Experimental Botany* 61: 4157-4168.

Zhou, X., Wang, G., Sutoh, K., Zhu, J.K. & Zhang, W. (2008). Identification of cold-inducible microRNAs in plants by transcriptome analysis. *Biochimica et Biophysica Acta* 1779: 780-788.

Zhou, X., Wang, G. & Zhang, W. (2007). UV-B responsive microRNA genes in Arabidopsis thaliana. *Molecular Systems Biology* 3: 103.

Zhu, J.K. (2002). Salt and drought stress signal transduction in plants. *Annual Review of Plant Biology* 53: 247-273.

Zilberman, D., Cao, X. & Jacobsen, S.E. (2003). ARGONAUTE4 control of locus-specific siRNA accumulation and DNA and histone methylation. *Science* 299: 716-719.

6

Cyclic Nucleotides and Nucleotide Cyclases in Plant Stress Responses

Fouad Lemtiri-Chlieh[1], Ludivine Thomas[1], Claudius Marondedze[1], Helen Irving[2] and Chris Gehring[1]
[1]*Division of Chemistry, Life Science and Engineering*
King Abdullah University of Science and Technology
[2]*Medicinal Chemistry and Drug Action, Monash Institute of Pharmaceutical Sciences*
Monash University
[1]*Saudi Arabia*
[2]*Australia*

1. Introduction

The cyclic nucleotides monophosphates (cNMP) and in particular adenosine 3′,5′-cyclic monophosphate (cAMP) and guanosine 3′,5′-cyclic monophosphate (cGMP) are cyclic catalytic products of adenosine 5'-triphosphate (ATP) and guanosine 5′-triphosphate (GTP) respectively. These cNMPs are universal second messengers with key roles in many and diverse physiological responses and processes in prokaryotes, and in both higher and lower eukaryotes.

Cyclic AMP is arguably one of the most extensively studied second messengers in animals, lower eukaryotes and bacteria where it has critical roles in signaling the metabolic status. In bacteria, cAMP is involved in the positive regulation of the *lac* operon where in an environment of a low glucose, cAMP accumulates and binds to the allosteric site of the cAMP receptor protein (CRP), a transcription activator protein. Once the CRP is in its active configuration, it binds to a *cis*-element upstream of the *lac* promoter and activates transcription. At high glucose concentrations, cAMP concentration decreases and CRP disengages from the *lac* operon promoter (Meiklejohn & Gralla, 1985). Cyclic AMP signaling is also critical for many aspects of the development of the slime mold *Dictyostelium discoideum* that grows unicellularly, but develops as multicellular organism (Kimmel & Firtel, 2004; McMains et al., 2008). Cyclic AMP has a role in chemotaxis and cAMP oscillations act as intracellular feedback loops in the transcriptional regulation of many regulatory pathways. Post aggregation, cAMP-dependent signals mediate cell sorting, pattern formation, and morphogenetic changes and cAMP receptor affinity can control wave dynamics, geometry and morphogenesis (Dormann et al., 2001). In animals, cAMP can be produced e.g. by G_s-coupled activation of adenylate cyclases (ACs) which triggers a signal cascade that includes the modification of Ca^{2+} channels, the phosphorylation of target proteins including enzymes involved in regulating glycogen metabolism and eventually the cAMP-dependent activation of the transcription factor, cAMP response element-binding (CREB) (Bolwell, 1995; Karin & Smeal, 1992). Cyclic AMP also plays a role in excitation-contraction where voltage-dependent potentiation of L-type Ca^{2+} channels is due to phosphorylation by cAMP-dependent protein kinase (Sculptoreanu et al., 1993). It has also been shown that bicarbonate can directly stimulate a mammalian soluble adenylyl cyclase (sAC) (Chen, 2000; Kaupp & Weyand, 2000) and that this sAC is most similar to adenylyl cyclases from cyanobacteria, and that bicarbonate regulation of cyclase activity is conserved in these early forms of life making it likely to be an ancient and fundamental regulatory component.

The other cyclic nucleotide under investigation here, cGMP, also has many and diverse roles in bacteria, and lower and higher eukaryotes. In bacteria, the work on cGMP undertaken in the early 1970s stalled mainly because firstly, the levels of cGMP are significantly lower than those of cAMP and secondly, no specific binding target for cGMP could be identified (for review see Linder, 2010). More recently, a specific bacterial guanylyl cyclase (GC; Cya2) from unicellular cyanobacterium *Synechocystis* sp. PCC 6803 has been identified and biochemically and structurally characterized (Ochoa De Alda et al., 2000). It was reported that Cya2 has the catalytic requirements for activity and purine specificity of being a GC and insertional mutagenesis of *cya2* caused a marked reduction in cGMP content without altering the cAMP content. Currently an increasing body of literature is accumulating that reports cGMP dependence of many bacterial processes for instance the control of cyst development in the α-proteobacterium *Rhodospirillum centenum* (Marden et al., 2011). In lower eukaryotes, cGMP is also implicated in signal transduction (Rericha & Parent, 2008; Shpakov & Pertseva, 2008) and *Dictyostelium discoideum* GCs, which interestingly share topological similarities with mammalian ACs, have been reported (Roelofs et al., 2001a; Roelofs et al., 2001b). In vertebrates, cGMP has been implicated in broad range of physiological processes, many of which are linked to specific receptors (Garbers, 1999). The processes include retinal phototransduction (Azadi et al., 2010; Luo et al., 2008) and signaling by the homeostasis regulating Atrial Natriuretic Peptide (Acher, 2002; Chinkers et al., 1989). Apart from the particulate membrane bound receptor GCs, there is a second class, the soluble GCs located in the cytosol (Arnold et al., 1977; Gao, 2010) that are involved in nitric oxide (NO) signaling.

While there is a large body of literature on cyclic nucleotide signaling in lower and higher eukaryotes, both dealing with biological functions and mechanisms of action, the acceptance that cyclic nucleotides play a role in plant signaling was delayed and not without controversy since the levels of cAMP and cGMP in plants appear to be generally lower than in animals or lower eukaryotes (for review see Bolwell, 1995; Gehring, 2010; Newton & Smith, 2004). Another and possibly more important reason for the reluctant acceptance of these signaling molecules in higher plants is that until quite recently, there was no genetic or molecular evidence of nucleotide cyclases (NCs) in higher plants. Here we undertake the review of the literature on biological and molecular aspects of cyclic nucleotide signaling in plants and higher plants in particular. In addition, we will examine the history of cyclic NCs in higher plants and the methods developed that led to their discovery. Finally, we will apply these search methods to generate and evaluate a list of candidate NCs and propose how they might impact on stress signaling in plants.

2. The discovery of adenosine cyclic monophosphate in plants

Perhaps the most convincing data towards directly establishing a specific function for cAMP came from whole-cell patch-clamp current recordings in *Vicia faba* mesophyll protoplasts that revealed that the outward K^+-current increased in a dose-dependent fashion following intracellular application of cAMP - and not AMP, cGMP or GMP - and indirect evidence indicated that this modulation occurred through a cAMP-regulated protein kinase (Li et al., 1994). Furthermore, cAMP-dependent up-regulation of a calcium-permeable conductance activated by hyperpolarization was also reported in guard cells as well as mesophyll cells of *Arabidopsis thaliana* and *Vicia faba* (Lemtiri-Chlieh & Berkowitz, 2004). Despite this compelling evidence, the history of cAMP function in plants has not been free of controversy and much more is needed before we have a clear picture of the generation and modes of action of cAMP in plant physiology and plant stress responses in particular.

2.1 Cyclic AMP in plants: the long road to acceptance

The question of whether or not cAMP exists and functions in plants was fiercely debated for almost three decades starting in the late 1960s and this during a time when cAMP was accepted as second messenger in animals, lower organisms and prokaryotes. Early published experimental data and various claims from several laboratories suggesting comparable roles for cAMP in higher plants were highly criticized (i) as lacking specificity in the effects elicited by cyclic nucleotides, (ii) uncertainty about cyclic nucleotide-generating enzymes, (iii) ambiguous identification of endogenous putative cyclic nucleotides largely due to inadequate chromatographic identification (Amrhein, 1977; Newton & Smith, 2004). For unequivocal proof of cAMP in plants and its role as a second messenger, one would have had to show for example phosphorylation response of proteins in response to elevated cAMP levels, unambiguously identified cAMP-dependent enzyme function and not least the presence of cAMP itself. Early papers were based on observations of various physiological and metabolical responses after the exogenous application of cAMP, cAMP analogues and phosphodiesterase (PDE) inhibitors known at the time, but these indirect pharmacological approaches were used without sufficient knowledge of the fundamental metabolic mechanisms in plants (Newton et al., 1999).

One of the reports disputing the existence of cAMP in higher plants was a review entitled "Evidence against the occurrence of adenosine 3':5'-cyclic monophosphate in higher plants" (Amrhein, 1974). In it, the author strongly disagreed with the then published literature attempting to demonstrate the occurrence of cAMP in tissues of higher plants including the work by numerous authors (Azhar & Krishna Murti, 1971; Becker & Ziegler, 1973; Brown & Newton, 1973; Janistyn, 1972; Kessler, 1972; Narayanan et al., 1970; Ownby et al., 1973; Pollard & Venere, 1970; Pradet et al., 1972; Raymond et al., 1973; Salomon & Mascarenhas, 1971, 1972; Wellburn et al., 1973). The main reason for disputing the presence of cAMP *in planta* was the low sensitivity of the methods employed to detect cAMP at the picomole level. For example, it was reported that a radiolabeled product derived from the incubation of 8-^{14}C-adenine of barley (*Hordeum vulgare*) seedlings could not be chromatographically separated from cAMP since the resolving power of the chromatographic system used was insufficient to resolve cAMP from the RNA catabolic intermediate, 2',3'-cAMP (Pollar & Venere, 1970). An attempt to demonstrate the chromatographed product as cAMP was carried out by measuring the hydrolysis of cAMP to AMP by cAMP PDE (Narayanan et al., 1970). This method also yielded inconclusive results because PDE had not been confirmed at that time to have specificity for cAMP. Some publications disputing the existence of cAMP in higher plants include explicit and categoric statements such as " ... cyclic AMP does not mediate the action of gibberellic acid" (Keates, 1973).

Skepticism about the occurrence of cAMP in plants on the other hand spurred the quest to develop scientific approaches that would eventually generate clear and direct evidence for cAMP and its role in plants. Since the mid-1980s, more reports discussing the presence and potential functions of cyclic nucleotides in plants have been published. The potential roles of cAMP in plants include the regulation of ion channels (Bolwell, 1995) and ion transport in *Arabidopsis thaliana* (Anderson et al., 1992; Trewavas, 1997), activation of phenylalanine ammonia lyase (Bolwell, 1992), cAMP-dependent signal transduction pathways and the cell cycle progression in tobacco BY-2 cells (Ehsan et al., 1998). Cyclic AMP has also been shown to play a role in stimulating protein kinase activity in rice (*Oryza sativa*) leaves (Komatsu & Hirano, 1993), and more recently, to eliciting stress responses and plant defense (Choi & Xu, 2010). For example, increased levels of cAMP coincide with the early stages of the response to phytoalexins and mediate the production of 6-methoxymellein and the activation of calcium uptake into cultured carrot (*Daucus carota*) cells (Kurosaki & Nishi, 1993; Kurosaki et al., 1987).

It is noteworthy that in the early 1970s, the same techniques for tissue extraction, purification and detection techniques employed in animals were applied to plant tissue without taking into account the particular characteristics of the (bio-)chemistry of plants. This led to disputes that were only resolved in the late 1970s when experimental procedures to analyze cAMP were adapted to suit the plant cellular and extracellular environments. Since then a number of research groups reported more reproducible data, for example, the activity of PDE to hydrolyze 3',5'-cAMP through enzyme inhibition reaction by methylxanthines (Brown et al., 1977) and the localization of AC activity in meristems of young pea (*Pisum sativum*) hypocotyl (Hilton & Nesius, 1978). Efforts to identify endogenous cAMP in plant cells coincided with the development of sequential chromatographic and electrophoretic methodologies for the extraction and isolation of cAMP (Brown & Newton, 1973) that led to improved cAMP recovery by co-chromatography. Together, these techniques enabled separation of cAMP from 2',3'-cAMP and other known naturally occurring adenine nucleotides. Yet, some scientists continued to dispute the claims arguing that the results were due to unidentified adenine compounds in higher plants that may have identical chromatographic properties as cAMP (Amrhein, 1974; Keates, 1973; Lin, 1974).

The elucidation of the structure and molecular conformation (Fig. 1) of cAMP was first established in animals (Sutherland & Rall, 1960) and helped to determine the chemical properties and biological function of this messenger. The molecular conformation and 3D structure of cAMP was established by X-ray crystallography and proton-NMR (Sundralingham, 1975).

2.2 Discovering cAMP in higher plants

The mode of action of cAMP as the second messenger for mammalian hormones had been accepted as early as the 1960s and were done long before any attempts to discover cAMP and cAMP functions in higher plants. Cyclic AMP was originally discovered as the intracellular mediator or activator of many peptides and catecholamine hormones in animals (Robison & Sutherland, 1971; Sutherland & Rall, 1960). Since then, cAMP has been implicated in signaling system that intercede cellular responses to imbalances of the external biosphere.

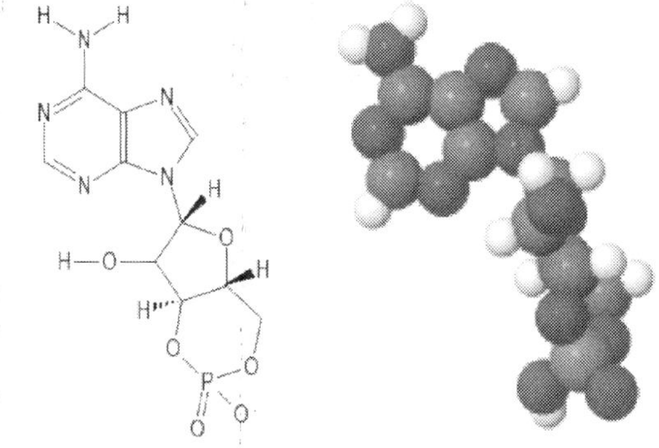

Fig. 1. Chemical and 3D structures of cAMP (Molecular formula: $C_{10}H_{12}N_5O_6P$)

The first functional clue that cAMP may play a role in hormone signaling in higher plants came from findings in barley aleurone layers suggesting that the nucleotide could substitute gibberellic acid (GA) in the induction of α-amylase synthesis (Duffus, 1969). The first evidence of the formation of cAMP in higher plants was demonstrated through the incorporation of ^{14}C-adenosine into cAMP by a method developed by Gilman (Gilman, 1970). The Gilman method is based on the competitive displacement of unlabelled nucleotide present in the complex containing the cAMP binding protein in the sample. However, incorporation of radioactively labeled adenine in plant tissues was observed to give rise to a complex with properties similar to cAMP (Azhar & Krishna Murti, 1971; Pollard, 1970; Salomon & Mascarenhas, 1971). Hence, the use of labeled products was not judged convincing enough to prove the synthesis of cAMP in plant cells, mainly due to the sensitivity limits, the specific activity of labeled precursor applied and the efficiency of conversion to cAMP. At the time, evidence for plant analogs of animal cAMP was limited and as such, endogenous cAMP could not easily be detected. Nevertheless, exogenous application of un-physiologically high levels of cAMP were shown to mimic some biological activities of auxin (Pollard & Venere, 1970) and GA (Kessler, 1972). It was later confirmed that indeed, cAMP is capable of entering plant cells causing an increase of the endogenous level of cAMP by a factor of 30 (Wiedmaier & Kull, 1976). However, these 'pharmacological' studies lacked biological significance and did not prove the existence of cAMP in higher plants.

Based on the specific activity of ATP, the substrate for cAMP formation, the upper limits of cAMP concentrations were estimated to be in the range of pmoles/g fresh weight. For instance, 0.04 pmole/mg protein in barley aleurone layers (Keates, 1973), 7-11 pmoles/g fresh weight in oat (*Avena sativa*) coleoptiles (Ownby et al., 1975) and 0.37 pmoles/g fresh weight in lettuce (*Lactuca sativa*) seedlings (Bressan et al., 1976). In the latter study, incorporation of adenosine-8-^{4}C into compounds, with chemically related properties to cAMP, was measured in germinating lettuce seeds. In addition to tracing radioactively labeled adenine, a modified Gilman binding assay and protein kinase activation assay were used to measure endogenous levels of cAMP in coleoptile leaf segments of oats, tubers of potato (*Solanum tuberosum*), callus of tobacco (*Nicotiana tabacum*), and germinating seeds of lettuce (Bressan et al., 1976). The binding assay proved reliable for mouse and rat liver analyses, but not sensitive or specific enough for plant tissues as it also responded to various components from lettuce and potato tissues that were chromatographically similar but not identical to cAMP. While the protein kinase activation assay was much more specific, it exhibited positive response to some compounds not chromatographically identical to cAMP. In this study, the concentrations of cAMP found in the plant tissues tested were below the detection limit (Bressan et al., 1976), a problem also encountered in other studies (Amrhein, 1974; Niles & Mount, 1974). Additionally, other authors claimed that cAMP detected in plants was a result of bacterial infection or contamination (Bonnafous et al., 1975), a claim that was refuted by the fact that bacteria contributed to less than 0.1% of the cAMP in plant cells (Ashton & Polya, 1978). By and large, reported cAMP levels were lower than those in animal tissues that typically range from 100 to 500 pmoles/g of fresh weight or 0.5 to 2.6 pmoles/mg protein (Robison et al., 1968). A comprehensive listing of cAMP concentrations in various plants are reported elsewhere (Newton & Brown, 1986).

The ubiquitous presence of cAMP in plant tissues assumes the definite existence of NCs and the likely presence of cyclic nucleotide PDEs. At present, there is hardly any doubt about the presence of adenylate cyclases and cAMP-dependent signaling system in plants. However, the role of these systems in various physiological and biochemical processes is yet to be properly elucidated. Cyclic AMP formation is catalyzed by adenylate cyclases (E.C.4.6.1.1) that convert ATP to cAMP and pyrophosphate (Helmreich et al., 1976), while the enzyme cAMP PDE (E.C.3.1.4.17) hydrolyzes 3',5'-cAMP to 5'-AMP (Robison & Sutherland, 1971). Endogenously applied cAMP can be used as an indicator of functional activity for adenylate cyclase signaling system (Lomovatskaya et al., 2011). The presence and localization of AC activity was first experimentally demonstrated in meristems of young pea hypocotyls (Hilton & Nesius, 1978) making use of adenylyl-imidodiphosphate as a substrate specific for the enzyme and lead nitrate as a precipitating product that can be visualized with electron microscopy and quantitated *in situ*. Lead precipitate was found to be localized in distinct areas bound to the smooth endoplasmic reticulum in differentiated cells of the pea root cap (Hilton & Nesius, 1978). Later, the use of mass spectrometric techniques explicitly identified the reaction product of AC activity in pea for the first time, thus erasing doubts casted on the methodology (Pacini et al., 1993). In this study, plant extracts from roots of pea seedlings and incubated with ATP and Mg^{2+} produced cAMP as measured by tandem mass spectrometry and hence are the first experimental proof for AC activity in higher plants. Unambiguous cAMP synthesis was later confirmed by mass spectrometric analysis of alfalfa (*Medicago sativa*) cell cultures after exposure to the glycoprotein elicitor of the phytopathogenic fungus *Verticilium alboatrum* (Cooke et al., 1994). Cyclic AMP synthesis by adenylyl cyclases was also reported to occur in apical hook plasma membrane from bean (*Phaseolus vulgaris*) (Roef et al., 1996). This study exploited the use of polyclonal antibodies raised in chicken against an adenosine 3',5' monophosphate-diphtheria toxoid antigen construct in developing an immunoaffinity purification procedure to measure cAMP and hence AC activity.

Technological advances providing an even better reproducibility of quantitative data brought the detection limit of cyclic nucleotides down to 25 fmoles in plant extracts by combining liquid chromatography and electrospray ionization (ESI) mass spectrometry (Witters et al., 1996; Witters et al., 1997; Witters et al., 1999). Due to the high sensitivity of ESI, it has become the technique of choice for the analysis of polar biomolecules. Other methods currently in use to detect cNMP concentrations in plants include the Gilman method (Gilman, 1970), radio-immune assays (Rosenberg et al., 1982) and the bioluminescent method (Isner & Maathuis, 2011; Nikolaev & Lohse, 2006). The radio-immune assay is based on the primary binding of standard antigen with antibodies followed by unlabeled antigen that competitively replace the radioactively labeled antigen, while the bioluminescent method permits the monitoring of cAMP levels in the living system, however, the use of the latter requires fluorophores with different spectrum properties than chlorophyll and phenols. A recent improved enzyme immunoassay method for determining cAMP concentration in plant tissue was proposed (Lomovatskaya et al., 2011) and allows for the detection of cAMP down to 5 pM, and this is about 10 times more sensitive than previously reported methods.

Enzymes that hydrolyze cAMP - cAMP PDEs - were first reported in pea seedlings (Lieberman & Kunishi, 1969), but apparently, such an activity was not detected in barley (Vandepeute et al., 1973). It has been demonstrated that plant PDEs differ from their animal counterpart in that they have several isoforms, each having its own substrate specificity. One isoform extracted from pea was shown to have an acidic isoelectric point and a substrate preference for 2′,3′-cAMP while its isomer preferred 3′,5′-cAMP (Lin & Varner, 1972). Partially purified PDE isoforms from scarlet runner bean (*Phaseolus coccineus*) showed similar enzymatic properties to those of mammalian PDEs (Brown et al., 1975, 1977). A multitude of PDEs have been reported in species like potato (Ashton & Polya, 1975), purslane (*Portuluca oleracea*) (Endress, 1979) and carrot (Kurosaki & Kaburaki, 1995; for review see Newton & Smith, 2004).

In the early 1980s, three cAMP-responsive protein kinases were identified in duckweed (*Lemna paucicostata*) and they catalyze the phosphorylation of histones (for review see Kato et al., 1983). Later, a cAMP-dependent protein kinase A (PKA) was reported in maize (*Zea mays*) and coconut (*Cocos nucifera*) (Janistyn, 1988, 1989); in addition, PKA activity was also detected in petunia (*Petunia hybrid* var. Old Glory Blue) (Polya et al., 1991) and rice (Komatsu & Hirano, 1993). Plant cAMP was shown to play a physiological role in regulating gene expression through eliciting dissociation of PKA (Inamdar et al., 1991). In this study, an isolated cDNA clone from *Vicia faba* with close resemblance to the animal cAMP response element-binding protein (CREB) was used to provide evidence for the presence of an analogous CREB system. In principle, PKA is a tetramer made up of two isoforms, and once cAMP binds to the regulatory subunits, it elicits dissociation of PKA yielding a regulatory dimer and two free catalytic subunits. Each of the regulatory dimer subunit is capable of binding two molecules of cAMP. In the free state, the catalytic subunits are active and able to phosphorylate other proteins thus altering their activity by increasing the surface charge. The active PKA catalytic subunits have also been shown to migrate to the nucleus allowing phosphorylation of the transcription factor CREB (De Cesare et al., 1999; Inamdar et al., 1991).

In summary, the existence of cAMP in higher plants has now been established using advanced analytical tools. In addition to presence of the cAMP, we also have conclusive evidence for the presence of PDEs and cAMP-binding proteins and a number of cAMP-dependent physiological responses in higher plants. Systematic studies of the function of cAMP are emerging and they include the study of the role of cAMP in ion transport e.g. in *Arabidopsis thaliana* (Anderson et al., 1992), in signal transduction and in cell cycle progression in tobacco BY-2 cells (Ehsan et al., 1998), in plant defence response (Kurosaki et al., 1987), in cAMP activation of phenylalanine ammonia lyase (Bolwell, 1992) and in stress responses (Choi & Xu, 2010; Meier & Gehring, 2006). Overall, this shows that regardless of the low and seemingly un-physiological levels of cAMP in plants as compared to animals, the perception that plants have also a functional cAMP-dependent signal system remains alive and awaits detailed elucidation.

3. The discovery of guanosine cyclic monophosphate in plants

3.1 Early evidence for cGMP in plants

Not long after the identification and characterization of cAMP, another cyclic nucleotide, cGMP, was synthesized *in vitro* (Smith et al., 1961) and subsequently degraded in an enzymatic reaction similar to the hydrolysis of cAMP (Drummond & Perrotty, 1961). Two years later, endogenous cGMP was detected in rat (*Rattus norvegicus*) urine (Ashman et al., 1963). In 1964, an enzyme capable of breaking the 3′,5′-phosphodiester bond of cGMP was characterized (Kuriyama et al., 1964). The same laboratory confirmed the detection of cGMP in rat urine and proposed that the cGMP synthesis is catalyzed by a cyclase (Price et al., 1967). The chemical and 3D structures of cGMP are shown in Fig. 2. In 1966, a PDE hydrolysing cAMP as well as cGMP was partially purified from dog (*Canis lupus*) heart (Nair, 1966) and by 1969, guanylyl cyclases (GCs) and PDEs were recognized as the enzymes responsible for the synthesis and hydrolysis, respectively, of cGMP. Unlike the ACs that synthesize cAMP, GCs are mostly detected in the soluble fractions of animal tissue homogenates. The activity of GC was barely detected in the presence of Mg^{2+}, whereas it was stimulated by a factor of 10 upon replacement of Mg^{2+} by Mn^{2+} (Hardman & Sutherland, 1969). Seven families of PDEs were reported, of which at least two are cGMP-dependent (Beavo, 1995). In the early seventies, further insights into cGMP-specific PDEs were obtained from different rat and bovine (*Bos taurus*) tissues (Beavo et al., 1970; Cheung, 1971; Kakiuchi et al., 1971; Thompson & Appleman, 1971). That time also saw the isolation of PDEs from various plant sources and tissues including pea seedlings (Lin & Varner, 1972), soybean (*Glycine max*) calluses (Brewin & Northcot, 1973), bean seedlings (Brown et al., 1979) and spinach (*Spinacia oleracea*) chloroplasts (Brown et al., 1980). To date, it is accepted that upon activation, GCs convert cytosolic GTP into cGMP and that hydrolysis of 3′,5′-cGMP to 5′-GMP is catalyzed by the enzyme family of PDEs (Hofmann et al., 2002; Reggiani, 1997). The latter enzymes are valuable as pharmaceutical targets for the development of several commercial drugs such as Sildenafil (Viagra) that keeps cellular cGMP levels elevated (Ghofrani et al., 2006). At least in animals, cellular levels of cGMP are regulated by the rate of synthesis and accumulation of the GCs. It is also regulated by the availability of the GTP substrate and Mg^{2+} co-factor, by the cGMP release from cells (transport and extrusion) and the rate of degradation by PDEs (Murad, 2006).

Although the importance of cGMP as a second messenger was recognized in the early 1970s in animals, appreciation of its significance in plants was initially rather slow. This was in part due to the controversy surrounding the existence of cAMP in plants. The occurrence of cGMP and other cyclic nucleotides including uridine 3′,5′-cyclic monophosphate (cUMP), cytidine 3′,5′-cyclic monophosphate (cCMP), inosine 3′ ,5′-cyclic monophosphate (cIMP) and 2′-deoxythymidine 3′,5′-cyclic monophosphate (c-dTMP) remained a matter of speculation and controversy until unambiguous demonstration by mass spectrometry in pea roots was achieved at the end of the 1980s (Newton et al., 1989).

While the occurrence of cGMP was further confirmed in several plant species and tissues including pea roots (Haddox et al., 1974) and bean seedlings (Newton et al., 1984), it was not fully accepted as a *bona fide* second messenger in plants before 1994 when three key articles were published. In the first, a significant and rapid increase in cGMP levels in response to nitric oxide (NO) exposure was detected in spruce pine (*Picea abies*) needles, thus suggesting the presence of NO-dependent guanylyl cyclases much like in animals (Pfeiffer et al., 1994). In another series of reports, cGMP was linked to light responses and chloroplast biogenesis (Bowler & Chua, 1994; Neuhaus et al., 1997).

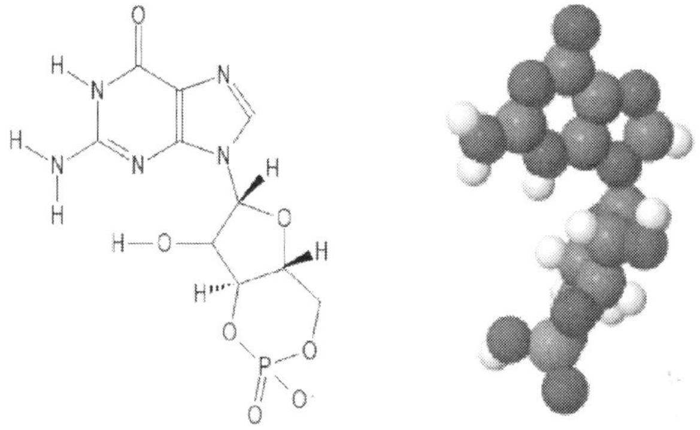

Fig. 2. Chemical and 3D structures cGMP (Molecular formula: $C_{10}H_{12}N_5O_7P$)

3.2 Methods for the detection of cGMP in plants

Cyclic GMP, much like cAMP, occurs at comparatively low concentrations in plant tissues and cells. As stated above, early research was hampered by the lack of sensitive and specific methods for measuring their small amounts and the requirement of complicated purification (and enrichment) methods prior to assaying (Brooker et al., 1968; Goldberg et al., 1969; Hardman et al., 1966). Cyclic GMP was first detected by measuring the radioactive molecule after injection of ^{32}P labeled substrate (Ashman et al., 1963). Three types of methods for measuring cGMP rapidly emerged: enzymatic methods (Kuo & Greengard, 1970), protein binding assay (Gilman, 1970) and immunoassay. The latter is based on immunological detection and ^{125}I-iodinated tracers. Two main techniques, the radioimmunoassay (Steiner et al., 1969) and enzyme immunoassay (Harper & Brooker, 1975) were developed subsequently and allowed determination of concentrations at the femtomole level (10^{-15} mol) with reasonable confidence. Nevertheless, unknown compounds from plant extracts may also cross-react, leading to inconsistencies. Thus, HPLC-mass spectrometric confirmation of the identity of the compound is advised (Brown & Newton, 1992). Other methods are based on the derivatization of cGMP and the successive purification by HPLC combined with UV-detection (Ohba et al., 2001; Soda et al., 2001). Recently, the combination of HPLC with mass spectrometry detection proved superior in terms of selectively and sensitivity (Cordell et al., 2008; Martens-Lobenhoffer et al., 2010; Witters et al., 1996). Additional information on the various assays and techniques developed to date is reviewed elsewhere (Schmidt, 2009) and most recently, an *in vivo* method for the detection of cGMP in real time in a non-invasive manner was described (Isner & Maathuis, 2011).

What are the cGMP concentrations in plants? For example, in pine (*Pinus densiflora*) pollen, cGMP was detected at the level of 60 pmol/g fresh weight (Takahashi et al., 1978). In maize seedlings, cGMP was estimated to be between 35 and 72 pmol/g fresh weight by GC-MS (Janistyn, 1983). Concentrations of cGMP ranging from 0.4 to 20 nmol/g fresh weight were detected in roots of bean (Haddox et al., 1974). In tobacco pith parenchyma, cGMP levels were estimated below 0.1 nmol/g dry weight (Lundeen et al., 1973). However, in the medicinal plants, *Evodia rutaecarpa, E. officinalis* and *Zizyphus jujuba*, levels as high as 10-50 nmol/g dry weight were reported (Cyong & Takahashi, 1982; Cyong et al., 1982). The rather wide range in cGMP concentrations is probably due, at least in part, to the presence of interfering substances and/or secondary metabolites that may have interfered with the assays. By combining gas chromatography and mass spectrometry on maize seedlings, 35-72 pmol/g fresh weight cGMP was measured (Janistyn, 1983). It is however noteworthy that considerable increases in cGMP levels occur in response to environmental cues such as light (Brown et al., 1989), NO (Pfeiffer et al., 1994) or pathogen attack (Meier et al., 2009), and these increases are a pointer to the functional importance of cGMP.

3.3 Biological functions of cGMP *in planta*

Research of the past few decades has connected the second messenger cGMP to a wide range of cellular and physiological processes in vertebrates and invertebrates, as well as bacteria, fungi, and algae. Receptor proteins can function as GCs and signal via cGMP (Chinkers et al., 1989) activating cGMP-dependent protein kinases (PKG, also abbreviated cGK), members of the PDE family and nucleotide-gated ion channels (CNGC) (Biel, 2009; Craven & Zagotta, 2006; Lincoln & Cornwell, 1993; Murad, 2006; Schmidt, 2009; Zimmerman, 2001), as well as transcription (Penson et al., 1996; Maathuis, 2006).

In invertebrates, cGMP controls homeotaxis of aggregating cells in the amoeba *Dictyostelium discoideum* (Bosgraaf et al., 2002) and modulates the expression of antimicrobial peptides in *Drosophila melanogaster* (Davies, 2006). In vertebrates, the signaling molecule is involved in differentiation, growth, apoptosis, regulation of electrolytes and fluid homeostasis (Kuhn, 2004; Steinbrecher et al., 2002) and in mammals, cGMP regulates a number of intracellular processes including neutrophil activation, phototransduction system of the retina (Cobbs & Pugh, 1985; Dhallan et al., 1992; Fesenko et al., 1985; Matthews et al., 1985; Yau & Nakatani, 1985) and natriuresis (Guillemant et al., 1994), and induces the mediation of smooth muscle relaxation in response to NO (Moncada et al., 1992). The role of cGMP on PDEs can also lead to alteration in the rate of hydrolysis of cAMP (Corbin & Francis, 1999).

In plants, growing evidence points to a central role of cGMP in a wide range of cellular and physiological processes including abiotic and biotic stress-response signaling (Donaldson et al., 2004; Durner et al., 1998; Gottig et al., 2008; Ma et al., 2009), the gating of ion channels (Anderson et al., 1992; Hoshi, 1995; Schuurink et al., 1998; Sentenac et al., 1992), light signal transduction (Bowler & Chua, 1994; Neuhaus et al., 1997; Neuhaus et al., 1993) and hormone signal transduction (Meier et al., 2009; Penson et al., 1996; Pharmawati et al., 1998; Pharmawati et al., 2001). Besides, treatment of Arabidopsis roots with the cell permeant cGMP analog (8-Br-cGMP) revealed that cGMP-mediated processes affect the transcriptome in general, and in particular lead to a transcriptional increase in transporter encoding genes. The transcriptional activation of non-selective ion channels and cation:proton antiporter encoding genes suggests a link between cGMP and the long-term modulation of monovalent cation (Na^+ and K^+) fluxes (Maathuis, 2006).

In 1992, two K^+ channels, KAT1 and AKT1, were determined in *Arabidopsis thaliana* (Anderson et al., 1992; Sentenac et al., 1992). While in mammals, the K^+ channels function in controlling the excitability of nerve and muscle cells (Hille, 1992), in plants and fungi, they are believed to be involved in turgor-mediated growth, osmotic adjustment, cell movement, and mineral nutrition (Gustin et al., 1986; Schroeder & Hedrich, 1989). Presence of a cGMP-binding motif in the C-terminal region suggested that cGMP might affect the channels (Schuurink & Jones, 1995), a role later confirmed (Hoshi, 1995), therefore linking the second messenger to the regulation of plant ion and water homeostasis.
In 1994, using the phytochrome A-deficient *aurea* tomato (*Lycopersicon esculentum*) mutant, phytochrome A, the photoreceptor responsible for regulating a wide range of morphogenesis responses such as flowering, seed germination or diurnal rhythms, was shown to utilize three different signal transduction pathways, dependent upon calcium and/or cGMP, to activate genes in the light (Bowler & Chua, 1994). In the first pathway, cGMP stimulates chalcone synthase and ferredoxin $NADP^+$ oxidoreductase, which in turn triggers anthocyanin synthesis (Bowler & Chua, 1994). The second pathway is dependent on calcium and calcium-activated calmodulin, whereas in the third one, both calcium and cGMP are required for the transcriptional activation of genes encoding photosystem I and cytochrome b6f complexes. Overall the combination of these three signaling systems mediates the development of chloroplasts and the biosynthesis of anthocyanin (Neuhaus et al., 1997).

3.4 Cellular signaling with cGMP

Cyclic GMP is becoming increasingly recognized as an important messenger in plant hormone signal transduction. It was linked for the first time to phytohormones while studying barley aleurone layers (Penson et al., 1996; Penson et al., 1997). Gibberellic acid (GA) was also shown to cause an increase in cytosolic cGMP, which in turn is essential for the induction of synthesis and secretion of $\propto$-amylase, the enzyme that breaks down starch. Auxin and kinetin, both of which cause stomatal opening, also show signaling via cGMP (Cousson & Vavasseur, 1998; Pharmawati et al., 2001), the former possibly through triggering the synthesis of cADP-ribose, which then influences Ca^{2+} fluxes. Abscisic acid (ABA) has been shown to induce stomatal closure through down-stream intermediates that include NO (Garcia-Mata & Lamattina, 2001) and hydrogen peroxide (McAinsh et al., 1996; Pei et al., 2000; Zhang et al., 2001). Recently, it came to light that germination of Arabidopsis seeds is modulated by cGMP levels, which in turn are dependent upon GA (Teng et al., 2010). However, the downstream effectors of cGMP triggering seed germination remain unknown. Taken together, it appears that cGMP is an effector of a number of phytohormones, leading to the downstream signaling mechanisms necessary for plants growth, development and responses to abiotic and biotic stresses.
In animals, NO acts as a highly instable signaling hormone in the cardiovascular system and the brain (Ignarro, 1999) acting locally and signaling via binding to and stimulating GCs (Arnold et al., 1977) giving rise to cGMP transients (Furchgott & Vanhoutte, 1989; Ignarro, 1991). Roles of NO have mainly been explored in animals where NO has been implicated in a number of physiological and pathological responses such as neurotransmission, smooth muscle relaxation, immunity, and apoptosis (Ignarro, 2000; Jaffrey & Snyder, 1995; Lloyd-Jones & Bloch, 1996; Moncada et al., 1991; Wink & Mitchell, 1998). Downstream effectors of NO includes activation of GCs, transcription factors and ion-channels (Melino et al., 1997;

Stamler, 1994). Detection of NO in plants came in 1994, when cGMP production was shown to be stimulated by NO in spruce needles (Pfeiffer et al., 1994). In 1998, it was confirmed that NO acts indirectly through activating the soluble form of GCs in plants and thus similarly to mammals (Martens-Lobenhoffer et al., 2010). Following stimulation by NO, cGMP induces expression of phenylalanine ammonia lyase (Durner et al., 1998), a key enzyme in the phenylpropanoid biosynthetic pathway. This stimulation is also believed to be implicated in the biosynthesis of salicylic acid (Dixon & Paiva, 1995; Mauch-Mani & Slusarenko, 1996; Pallas et al., 1996). In addition, a number of publications have revealed NO as an ubiquitous plant signaling molecule that also plays a role in growth and development including induction of leaf expansion, root growth as well as cell death and defense against pathogens (Durner et al., 1998; Gouvea et al., 1997; Leshem, 1996; Neill et al., 2008). Nitric oxide has also been linked to the induction, stimulation or inhibition of ABA, GA and ethylene (Beligni et al., 2002; Leshem, 1996; Neill et al., 2002). In analogy with the mammalian system where the majority of NO effects are accomplished through elevation in cGMP levels (Bogdan, 2001; Mayer & Hemmens, 1997), cGMP was demonstrated to be required in NO-induced programmed cell death in Arabidopsis (Clarke et al., 2000).

Cyclic GMP is also a second messenger in plant natriuretic peptide (PNP) responses. PNPs are small peptidic molecules with biological activity at nanomolar concentrations (Gehring, 1999; Gehring & Irving, 2003) that induce rapid and transient increases in cellular cGMP levels (Pharmawati et al., 1998; Pharmawati et al., 2001; Wang et al., 2007) and modulate K^+, Na^+ and H^+ net fluxes (Ludidi et al., 2004; Pharmawati et al., 1999) thus linking cGMP effects to the regulation of cellular homeostasis. PNPs have also been implicated in stress defense and possibly play a role in the changes of host homeostasis, which in turn are a part of plant defense against biotrophic bacteria (Gottig et al., 2009).

4. Cyclic nucleotide gated channels – structure and function

At the turn of the 20th century, one could read statements such as "*Corpora non agunt nisi fixata*" (drugs do not act unless they are bound). John N. Langley in England and Paul Ehrlich in Germany are usually credited for developing the concept of "receptive substance"; that could also act as a transducing engine. Fifty years later, Earl W. Sutherland (in the USA), studying the hyperglycemic response in hepatocytes, discovered that addition of epinephrine and/or glucagon in the presence of ATP and Mg^{2+} to the so called "particulate material" (i.e. fraction containing fragments of plasma membrane) obtained from liver slices, but not to "the soluble system", resulted in the generation of a newly synthesized heat-stable factor that was identified soon after to be cAMP. Moreover, cAMP was able to mimic the effects of the hormones on the liver glycogen phosphorylase when added on its own to the soluble fraction (Sutherland et al., 1968). Sutherland and his team paved the way for the fields of secondary messenger systems, G proteins and cell signaling and thus, both cAMP and cGMP became the focus of extensive studies. They were found to be ubiquitous across kingdoms - however, with considerable ambiguity in higher plants (discussed above) - and to regulate a vast array of cellular functions by controlling the activity of numerous proteins downstream of hormone-receptor interaction. Amongst the first proteins to be found regulated by cyclic nucleotides were the protein kinases (PKA and PKG types activated by cAMP and cGMP, respectively). When activated, these kinases phosphorylate an array of other cellular targets including other kinases, phosphatases, gene transcription factors and an ever-growing list of ion channels (Gray et al., 1998; Montminy,

1997; Shipston, 2001; Thevelein & De Winde, 1999; Trautwein & Hescheler, 1990). Note that while the activity of some ion channels can be enhanced by cAMP and/or cAMP-dependent PKA treatment, the activity of others can be down-regulated following the same treatment (Milhaud et al., 1998; Osterrieder et al., 1982; Shuster et al., 1985; Siegelbaum et al., 1982).

By the mid 1980's, Fesenko and co-workers (Fesenko et al., 1985) published a seminal paper where they showed that cationic channels from the retinal outer segments could be directly gated 'open' (no phosphorylation reaction required) by the addition of cGMP. Nowadays, the rod outer segment channel is classically described as part of a special class of channels: the cyclic nucleotide gated channels or CNGCs. These CNGCs form a distinct branch within the superfamily of pore-loop cation channels (Hofmann et al., 2005; Yu et al., 2005). The family of mammalian CNGCs comprises six homologous members, which are classified as A subunits (CNGA1-4) and B subunits (CNGB1 and CNGB3). Based on the detection of their respective mRNAs, CNGCs were found in the heart, brain, muscle, liver, kidney, and testes (Kaupp & Seifert, 2002; Zimmerman, 2001) but that the analysis of CNGC knock-out mice did not provide convincing evidence for relevant physiological function raising questions about CNGC expression in non-sensory system cells (Biel & Michalakis, 2009).

Two cyclic binding domains (CNB), named S1 and S2, were originally shown in the regulatory subunits of both RI and RII monomers of the bovine PKA (Corbin & Francis, 1999). The bacterial protein catabolite activator protein (CAP), also contains this CNB domain, which is linked to a DNA binding domain, and mediates gene transcription in the absence of glucose. The elucidation of many bacterial genomes indicates that this domain is much more diverse. There are literally hundreds of proteins that share this common ancestral structural domain made of about 120 amino acid residues and comprising both β-strands and α-helical elements (Berman et al., 2005; Wu et al., 2004). So, besides the mammalian CNGC, this CNB structure is found in a protein that binds calcium in the cytosol: calmodulins (CaM), in channels like the hyperpolarization-activated cyclic nucleotide-gated channels (HCN), Eag-like K^+-channels (Berman et al., 2005; Biel, 2009; Biel & Michalakis, 2009; Craven & Zagotta, 2006). This feature is also shared by K^+-channels from higher plants; indeed KAT1 and AKT1 were shown to contain the structural domain and/or to be regulated by cGMP (Gaymard et al., 1996; Hoshi, 1995; Sentenac et al., 1992).

In 1998, Schuurink and colleagues demonstrated the existence of CNG-like channels in *Arabidopsis thaliana* plants (Schuurink et al., 1998). They employed a strategy whereby 'CaM conjugated to horseradish peroxidase', was used to screen a complementary DNA expression library for CaM binding proteins in barley aleurones. One of the cDNAs obtained by this screen was shown to be a unique protein of 702 residues; the molecular architectural structure of the deduced translation product resembled that of the superfamily of genes encoding the voltage-gated ion channels. The *Arabidopsis thaliana* genome was discovered to encode 20 putative members of the CNGC family and homologs of these proteins were also found in major crop plants like rice, maize and cotton (*Gossypium spp.*). Plant CNGCs were classified into four main groups: Group I, II, III and IV (group IV is subdivided in two groups: A & B). For more details, the reader is encouraged to consult the following excellent reviews on this subject: (Demidchik et al., 2002; Dietrich et al., 2010; Kaplan et al., 2007; Martinez-Atienza et al., 2007; Mäser et al., 2001; Sherman & Fromm, 2009; Talke et al., 2003; Very & Sentenac, 2002; Ward et al., 2009; White et al., 2002).

Biochemical (Peng et al., 2004; Weitz et al., 2002; Zhong et al., 2002) and biophysical data (Zheng et al., 2002; Zheng & Zagotta, 2004) as well as patch clamp recordings from CNGCs in animal native sensory neurons and those obtained from heterologous co-expression of distinct subunit combinations (Chen et al., 1993; Kaupp et al., 1989; Korschen et al., 1995; Matthews & Watanabe, 1987; Torre et al., 1992), all strongly suggest that animal CNGC subunits must assemble into tetrameric complexes with stoichiometries that are variable from one type of sensory neuron to another (reviewed in Biel, 2009). For instance, the CNGC expressed in cilia of olfactory sensory neurons consists of three different subunits: CNGA2, CNGA4, and a short isoform of the CNGB1 subunit (CNGB1b) (2:1:1 stoichiometry). The CNG channel of rod photoreceptors consists of the CNGA1 subunit and a long isoform of the CNGB1 subunit (CNGB1a) (3:1 stoichiometry), while the cone photoreceptor channel consists of the CNGA3 and CNGB3 subunits (2:2 stoichiometry). The hydrophobic core of each of the subunits consists of the classic six putative α-helical membrane-spanning segments (named S1 to S6), a re-entrant pore (P) loop between S5 and S6, and hydrophobic N- and C-termini regions, both projecting into the cytosol (Fig. 3). In animals, CNGCs have the CaM binding- and CNB domains (CaMBD and CNBD, respectively) found in the opposite N- and C termini (Fig. 3A). However in plants, both CaM and CNB domains co-exist at the carboxyl terminus (Fig. 3B). Three-dimensional modeling of the CNBD structure of an *Arabidopsis thaliana* CNGC (AtCNGC2) revealed that cAMP binding to CNBD occurs in a pocket formed by a β barrel structure appressed against a shorter (relative to animal CNBD) αC helix (Hua et al., 2003b).

All channels from the voltage-gated channels superfamily including Na_v, K_v and HCN contain a highly positively charged S4 helix: the sensor for the allosteric modulation of gating by membrane voltage (Flynn et al., 2001). S4 is classically described as carrying three to nine regularly spaced arginine or lysine residues at every third position and these residues may confer voltage sensitivity to the channels (Biel, 2009; Jan & Jan, 1992). But in CNGCs it was proposed that glutamate residues in the vicinity of S4 contribute negative charges that may neutralize positively charged arginine and lysine residues leaving the S4 domain with a lower net positive charge with respect to K_v channels and rendering it less sensitive to voltage (Wohlfart et al., 1992). Moreover, it was found that the S4–S5 linker is shorter (11 amino acids instead of 20) and probably forms a loop in CNGCs. This is in contrast to related K_v and HCN channels, where the S4–S5 linker forms an α-helical structure and thus a rigid connection between the voltage sensor domain and the pore lining domain S5 (Ohlenschlager et al., 2002). Consequently, this causes the motion of the S4 domain to be less efficiently coupled to the S6 domain as in K_v and HCN channels (Anselmi et al., 2007). CNG current recordings in whole cell mode show no voltage-dependent inactivation and analysis of their current-voltage (I-V) relations shows only a weak voltage-dependence (especially at saturating or near-saturating ligand concentration and in the absence of divalent cations; Haynes & Yau, 1985; Picones & Korenbrot, 1992b; Zagotta & Siegelbaum, 1996). The weak rectification seen in I-V relations from recordings made in intact animal cells may originate from voltage-dependent channel block by Ca^{2+} and Mg^{2+} (Zimmerman & Baylor, 1992).

Much of our current understanding of the molecular mechanisms of channel ion conduction is in large part due to the elucidation, by MacKinnon and co-workers, of the X-ray crystal structure of KcsA, a K^+-channel from *Streptomyces lividans* (Doyle et al., 1998). The molecular architecture of the putative pore region (P-loop) between S5 and S6 forms the basis of ion permeation and selectivity. In all the P-loop-containing protein family of channels including animal and plant CNGCs, this structure (Fig. 3) dips into the membrane bilayer as an α-helix (pore helix) and exits back extracellularly as an uncoiled strand (Biel, 2009; Flynn et al., 2001; Hua et al., 2003b). In K^+-channels, a highly conserved amino acid triplet, GYG, seems to be a pre-requisite for K^+ selectivity (Doyle et al., 1998). Sequence alignments of the pore helix shows that both animal and plant CNGCs lack the GYG motif. Animal CNGC1 and CNGC2 have a GXET triplet; where the crucial glutamic acid (E) in place of the tyrosine (Y), confers nonselective cation permeation properties (Doyle et al., 1998; Flynn et al., 2001; Zagotta & Siegelbaum, 1996). CNGCs discriminate poorly among monovalent cations with no significant permeability to anions (example of selectivity sequence in rods: $Li^+ \geq Na^+ \geq K^+ > Rb^+ > Cs^+$), with Ca^{2+} and Mg^{2+} being more permeable than the other cations (Yau & Baylor, 1989; Zagotta & Siegelbaum, 1996; Zimmerman, 2001). In the absence of divalent cations, the single channel conductance of animal CNGCs can be as high as tens of picoSiemens (pS; at least two conductance states have been resolved in rods: ~ 8 - 10 pS and ~ 25 - 30 pS; and in cones and olfactory cells: 45 to 50 pS with apparent sub-conductance states; Haynes et al., 1986; Ildefonse & Bennett, 1991; Matthews, 1987; Ruiz & Karpen, 1997; Taylor & Baylor, 1995; Zimmerman & Baylor, 1992).

To study the biophysical properties of CNGCs (or any other channel for that matter), electrophysiologists took advantage of an elegant and powerful technique that allows heterologous expression of the clones (cDNA and/or mRNA) in a suitable system (HEK cells or oocytes). The end product (ion channel) should in theory 'correctly' fold and get inserted into the membrane to allow further analysis using the patch- or any another voltage clamp technique. Unfortunately, this technique, which by the way yielded important key biophysical information for animal CNGCs (and has also being used successfully to study plant KAT1 and AKT1 ion channels) turned out to be surprisingly unproductive for the electrophysiological study of plant CNGCs. Witness of this statement is the rather scarce number of reports related to this issue (see recent reviews from Dietrich et al., 2010; Ward et al., 2009). Speculations were advanced that maybe the functional analysis of these plant CNGCs upon expression in heterologous systems could be hindered by interaction of the plant protein with regulatory systems/molecules present in the host cell (Ali et al., 2006; Very & Sentenac, 2002). To date, only two groups have produced functional data with some electrophysiological characterization of plant CNGCs (Balague et al., 2003; Hua et al., 2003a; Hua et al., 2003b; Leng et al., 2002; Leng et al., 1999). The *Arabidopsis thaliana* channel AtCNGC2 was the first cloned plant channel shown to be permeable to Ca^{2+} (Leng et al., 1999). A few years later, AtCNGC4 was found to be permeable to both Na^+ and K^+ as is the case for animal CNGCs (Balague et al., 2003). AtCNGC2 was also found to discriminate poorly between all monovalent cations tested (K^+, Li^+, Rb^+ and Cs^+) except for Na^+. Indeed, AtCNGC2, unlike any animal or plant CNGCs cloned to date, is impermeable to Na^+ (Hua et al., 2003a). A molecular analysis of what is presumed to form the selectivity filter in the P-loop indicates that AtCNGC2 has an "AND" triplet that differs from the selectivity filter motifs of all other known CNG channels. The same authors found that specific amino acids within the AtCNGC2 pore selectivity filter, namely Asn-416 and Asp-417, facilitated K^+ permeation over Na^+ (Hua et al., 2003b). They proposed that the "AND" motif can confer selectivity against Na^+, as does the "GYG" motif in K^+-selective channels (Doyle et al., 1998; Jiang & MacKinnon, 2000). Note that most plant CNGCs, for instance AtCNGC1, or their homologs from other plants like NtCBP4 or HvCBT1 have a GQN triplet. It was suggested that glutamine (Q) could equally compromise the K^+ selectivity in animal cation channels, as does the glutamic acid (E) residue (Kerr & Sansom, 1995; Leng et al., 2002).

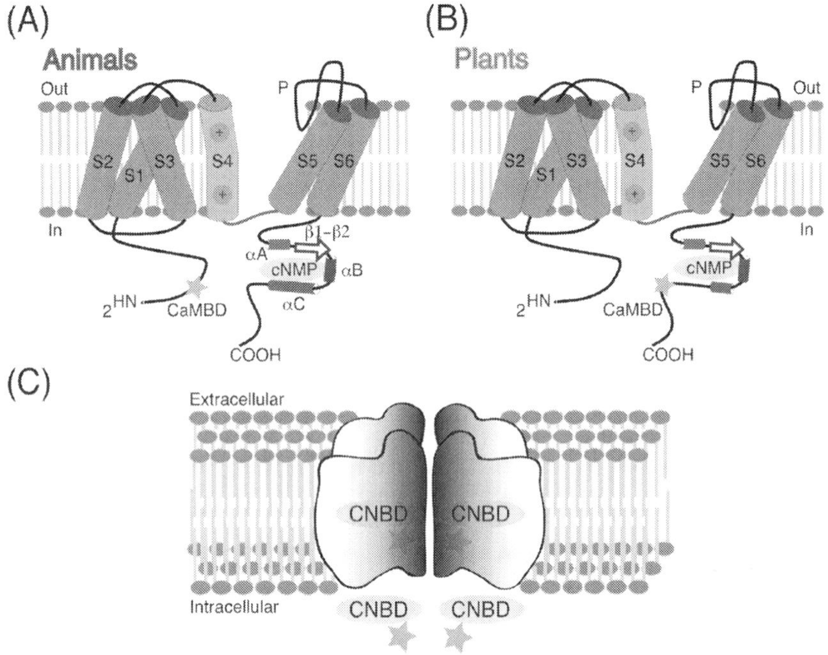

Fig. 3. Animal and plant CNGC membrane topologies and domain structure models. (A) Animal and (B) plant CNGCs share the common six α-helical transmembrane segments (S1-S6) and a pore loop (P) between S5 and S6, which is thought to contain the selectivity filter. The S4 segment is composed of a series of positively charged residues that forms the voltage sensor in all voltage-dependent channels, but in CNGCs, S4 is believed to carry a diminished net positive charge resulting in weaker voltage sensitivity. The N- and C-termini of plant and animal CNGCs extend into the cell cytosol. In the cytosolic C terminus, all CNGCs subunits carry a CNBD that is functionally coupled to the transmembrane channel core via the C-linker domain which is formed by three α-helices (A-C) and two β sheets (β1 and β2) forming a β barrel between the αA and αB helices. In sensory neurons, binding of cNMPs to the CNBD is thought to produce an allosteric conformational change that increases the open probability of the channel pore. The CaMBD is located at the N-terminus in animal CNGCs while in plants, it is located at the C-terminus, beginning at the truncated αC helix. (C) Four subunits assemble to form functional tetramers in animal CNGCs. Plant CNGCs are also predicted to form tetramers with each subunit associated to one CNBD and one CaMBD (figure adapted and modified from Biel, 2009 and Hua et al., 2003b).

In addition to their direct activation by cNMPs, another distinguishing feature of CNGC in native (animal) membranes is their remarkable fast gating kinetics (open-shut transitions) when recording from toad rods in cell-attached or excised patch modes (Matthews & Watanabe, 1987). This rapid flickery behaviour occurs even in the absence of any added cations like Ca^{2+} and Mg^{2+} that are known to result in flicker block mode (Zimmerman, 2001). However, when the channels were purified and reconstituted, or cloned and heterologously expressed, the channels performed like any other typical channels, i.e. they had slower gating kinetics (Kaupp et al., 1989). It was found that the flickery gating pattern from the native membranes resulted partly from the presence of at least one more type of subunit that was missing in the original reconstitution and expression studies (Chen et al., 1993; Korschen et al., 1995).

To date, there are only few reports in the literature showing any kind of modulation (i.e. direct or indirect) by cNMPs of an ion channel in a native plant membrane. It was suggested (Li et al., 1994) that cAMP, in the presence of 3-isobutyl-1-methylxanthine (IBMX), a cyclic mononucleotide phosphodiesterase inhibitor, modulated an outward K^+-current ($I_{K,out}$) in *Vicia faba* mesophyll cells indirectly through a PKA. Although the amount of cAMP used in the paper was high (1.5 mM), this modulation was specific and could not be reproduced by cGMP or other non-cyclic nucleotides (AMP or GMP). In addition, this modulation was affected with all the known usual suspects of the PKA pathway: PKI and Rp-cAMP, both inhibitors of the PKA, also inhibited $I_{K,out}$ and more importantly, the catalytic subunit of PKA enhanced the magnitude of $I_{K,out}$ in the absence of any added cAMP. In contrast to this, it was also shown that cAMP added alone did not affect ion channels in Vicia guard cells, but if added after ABA and/or internal Ca^{2+} treatments, cAMP was able to antagonize their well-characterized inhibitory effect on the inward K^+-current ($I_{K,in}$) (Jin & Wu, 1999). Other lines of evidence that suggest a role for cNMPs in plants come from studies of the modulation of stomatal aperture and/or internal free cytosolic Ca^{2+} levels (Jin & Wu, 1999; Kurosaki et al., 1994; Moutinho et al., 2001; Pharmawati et al., 1998; Pharmawati et al., 2001; Volotovski et al., 1998). The work of Volotovski and coworkers (1998) clearly and unambiguously demonstrated that physiological responses such as protoplast swelling and change in cell Ca^{2+} homeostasis occurred in response to cAMP and cGMP. Indeed, both cNMP-evoked elevation in cytosolic Ca^{2+} concentration ($[Ca^{2+}]_{cyt}$) and other pharmacological compounds such as forskolin (a direct activator of AC), IBMX (a cAMP PDE inhibitor) and zaprinast (a cGMP PDE inhibitor) could all mimic the effect of cNMPs on $[Ca^{2+}]_{cyt}$. The elevation in $[Ca^{2+}]_{cyt}$ was suppressed by verapamil, suggesting to the authors that a significant proportion of the $[Ca^{2+}]_{cyt}$ response occurs via activation of verapamil-sensitive Ca^{2+}-channels. Having said that, the authors could not exclude that internal Ca^{2+} stores could also be involved in the cNMP-elevation of $[Ca^{2+}]_{cyt}$. All these reports point to an indirect effect of cNMPs on plant cells through PKA or other cross talk mechanisms. In addition, support for a role of cNMPs in regulating plant ion channels comes indirectly from work done on the regulatory effects of a PNP hormone that is thought to operate through a cGMP-dependent pathway (Pharmawati et al., 2001; Wang et al., 2007). Using MIFE, a non-invasive microelectrode technique for measurements of net ion fluxes, it has been found that NP and cGMP activates non-selective cation channels (NSCCs) in maize root conductive tissue (Pharmawati et al., 1999). It is important to note that although, these authors referred to a direct effect of cGMP on NSCCs, they used that term to distinguish between NP and cGMP action. Indeed, unlike NP, cGMP was found not to stimulate the plasma membrane H^+-pump. In the same line of indirect evidence, NO which was shown in plant cells to dramatically and transiently increase cGMP levels (Dubovskaya et al., 2011; Durner et al., 1998), was found to induce stomatal closure and confer increased tolerance to plant water deficits (Garcia-Mata & Lamattina, 2001). Nitric oxide raised $[Ca^{2+}]_{cyt}$ by promoting Ca^{2+} release from intracellular stores. Antagonists of GCs and cyclic ADP ribose-dependent endomembrane Ca^{2+}-channels blocked the increase of free cytosolic Ca^{2+}. It was proposed that a cGMP-dependent cascade might be the origin of this effect (Garcia-Mata et al., 2003). Recently, the evidence of cGMP acting upstream of $[Ca^{2+}]_{cyt}$ increases following either H_2O_2 and/or NO treatment was provided (Dubovskaya et al., 2011). Also, H_2O_2 and NO were found to act downstream of ABA to close the stomata and the changes in NO synthesis were induced by H_2O_2, resulting in increases in [cGMP] and $[Ca^{2+}]_{cyt}$.

A direct effect of cGMP (not cAMP) on the plant K^+ channels KAT1 (Hoshi, 1995) and AKT1 (Gaymard et al., 1996) has been proposed. However, in these studies a cGMP-ATP antagonism was described and indirect effects through possible kinases could not be ruled out (Gaymard et al., 1996). The effect of cGMP on these channels was to shift the activation voltage to more negative potentials, thus cGMP effectively reduced current through these channels over the physiological range of voltages. The first clear-cut evidence for a direct, effect of cNMP on any plant ion channel came from the work of Maathuis and Sanders (2001). They demonstrated cNMP inhibition of a root voltage-insensitive channel (VIC) that showed weak selectivity amongst monovalent cations and which probably provides a pathway for Na^+ uptake in root cells (Maathuis & Sanders, 2001). They suggested that this NSCC could be the target for the direct effect of cNMP seen in excised inside-out patches, and hence, be considered as candidate for the presence of functional CNGCs *in planta*. This effect was somehow unexpected since most animal and plant CNGCs, native and/or heterologously expressed, are rather activated by addition of cNMPs not deactivated. It was argued that cNMPs might exert their attenuating effect on Na^+ toxicity (Essah et al., 2003; Maathuis & Sanders, 2001; Rubio et al., 2003) not only by decreasing root cell NSCC activity and thus Na^+ uptake but also by activating CNGCs, like *At*CNGC2 involved in Ca^{2+} signaling (Leng et al., 1999), thus providing a possible role for CNGCs in plant salinity tolerance (Talke et al., 2003). There is also strong evidence to support a direct effect of cAMP in the shoot. Indeed, cAMP (or its membrane permeable derivative, dibutyryl cAMP) regulated a calcium conductance in both whole-cell mode and excised patches from *Vicia faba* guard and mesophyll cells (Lemtiri-Chlieh & Berkowitz, 2004). In this work, the authors showed that addition of cAMP consistently activated a channel with a 13 pS chord conductance. It was concluded that cAMP-activated calcium current seen in the whole protoplast configuration could be explained by a direct effect of cAMP binding on the Ca^{2+}-channel itself or a close entity to it. Like animal native CNGCs (Matthews & Watanabe, 1987), cAMP induced a rapid flickery gating behavior of the plant channel, but unlike them, cAMP binding was not the exclusive activator. Voltage (hyperpolarization) also activated the channel even in the absence of any added cAMP, resembling a feature of the mammalian HCN channels (Biel & Michalakis, 2009; Craven & Zagotta, 2006). However, CNGCs in the photoreceptor outer segments from striped bass, exhibit a very low level of activity even in the absence of added CNs, and the kinetics of this activity were the same as those measured in the presence of cGMP (Picones & Korenbrot, 1992a).

5. Evidence of nucleotide cyclases in higher plants

Given that the role of cNMP is recognized in many biological processes as both essential and sufficient to transduce signals and/or elicit physiological responses ranging from protein phosphorylation to transcriptional activation of specific genes, it is somewhat surprising how little we still know about NCs, the enzymes that catalyze the conversion of nucleotide triphosphate to cNMP and pyrophosphate. This gap in our understanding is particularly noticeable in higher plants.

In late 1997, a potentially highly influential paper was published in "Nature" linking an AC to the signaling of the plant hormone auxin in tobacco. The AC gene (*axi141*) was claimed to have AC activity and protoplasts transformed with it and treated with the AC stimulating compound forskolin could grow in the absence of cAMP, thus establishing the messenger as necessary and sufficient for auxin-dependent growth. However, the paper has been retracted since *axi141* does not confer auxin-independent growth and we do not know if indeed the reported data on AC activity are factual. To-date, the only annotated and experimentally confirmed cAMP generating molecule in plants is a maize pollen AC required for polarized pollen tube growth, which in turn depends on cAMP (Moutinho et al., 2001). The Arabidopsis orthologue of this protein (At3g14460) is annotated as a disease resistance protein belonging to the nucleotide-binding site-leucine-rich repeat (NBS-LRR) family used for pathogen sensing with a role in defense responses and apoptosis (DeYoung & Innes, 2006). NBS-LRR proteins directly bind pathogen proteins and associate with either a modified host protein or a pathogen protein, leading to conformational changes in the amino terminal and LRR domains of NBS-LRR proteins that are thought to promote the exchange of ADP for ATP by the NBS domain. It is thus conceivable that NBS-LRR downstream signaling is enabled by cAMP.

The discovery of the first GC (AtGC1; At5g05930) in higher plants is even more recent (Ludidi & Gehring, 2003) and was made possible by the deduction of a 14 amino acid (AA) long search term (GC core motif, see Fig. 5) based on an alignment of conserved and functionally assigned amino acids in the catalytic centre of annotated type III GCs from lower and higher eukaryotes (Liu et al., 1997; McCue et al., 2000). It has since been shown that the Arabidopsis brassinosteroid receptor (AtBRI) (Kwezi et al., 2007), a stress responsive wall associated kinase-like molecule (AtWAKL10; At1g79680) (Meier et al., 2010) and the Arabidopsis peptide signaling molecule (Pep1) receptor (AtPepR1; At1g73080) (Qi et al., 2010) also contain functional GC domains and this is commensurate with the presence of the amino acid residues essential for catalysis. The discovery of increasing complexities in the molecular architecture of higher plant NCs and in particular GCs (Fig. 4) is entirely compatible with findings in the single celled green alga *Chlamydomonas reinhardtii*, where catalytic NC domains appear in >20 different domain combinations including H-NOX, periplasmic binding protein, GAF-like, protein kinase-like domain, ATPase domain of HSP90, ribonuclease-H domain, G protein-coupled receptors and cysteine proteinase (Meier et al., 2007).

If we assume that many of these multiple domain proteins operate as multifunctional units, they may well have highly diverse and complex roles in cellular signaling (Meier et al., 2007). There are potentially several signaling modes that could be in operation in these complex multi-domain NCs, one of which is intramolecular cross-talk where the GCs activated in a stimulus or ligand specific way are generating a cytosolic cGMP signature which in turn modifies the activity of a second domain (e.g. kinase) and thereby enabling intra-molecular cross-talk.

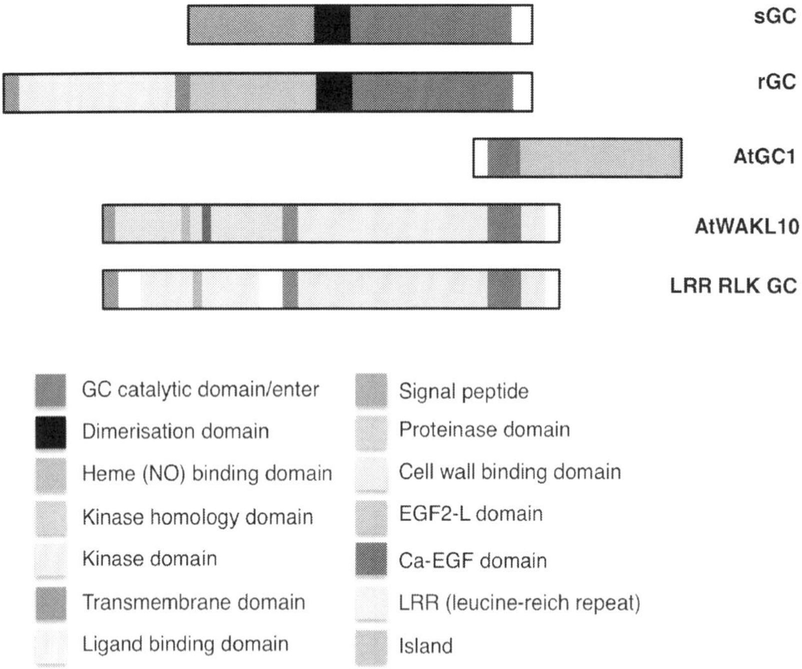

Fig. 4. Guanylyl cyclase domain organisations. Soluble GCs typically contain heme-binding domains allowing for NO dependent activation while reticulate GCs (rGC) contain transmembrane domains and extracellular ligand domains. The signal peptides direct the peptide to the cell membrane. AtWAKL10 contains an extracellular region that is tightly associated with the cell wall and contains several epidermal growth factor (EGF) repeats that may act as ligand binding domains (Meier et al., 2010). The AtWAKL10 protein has previously been predicted to contain an extracellular calcium binding EGF-like domain and a degenerate EGF2-like domain (He et al., 1999).

6. In search of novel nucleotide cyclases in higher plants

It has long been evident that BLAST searches using annotated or experimentally proven prokaryote or lower and higher eukaryote AC and GC sequences do not identify plausible plant candidate NC (Ludidi & Gehring, 2003) with low e-values since the degree of conservation in the NC catalytic center of these multi-domain enzymes is too low. Instead, we have argued that the 14 AA catalytic centre (not including the metal binding residues) may be sufficiently conserved to identify NCs in higher plants. For this reason, we have deduced a rational query motif based on specific functional residues in the GC catalytic center, which has yielded seven candidates (Fig. 5) (Ludidi & Gehring, 2003), one of which (AtGC1) was functionally tested and proved to be the first GC in higher plants. Recently, a homologue of this gene was identified in Morning Glory (*Pharbitus nil*) and shown to have catalytic activity *in vitro* as well as being a potential element of the light signalling responses *in vivo* (Szmidt-Jaworska et al., 2009).

The original GC motif:

```
--[RKS][YFW][CTGH][VIL][FV]X[DNA]X[VIL]X(4)[KR]--
   1     2     3     4   5  6  7  8  9 10-13 14
```

The relaxed GC motif:

```
--[RKS][YFW][CTGH][VIL][FV]X(3)[VIL]X(4)[KR]X(1,2)[D]--
   1     2     3     4   5         9 10-13 14
```

The relaxed AC motif:

```
--[RKS]X[DE]X(9,11)[KR]X(1,3)[DE]--
   1     3         12/14
```

The extended AC motif:

```
--[RK][YFW][DE][VIL][FV]X(8)[KR]X(1,3)[DE]--
  1    2    3    4    5       14
```

Fig. 5. Guanylyl cyclase and adenylyl cyclase motifs employed in the identification of nucleotide cyclases in higher plants. In the original GC motif the residue (red) in position 1 does the hydrogen bonding with the guanine, the amino acid in position 3 confers substrate specificity and the residue in position 14 stabilises the transition (GTP/cGMP). The Mg^{2+}/Mn^{2+}-binding site is in the C-terminal (green). In the derived AC motifs position 3 (magenta) has been substituted to [DE] to allow for ATP binding.

In a further step, we used site directed mutagenesis to reduce stringency of the 14 AA GC core motif but added additional AAs to the search term to include the C-terminal Mg^{2+}/Mn^{2+}-binding site and the N-terminal PPi-binding site and this has yielded a further 27 candidate GCs in Arabidopsis, two of which (AtBRI1 and AtPepR1) have already been shown to be catalytically active *in vitro*. We have also undertaken a gene ontology analysis (Al-Shahrour et al., 2004) of these genes with a view to determine if this group is enriched for specific functions and found a significant over-representation ($p < 1e^{-5}$) in the Fatigo+ (level 4) categories of phosphorus metabolic processes, protein metabolic process, cellular macromolecular metabolic process and biopolymer metabolic process. It is noteworthy that several are annotated as leucine rich repeat receptor like kinases (LRR RLKs) and that the GC domain identified lies within the cytosolic kinase domain. Furthermore, three genes of this group (CLAVATA1 receptor (At4g20270) and ERECTA (At5g07180 and At2g26330)) have roles in plant development and meristem development in particular while one candidate (At3g46400) has similarity to a light repressible protein kinase. The key challenge is to elucidate GC activity of the untested GCs *in vitro* and *in vivo* to establish the role(s) of cGMP in downstream signaling which will yield important new insights into this family of plant proteins.

As mentioned above, currently the only annotated and experimentally confirmed AC in plants is a maize pollen protein, which can generate cAMP that in turn acts as second messenger in polarized pollen tube growth (Moutinho et al., 2001). The *Arabidopsis thaliana* orthologue of this protein (At3g14460) is annotated as disease resistance protein and belongs to the NBS-LRR family that is used for pathogen sensing and also has a role in defense responses and apoptosis (DeYoung & Innes, 2006). NBS-LRR proteins directly bind pathogen proteins and associate with either a modified host protein or a pathogen protein leading to conformational changes in the amino-terminal and LRR domains of NBS-LRR proteins that are thought to promote the exchange of ADP for ATP by the NBS domain. It is thus conceivable that NBS-LRR downstream signaling (DeYoung & Innes, 2006) is enabled by cAMP.

Given that cyclic nucleotides have important and diverse roles in plant signaling via cyclic nucleotide-responsive protein kinases, -binding proteins and -gated ion channels (Newton & Smith, 2004), it is unlikely that a single AC or GC can account for all cAMP and cGMP dependent processes in higher plants. In line with this hypothesis is the fact that a number of Arabidopsis molecules with highly variable domain combinations and experimentally confirmed GC activity have recently been reported (Kwezi et al., 2007; Ludidi & Gehring, 2003; Meier et al., 2010). It is likely that what is true for GCs will also hold true for ACs. This then leaves us with the task to identify them, a task made complicated by the fact that BLAST searches with known and/or experimentally confirmed NCs from lower or higher eukaryotes do not return plausible candidate molecules. We also note that Prosite signatures for class one and class two ACs (EYFG[SA]X(2)LWXLYK and YRNXW[NS]E[LIVM]RTLHFXG, respectively) are not present in the *Arabidopsis thaliana* proteome even if we allow two mismatches.

Since in Arabidopsis, GCs have been identified with a catalytic motif search (Fig. 5), it was reasoned that a similar approach might lead to the discovery of novel ACs. In the modified AC search motifs the amino acid residues that confer substrate specificity (position 3 in Fig. 5) are substituted to [DE]. Consequently, the AC core motif within the catalytic centre consists of the functionally assigned residue that does the hydrogen bonding with the adenine (position 1), the amino acid that confers substrate specificity for ATP (position 3) and the amino acid that stabilizes the transition state from ATP to cAMP ([K,R], position 12-14). Additional diagnostic residues are the Mg^{2+}/Mn^{2+} binding amino acid [D,E], usually 1 – 3 amino acids removed from C-terminal of the transition state stabilizing residue. Such a relaxed AC motif ([RKS]X[DE]X(9,11)[KR]X(1,3)[DE]) (Fig. 5) is indeed present in a maize AC (AJ307886.1) which is the only experimentally tested AC in plants. It is also present in the *Sorghum bicolor* ortholog (gb|EER90437.1) and the related ($2e^{-70}$) Arabidopsis NBS-LRR class protein (At3g14460).

In many annotated GCs and all experimentally confirmed GCs in Arabidopsis (Kwezi et al., 2007; Ludidi & Gehring, 2003; Meier et al., 2010; Szmidt-Jaworska et al., 2009), the position between the assigned residue that does the hydrogen bonding (position 1) and the amino acid that confers substrate specificity (position 3) of the core motif is [YFW] and this is also the case in the confirmed maize AC which incidentally shares some similarity ($5e^{-05}$) with an annotated *Neurospora crassa* AC (XP_965280.1).

In Arabidopsis (TAIR: www.arabidopsis.org) there are currently only three annotated but functionally unconfirmed ACs (At1g26190, At1g73980 and At2g11890) and they all do contain a relaxed AC motif ([RKS]X[DE]X(9,11)[KR]X(1.3)[DE]) (Fig. 5). The first, a phosphoribulokinase/uridine kinase family protein shows similarity ($3e^{-108}$) to an AC domain-containing protein from the cellular slime mold *Polysphondylium pallidum* PN500. This protein contains a conserved CYTH-like domain typical for the superfamily of enzymes that hydrolyze triphosphate-containing substrates and requires metal cations as cofactors. The term CYTH derives from bacterial class IV adenylyl cyclases (CyaB) and thiamine triphosphatase and the domain occur in RNA triphosphatases, membrane-associated polyphosphate polymerases, tripolyphosphatases, nucleoside triphosphatases, nucleoside tetraphosphatases and other proteins with unknown functions. Furthermore, searches initiated from the C-terminal region to the uridine kinase from *Oryza sativa* identified archaeal CyaB homologs (Iyer & Aravind, 2002). The second candidate AC (At1g73980) has a similar domain organization and high homology ($6e^{-109}$) to a *Dictyostelium discoideum* AX4 AC domain containing protein. The third has only some similarity to non-plant proteins, one being a *Trichomonas vaginalis* G3 AC family protein ($2e^{-04}$). In addition, there is one putative Arabidopsis AC (At3g21465) annotated at NCBI (http://www.ncbi.nlm.nih.gov/protein/51968402). It does contain the core motif, but has no annotated domains or known functions and does not share any similarity with annotated and/or experimentally confirmed ACs but appears to be transcriptionally up-regulated in response to biotic stress.

While the presence of the relaxed AC motif may prove useful as supporting criterion for the identification of candidate ACs, it is not stringent enough to identify candidate ACs *ab initio* with any reasonable degree of confidence. In order to achieve this, we have proposed to use an extended AC motif that accounts for the specificity for ATP rather than GTP binding and with the C-terminal metal binding residue ([RK][YFW][DE][VIL][FV]X(8)[KR]X(1,3)[DE]) (Fig. 5). This extended motif retrieves nine putative Arabidopsis ACs (see Table 1).

The candidate ACs include two F-box proteins (At3g28223 and At4g39756) and a toll interleukin receptor (TIR) NBS-LRR (At3g04220). In the former, the F-box domains (cyclin like; IPR001810) have a role in protein-protein interactions and have also been associated with cellular functions, such as signal transduction and the regulation of the cell cycle that in turn is linked to both auxin responses and changes in cellular cAMP content (Ehsan et al., 1998; Leyser, 1998; Mohanty et al., 2001). Given this association one might be tempted to speculate that *axi*, an auxin independence conferring gene could encode an AC, particularly since the Arabidopsis axi1 protein does contain the AC core motif as well as the C-terminal metal binding residue.

With regards to the latter, At3g04220, we note that LRR proteins with AC domains or AC-like domains have been reported (Suzuki et al., 1990). Further, the maize AC (Moutinho et al., 2001) is structurally similar to plant TIR-NBS-LRR type disease resistance proteins (e.g. ADB66335.1, *Populus trichocarpa*, $4e^{-76}$) and At3g04220 also contains a P-loop NTase signature which also occurs in signal transduction ATPases with numerous domains (STAND) that in turn include ACs (Leipe et al., 2004). If one queries "Panther" (A search tool for a library of protein families and subfamilies indexed by function (http://www.pantherdb.org; Thomas et al., 2003) and look for ACs in *Arabidopsis thaliana*, one retrieves 83 entries with the GO (gene ontology) biological process categories: immune system process, cell surface receptor linked signal transduction, intracellular signaling cascade, nucleotide and nucleic acid metabolic process and signal transduction. Many of the retrieved sequences are in fact TIR-NBS-LLRs and Coiled-Coil (CC)-NBS-LLRs that also have a role in plant resistance.

Arabidopsis thaliana proteins containing the AC search motif:		
[RK][YFW][DE][VIL][FV]X(8)[KR]X(1,3)[DE]		
At1g25240	KWEIFEDDFCFTCKDIKE	Epsin N-terminal homology
At1g62590	KFDVVISLGEKMQR—-LE	Pentatricopeptide (PPR) protein
At1g68110	KWEIFEDDYRCFDR-—KD	Epsin N-terminal homology
At2g34780	KFEIVRARNEELKK-EME	MATERNAL EFFECT EMBRYO ARREST 22
At3g02930	KFEVVEAGIEAVQR—-KE	Chloroplast protein
At3g04220	KYDVFPSFRGEDVR--KD	TIR-NBS-LRR class protein
At3g18035	KFDIFQEKVKEIVKVLKD	Linker histone-like prot. - HNO4
At3g28223	KWEIVSEISPACIKSGLD	F-box protein
At4g39756	KWDVVASSFMIERK--CE	F-box protein

Table 1. Candidate adenylyl cyclase molecules.

If we were to allow an [S] in position one, 10 more sequences are identified and they include a Type IIA (sarco/endoplasmic reticulum Ca, SERCA-type) Ca^{2+}-ATPase (At4G00900) that catalyzes the efflux of calcium from the cytoplasm and is transcriptionally up-regulated in response to syringolin and flagellin 22 (For expression analyses see https://www.genevestigator.com/gv/index.jsp; Zimmermann et al., 2005). Syringolin A is a molecular determinant secreted by *Pseudomonas syringae* pv *syringae* that is perceived by non-host plant species e.g. rice. It is recognized by wheat and does induce the accumulation of gene transcripts and increases protection against powdery mildew when applied before inoculation. By doing so, it essentially eradicates powdery mildew from infected wheat if applied after inoculation and therefore has a curative effect. Syringolin A has no fungicidal activity against a variety of fungi and its action on wheat cannot be mimicked by the fungicide cyprodinil and has therefore been proposed as counteracting the suppression of host defense reactions (Wäspi et al., 2001). Flagellin 22 (flg22) designates the conserved N-terminal part of flagellin that activates plant defense mechanisms in *Arabidopsis thaliana* and functions via the receptor-like-kinase, flagellin-sensitive-2 (FLS2). Mitogen-activated-protein-kinases (MAPK) are key signaling compounds that mediate the transcriptional regulation of > 900 flg22 responsive genes. While Ca^{2+} has long been recognized as an essential signal in plant defense responses, the mechanisms by which Ca^{2+} signals are sensed and translated into early microbe-associated molecular patterns (MAMPs) signals remain little understood. However, recently, four calcium-dependent protein kinases (CDPKs) were shown to be Ca^{2+}-sensor protein kinases and to be critical for transcriptional reprogramming in plant innate immune signaling (Boudsocq et al., 2010). It was reported that CDPK and MAPK cascades act differentially in MAMP-mediated regulatory programs that control early genes involved in the synthesis of defense peptides and metabolites, cell wall modifications and redox signaling. Double, triple and quadruple cAMP-dependent protein kinase mutants (*cpk* mutant) display progressively diminished oxidative burst and

gene activation induced by flg22, as well as compromised pathogen defense (Boudsocq et al., 2010).
Finally, we have also applied our AC/GC search strategies to the Rice proteome and identified 40 candidate GCs and 25 ACs and many of the putative GC genes are in families with either publications documenting roles in pathogen signalling, or annotated as such. In addition, there are also families of development-related LRR-RLKs, similar to BRI1 and CLAVATA1 and STRUBBELIG-receptor LRR-RLKs involved in immune and self/non-self signaling (Alcazar et al., 2010) and/or signaling during development (Eyuboglu et al., 2007).

7. Cyclic nucleotide cyclases and cyclic nucleotides in plant stress responses

Recently, cyclic nucleotides have been shown to play an important role in pathogen defense (Ma et al., 2009; Meier et al., 2009). Cyclic AMP was shown to be elevated at the initial site of infection initiating pathogen-related cytosolic Ca^{2+} signaling (Ma et al., 2009). Further, recent research (Qi et al., 2010) demonstrated cyclic nucleotides such as cAMP and cGMP are activating ligands for plant CNGCs, including CNGC2 and are thus linked to Ca^{2+} transport and hence Ca^{2+} signalling. Activation of channels by cGMP within a cell may be confined to microdomains near the plasma membrane and so provide spatial resolution of the signal. In addition, the role AtPeps, danger-associated molecular pattern (DAMPs) molecules play in pathogen-defense signaling cascades, and the guanylyl cyclase activity of the AtPep receptor AtPepR1 (Qi et al., 2010) provided a model for linking pathogen perception at the cell surface to intracellular Ca^{2+} signaling and immune responses in plants (Ma et al., 2009; Ryan et al., 2006). This implies that both cAMP and cGMP have specific roles in plant-defense signal transduction cascades, suggesting that specific cyclic nucleotide signatures generated in response to biotic (Ma et al., 2009; Meier et al., 2009) and abiotic (Donaldson et al., 2004) stresses act as second messengers in signaling cascades that critically depend on CNGCs (Kaplan et al., 2007; Talke et al., 2003). This is a concept that is in accord with the suggestion that the largest group of cNMP targets in plant cells are the CNGCs (Kaplan et al., 2007) that have a key role in the control of ion homeostasis and defense against biotic and abiotic stress (Kaplan et al., 2007; Ma et al., 2010).
Further evidence for the role of cGMP in pathogen response has been provided by a direct pathogen dependent increase in tissue cGMP levels in *planta* (Meier et al., 2009). Here, cGMP accumulation in *Arabidopsis thaliana* leaves was measured after inoculation with virulent (DC3000) and avirulent (AvirB) *Pseudomonas syringae* strains and caused marked and sustained increases in cGMP levels in response to the avirulent strain only, while the virulent strain caused smaller increases with a delayed onset. The earlier induction of cGMP by the avirulent strain is compatible with the instant recognition of specific pathogen avirulent (*avr*) gene encoded molecules that together with their corresponding *R* genes in plants triggers activation of plant defense responses. In the delayed response case for virulent DC3000 strain, pathogens are able to grow and spread due to the absence of specific pathogen recognition by plant *R* gene products. In Arabidopsis cell cultures, cGMP has been implicated as required for NO-induced cell death in response to challenge by avirulent bacterial pathogens (Clarke et al., 2000). In tobacco, cGMP has also been implicated in NO-dependent defense responses and required for induction of expression of defense-related genes, pathogenesis-related 1 gene (*PR1*) and the phenylalanine ammonia lyase gene (*PAL*) (Durner et al., 1998; Klessig et al., 2000).

Moreover, specific and transient increases in intracellular cGMP levels have also been reported in response to NaCl, drought stress (Donaldson et al., 2004) and ozone (Pasqualini et al., 2008). Cyclic GMP levels in *Arabidopsis thaliana* have been demonstrated to be time dependent and differing in response to salt and osmotic stress (Donaldson et al., 2004). Salt stress triggers an osmotic conduit that is independent of calcium concentration and the ionic response pathway triggered by high NaCl is calcium-dependent. These results suggest that cGMP also plays a complex role in abiotic stress responses (Donaldson et al., 2004). Remarkably, cGMP increases were observed two hours following ozone treatment in tobacco leaf, suggesting that cGMP is essential for the induction of some late ozone-dependent pathways but not critical in the early signal responses (Pasqualini et al., 2008). In this study, it was shown that early response to ozone and NO caused transcriptional activation of the scavenger coding for proteins *AOX1, GPX* and *ACS2* and were cGMP independent, but the early response of *PAL*a and late response of *PR1a* showed critical dependence on cGMP. In this context, it is noteworthy that the early response of *PALa* was observed two hours after fumigation coinciding with increase in cGMP transients.

Cyclic AMP may also have an important role in abiotic stress responses and particularly to salt stress since VICs in *Arabidopsis thaliana* roots have been reported to have open probabilities that are sensitive to micromolar concentrations of cAMP or cGMP at the cytoplasmic side of the plasma membrane. Here, presence of permeable cyclic nucleotides during growth improved plant tolerance to salinity that corresponded to a reduction in sodium accumulation (Maathuis & Sanders, 2001). This may suggest that plants contain a cyclic nucleotide-based signaling response that directly affects sodium transport across VICs in response to salinity stress.

Nucleotide cyclases have also been implicated in playing a critical role in biotic stress. An experimentally confirmed putative Arabidopsis AC (At3g21465) has been shown to be transcriptionally up-regulated in response to biotic stress (Gehring, 2010). The AC F-box domains (cyclin like; IPR001810) have a role in protein-protein interactions and are also associated with signal transduction and cell cycle regulation that are linked to both auxin responses and changes in cellular cAMP content (Ehsan et al., 1998; Leyser, 1998; Mohanty et al., 2001). In addition, microarray analysis of the AtWAKL10 (At1g79680) (Zimmermann et al., 2005) which is also a GC (Ludidi & Gehring, 2003), supports a role for GCs and hence cGMP in plant pathogen signaling. In summary, AC and GC are increasingly understood to have a role in deciphering pathogen perception, response signalling and downstream pathogen defense responses in the plant immune signal transduction cascade (Fig. 6), a view that is supported by older studies where the application of fungal extracts was shown to induce elevation in the endogenous levels of cAMP in cultured French bean cells, alfalfa and carrot (Bolwell, 1992; Cooke et al., 1994; Kurosaki et al., 1987).

Finally, it is noteworthy that cNMPs have a role in both abiotic and biotic stress responses and this is perhaps not surprising given that in biotic interactions pathogens can, directly or indirectly, cause homeostatic disturbances that in turn lead to conditions that amount to abiotic stress. Interestingly, some of the host homeostatic modifications are part of the defense response to pathogens (Garavaglia et al., 2010; Gottig et al., 2008, 2009) where the aim of the host is either to starve the biotroph pathogen by restricting access to water and nutrients or to dehydrate the pathogen by sharply increasing the apoplastic osmotic pressure.

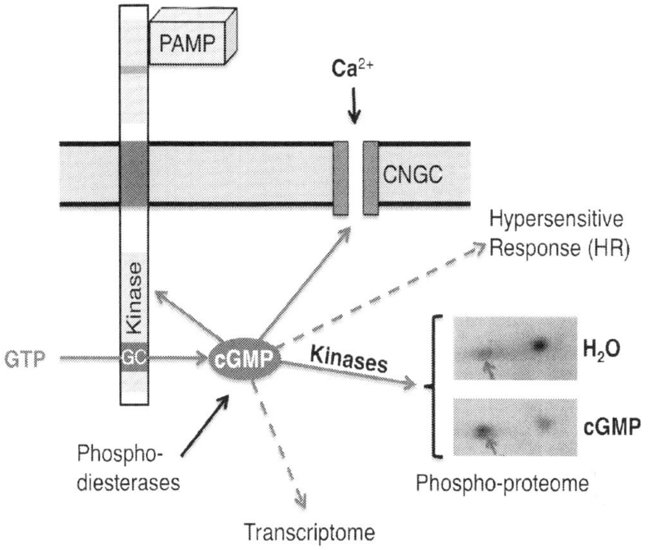

Fig. 6. Guanylyl cyclase mediated signal transduction. A receptor type GC (the domain organisation is annotated in the legend to Fig. 4) is activated by a ligand, here PAMP (Pathogen-associated molecular pattern) and this in turn activates the cytosolic GC domain that generates cGMP that causes Ca^{2+} influx through the regulation of CNGCs (cyclic nucleotide gated channels). Cyclic GMP, possibly together with Ca^{2+} and calmodulin, activates protein kinases which result in a rapid change of the phospho-proteome; cGMP also plays a critical role in the transcriptional regulation of cGMP-dependent gene transcription (Bastian et al., 2010) and the complex processes leading to the "hypersensitive response" (HR). Solid lines represent direct interactions and dashed lines indicate indirect actions that are only partly understood. Additional soluble GCs may also contribute to raises in cGMP levels (not show in this model).

8. Out-look

Given that our understanding of the structural features that enable NCs activity in higher plants is growing, we can expect significant progress in the discovery and experimental confirmation of novel NCs in higher plants in the foreseeable future. This will afford a better understanding of the role of both cAMP and cGMP as second messengers in plant development, responses to environmental stimuli and/or hormones and in particular stress responses. In addition, we will be likely to see cAMP- and cGMP-dependent transcriptomes, (phospho-)proteomes and metabolomes that, together with studies in mutants (e.g. affected in their ability to respond to stress), will afford new and fundamental insights into plant stress signaling. Finally, given the central role of the cNMPs and hence NCs in homeostasis regulation and stress responses, it is conceivable that some of the NCs will be targets for rational bioengineering strategies that will eventually deliver crop plants with improved abiotic stress resistance.

9. References

Acher, R. (2002). Water homeostasis in the living: molecular organization, osmoregulatory reflexes and evolution. *Annales d'Endocrinologie,* 63, 3, 197-218, ISSN 0003-4266

Al-Shahrour, F., Diaz-Uriarte, R. & Dopazo, J. (2004). FatiGO: a web tool for finding significant associations of Gene Ontology terms with groups of genes. *Bioinformatics,* 20, 4, 578-580, ISSN 0305-1048

Alcazar, R., Garcia, A.V., Kronholm, I., de Meaux, J., Koornneef, M., Parker, J.E. & Reymond, M. (2010). Natural variation at Strubbelig Receptor Kinase 3 drives immune-triggered incompatibilities between Arabidopsis thaliana accessions. *Nature Genetics,* 42, 12, 1135-1139, ISSN 1546-1718

Ali, R., Zielinski, R.E. & Berkowitz, G.A. (2006). Expression of plant cyclic nucleotide-gated cation channels in yeast. *Journal of Experimental Botany,* 57, 1, 125-138, ISSN 0022-0957

Amrhein, N. (1974). Evidence against the occurrence of adenosine 3':5'-cyclic monophosphate in higher plants. *Planta,* 118, 241-258

Amrhein, N. (1977). Current status of cyclic-AMP in higher-plants. *Annual Review of Plant Physiology and Plant Molecular Biology,* 28, 123-132, ISSN 0066-4294

Anderson, J.A., Huprikar, S.S., Kochian, L.V., Lucas, W.J. & Gaber, R.F. (1992). Functional expression of a probable *Arabidopsis thaliana* potassium channel in *Saccharomyces cerevisiae. Proceedings of the National Academy of Sciences of the United States of America,* 89, 9, 3736-3740, ISSN 0027-8424

Anselmi, C., Carloni, P. & Torre, V. (2007). Origin of functional diversity among tetrameric voltage-gated channels. *Proteins-Structure Function and Bioinformatics,* 66, 1, 136-146, ISSN 0887-3585

Arnold, W., Mittal, C., Katsuki, S. & Murad, F. (1977). Nitric oxide activates guanylate cyclase and increases guanosine 32:52-cyclic monophosphate levels in various tissue preparations. *Proceedings of the National Academy of Sciences of the United States of America,* 78, 3203–3207

Ashman, D.F., Lipton, R., Melicow, M.M. & Price, T.D. (1963). Isolation of adenine 3',5'-monophosphate and guanosine 3',5'-monohosphate from rat urine. *Biochemistry and Biophysics Research Communications,* 11, 330-334

Ashton, A.R. & Polya, G.M. (1975). Higher-plant cyclic nucleotide phosphodiesterases - resolution, partial-purification and properties of 3 phosphodiesterases from potato-tuber. *Biochemical Journal,* 149, 2, 329-339, ISSN 0264-6021

Ashton, A.R. & Polya, G.M. (1978). Cyclic adenosine 3':5'-monophosphate in axenic rye grass endosperm cell cultures. *Plant Physiology,* 61, 5, 718-722, ISSN 1532-2548

Azadi, S., Molday, L.L. & Molday, R.S. (2010). RD3, the protein associated with Leber congenital amaurosis type 12, is required for guanylate cyclase trafficking in photoreceptor cells. *Proceedings of the National Academy of Sciences of the United States of America,* 107, 49, ISSN 21158-21163

Azhar, S. & Krishna Murti, C.R. (1971). Effect of Indole-3-acetic acid on synthesis of cyclic 3'-5' adenosine phosphate by Bengal gram seeds. *Biochemical and Biophysical Research Communications,* 43, 1, 58-64, ISSN 0006-291X

Balague, C., Lin, B., Alcon, C., Flottes, G., Malmstrom, S., Kohler, C., Neuhaus, G., Pelletier, G., Gaymard, F. & Roby, D. (2003). HLM1, an essential signaling component in the hypersensitive response, is a member of the cyclic nucleotide-gated channel ion channel family. *Plant Cell,* 15, 2, 365-379, ISSN 1040-4651

Bastian, R., Dawe, A., Meier, S., Ludidi, N., Bajic, V.B. & Gehring, C. (2010). Gibberelic acid and cGMP-dependent transcriptional regulation in *Arabidopsis thaliana. Plant Signaling & Behavior,* 5, 3, 224-232, doi: 10.4161/psb.5.3.10718

Beavo, J.A. (1995). Cyclic-nucleotide phosphodiesterases: Functional implications of multiple isoforms. *Physiological Reviews,* 75, 4, 725-748, ISSN 0031-9333

Beavo, J.A., Hardman, J.G. & Sutherland, E.W. (1970). Hydrolysis of cyclic guanosine and adenosine 3',5'-monophosphates by rat and bovine tissues. *Journal of Biological Chemistry,* 245, 21, 5649-5655, ISSN 0021-9258

Becker, D. & Ziegler, H. (1973). Cyclic adenosine-3'-5'-monophosphate in translocation tissues of plants. *Planta,* 110, 1, 85-89, ISSN 0032-0935

Beligni, M.V., Fath, A., Bethke, P.C., Lamattina, L. & Jones, R.L. (2002). Nitric oxide acts as an antioxidant and delays programmed cell death in barley aleurone layers. *Plant Physiology,* 129, 4, 1642-1650, ISSN 0032-0889

Berman, H.M., Ten Eyck, L.F., Goodsell, D.S., Haste, N.M., Kornev, A. & Taylor, S.S. (2005). The cAMP binding domain: an ancient signaling module. *Proceedings of the National Academy of Sciences of the United States of America,* 102, 1, 45-50, ISSN 0027-8424

Biel, M. (2009). Cyclic nucleotide-regulated cation channels. *The Journal of Biological Chemistry,* 284, 14, ISSN 9017-9021

Biel, M. & Michalakis, S. (2009). Cyclic nucleotide-gated channels. *Handbook of Experimental Pharmacology,* 191, 111-136, ISSN 0171-2004

Bogdan, C. (2001). Nitric oxide and the regulation of gene expression. *Trends in Cell Biology,* 11, 2, 66-75, ISSN 0962-8924

Bolwell, G.P. (1992). A role for phosphorylation in the down-regulation of phenylalanine ammonia-lyase in suspension-cultured cells of French bean. *Phytochemistry,* 31, 12, 4081-4086, ISSN 0031-9422

Bolwell, G.P. (1995). Cyclic AMP, the reluctant messenger in plants. *Trends in Biochemical Science,* 20, 12, 492-495, ISSN 0968-0004

Bonnafous, J.C., Olive, J.L., Borgna, J.L. & Mousseron-Cadet, M. (1975). L'AMP cyclique dans les graines et les plantules d'orge et la contamination bacterienne ou fongique. *Biochimie,* 57, 661-663

Bosgraaf, L., Russcher, H., Snippe, H., Bader, S., Wind, J. & Van Haastert, P.J.M. (2002). Identification and characterization of two unusual cGMP-stimulated phoshodiesterases in Dictyostelium. *Molecular Biology of the Cell,* 13, 11, 3878-3889, ISSN 1059-1524

Boudsocq, M., Willmann, M.R., McCormack, M., Lee, H., Shan, L., He, P., Bush, J., Cheng, S.H. & Sheen, J. (2010). Differential innate immune signalling via Ca^{2+} sensor protein kinases. *Nature,* 464, 7287, 418-422, ISSN 1476-4687

Bowler, C. & Chua, N.H. (1994). Emerging themes of plant signal transduction. *Plant Cell,* 6, 1529-1541, ISSN 1040-4651

Bressan, R.A., Ross, C.W. & Vandepeute, J. (1976). Attempts to detect cyclic adenosine 3'-5'-monophosphate in higher-plants by 3 assay methods. *Plant Physiology,* 57, 1, 29-37, ISSN 0032-0889

Brewin, N.J. & Northcot, D.H. (1973). Partial-purification of a cyclic-AMP phosphodiesterase from soybean callus - isolation of a non-dialyzable inhibitor. *Biochimica et Biophysica Acta,* 320, 1, 104-122, ISSN 0006-3002

Brooker, G., Thomas, L.J. & Appleman, M.M. (1968). Assay of adenosine 3',5'-cyclic monophosphate and guanosine 3',5'-cyclic monophosphate in biological materials by enzymatic radioisotopic displacement. *Biochemistry,* 7, 12, 4177-4181, ISSN 0006-2960

Brown, E.G., Alnajafi, T. & Newton, R.P. (1975). Partial-purification of adenosine 3' - 5'-cyclic monophosphate phosphodiesterase from *Phaseolus vulgaris* L. associated activator and inhibitors. *Biochemical Society Transactions,* 3, 3, 393-395, ISSN 0300-5127

Brown, E.G., Alnajafi, T. & Newton, R.P. (1977). Cyclic nucleotide phosphodiesterase activity in *Phaseolus vulgaris. Phytochemistry,* 16, 9, 1333-1337, ISSN 0031-9422

Brown, E.G., Edwards, M.J., Newton, R.P. & Smith, C.J. (1979). Plurality of cyclic nucleotide phosphodiesterase in *Spinacea oleracea,* subcellular-distribution, partial-purification, and properties. *Phytochemistry,* 18, 12, 1943-1948, ISSN 0031-9422

Brown, E.G., Edwards, M.J., Newton, R.P. & Smith, C.J. (1980). Cyclic-nucleotide phosphodiesterases of spinach-chloroplasts and microsomes. *Phytochemistry,* 19, 1, 23-30, ISSN 0031-9422

Brown, E.G. & Newton, R.P. (1973). Occurrence of adenosine 3'-5'-cyclic monophosphate in plant tissues. *Phytochemistry,* 12, 11, 2683-2685, ISSN 0031-9422

Brown, E.G. & Newton, R.P. (1992). Analytical procedures for cyclic-nucleotides and their associated enzymes in plant tissues *Phytochemical Analysis,* 3, 1, 1-13, ISSN 0958-0344

Brown, E.G., Newton, R.P., Evans, D.E., Walton, T.J., Younis, L.M. & Vaughan, J.M. (1989). Influence of light on cyclic-nucleotide metabolism in plants - Effect of dibutyryl cyclic-nucleotides on chloroplast components. *Phytochemistry,* 28, 10, 2559-2563, ISSN 0031-9422

Chen, T.Y., Peng, Y.W., Dhallan, R.S., Ahamed, B., Reed, R.R. & Yau, K.W. (1993). A new subunit of the cyclic nucleotide-gated cation channel in retinal rods. *Nature,* 362, 6422, 764-767, ISSN 0028-0836

Chen, Y., Cann, M.J., Litvin, T.N., Iourgenko, V., Sinclair, M.L., Levin, L.R. & Buck, J. (2000). Soluble adenylyl cyclase as an evolutionarily conserved bicarbonate sensor. *Science,* 289, 5479, 625-628, ISSN 0036-8075

Cheung, W.Y. (1971). Cyclic 3',5'-nucleotide phosphodiesterase - Effect of divalent cations. *Biochimica et Biophysica Acta,* 242, 2, 395-409, ISSN 0006-3002

Chinkers, M., Garbers, D.L., Chang, M.-S., Lowe, D.G., Chin, H., Goeddel, D.V. & Schulz, S. (1989). A membrane form of guanylate cyclase is an atrial natriuretic peptide receptor. *Nature,* 338, 6210, 78-83

Choi, Y.E. & Xu, J.R. (2010). The cAMP signaling pathway in *Fusarium verticillioides* is important for conidiation, plant infection, and stress responses but not fumonisin production. *Molecular Plant-Microbe Interactions,* 23, 4, 522-533, ISSN 0894-0282

Clarke, A., Desikan, R., Hurst, R.D., Hancock, J.T. & Neill, S.J. (2000). NO way back: nitric oxide and programmed cell death in Arabidopsis thaliana suspension cultures. *Plant Journal,* 24, 5, 667-677, ISSN 0960-7412

Cobbs, W.H. & Pugh, E.N. (1985). Cyclic-GMP can increase rod outer-segment light-sensitive current 10-fold without delay of excitation. *Nature,* 313, 6003, 585-587, ISSN 0028-0836

Cooke, C.J., Smith, C.J., Walton, T.J. & Newton, R.P. (1994). Evidence that cyclic-AMP is involved in the hypersensitive response of *Medicago sativa* to a fungal elicitor. *Phytochemistry,* 35, 4, 889-895, ISSN 0031-9422

Corbin, J.D. & Francis, S.H. (1999). Cyclic GMP phosphodiesterase-5: Target of sildenafil. *Journal of Biological Chemistry*, 274, 20, 13729-13732, ISSN 0021-9258

Cordell, R.L., Hill, S.J., Ortori, C.A. & Barrett, D.A. (2008). Quantitative profiling of nucleotides and related phosphate-containing metabolites in cultured mammalian cells by liquid chromatography tandem electrospray mass spectrometry. *Journal of Chromatography B-Analytical Technologies in the Biomedical and Life Sciences*, 871, 1, 115-124, ISSN 1570-0232

Cousson, A. & Vavasseur, A. (1998). Putative involvement of cytosolic Ca^{2+} and GTP-binding proteins in cyclic-GMP-mediated induction of stomatal opening by auxin in *Commelina communis* L. *Planta*, 206, 2, 308-314, ISSN 0032-0935

Craven, K.B. & Zagotta, W.N. (2006). CNG and HCN channels: Two peas, one pod. *Annual Review of Physiology*, 68, 375-401, ISSN 0066-4278

Cyong, J.C. & Takahashi, M. (1982). Identification of guanosine-3':5'-monophosphate in the fruits of *Zizyphus jujuba*. *Phytochemistry*, 21, 1871-1874, ISSN 0031-9422

Cyong, J.C., Takahashi, M., Hanabusa, K. & Otsuka, Y. (1982). Guanosine 3',5'-monophosphate in fruits of *Evodia rutaecarpa* and *E. officinalis*. *Phytochemistry*, 21, 3, 777-778, ISSN 0031-9422

Davies, S.A. (2006). Signalling via cGMP: Lessons from Drosophila. *Cellular Signalling*, 18, 4, 409-421, ISSN 0898-6568

De Cesare, D., Fimia, G.M. & Sassone-Corsi, P. (1999). Signaling routes to CREM and CREB: plasticity in transcriptional activation. *Trends in Biochemical Sciences*, 24, 7, 281-285, ISSN 0968-0004

Demidchik, V., Davenport, R.J. & Tester, M. (2002). Nonselective cation channels in plants. *Annual Review in Plant Biology*, 53, 67-107, ISSN 1543-5008

DeYoung, B.J. & Innes, R.W. (2006). Plant NBS-LRR proteins in pathogen sensing and host defense. *Nature Immunology*, 7, 12, 1243-1249, ISSN 1529-2908

Dhallan, R.S., Macke, J.P., Eddy, R.L., Shows, T.B., Reed, R.R., Yau, K.W. & Nathans, J. (1992). Human rod photoreceptor cGMP-gated channel amino acid sequence, gene structure, and functional expression. *Journal of Neuroscience*, 12, 8, 3248-3256, ISSN 0270-6474

Dietrich, P., Anschutz, U., Kugler, A. & Becker, D. (2010). Physiology and biophysics of plant ligand-gated ion channels. *Plant Biology*, 12, 80-93, ISSN 1435-8603

Dixon, R.A. & Paiva, N.L. (1995). Stress-induced phenylpropanoid metabolism. *The Plant Cell*, 7, 1085-1097, ISSN 1040-4651

Donaldson, L., Ludidi, N., Knight, M.R., Gehring, C. & Denby, K. (2004). Salt and osmotic stress cause rapid increases in *Arabidopsis thaliana* cGMP levels. *Federation of European Biochemical Societies Letters*, 569, 1-3, 317-320, ISSN 0014-5793

Dormann, D., Kim, J.Y., Devreotes, P.N. & Weijer, C.J. (2001). cAMP receptor affinity controls wave dynamics, geometry and morphogenesis in Dictyostelium. *Journal of Cell Science*, 114, Pt 13, 2513-2523, ISSN 0021-9533

Doyle, D.A., Cabral, J.M., Pfuetzner, R.A., Kuo, A.L., Gulbis, J.M., Cohen, S.L., Chait, B.T. & MacKinnon, R. (1998). The structure of the potassium channel: Molecular basis of K^+ conduction and selectivity. *Science*, 280, 5360, 69-77, ISSN 0036-8075

Drummond, G. & Perrotty, S. (1961). Enzymatic hydrolysis of adenosine 3',5'-phosphoric acid. *Journal of Biological Chemistry*, 236, 4, 1126-1129, ISSN 0021-9258

Dubovskaya, L.V., Bakakina, Y.S., Kolesneva, E.V., Sodel, D.L., McAinsh, M.R., Hetherington, A.M. & Volotovski, I.D. (2011). cGMP-dependent ABA-induced stomatal closure in the ABA-insensitive Arabidopsis mutant abi1-1. *New Phytologist*, no. doi: 10.1111/j.1469-8137.2011.03661.x

Duffus, C.M. (1969). Alpha-amylase activity in developing barley grain and its dependence on gibberellic acid. *Phytochemistry*, 8, 7, 1205-1209, ISSN 0031-9422

Durner, J., Wendehenne, D. & Klessig, D.F. (1998). Defense gene induction in tobacco by nitric oxide, cyclic GMP, and cyclic ADP-ribose. *Proceedings of the National Academy of Sciences of the United States of America*, 95, 17, 10328-10333, ISSN 0027-8424

Ehsan, H., Reichheld, J.P., Roef, L., Witters, E., Lardon, F., Van Bockstaele, D., Van Montagu, M., Inze, D. & Van Onckelen, H. (1998). Effect of indomethacin on cell cycle dependent cyclic AMP fluxes in tobacco BY-2 cells. *Federation of European Biochemical Societies Letters*, 422, 2, 165-169, ISSN 0014-5793

Endress, R. (1979). Allosteric regulation of phosphodiesterase from Portulaca callus by cGMP and papaverin. *Phytochemistry*, 18, 1, 15-20, ISSN 0031-9422

Essah, P.A., Davenport, R. & Tester, M. (2003). Sodium influx and accumulation in Arabidopsis. *Plant Physiology*, 133, 1, 307-318, ISSN 0032-0889

Eyuboglu, B., Pfister, K., Haberer, G., Chevalier, D., Fuchs, A., Mayer, K.F. & Schneitz, K. (2007). Molecular characterisation of the STRUBBELIG-RECEPTOR FAMILY of genes encoding putative leucine-rich repeat receptor-like kinases in Arabidopsis thaliana. *BMC Plant Biology*, 7, 16, ISSN 1471-2229

Fesenko, E.E., Kolesnikov, S.S. & Lyubarsky, A.L. (1985). Induction by cyclic-GMP of cationic conductance in plasma-membrane of retinal rod outer segment. *Nature*, 313, 6000, 310-313, ISSN 0028-0836

Flynn, G.E., Johnson, J.P. & Zagotta, W.N. (2001). Cyclic nucleotide-gated channels: Shedding light on the opening of a channel pore. *Nature Reviews Neuroscience*, 2, 9, 643-651, ISSN 1471-0048

Furchgott, R.F. & Vanhoutte, P.M. (1989). Endothelium-derived relaxing and contracting factors. *The Federation of American Societies for Experimental Biology Journal*, 3, 9, 2007-2018, ISSN 0892-6638

Gao, Y. (2010). The multiple actions of NO. *Pflügers Archiv European Journal of Physiology*, 459, 6, 829-839, ISSN 0031-6768

Garavaglia, B.S., Thomas, L., Gottig, N., Dunger, G., Garofalo, C.G., Daurelio, L.D., Ndimba, B., Orellano, E.G., Gehring, C. & Ottado, J. (2010). A eukaryotic-acquired gene by a biotrophic phytopathogen allows prolonged survival on the host by counteracting the shut-down of plant photosynthesis. *Plos One*, 5, 1, e8950, ISSN 1932-6203

Garbers, D.L. (1999). The guanylyl cyclase receptors. *Methods*, 19, 4, 477-484, ISSN 1046-2023

Garcia-Mata, C., Gay, R., Sokolovski, S., Hills, A., Lamattina, L. & Blatt, M.R. (2003). Nitric oxide regulates K^+ and Cl^- channels in guard cells through a subset of abscisic acid-evoked signaling pathways. *Proceedings of the National Academy of Science of the USA*, 100, 19, 11116-11121, ISSN 0027-8424

Garcia-Mata, C. & Lamattina, L. (2001). Nitric oxide induces stomatal closure and enhances the adaptive plant responses against drought stress. *Plant Physiology*, 126, 3, 1196, ISSN 0032-0889

Gaymard, F., Cerutti, M., Horeau, C., Lemaillet, G., Urbach, S., Ravallec, M., Devauchelle, G., Sentenac, H. & Thibaud, J.B. (1996). The baculovirus/insect cell system as an alternative to Xenopus oocytes. First characterization of the AKT1 K^+ channel from *Arabidopsis thaliana*. *Journal of Biological Chemistry*, 271, 37, 22863-22870, ISSN 0021-9258

Gehring, C. (2010). Adenyl cyclases and cAMP in plant signaling - past and present. *Cell Communication and Signaling,* 8, 15, ISSN 1478-811X

Gehring, C.A. (1999). Natriuretic peptides - A new class of plant hormone? *Annals of Botany,* 83, 4, 329-334, ISSN 0305-7364

Gehring, C.A. & Irving, H.R. (2003). Natriuretic peptides - a class of heterologous molecules in plants. *International Journal of Biochemistry & Cell Biology,* 35, 9, 1318-1322, ISSN 1357-2725

Ghofrani, H.A., Osterloh, I.H. & Grimminger, F. (2006). Sildenafil: from angina to erectile dysfunction to pulmonary hypertension and beyond. *Nature Reviews Drug Discovery,* 5, 8, 689-702, ISSN 1474-1776

Gilman, A.G. (1970). A protein binding assay for adenosine 3':5'-cyclic monophosphate. *Proceedings of the National Academy of Science of the United States of America,* 67, 1, 305-312

Goldberg, N.D., Dietz, S.B. & O'Toole, A.G. (1969). Cyclic guanosine 3',5'-monophosphate in mammalian tissues and urine. *Journal of Biological Chemistry,* 244, 16, 4458-4466, ISSN 0021-9258

Gottig, N., Garavaglia, B.S., Daurelio, L.D., Valentine, A., Gehring, C., Orellano, E.G. & Ottado, J. (2008). *Xanthomonas axonopodis* pv. citri uses a plant natriuretic peptide-like protein to modify host homeostasis. *Proceeding of the National Academy of Science of the United States of America,* 105, 47, 18631-18636, ISSN 1091-6490

Gottig, N., Garavaglia, B.S., Daurelio, L.D., Valentine, A., Gehring, C., Orellano, E.G. & Ottado, J. (2009). Modulating host homeostasis as a strategy in the plant-pathogen arms race. *Communicative & Integrative Biology,* 2, 2, 89-90, ISSN 1942-0889

Gouvea, C., Souza, J.F., Magalhaes, A.C.N. & Martins, I.S. (1997). NO-releasing substances that induce growth elongation in maize root segments. *Plant Growth Regulation,* 21, 3, 183-187, ISSN 0167-6903

Gray, P.C., Scott, J.D. & Catterall, W.A. (1998). Regulation of ion channels by cAMP-dependent protein kinase and A-kinase anchoring proteins. *Current Opinion in Neurobiology,* 8, 3, 330-334, ISSN 0959-4388

Guillemant, J., Oberlin, F., Carayon, A., Bourgeois, P. & Guillemant, S. (1994). Natriuresis and cyclic GMP stimulation by parathyroid hormone 1-34 in man. *Journal of Bone Mineral and Metabolism,* 12, 1, S45-S48

Gustin, M.C., Martinac, B., Saimi, Y., Culbertson, M.R. & Kung, C. (1986). Ion channels in yeast. *Science,* 233, 4769, 1195-1197, ISSN 0036-8075

Haddox, M.K., Stephens, J.H. & Goldberg, N.D. (1974). Cyclic GMP in meristematic and elongating regions of bean root. *Federation Proceedings,* 33, 3, 522-526, ISSN 0014-9446

Hardman, J.G., Davis, J.W. & Sutherland, E.W. (1966). Measurement of guanosine 3',5'-monophosphate and other cyclic nucleotides. Variations in urinary excretion with hormonal state of rat. *Journal of Biological Chemistry,* 241, 20, 4812-4815, ISSN 0021-9258

Hardman, J.G. & Sutherland, E.W. (1969). Guanyl cyclase, an enzyme catalyzing the formation of guanosine 3',5'-monophosphate from guanosine triphosphate. *Journal of Biological Chemistry,* 244, 23, 6363-6370, ISSN 0021-9258

Harper, J.F. & Brooker, G. (1975). Femtomole sensitive radioimmunoassay for cyclic-AMP and cyclic-GMP after 2'0 acetylation by acetic-anhydride in aqueous solution. *Journal of Cyclic Nucleotide Research,* 1, 4, 207-218, ISSN 0095-1544

Haynes, L.W. & Yau, K.W. (1985). Cyclic GMP-sensitive conductance in outer segment membrane of Catfish cones. *Nature,* 317, 6032, 61-64, ISSN 0028-0836

Haynes, L.W., Kay, A.R. & Yau, K.W. (1986). Single cyclic GMP-activated channel activity in excised patches of rod outer segment membrane. *Nature,* 321, 6065, 66-70, ISSN 0028-0836

He, Z.H., Cheeseman, I., He, D. & Kohorn, B.D. (1999). A cluster of five cell wall-associated receptor kinase genes, Wak1-5, are expressed in specific organs of Arabidopsis. *Plant Molecular Biology,* 39, 6, 1189-1196, ISSN 0167-4412

Helmreich, E.J., Zenner, H.P. & Pfeuffer, T. (1976). Signal transfer from hormone receptor to adenylate cyclase. *Current Topics in Cellular Regulation,* 10, 41-87, ISSN 0070-2137

Hille, B. (1992). *Ionic channels of excitable membranes* (Second Edition). Sunderland, Massachusetts, USA. Sinauer Associates, Inc.

Hilton, G.M. & Nesius, K.K. (1978). Localization of adenyl-cyclase in meristems of young pea hypocotyls. *Physiologia Plantarum,* 42, 1, 49-52, ISSN 0031-9317

Hofmann, A., Grella, M., Botos, I., Filipowicz, W. & Wlodawer, A. (2002). Crystal structures of the semireduced and inhibitor-bound forms of cyclic nucleotide phosphodiesterase from *Arabidopsis thaliana*. *Journal of Biological Chemistry,* 277, 2, 1419-1425, ISSN 0021-9258

Hofmann, F., Biel, M. & Kaupp, U.B. (2005). International Union of Pharmacology. LI. Nomenclature and structure-function relationships of cyclic nucleotide-regulated channels. *Pharmacological Reviews,* 57, 4, 455-462, ISSN 0031-6997

Hoshi, T. (1995). Regulation of voltage-dependence of the KAT1 channel by intracellular factors *Journal of General Physiology,* 105, 3, 309-328, ISSN 0022-1295

Hua, B.G., Mercier, R.W., Leng, Q. & Berkowitz, G.A. (2003a). Plants do it differently. A new basis for potassium/sodium selectivity in the pore of an ion channel. *Plant Physiology,* 132, 3, 1353-1361, ISSN 0032-0889

Hua, B.G., Mercier, R.W., Zielinski, R.E. & Berkowitz, G.A. (2003b). Functional interaction of calmodulin with a plant cyclic nucleotide gated cation channel. *Plant Physiology and Biochemistry,* 41, 11-12, 945-954, ISSN 0981-9428

Ignarro, L.J. (1991). Signal transduction mechanisms involving nitric oxide. *Biochemical Pharmacology,* 41, 4, 485-490, ISSN 0006-2952

Ignarro, L.J. (1999). Nitric oxide: A unique endogenous signaling molecule in vascular biology. *Angewandte Chemie-International Edition,* 38, 13-14, 1882-1892, ISSN 1433-7851

Ignarro, L.J. (2000). *Nitric oxide. Biology and Pathobiology.* San Diego: Academic Press.

Ildefonse, M. & Bennett, N. (1991). Single-channel study of the cGMP-dependent conductance of retinal rods from incorporation of native vesicles into planar lipid bilayers. *Journal of Membrane Biology,* 123, 2, 133-147, ISSN 0022-2631

Inamdar, N.M., Ehrlich, K.C. & Ehrlich, M. (1991). Cpg methylation inhibits binding of several sequence-specific DNA-binding proteins from pea, wheat, soybean and cauliflower. *Plant Molecular Biology,* 17, 1, 111-123, ISSN 0167-4412

Isner, J.C. & Maathuis, F.J.M. (2011). Measurement of cellular cGMP in plant cells and tissues using the endogenous fluorescent reporter *Flinc*G. *Plant Journal,* 65, 2, 329-334, ISSN 0960-7412

Iyer, L.M. & Aravind, L. (2002). The catalytic domains of thiamine triphosphatase and CyaB-like adenylyl cyclase define a novel superfamily of domains that bind organic phosphates. *BMC Genomics,* 3, 1, 33, ISSN 1471-2164

Jaffrey, S.R. & Snyder, S.H. (1995). Nitric oxide: A neural messenger. *Annual Review of Cell and Developmental Biology,* 11, 417-440, ISSN 1081-0706

Jan, L.Y. & Jan, Y.N. (1992). Structural elements involved in specific K^+ channel functions. *Annual Review of Physiology,* 54, 537-555, ISSN 0066-4278

Janistyn, B. (1972). 3-Indole acetic-acid induced efflux of nucleotides concomitant with an increased synthesis of adenosine-3',5'-monophosphate (cAMP) in maize coleoptile sections. *Zeitschrift Fur Naturforschung Part B-Chemie Biochemie Biophysik Biologie und Verwandten Gebiete,* B 27, 3, 273-276

Janistyn, B. (1983). Gas chromatographic-mass spectroscopic identification and quantification of cyclic guanosine-3':5' monophosphate in maize seedlings (*Zea mays*). *Planta,* 159, 4, 382-385, ISSN 0032-0935

Janistyn, B. (1988). Stimulation by manganese(II) sulfate of a cAMP-dependent protein-kinase from *Zea mays* seedlings. *Phytochemistry,* 27, 9, 2735-2736, ISSN 0031-9422

Janistyn, B. (1989). cAMP promoted protein-phosphorylation of dialyzed coconut milk. *Phytochemistry,* 28, 2, 329-331, ISSN 0031-9422

Jiang, Y.X. & MacKinnon, R. (2000). The barium site in a potassium channel by X-ray crystallography. *Journal of General Physiology,* 115, 3, 269-272, ISSN 0022-1295

Jin, X.C. & Wu, W.H. (1999). Involvement of cyclic AMP in ABA- and Ca^{2+}-mediated signal transduction of stomatal regulation in *Vicia faba. Plant and Cell Physiology,* 40, 11, 1127-1133, ISSN 0032-0781

Kakiuchi, S., Yamazaki, R. & Teshima, Y. (1971). Cyclic 3',5'-nucleotide phosphodiesterase. IV. Enzymes with different properties form brain. *Biochemical and Biophysical Research Communications,* 42, 5, 968-974, ISSN 0006-291X

Kaplan, B., Sherman, T. & Fromm, H. (2007). Cyclic nucleotide-gated channels in plants. *Federation of European Biochemical Societies Letters,* 581, 12, 2237-2246, ISSN 0014-5793

Karin, M. & Smeal, T. (1992). Control of transcription factors by signal transduction pathways: the beginning of the end. *Trends in Biochemical Sciiences,* 17, 10, 418-422, ISSN 0968-0004

Kato, R., Uno, I., Ishikawa, T. & Fujii, T. (1983). Effects of Cyclic-Amp on the Activity of Soluble-Protein Kinases in Lemna-Paucicostata. *Plant and Cell Physiology,* 24, 5, 841-848, ISSN 0032-0781

Kaupp, U.B., Niidome, T., Tanabe, T., Terada, S., Bonigk, W., Stuhmer, W., Cook, N.J., Kangawa, K., Matsuo, H., Hirose, T., et al. (1989). Primary structure and functional expression from complementary DNA of the rod photorecepetor cyclic-GMP-gated channel. *Nature,* 342, 6251, 762-766, ISSN 0028-0836

Kaupp, U.B. & Seifert, R. (2002). Cyclic nucleotide-gated ion channels. *Physiological Reviews,* 82, 3, 769-824, ISSN 0031-9333

Kaupp, U.B. & Weyand, I. (2000). Cell biology. A universal bicarbonate sensor. *Science,* 289, 5479, 559-560, ISSN 0036-8075

Keates, R.A.B. (1973). Evidence that cyclic-AMP does not mediate action of gibberellic acid. *Nature,* 244, 5415, 355-357, ISSN 0028-0836

Kerr, I.D. & Sansom, M.S.P. (1995). Cation selectivity in ion channels. *Nature,* 373, 6510, 112-112, ISSN 0028-0836

Kessler, B. (1972). Hormonal and environmental modulation of gene expression in plant development. *Proceedings of the Australian Biochemical Society,* 5, Nmay, P16, ISSN 0067-1703

Kimmel, A.R. & Firtel, R.A. (2004). Breaking symmetries: regulation of Dictyostelium development through chemoattractant and morphogen signal-response. *Current Opinion Genetics & Development*, 14, 5, 540-549, ISSN 0959-437X

Klessig, D.F., Durner, J., Noad, R., Navarre, D.A., Wendehenne, D., Kumar, D., Zhou, J.M., Shah, J., Zhang, S.Q., Kachroo, P., et al. (2000). Nitric oxide and salicylic acid signaling in plant defense. *Proceedings of the National Academy of Sciences of the United States of America*, 97, 16, 8849-8855, ISSN 0027-8424

Komatsu, S. & Hirano, H. (1993). Protein kinase activity and protein phosphorylation in rice (*Oryza sativa*. L) leaf. *Plant Science*, 94, 1-2, 127-137, ISSN 0168-9452

Korschen, H.G., Illing, M., Seifert, R., Sesti, F., Williams, A., Gotzes, S., Colville, C., Muller, F., Dose, A., Godde, M., et al. (1995). A 240 kDa protein represents the complete beta-subunit of the cyclic nucleotide-gated channel from rod photoreceptor. *Neuron*, 15, 3, 627-636, ISSN 0896-6273

Kuhn, M. (2004). Molecular physiology of natriuretic peptide signalling. *Basic Research in Cardiology*, 99, 2, 76-82, ISSN 0300-8428

Kuo, J.F. & Greengard, P. (1970). Cyclic nucleotide-dependent protein kinases. VI. Isolation and parial purification of a protein kinase activated by guanosine 3',5'-monophosphate. *The Journal of Biological Chemistry*, 245, 10, 2493-2498

Kuriyama, Y., Koyama, J. & Egami, F. (1964). Digestion of chemically synthesized polyguanylic acids by ribonuclease T1 and spleen phosphodiesterase. *Seikagaku*, 36, 135-139

Kurosaki, F. & Kaburaki, H. (1995). Phosphodiesterase isoenzymes in cell-extracts of cultured carrot. *Phytochemistry*, 40, 3, 685-689, ISSN 0031-9422

Kurosaki, F., Kaburaki, H. & Nishi, A. (1994). Involvement of plasma membrane-located calmodulin in the response decay of cyclic nucleotide-gated cation channel of cultured carrot cells. *Federation of European Biochemical Societies Letters*, 340, 3, 193-196, ISSN 0014-5793

Kurosaki, F. & Nishi, A. (1993). Stimulation of calcium influx and calcium cascade by cyclic AMP in cultured carrot cells. *Archives of Biochemistry and Biophysics*, 302, 1, 144-151, ISSN 0003-9861

Kurosaki, F., Tsurusawa, Y. & Nishi, A. (1987). The elicitation of phytoalexins by Ca^{2+} and cyclic-AMP in carrot cells. *Phytochemistry*, 26, 7, 1919-1923, ISSN 0031-9422

Kwezi, L., Meier, S., Mungur, L., Ruzvidzo, O., Irving, H. & Gehring, C. (2007). The Arabidopsis thaliana brassinosteroid receptor (AtBRI1) contains a domain that functions as a guanylyl cyclase in vitro. *Plos One*, 2, 5, e449, ISSN 1932-6203

Leipe, D.D., Koonin, E.V. & Aravind, L. (2004). STAND, a class of P-loop NTPases including animal and plant regulators of programmed cell death: multiple, complex domain architectures, unusual phyletic patterns, and evolution by horizontal gene transfer. *Journal of Molecular Biology*, 343, 1, 1-28, ISSN 0022-2836

Lemtiri-Chlieh, F. & Berkowitz, G.A. (2004). Cyclic adenosine monophosphate regulates calcium channels in the plasma membrane of Arabidopsis leaf guard and mesophyll cells. *Journal of Biological Chemistry*, 279, 34, 35306-35312, ISSN 0021-9258

Leng, Q., Mercier, R.W., Hua, B.G., Fromm, H. & Berkowitz, G.A. (2002). Electrophysiological analysis of cloned cyclic nucleotide-gated ion channels. *Plant Physiologiy*, 128, 2, 400-410, ISSN 0032-0889

Leng, Q., Mercier, R.W., Yao, W. & Berkowitz, G.A. (1999). Cloning and first functional characterization of a plant cyclic nucleotide-gated cation channel. *Plant Physiology,* 121, 3, 753-761, ISSN 0032-0889

Leshem, Y.Y. (1996). Nitric oxide in biological systems. *Plant Growth Regulation,* 18, 3, 155-159, ISSN 0167-6903

Leyser, O. (1998). Auxin signalling: protein stability as a versatile control target. *Current Biology,* 8, 9, R305-R307, ISSN 0960-9822

Li, W., Luan, S., Schreiber, S.L. & Assmann, S.M. (1994). Cyclic AMP stimulates K^+ channel activity in mesophyll cells of *Vicia faba* L. *Plant Physiology,* 106, 3, 957-961, ISSN 0032-0889

Lieberman, M. & Kunishi, A.T. (1969). *Cyclic 3',5'-adenosine monophosphate phosphodiesterase in tissues of pea seedlings.* Paper presented at the Abstracts 11th International Congress Botany Seattle.

Lin, P.P. (1974). Cyclic nucleotides in higher plants? *Advances in Cyclic Nucleotide Research,* 4, 0, 439-460, ISSN 0084-5930

Lin, P.P.C. & Varner, J.E. (1972). Cyclic nucleotide phosphodiesterase in pea-seedlings. *Biochimica et Biophysica Acta,* 276, 2, 454-474, ISSN 0006-3002

Lincoln, T.M. & Cornwell, T.L. (1993). Intracellular cyclic-GMP receptor proteins. *Federation of American Societies for Experimental Biology Journal,* 7, 2, 328-338, ISSN 0892-6638

Linder, J.U. (2010). cGMP production in bacteria. *Molecular and Cellular Biochemistry,* 334, 1-2, 215-219, ISSN 0300-8177

Liu, Y., Ruoho, A.E., Rao, V.D. & Hurley, J.H. (1997). Catalytic mechanism of the adenylyl and guanylyl cyclases: modeling and mutational analysis. *Proceedings of the National Academy of Sciences of the United States of America,* 94, 25, 13414-13419, ISSN 0027-8424

Lloyd-Jones, D.M. & Bloch, K.D. (1996). The vascular biology of nitric oxide and its role in atherogenesis. *Annual Review of Medicine,* 47, 365-375, ISSN 0066-4219

Lomovatskaya, L.A., Romanenko, A.S., Filinova, N.V. & Dudareva, L.V. (2011). Determination of cAMP in plant cells by a modified enzyme immunoassay method. *Plant Cell Reports,* 30, 1, 125-132, ISSN 0721-7714

Ludidi, N. & Gehring, C. (2003). Identification of a novel protein with guanylyl cyclase activity in *Arabidopsis thaliana. Journal of Biological Chemistry,* 278, 8, 6490-6494, ISSN 0021-9258

Ludidi, N., Morse, M., Sayed, M., Wherrett, T., Shabala, S. & Gehring, C. (2004). A recombinant plant natriuretic peptide causes rapid and spatially differentiated K^+, Na^+ and H^+ flux changes in *Arabidopsis thaliana* roots. *Plant Cell Physiology,* 45, 8, 1093-1098, ISSN 0032-0781

Lundeen, C.V., Wood, H.N. & Brauna, A.C. (1973). Intracellular levels of cyclic nucleotides during cell enlargement and cell division in excised tobacco pith tissues. *Differentiation,* 1, 4, 255-260

Luo, D.-G., Xue, T. & Yau, K.-W. (2008). How vision begins: An odyssey. *Proceedings of the National Academy of Sciences of the United States of America,* 105, 29, 9855-9862

Ma, W., Smigel, A., Verma, R. & Berkowitz, G.A. (2009). Cyclic nucleotide gated channels and related signaling components in plant innate immunity. *Plant Signaling & Behaviouriour,* 4, 4, 277-282, ISSN 1559-2324

Ma, W., Yoshioka, K., Gehring, C. & Berkowitz, G.A. (2010). The function of cyclic nucleotide-gated channels in biotic stress. *In: Ion Channels and Plant Stress Responses,*

Signaling and Communication in Plants. Eds. V. Demidchik and F. Maathuis, Springer-Verlag Berlin Heidelberg (Germany), pp. 159-174

Maathuis, F.J. & Sanders, D. (2001). Sodium uptake in Arabidopsis roots is regulated by cyclic nucleotides. *Plant Physiology*, 127, 4, 1617-1625, ISSN 0032-0889

Maathuis, F.J.M. (2006). cGMP modulates gene transcription and cation transport in Arabidopsis roots. *Plant Journal*, 45, 5, 700-711, ISSN 0960-7412

Marden, J.N., Dong, Q.A., Roychowdhury, S., Berleman, J.E. & Bauer, C.E. (2011). Cyclic GMP controls Rhodospirillum centenum cyst development. *Molecular Microbiology*, 79, 3, 600-615, ISSN 0950-382X

Martens-Lobenhoffer, J., Dautz, C. & Bode-Boeger, S.M. (2010). Improved method for the determination of cyclic guanosine monophosphate (cGMP) in human plasma by LC-MS/MS. *Journal of Chromatography B*, 878, 3-4, 487-491, ISSN 1570-0232

Martinez-Atienza, J., Van Ingelgem, C., Roef, L. & Maathuis, F.J. (2007). Plant cyclic nucleotide signalling: facts and fiction. *Plant Signaling & Behaviouriour*, 2, 6, 540-543, ISSN 1559-2324

Mäser, P., Thomine, S., Schroeder, J.I., Ward, J.M., Hirschi, K., Sze, H., Talke, I.N., Amtmann, A., Maathuis, F.J., Sanders, D., et al. (2001). Phylogenetic relationships within cation transporter families of Arabidopsis. *Plant Physiology*, 126, 4, 1646-1667, ISSN 0032-0889

Matthews, G. (1987). Single-channel recordings demonstrate that cGMP opens the light-sensitive ion channel of the rod receptor. *Proceedings of the National Academy of Sciences of the United States of America*, 84, 1, 299-302, ISSN 0027-8424

Matthews, G. & Watanabe, S. (1987). Properties of ion channels closed by light and opened by guanosine 3',5'-cyclic-monophosphate in toad retinal rods. *Journal of Physiology*, 389, 691-715, ISSN 0022-3751

Matthews, H.R., Torre, V. & Lamb, T.D. (1985). Effects on the photoresponse of calcium buffers and cyclic-GMP incorporated into the cytoplasm of retinal rods. *Nature*, 313, 6003, 582-585, ISSN 0028-0836

Mauch-Mani, B. & Slusarenko, A.J. (1996). Production of salicylic acid precursors is a major function of phenylalanine ammonia-lyase in the resistance of Arabidopsis to *Peronospora parasitica*. *Plant Cell*, 8, 2, 203-212, ISSN 1040-4651

Mayer, B. & Hemmens, B. (1997). Biosynthesis and action of nitric oxide in mammalian cells. *Trends in Biochemical Sciences*, 22, 12, 477-481, ISSN 0968-0004

McAinsh, M.R., Allen, G.J., Hetherington, A.M. & Sanders, D. (1996). The role of cyclic-ADP-ribose in stomatal guard cells. *Plant Physiology*, 111, 2, 692-692, ISSN 0032-0889

McCue, L.A., McDonough, K.A. & Lawrence, C.E. (2000). Functional classification of cNMP-binding proteins and nucleotide cyclases with implications for novel regulatory pathways in *Mycobacterium tuberculosis*. *Genome Research*, 10, 2, 204-219, ISSN 1088-9051

McMains, V.C., Liao, X.H. & Kimmel, A.R. (2008). Oscillatory signaling and network responses during the development of *Dictyostelium discoideum*. *Ageing Research Reviews*, 7, 3, 234-248, ISSN 1872-9649

Meier, S. & Gehring, C. (2006). Emerging roles in plant biotechnology for the second messenger cGMP - guanosine 3',5'-cyclic monophosphate. *African Journal of Biotechnology*, 5, 19, 1687-1692, ISSN 1684-5315

Meier, S., Madeo, L., Ederli, L., Donaldson, L., Pasqualini, S. & Gehring, C. (2009). Deciphering cGMP signatures and cGMP-dependent pathways in plant defence. *Plant Signaling & Behaviouriour,* 4, 4, 307-309, ISSN 1559-2324

Meier, S., Ruzvidzo, O., Morse, M., Donaldson, L., Kwezi, L. & Gehring, C. (2010). The Arabidopsis wall associated kinase-like 10 gene encodes a functional guanylyl cyclase and is co-expressed with pathogen defense related genes. *Plos One,* 5, 1, e8904, ISSN 1932-6203

Meier, S., Seoighe, C., Kwezi, L., Irving, H. & Gehring, C. (2007). Plant nucleotide cyclases: an increasingly complex and growing family. *Plant Signaling & Behaviour,* 2, 6, 536-539, ISSN 1559-2324

Meiklejohn, A.L. & Gralla, J.D. (1985). Entry of RNA polymerase at the lac promoter. *Cell,* 43, 3 Pt 2, 769-776, ISSN 0092-8674

Melino, G., Bernassola, F., Rossi, A., Catani, V. & Knight, R.A. (1997). S-nitrosylation of caspases and tissue-transglutaminase regulates death by apoptosis of necrosis. *Anticancer Research,* 17, 5C, 3961-3962, ISSN 0250-7005

Milhaud, D., Michel, J., Bockaert, J. & Fagni, L. (1998). cAMP-mediated long-term modulation of voltage-dependent K^+ channels in cultured colliculi neurons. *Pflügers Archives- European Journal of Physiology,* 437, 1, 74-78, ISSN 0031-6768

Mohanty, S., Lee, S., Yadava, N., Dealy, M.J., Johnson, R.S. & Firtel, R.A. (2001). Regulated protein degradation controls PKA function and cell-type differentiation in Dictyostelium. *Genes & Development,* 15, 11, 1435-1448, ISSN 0890-9369

Moncada, S., Marletta, M.A., Hibbs, J.B. & Higgs, E.A. (1992). *The biology of nitric oxide, Parts I-II.* London: Portland Press

Moncada, S., Palmer, R.M.J. & Higgs, E.A. (1991). Nitric-oxide: Physiology, pathophysiology, and pharmacology. *Pharmacological Reviews,* 43, 2, 109-142, ISSN 0031-6997

Montminy, M. (1997). Transcriptional regulation by cyclic AMP. *Annual Review of Biochemistry,* 66, 807-822, ISSN 0066-4154

Moutinho, A., Hussey, P.J., Trewavas, A.J. & Malho, R. (2001). cAMP acts as a second messenger in pollen tube growth and reorientation. *Proceedings of the National Academy of Sciences of the United States of America,* 98, 18, 10481-10486, ISSN 0027-8424

Murad, F. (2006). Shattuck Lecture: Nitric oxide and cyclic GMP in cell signaling and drug development. *New England Journal of Medicine,* 355, 19, 2003-2011, ISSN 0028-4793

Nair, K.G. (1966). Purification and properties of 3',5'-cyclic nucleotide phosphodiesterase form dog heart. *Biochemistry,* 5, 1, 150-157, ISSN 0006-2960

Narayanan, A., Vermeersch, J. & Pradet, A. (1970). Enzymatic determination of cyclic adenosine 3', 5'-monophosphate in the seeds of lettuce, variety "Reine de mai". *Comptes Rendus Hebdomadaires des Seances de l' Academie des Sciences. D,* 271, 25, 2406-2407, ISSN 0567-655X

Neill, S., Bright, J., Desikan, R., Hancock, J., Harrison, J. & Wilson, I. (2008). Nitric oxide evolution and perception. *Journal of Experimental Botany,* 59, 1, 25-35, ISSN 0022-0957

Neill, S.J., Desikan, R., Clarke, A. & Hancock, J.T. (2002). Nitric oxide is a novel component of abscisic acid signaling in stomatal guard cells. *Plant Physiology,* 128, 1, 13-16, ISSN 0032-0889

Neuhaus, G., Bowler, C., Hiratsuka, K., Yamagata, H. & Chua, N.H. (1997). Phytochrome-regulated repression of gene expression requires calcium and cGMP. *EMBO Journal,* 16, 10, 2554-2564, ISSN 0261-4189

Neuhaus, G., Bowler, C., Kern, R. & Chua, N.H. (1993). Calcium/calmodulin-dependent and calcium/calmodulin-independent phytochrome signal-transduction pathways. *Cell,* 73, 5, 937-952, ISSN 0092-8674

Newton, R.P. & Brown, E.G. (1986). The biochemistry and physiology of cyclic AMP in higher plants, In *Receptors in plants and cellular slime moulds,* Garrod, D.R. & Chadwick, C.M. (Eds.), 115-153. Cambridge: Cambridge University Press.

Newton, R.P., Chiatante, D., Ghosh, D., Brenton, A.G., Walton, T.J., Harris, F.M. & Brown, E.G. (1989). Identification of cyclic-nucleotide constituents of meristematic and non-meristematic tissue of *Pisum sativum* roots. *Phytochemistry,* 28, 9, 2243-2254, ISSN 0031-9422

Newton, R.P., Kingston, E.E., Evans, D.E., Younis, L.M. & Brown, E.G. (1984). Occurrence of guanosine 3',5'-cyclic-monophosphate (cyclic-GMP) and associated enzyme-systems in *Phaseolus vulgaris. Phytochemistry,* 23, 7, 1367-1372, ISSN 0031-9422

Newton, R.P., Roef, L., Witters, E. & Van Onckelen, H. (1999). Cyclic nucleotides in higher plants: the enduring paradox. *New Phytologist,* 143, 3, 427-455, ISSN 0028-646X

Newton, R.P. & Smith, C.J. (2004). Cyclic nucleotides. *Phytochemistry,* 65, 17, 2423-2437, ISSN 0031-9422

Nikolaev, V.O. & Lohse, M.J. (2006). Monitoring of cAMP synthesis and degradation in living cells. *Physiology,* 21, 86-92, ISSN 1548-9213

Niles, R.M. & Mount, M.S. (1974). Failure to detect cyclic 3',5'-adenosine-monophosphate in healthy and crown gall tumorous tissues of *Vicia faba. Plant Physiology,* 54, 3, 372-373, ISSN 0032-0889

Ochoa De Alda, J.A., Ajlani, G. & Houmard, J. (2000). Synechocystis strain PCC 6803 cya2, a prokaryotic gene that encodes a guanylyl cyclase. *Journal of Bacteriology,* 182, 13, 3839-3842, ISSN 0021-9193

Ohba, Y., Soda, K. & Zaitsu, K. (2001). A sensitive assay of human blood platelet cyclic nucleotide phosphodiesterase activity by HPLC using fluorescence derivatization and its application to assessment of cyclic nucleotide phosphodiesterase inhibitors. *Biological & Pharmaceutical Bulletin,* 24, 5, 567-569, ISSN 0918-6158

Ohlenschlager, O., Hojo, H., Ramachandran, R., Gorlach, M. & Haris, P.I. (2002). Three-dimensional structure of the S4-S5 segment of the shaker potassium channel. *Biophysical Journal,* 82, 6, 2995-3002, ISSN 0006-3495

Osterrieder, W., Brum, G., Hescheler, J., Trautwein, W., Flockerzi, V. & Hofmann, F. (1982). Injection of subunits of cyclic AMP-dependent protein kinase into cardiac myocytes modulates Ca^{2+}. *Nature* , 128, 5874, 576-578, ISSN 0028-0836

Ownby, J.D., Ross, C.W., Bressan, R.A. & Key, J.L. (1973). Studies on presence of cyclic-AMP in plants. *Plant Physiology,* 51, 14-14, ISSN 0032-0889

Ownby, J.D., Ross, C.W. & Key, J.L. (1975). Studies on presence of adenosine cyclic 3',5'-monophosphate in oat coleoptiles. *Plant Physiology,* 55, 2, 346-351, ISSN 0032-0889

Pacini, B., Petrigliano, A., Diffley, P., Paffetti, A., Brown, E.G., Martelli, P., Trabalzini, L., Bovalini, L., Lusini, P. & Newton, R.P. (1993). Adenylyl-cyclase activity in roots of *Pisum sativum. Phytochemistry,* 34, 4, 899-903, ISSN 0031-9422

Pallas, J.A., Paiva, N.L., Lamb, C. & Dixon, R.A. (1996). Tobacco plants epigenetically suppressed in phenylalanine ammonia-lyase expression do not develop systemic acquired resistance in response to infection by tobacco mosaic virus. *Plant Journal,* 10, 281-293, ISSN 0960-7412

Pasqualini, S., Meier, S., Gehring, C., Madeo, L., Fornaciari, M., Romano, B. & Ederli, L. (2008). Ozone and nitric oxide induce cGMP-dependent and -independent transcription of defence genes in tobacco. *New Phytologist,* 181, 860–870, ISSN 1469-8137

Pei, Z.M., Murata, Y., Benning, G., Thomine, S., Klusener, B., Allen, G.J., Grill, E. & Schroeder, J.I. (2000). Calcium channels activated by hydrogen peroxide mediate abscisic acid signalling in guard cells. *Nature,* 406, 6797, 731-734, ISSN 0028-0836

Peng, C.H., Rich, E.D. & Varnum, M.D. (2004). Subunit configuration of heteromeric cone cyclic nucleotide-gated channels. *Neuron,* 42, 3, 401-410, ISSN 0896-6273

Penson, S.P., Schuurink, R., Shartzer, S.F. & Jones, R.L. (1997). Cyclic GMP and its possible targets in barley aleurone cells. *Plant Physiology,* 114, 1392, ISSN 1040-4651

Penson, S.P., Schuurink, R.C., Fath, A., Gubler, F., Jacobsen, J.V. & Jones, R.L. (1996). cGMP is required for gibberellic acid-induced gene expression in barley aleurone. *The Plant Cell,* 8, 2325-2333, ISSN 1040-4651

Pfeiffer, S., Janistyn, B., Jessner, G., Pichorner, H. & Ebermann, R. (1994). Gaseous nitric-oxide stimulates guanosine-3',5'-cyclic monophosphate (cGMP) formation in spruce needles *Phytochemistry,* 36, 2, 259-262, ISSN 0031-9422

Pharmawati, M., Billington, T. & Gehring, C.A. (1998). Stomatal guard cell responses to kinetin and natriuretic peptides are cGMP-dependent. *Cellular and Molecular Life Sciences,* 54, 3, 272-276, ISSN 1420-682X

Pharmawati, M., Maryani, M.M., Nikolakopoulos, T., Gehring, C.A. & Irving, H.R. (2001). Cyclic GMP modulates stomatal opening induced by natriuretic peptides and immunoreactive analogues. *Plant Physiology and Biochemistry,* 39, 5, 385-394, ISSN 0981-9428

Pharmawati, M., Shabala, S.N., Newman, I.A. & Gehring, C.A. (1999). Natriuretic peptides and cGMP modulate K+, Na+, and H+ fluxes in *Zea mays* roots. *Molecular Cell Biology Research Communications,* 2, 1, 53-57, ISSN 1522-4724

Picones, A. & Korenbrot, J.I. (1992a). cGMP-dependent conductance of cone outer segment membrane is active even in the absence of nucleotide. *Biophysical Journal,* 61, 2 Pt 2, A427, ISSN 0006-3495

Picones, A. & Korenbrot, J.I. (1992b). Permeation and interaction of monovalent cations with the cGMP-gated channel of cone photoreceptors. *Journal of General Physiology,* 100, 4, 647-673, ISSN 0022-1295

Pollard, C.J. (1970). Influence of gibberellic acid on the incorporation of 8-14C adenine into adenosine 3',5'-cyclic phosphate in barley aleurone layers. *Biochimica et Biophysica Acta,* 201, 511-512, ISSN 0006-3002

Pollard, C.J. & Venere, R.J. (1970). Gibberellin action and 3',5' cyclic AMP metabolism in aleurone layers. *Federation Proceedings,* 29, 2, A670, ISSN 0014-9446

Polya, G.M., Chung, R. & Menting, J. (1991). Resolution of a higher plant protein-kinase similar to the catalytic subunit of cyclic AMP-dependent protein-kinase. *Plant Science,* 79, 1, 37-45, ISSN 0168-9452

Pradet, A., Raymond, P. & Narayana.A. (1972). Confirmation of presence of cyclic AMP in lettuce seeds. *Comptes Rendus Hebdomadaires des Seances de l' Academie des Sciences Serie D,* 275, 18, 1987-1988

Price, T.D., Ashman, D.F. & Melicow, M.M. (1967). Organophosphates of urine including adenosine 3',5'-monophosphate and guanosine 3',5'-monophosphate. *Biochimica et Biophysica Acta,* 138, 3, 452-465, ISSN 0006-3002

Qi, Z., Verma, R., Gehring, C., Yamaguchi, Y., Zhao, Y.C., Ryan, C.A. & Berkowitz, G.A. (2010). Ca^{2+} signaling by plant Arabidopsis thaliana Pep peptides depends on AtPepR1, a receptor with guanylyl cyclase activity, and cGMP-activated Ca^{2+} channels. *Proceedings of the National Academy of Sciences of the United States of America,* 107, 49, 21193-21198, ISSN 0027-8424

Raymond, P., Narayanan, A. & Pradet, A. (1973). Evidence for the presence of 3', 5'-cyclic AMP in plant tissues. *Biochemical and Biophysical Research Communications,* 53, 4, 1115-1121, ISSN 0006-291X

Reggiani, R. (1997). Alteration of levels of cyclic nucleotides in response to anaerobiosis in rice seedlings. *Plant and Cell Physiology,* 38, 6, 740-742, ISSN 0032-0781

Rericha, E.C. & Parent, C.A. (2008). Steering in quadruplet: the complex signaling pathways directing chemotaxis. *Science Signaling,* 1, 22, pe26, ISSN 1937-9145

Robison, G.A., Butcher, R.W. & Sutherland, E.W. (1968). Cyclic AMP. *Annual Review of Biochemistry,* 37, 149-174, ISSN 0066-4154

Robison, G.A. & Sutherland, E.W. (1971). Cyclic AMP and function of eukaryotic cells: An introduction. *Annals of the New York Academy of Sciences,* 185, 5-9, ISSN 0077-8923

Roef, L., Witters, E., Gadeyne, J., Marcussen, J., Newton, R.P. & Van Onckelen, H.A. (1996). Analysis of 3',5'-cAMP and adenylyl cyclase activity in higher plants using polyclonal chicken egg yolk antibodies. *Analytical Biochemistry,* 233, 2, 188-196, ISSN 0003-2697

Roelofs, J., Meima, M., Schaap, P. & Van Haastert, P.J. (2001a). The Dictyostelium homologue of mammalian soluble adenylyl cyclase encodes a guanylyl cyclase. *EMBO Journal,* 20, 16, 4341-4348, ISSN 0261-4189

Roelofs, J., Snippe, H., Kleineidam, R.G. & Van Haastert, P.J. (2001b). Guanylate cyclase in *Dictyostelium discoideum* with the topology of mammalian adenylate cyclase. *Biochemical Journal,* 354, Pt 3, 697-706, ISSN 0264-6021

Rosenberg, N., Pines, M. & Sela, I. (1982). Adenosine 3'-5'-cyclic monophosphate: Its release in a higher plant by an exogeneous stimulus as detected by radioimmunoassay. *Federation of European Biochemical Societies Letters,* 137, 1, 105-107, ISSN 0014-5793

Rubio, F., Flores, P., Navarro, J.M. & Martinez, V. (2003). Effects of Ca^{2+}, K^{+} and cGMP on Na^{+} uptake in pepper plants. *Plant Science,* 165, 5, 1043-1049, ISSN 0168-9452

Ruiz, M. & Karpen, J.W. (1997). Single cyclic nucleotide-gated channels locked in different ligand-bound states. *Nature,* 389, 6649, 389-392, ISSN 0028-0836

Ryan, M.M., Lockstone, H.E., Huffaker, S.J., Wayland, M.T., Webster, M.J. & Bahn, S. (2006). Gene expression analysis of bipolar disorder reveals downregulation of the ubiquitin cycle and alterations in synaptic genes. *Molecular Psychiatry,* 11, 10, 965-978, ISSN 1359-4184

Salomon, D. & Mascarenhas, J.P. (1971). Effect of cyclic 3',5'-adenosine monophosphate on abscission of Coleus petioles. *Zeitschrift Fur Pflanzenphysiologie,* 65, 5, 385-388, ISSN 0044-328X

Salomon, D. & Mascarenhas, J.P. (1972). Time course of synthesis of cyclic AMP in Avena coleoptile sections in response to auxin. *Plant Physiology,* 49, 30, ISSN 0032-0889

Schmidt, P.M. (2009). Biochemical detection of cGMP from past to present: An overview. *Handbook of Experimental Pharmacology,* 195-228, ISSN 0171-2004

Schroeder, J.I. & Hedrich, R. (1989). Involvement of ion channels and active-transport in osmoregulation and signaling of higher-plant cells. *Trends in Biochemical Sciences,* 14, 5, 187-192, ISSN 0968-0004

Schuurink, R.C. & Jones, R.L. (1995). A putative CaM and cyclic-nucleotide-regulated ion-channel from barley. *Journal of Cellular Biochemistry,* 19A,153, ISSN 0730-2312

Schuurink, R.C., Shartzer, S.F., Fath, A. & Jones, R.L. (1998). Characterization of a calmodulin-binding transporter from the plasma membrane of barley aleurone. *Proceedings of the National Academy of Sciences of the United States of America,* 95, 4, 1944-1949, ISSN 0027-8424

Sculptoreanu, A., Scheuer, T. & Catterall, W.A. (1993). Voltage-dependent potentiation of L-type Ca^{2+} channels due to phosphorylation by cAMP-dependent protein kinase. *Nature,* 364, 6434, 240-243, ISSN 0028-0836

Sentenac, H., Bonneaud, N., Minet, M., Lacroute, F., Salmon, J.M., Gaymard, F. & Grignon, C. (1992). Cloning and expression in yeast of a plant potassium-ion transport-system. *Science,* 256, 5057, 663-665, ISSN 0036-8075

Sherman, T. & Fromm, H. (2009). Physiological roles of cyclic nucleotide gated channels in plants. *Signaling in Plants,* 91-106, ISSN 1867-9048

Shipston, M.J. (2001). Alternative splicing of potassium channels: a dynamic switch of cellular excitability. *Trends in Cell Biology,* 11, 9, 353-358, ISSN 0962-8924

Shpakov, A.O. & Pertseva, M.N. (2008). Signaling systems of lower eukaryotes and their evolution. *International Review of Cell and Molecular Biology,* 269, 151-282, ISSN 1937-6448

Shuster, M.J., Camardo, J.S., Siegelbaum, S.A. & Kandel, E.R. (1985). Cyclic AMP-dependent protein-kinase closes the serotonin-sensitive K^+ channels of aplysia sensory neurons in cell-free membrane patches. *Nature,* 313, 6001, 392-395, ISSN 0028-0836

Siegelbaum, S.A., Camardo, J.S. & Kandel, E.R. (1982). Serotonin and cyclic-AMP close single K^+ channels in aplysia sensory neurons. *Nature,* 299, 5882, 413-417, ISSN 0028-0836

Smith, M., Drummond, G.I. & Khorana, H.G. (1961). Cyclic phosphates. IV. Ribonucleoside 3',5'-cyclic phosphates. A general method of synthesis and some properties. *Journal of the American Chemical Society,* 83, 698-706, ISSN 0002-7863

Soda, K., Ohba, Y. & Zaitsu, K. (2001). Assay of human platelet guanylate cyclase activity by high-performance liquid chromatography with fluorescence derivatization. *Journal of Chromatography B,* 752, 1, 55-60, ISSN 0378-4347

Stamler, J.S. (1994). Redox signalling: Nitrosylation and related target interactions of nitric-oxide. *Cell,* 78, 6, 931-936, ISSN 0092-8674

Steinbrecher, K.A., Wowk, S.A., Rudolph, J.A., Witte, D.P. & Cohen, M.B. (2002). Targeted inactivation of the mouse guanylin gene results in altered dynamics of colonic epithelial proliferation. *American Journal of Pathology,* 161, 6, 2169-2178, ISSN 0002-9440

Steiner, A.L., Kipnis, D.M., Utiger, R. & Parker, C. (1969). Radioimmunoassay for measurement of adenosine 3',5'-cyclic phosphate. *Proceedings of the National Academy of Sciences of the United States of America,* 64, 1, 367-373, ISSN 0027-8424

Sundralingham, M. (1975). Structure and function of nucleosides, nucleotides and their analogues as determined by X-ray diffraction. *Annals of the New York Academy of Science,* 255, 3-42

Sutherland, E.W. & Rall, T.W. (1960). The relation of adenosine-3',5'-phosphate and phosphorylase to the actions of catecholamines and other hormones. *Pharmacological Reviews*, 12, 3, 265-299, ISSN 0031-6997

Sutherland, E.W., Robison, G. & Butcher, R.W. (1968). Some aspects of the biological role of adenosine 3', 5'-monophosphate (cyclic AMP). *Circulation*, 37, 2, 279-306, ISSN 0009-7322

Suzuki, N., Choe, H.R., Nishida, Y., Yamawaki-Kataoka, Y., Ohnishi, S., Tamaoki, T. & Kataoka, T. (1990). Leucine-rich repeats and carboxyl terminus are required for interaction of yeast adenylate cyclase with RAS proteins. *The Proceedings of the National Academy of Sciences of the United States of America*, 87, 22, 8711-8715, ISSN 0027-8424

Szmidt-Jaworska, A., Jaworski, K., Pawelek, A. & Kocewicz, J. (2009). Molecular cloning and characterization of a guanylyl cyclase, PNGC-1, involved in light signaling in *Pharbitis nil*. *Journal of Plant Growth Reguation*, 28, 4, 367-380, ISSN 0721-7595

Takahashi, N., Ejiri, S. & Katsumata, T. (1978). Changes of cyclic-AMP and cyclic-GMP levels during germination of pine pollen. *Agricultural and Biological Chemistry*, 42, 8, 1605-1606, ISSN 0002-1369

Talke, I.N., Blaudez, D., Maathuis, F.J. & Sanders, D. (2003). CNGCs: prime targets of plant cyclic nucleotide signalling? *Trends in Plant Science*, 8, 6, 286-293, ISSN 1360-1385

Taylor, W.R. & Baylor, D.A. (1995). Conductance and kinetics of single cGMP-activated channels in Salamander rod outer segments. *Journal of Physiology*, 483, 3, 567-582, ISSN 0022-3751

Teng, Y., Xu, W. & Ma, M. (2010). cGMP is required for seed germination in *Arabidopsis thaliana*. *Journal of Plant Physiology*, 167, 11, 885-889, ISSN 0176-1617

Thevelein, J.M. & De Winde, J.H. (1999). Novel sensing mechanisms and targets for the cAMP-protein kinase A pathway in the yeast *Saccharomyces cerevisiae*. *Molecular Microbiology*, 33, 5, 904-918, ISSN 1365-2958

Thomas, P.D., Campbell, M.J., Kejariwal, A., Mi, H., Karlak, B., Daverman, R., Diemer, K., Muruganujan, A. & Narechania, A. (2003). PANTHER: a library of protein families and subfamilies indexed by function. *Genome Research*, 13, 9, 2129-2141, ISSN 1088-9051

Thompson, W.J. & Appleman, M.M. (1971). Characterization of cyclic nucleotide phosphodiesterases of rat tissues. *Journal of Biological Chemistry*, 246, 10, 3145-3150, ISSN 0021-9258

Torre, V., Straforini, M., Sesti, F. & Lamb, T.D. (1992). Different channel-gating properties of two classes of cyclic GMP-activated channel in vertebrate photoreceptors. *Proceedings of the Royal Society of London Series B-Biological Sciences*, 250, 1329, 209-215, ISSN 0962-8452

Trautwein, W. & Hescheler, J. (1990). Regulation of cardiac L-type calcium current by phosphorylation and G proteins. *Annual Review of Physiology*, 52, 1, 257-274, ISSN 0066-4278

Trewavas, A.J. (1997). Plant cyclic AMP comes in from the cold. *Nature*, 390, 6661, 657-658, ISSN 0028-0836

Vandepeute, J., Huffaker, R.C. & Alvarez, R. (1973). Cyclic nucleotide phosphodiesterase activity in barley seeds. *Plant Physiology*, 52, 3, 278-282, ISSN 0032-0889

Very, A.A. & Sentenac, H. (2002). Cation channels in the Arabidopsis plasma membrane. *Trends in Plant Science*, 7, 4, 168-175, ISSN 1360-1385

Volotovski, I.D., Sokolovsky, S.G., Molchan, O.V. & Knight, M.R. (1998). Second messengers mediate increases in cytosolic calcium in tobacco protoplasts. *Plant Physiology*, 117, 3, 1023-1030, ISSN 0032-0889

Wang, Y., Gehring, C., Cahill, D. & Irving, H. (2007). Plant natriuretic peptide active site determination and effects on cGMP and cell volume regulation. *Functional Plant Biology*, 34, 7, 645-653, ,ISSN 1445-4408

Ward, J.M., Mäser, P. & Schroeder, J.I. (2009). Plant ion channels: gene families, physiology, and functional genomics analyses. *Annual Review of Physiology*, 71, 59-82, ISSN 1545-1585

Wäspi, U., Schweizer, P. & Dudler, R. (2001). Syringolin reprograms wheat to undergo hypersensitive cell death in a compatible interaction with powdery mildew. *Plant Cell*, 13, 1, 153-161, ISSN 1040-4651

Weitz, D., Ficek, N., Kremmer, E., Bauer, P.J. & Kaupp, U.B. (2002). Subunit stoichiometry of the CNG channel of rod photoreceptors. *Neuron*, 36, 5, 881-889, ISSN 0896-6273

Wellburn, A.R., Ashby, J.P. & Wellburn, F.A. (1973). Occurrence and biosynthesis of adenosine 3',5'-cyclic monophosphate in isolated Avena etioplasts. *Biochimica et Biophysica Acta*, 320, 2, 363-371, ISSN 0006-3002

White, P.J., Bowen, H.C., Demidchik, V., Nichols, C. & Davies, J.M. (2002). Genes for calcium-permeable channels in the plasma membrane of plant root cells. *Biochimica et Biophysica Acta*, 1564, 2, 299-309, ISSN 0006-3002

Wiedmaier, J. & Kull, U. (1976). Evidence of cyclic-AMP uptake into plant cells. *Naturwissenschaften*, 63, 3, 147-147, ISSN 0028-1042

Wink, D.A. & Mitchell, J.B. (1998). Chemical biology of nitric oxide: Insights into regulatory, cytotoxic, and cytoprotective mechanisms of nitric oxide. *Free Radical Biology and Medicine*, 25, 4-5, 434-456, ISSN 0891-5849

Witters, E., Roef, L., Newton, R.P., VanDongen, W., Esmans, E.L. & Van Onckelen, H.A. (1996). Quantitation of cyclic nucleotides in biological samples by negative electrospray tandem mass spectrometry coupled to ion suppression liquid chromatography. *Rapid Communications in Mass Spectrometry*, 10, 2, 225-231, ISSN 0951-4198

Witters, E., VanDongen, W., Esmans, E.L. & VanOnckelen, H.A. (1997). Ion-pair liquid chromatography electrospray mass spectrometry for the analysis of cyclic nucleotides. *Journal of Chromatography B*, 694, 1, 55-63, ISSN 0378-4347

Witters, E., Vanhoutte, K., Dewitte, W., Machackova, I., Benkova, E., Van Dongen, W., Esmans, E.L. & Van Onckelen, H.A. (1999). Analysis of cyclic nucleotides and cytokinins in minute plant samples using phase-system switching capillary electrospray-liquid chromatography-tandem mass spectrometry. *Phytochemical Analysis*, 10, 3, 143-151, ISSN 0958-0344

Wohlfart, P., Haase, W., Molday, R.S. & Cook, N.J. (1992). Antibodies against synthetic peptides used to determine the topology and site of glycosylation of the cGMP-gated channel from bovine rod photoreceptors. *Journal of Biological Chemistry*, 267, 1, 644-648, ISSN 0021-9258

Wu, J., Brown, S., Xuong, N.H. & Taylor, S.S. (2004). Rl alpha subunit of PKA: A cAMP-free structure reveals a hydrophobic capping mechanism for docking cAMP into site B. *Structure*, 12, 6, 1057-1065, ISSN 0969-2126

Yau, K.W. & Baylor, D.A. (1989). Cyclic GMP-activated conductance of retinal photoreceptor cells. *Annual Review of Neuroscience*, 12, 289-327, ISSN 0147-006X

Yau, K.W. & Nakatani, K. (1985). Light-suppressible, cyclic GMP-sensitive conductance in the plasma-membrane of a truncated rod outer segment. *Nature*, 317, 6034, 252-255, ISSN 0028-0836

Yu, F.H., Yarov-Yarovoy, V., Gutman, G.A. & Catterall, W.A. (2005). Overview of molecular relationships in the voltage-gated ion channel superfamily. *Pharmacological Reviews*, 57, 4, 387-395, ISSN 0031-6997

Zagotta, W.N. & Siegelbaum, S.A. (1996). Structure and function of cyclic nucleotide-gated channels. *Annual Review of Neuroscience*, 19, 235-263, ISSN 0147-006X

Zhang, X., Zhang, L., Dong, F.C., Gao, J.F., Galbraith, D.W. & Song, C.P. (2001). Hydrogen peroxide is involved in abscisic acid-induced stomatal closure in *Vicia faba*. *Plant Physiology*, 126, 4, 1438-1448, ISSN 0032-0889

Zheng, J., Varnum, M.D. & Zagotta, W.N. (2002). Disruption of intersubunit interactions of cyclic nucleotide-gated channels by Ca-calmodulin measured by FRET. *Biophysical Journal*, 82, 1, 1337, ISSN 0006-3495

Zheng, J. & Zagotta, W.N. (2004). Stoichiometry and assembly of olfactory cyclic nucleotide-gated channels. *Neuron*, 42, 3, 411-421, ISSN 0896-6273

Zhong, H.M., Molday, L.L., Molday, R.S. & Yau, K.W. (2002). The heteromeric cyclic nucleotide gated channel adopts a 3A : 1B stoichiometry. *Nature*, 420, 6912, 193-198, ISSN 0028-0836

Zimmerman, A.L. (2001). *Cyclic nucleotide-gated ion channels*, In: Cell Physiology Source Book, Sperelakis, N., 795-805, Academic Press, ISBN 0-12-656976-2 San Diego

Zimmerman, A.L. & Baylor, D.A. (1992). Cation interactions within the cyclic GMP-activated channel of retinal rods from the tiger salamander. *Journal of Physiology*, 449, 759-783, ISSN 0022-3751

Zimmermann, P., Hennig, L. & Gruissem, W. (2005). Gene-expression analysis and network discovery using Genevestigator. *Trends in Plant Science*, 10, 9, 407-409, ISSN 1360-1385

Abiotic Stress-Induced Programmed Cell Death in Plants: A Phytaspase Connection

Alexander I. Tuzhikov[1], Boris B. Vartapetian[2]
Andrey B. Vartapetian[1] and Nina V. Chichkova[1]
[1]*A.N.Belozersky Institute of Physico-Chemical Biology*
Moscow State University
[2]*K.A.Timiryazev Institute of Plant Physiology, Russian Academy of Sciences*
Moscow
Russia

1. Introduction

The plants in the course of ontogenesis are frequently exposed to various unfavourable environmental stress factors (such as high and low temperatures, heavy metals, salt and anaerobic stress, biotic stresses, etc.). If extreme, these stress factors can bring about damage of agricultural crops and wild flora causing considerable environmental and economic losses worldwide.

In the course of evolution, and also due to targeted selection, many plant species have developed the capacity for adaptation to some extent to these unfavourable stresses. Under these circumstances adaptive processes in plants are aimed on sustainability of the plant cell and thus on survival of the whole organism. It is worth noting that plants have also developed alternative adaptive mechanisms which are aimed at selective death of the cells and tissues under abiotic and biotic stresses rather than on their survival. Such selective programmed cell death (PCD) eventually provides survival benefits for the whole organism under extreme conditions (Drew et al., 2000; Jackson & Armstrong, 1999).

This review focuses on abiotic stress-induced PCD in plants and highlights the role of a newly discovered plant protease, phytaspase, in this process. We describe the role of phytaspase in PCD execution, structure and properties of the enzyme, and its intriguing trafficking in response to stress stimuli. Parallels between the mechanisms of PCD in plants and animals are drawn, highlighting both similarities and differences of the apoptotic proteases involved and the strategies to keep them under control.

2. Abiotic stress-induced programmed cell death responses in plants

The two above mentioned distinct strategies of plant adaptation to unfavourable abiotic stresses are most pronounced under anaerobic stress (anoxia and hypoxia) during ontogenesis of rice (Vartapetian et al., 2008). Rice is a unique agricultural crop that readily inhabits waterlogged and flooded anaerobic soils, which occupy vast territories of our planet. In such soils anaerobic conditions are caused due to low solubility and slow diffusion of molecular oxygen in water. Like other higher plants, rice is an obligate aerobe and needs a permanent supply of environmental oxygen. However, in contrast to the other agricultural crops, the evolution of this species has resulted in the two above mentioned opposite adaptive strategies to hypoxia and anoxia. The first adaptive strategy is aimed at the survival of the cells and the whole rice seedling during germination under hypoxia and even anoxia, when respiration, which is the main energy-supplying mechanism of aerobes, is entirely arrested. Under these extreme conditions, which virtually preclude aerobic respiration of the plant cells, the rice seeds have developed the capacity to support not only the germinating ability but also the active growth of the seedling due to stimulation of cell anaerobic energy metabolism (Magneschi & Perata, 2009).

However, the rice plant has also developed the mechanism of adaptation to anaerobic stress, which is paradoxically aimed on selective death of the root and shoot cells (Drew et al., 2000). This mechanism based on the pivotal role of the PCD starts up when anaerobically sprouted seedlings, having pushed through the water layer and come in contact with atmospheric air, start forming the root system. The cell death results in the formation of continuous air cavities (aerenchyma), which readily enable the oxygen transport from overground aerated organs to the roots and survival of the whole plant under these unfavourable conditions.

PCD, as a way of cell demise, is operative in plant responses to various other abiotic stresses such as salinity, cold and heat stress, UV, oxidative stress, etc. Therefore stress-induced PCD responses are of considerable interest, not only from the fundamental perspective but also in the context of applied agriculture, ecology and environmental protection. Elucidation of the mechanisms involving specific enzyme systems in these processes lays foundation for future development of biotechnological methods enabling the creation of plants that are more tolerant to these stress conditions.

3. Aspartate-specific proteases in programmed cell death

PCD is a fundamental process that operates during tissue development and homeostasis of a multicellular organism. This process employs suicide molecular machinery which is activated in order to rapidly destroy the cell. PCD helps maintain tissue balance by providing the means to eliminate redundant cells. It also plays a crucial role during a response to various abiotic stress factors by dismantling a limited number of affected cells to prevent severe damage to the whole organism. In animal organisms, one of the best characterized forms of PCD is apoptosis (Thompson, 1995). Essential molecular components of apoptosis are caspases – a family of highly specific cysteine proteases which execute apoptosis by cleaving specific cellular protein substrates after aspartate residues within a specific (usually tetrapeptide) recognition site (Cohen, 1997). Caspases are stored in the cytoplasm as latent zymogens (inactive precursors) and are processed in response to a variety of death stimuli to generate a mature active enzyme. Caspases divide in two groups: initiator and effector caspases. Initiator caspases (e.g. caspases-2, -8, -9 and -10) cleave inactive zymogens of effector caspases, thus activating them. Effector caspases (e.g. caspases-3, -6 and -7) in turn cleave other protein substrates within the cell, among which are proteins and enzymes of a crucial importance for the cellular homeostasis (Chang & Yang, 2000). Due to the supremely important role that caspases play in apoptosis, it is imperative that their activation is strictly controlled. Caspase activity is tightly regulated by the caspase inhibitors, thereby inhibiting of the inappropriately activated caspases leads to suppression of apoptosis and survival of the cell (Brady & Duckett, 2009).

Animal and plant PCD have some features in common, such as nuclear DNA fragmentation (laddering) (Ryerson & Heath, 1996), release of cytochrome *c* from mitochondria, cytoplasm shrinkage (Levine et al., 1996), and cellular plasma membrane blabbing (Lam & del Pozo, 2000). However, bioinformatics studies of the plant genomes sequenced thus far have failed to reveal direct caspase homologues in plants. Plants do possess metacaspases - a family of distant caspase homologues, which may be involved in PCD, but possess specificity distinct from that of caspases (substrate cleavage after basic amino acid residues, Arg and Lys, not after Asp) (Aravind et al., 1999; Vercammen et al., 2006; Watanabe & Lam, 2005).

However, it is becoming increasingly evident, that plants employ the caspase-like (that is, Asp-specific) activity during the PCD. In various plant PCD models, application of synthetic peptide inhibitors based on the recognition sites of the specific members of the animal caspase family was reported to suppress PCD (Bonneau et al., 2008; del Pozo & Lam, 1998). Likewise, protein caspase inhibitors, such as p35 from baculovirus and IAP, having been synthesized in plants, counteract the PCD (Danon et al., 2004; del Pozo & Lam, 2003; Dickman et al., 2001). Conforming to these results, hydrolysis of peptide-based caspase substrates has been observed during the PCD in different plant systems. The absence of the caspase family genes in plant genomes therefore raises the possibility that plants possess functional, rather than structural, analogues of caspases.

Until recently, scarce information about plant PCD-related proteases with the 'caspase-like' activity was available, the unidentified putative caspase-like enzymes being named VEIDase, YVADase, VADase, DEVDase, etc., in accord with their peptide cleavage specificity (Bonneau et al., 2008). In rare cases, however, plant enzymes possessing some of the above mentioned 'caspase-like' activities have been identified. For example, vacuolar processing enzyme (VPE), which is an asparaginyl-specific cysteine protease distantly related to caspases, was shown to display an YVADase activity as well (Hatsugai et al., 2004; Rojo et al., 2004). Recently, a DEVDase activity was attributed to the *Arabidopsis* PBA1, a proteasome subunit (Hatsugai et al., 2009).

In this review, we will focus on the newly identified plant protease named phytaspase (for plant aspartate-specific protease) (Chichkova et al., 2010). Being structurally completely different from animal caspases, phytaspase possesses a 'caspase-like' cleavage specificity and its activity is essential for accomplishment of PCD induced by a variety of stress stimuli. Phytaspase thus appears to represent a functional analogue of the animal caspases.

4. Identification of the phytaspase

Nearly a decade ago, while studying a human protein prothymosin alpha (ProTa) it was noticed that ProTa is subject to caspase fragmentation in the course of apoptosis caused by a variety of death-inducing stimuli, including abiotic stresses (Evstafieva et al., 2000; Evstafieva et al., 2003). ProTa is a proliferation-related protein localizing in the nucleus where it is engaged in activity regulation of several stress-related transcription factors, such as p53 tumour suppressor and Nrf2 (Karapetian et al., 2005; Kobayashi et al., 2006; Zakharova et al., 2008). Caspase-3-mediated cleavage detaches a short C-terminal peptide of ProTa. This, however, exerts a profound effect on the fate of ProTa because this very region encompasses a nuclear localization signal (NLS) that drives protein import into the nucleus

(Rubtsov et al., 1997). As a result of caspase-mediated truncation, ProTa loses its ability to accumulate in the nucleus and, furthermore, becomes surface exposed by dying human cells (Evstafieva et al., 2003).

To learn whether some other proteins could follow this exciting scheme of relocalization in response to caspase-mediated fragmentation, a bioinformatics search for proteins containing a putative NLS lying close to a putative caspase cleavage site was performed. Among many hits obtained, one was particularly astonishing, predicting that a bacterial protein possesses an NLS and a caspase cleavage site nearby. It was realized however that this protein, VirD2, is encoded by *Agrobacterium tumefaciens*, a plant pathogen which causes the crown gall disease. Furthermore, one of the VirD2 functions is to guide a segment of bacterial DNA (T-DNA) into the plant cell nucleus to achieve transformation of the host cell (Pitzschke & Hirt, 2010). For this purpose, the bacterially encoded VirD2 protein is indeed equipped with an NLS positioned close to the C-terminus of the protein (Howard et al., 1992), which perfectly matched with the predicted one.

The second prediction came true as well. Caspase-3, the major human executioner caspase, has been shown to cleave VirD2 *in vitro* at a $TATD^{400}$ site close to the C-terminus of the protein, just upstream from the NLS. This observation has raised the possibility that such VirD2 fragmentation, if accomplished in plant cell by a plant caspase-like protease, could make sense to protect plants from transformation by limiting delivery of foreign DNA into the nucleus.

In planta studies involving a VirD2 C-terminal region (VirD2Ct)-based fluorescent reporter GFP-VirD2Ct have revealed that during the tobacco mosaic virus (TMV)-mediated PCD in *Nicotiana tabacum* leaves a caspase-like protease is activated which cleaves the VirD2 moiety at the same TATD motif (Chichkova et al., 2004). A substitution of D residue with A abolished the cleavage of GFP-VirD2Ct, testifying to the unique D-specificity of the newly discovered protease.

A synthetic peptide inhibitor designed based on the TATD-recognition motif has been shown to inactivate the enzyme and to suppress the development of the TMV-induced PCD in *Nicotiana tabacum* leaves, pointing to an essential role for this protease in PCD execution (Chichkova et al., 2004). The protease was named "phytaspase" in respect to its caspase-like activity, and phytaspase activity was detected in many plant species, both di- and monocotyledonous (Chichkova et al., 2008).

It was also found that healthy plant tissue wounding/disruption produces high level of phytaspase activity in extracts, indicating that mechanical stress, as well as PCD, causes rapid activation (or de-sequestration) of the enzyme.

5. Structure of phytaspases

Phytaspases from *Nicotiana tabacum* and *Oryza sativa* were purified, identified, and their cDNAs cloned (Chichkova et al., 2010). The enzymes from both species turned out to be homologous and belong to the S8A subtilisin-like (that is, Ser-dependent) protease family (as all known plant subtilases are). Alignment of phytaspases with known subtilases of bacterial and plant origin revealed a canonical triad of catalytic amino acid residues: Asp149, His220 and Ser537. Accordingly, mutating the predicted active site Ser^{537} produced an inactive enzyme. Analysis of the primary structure indicated that phytaspases are synthesized as precursor proteins, with the hydrophobic N-terminal signal peptide and the prodomain preceding the protease domain of the enzyme, which is typical for subtilases

(Schaller, 2004) - Figure 1. Phylogenetic comparison of tobacco and rice phytaspases with the *Arabidopsis* subtilisin-like proteins places phytaspases as a branch within the Subgroup 1 of *Arabidopsis* subtilases (Vartapetian et al., 2011).

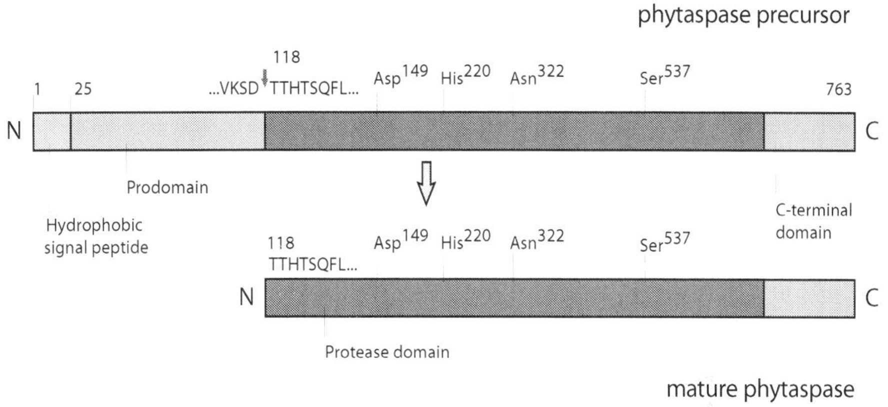

Fig. 1. Schematic representation of the pre-pro-phytaspase and the mature form of phytaspase. Amino acid numbering is given for tobacco phytaspase (MEROPS ID S08.150). Asp^{149}, His^{220} and Ser^{537} are active site residues; Asn^{322} is the oxyanion hole residue. Red arrow points to the prodomain cleavage site. Note that autocatalytic Asp-specific prodomain cleavage conforms to the caspase-like specificity of phytaspase.

6. Involvement of phytaspase in the abiotic stress response

In order to assess possible impact of phytaspase activity on plant development and response to stress factors, *Nicotiana tabacum* transgenic plants were constructed that either overproduced tobacco phytaspase, or possessed markedly decreased level of phytaspase gene expression due to the RNAi silencing. Notably, despite the significant variations in phytaspase activity levels achieved, under normal growth conditions neither overproduction, nor gene silencing has resulted in any discernible phenotype as compared to the wild type plants. Increased enzymatic activity could possibly undergo neutrally due to the tight regulation and a specific tissue localization of phytaspase under the normal conditions (described below). The silencing experiments suggest that functional redundancy may exist and another enzyme may compensate for the low level of phytaspase during plant growth and development.

Possible involvement of phytaspase in the PCD-related response to abiotic stresses, such as oxidative stress and osmotic stress, was assessed by treating leaf discs of transgenic tobacco plants with methyl viologen (MV) or with NaCl at elevated concentrations. Methyl viologen (paraquat) is known to be a toxic agent for both animals and plants. Upon contact with a cell, this redox-active compound intercalates into the intracellular electron transfer systems and causes uncontrollable formation of reactive oxygen species (ROS), leading to extensive damage of the macromolecules, including intracellular membrane components, and therefore destruction of the organelles of the cell. In plant cells, methyl viologen prevents the reduction of NADP to NADPH during photosynthesis, and thereby causes rapid elimination of the chloroplasts and bleaching of the leaf tissues (Fujibe et al., 2004). The oxidative stress caused by methyl viologen leads to rapid development of PCD in plant cells. The effect of NaCl at concentrations exceeding the physiological rate represents a form of abiotic stress, similar to salinization of soil. Exposure to relatively high concentrations of NaCl causes a misbalance of cellular ion concentrations, osmotic shock and oxidative stress. Salt-induced PCD response, as well as treatment with methyl viologen, leads to rapid bleaching of the tissues and loss of viability (Huh et al., 2002).

Treatment with these stress-inducing agents produced characteristic molecular manifestations of PCD in plants, such as ROS accumulation (H_2O_2 production) and release of cytochrome *c* from mitochondria, which preceded the morphological changes (Chichkova et al., 2010). Both molecular and morphological changes have been shown to develop more rapidly in the phytaspase overexpressing line of tobacco plants, comparing to the wild type plants in response to treatment with 2.5 mM MV and with 75 mM NaCl. At the same time, even being exposed to the higher concentrations of stress-inducing agents (10 mM MV and at 250 mM NaCl), the phytaspase gene knockdown plants showed considerably improved viability comparing to the wild type plants. Expression of the active rice phytaspase in the RNAi knockdown tobacco plants restored the wild type phenotype (bleaching and the H_2O_2 levels), while expression of the catalytically inactive mutant did not, indicating a role for phytaspase proteolytic activity in the PCD execution.

7. Cleavage specificity of phytaspases

Phytaspase, like animal caspases, displays a unique specificity towards the Asp residue in the P_1 position of a substrate. Such a stringent specificity is quite unusual for subtilases. Most of the subtilisin-like proteases either display limited specificity, or possess only partial substrate selectivity towards different groups of amino acid residues. For instance, cleavage by bacterial subtilisin A typically occurs adjacent to large uncharged residues (Bryan et al., 1986). Likewise, proteinase K, a typical member of the subtilisin-like protease family, displays broad specificity towards aliphatic and aromatic amino acid residues and is frequently employed for complete protein digestion (Kraus & Femfert, 1976). Plant subtilases are usually also devoid of strict cleavage specificity (Beers et al., 2000). However, remarkable phytaspase specificity demonstrates that the structure of plant subtilisin-like proteins is able to support different levels of substrate selectivity, including the most rigorous and specific ones. It is also evident that phytaspases are processive, rather than digestive, proteolytic enzymes.

Besides a requirement for D at the P_1 position of a substrate, phytaspases display selectivity towards a preceding amino acid motif. By using peptide-based fluorogenic substrates of caspases, it was found that Ac-VEID-AFC (AFC is 7-amino-4-trifluoromethylcoumarin) is the optimal substrate (among the ones tested) for both tobacco and rice phytaspases. Several other substrates, such as YVAD-, VAD-, IETD- and LEHD-AFC are also cleaved albeit somewhat less efficiently, whereas the DEVD-AFC is not cleaved by phytaspases at all. Consistent with these results, peptide aldehyde caspase inhibitors based on the same amino acid motifs could reversibly inactivate phytaspases, Ac-VEID-CHO being the most potent inhibitor, whereas Ac-DEVD-CHO failed to inhibit phytaspases (Chichkova et al., 2010). Interestingly, the STATD motif representing an efficient phytaspase cleavage site in the VirD2 protein turned out to be a rather poor phytaspase substrate at the peptide level. This suggests that the activity of phytaspase may not only depend on the amino acid sequence of the cleavage site, but also on the position, steric availability and (or) the overall context of the cleavage site within the protein molecule.

Of note, an unidentified subtilisin-like protease from oat (saspase) was reported to cleave various caspase peptide-based substrates as well, with the same exception of DEVD-AFC. However, cleavage specificities of phytaspases and saspases are markedly distinct, as VEID-AFC, an optimal phytaspase substrate was not cleaved by saspase at all (Coffeen & Wolpert, 2004).

Stringent hydrolytic specificity of phytaspases resembles that of animal and yeast subtilisin-like proteases called proprotein converting enzymes (convertases). Convertases belong to the S8B subfamily of subtilases which appear to be absent from plants (Tripathi & Sowdhamini, 2006). Convertases are involved in proteolytic processing of precursors to generate bioactive proteins and peptides (Seidah & Chretien, 1999). Unlike phytaspases, however, convertases introduce cleavage after a basic amino acid residue (Lys, Arg), not after Asp.

8. Phytaspase localization and processing

In healthy plant tissues, phytaspase was shown to accumulate in the apoplast in the processed form lacking the signal peptide and the prodomain. This was elucidated by biochemical fractionation and by tracing the phytaspase linked to a fluorescent protein in *N. tabacum* leaves (Chichkova et al., 2010) - Figure 2A. The apoplast separates the plasma membrane and the cellular wall and forms a continuous film of fluid on the cell exterior. It facilitates the transport of water and solutes across a tissue, and moreover, it is considered to play a key role in plant resistance to multiple abiotic stress factors, such as draught, low temperatures and salinization of soil. All of these stress conditions lead to a rapid alkalization of the apoplast with characteristic times of minutes after exposure to the stress factor. As long as the apoplastic fluid spreads from roots to the shoots of the plants, it seems to be capable of rapid transfer of the signal across the tissues (Felle et al., 2005). Apart from responding to various abiotic challenges, the apoplast also mediates plant hormone signaling and participates in plant organism homeostasis (Matsubayashi, 2003; Ryan et al., 2002). Secretory proteins are usually transferred and concentrate in the apoplast media. Many plant proteases undergo this secretory pathway, including those of the subtilisin-like proteases involved in plant response to stresses.

Application of brefeldin A, an inhibitor of secretion, resulted in retention of the phytaspase-mRFP fusion inside the plant cells indicating that the canonical secretory pathway mediates translocation of the de novo synthesized phytaspase to the apoplast (Chichkova et al., 2010). Interestingly, the brefeldin A-mediated arrest of secretion did not severely affect the processing of phytaspase suggesting that secretion is not required for this process. Although processing of phytaspase does not directly depend on the secretion, some molecular events during the transit through the secretory pathway may favour this process. In this case, proteolytic activity of phytaspase would remain shaded by the prodomain until phytaspase had reached the extracellular fluid, thus protecting the cell from undesirable induction of PCD.

It is thus evident that in healthy plant tissues phytaspase is constitutively processed and secreted into the apoplast. It turned out that the phytaspase processing (detachment of the prodomain) occurs autocatalytically, as revealed by the properties of the phytaspase Ser537Ala mutant. The loss of the enzymatic activity by the mutant led to its complete inability to detach the prodomain, which was demonstrated by Western blot assay (the unprocessed enzyme displays lower electrophoretic mobility comparing to the wild type protein), and further unambiguously ascertained by Edman degradation sequencing of the N-terminal region of the Ser537Ala mutant.

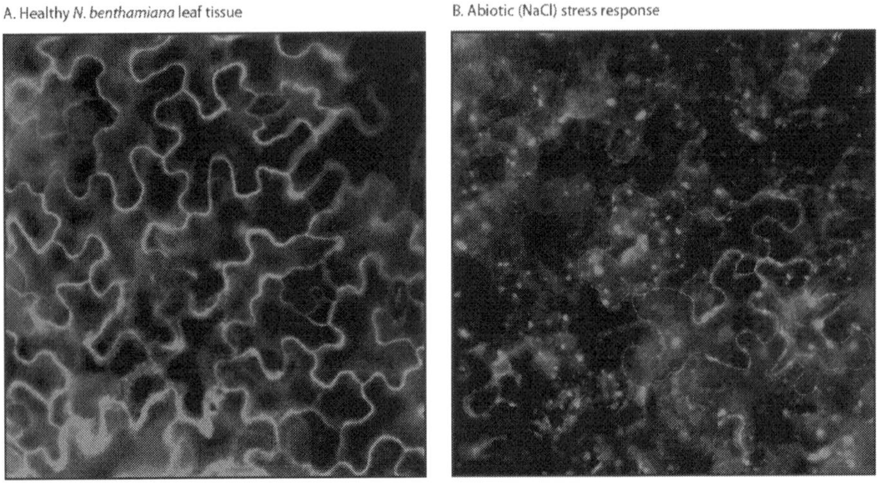

Fig. 2. Fluorescence microscopy visualization of phytaspase-EGFP in healthy (A) and osmotic stress-induced (B) *Nicotiana benthamiana* leaf tissues. PCD was induced in (B) by treatment with 300 mM NaCl for 6 h. The fluorescence colour has been changed to fit the colours given in Figures 1 and 3. White borders in (B) mark boundaries of the stressed cells, as visualized by the phase contrast.

In accordance with the phytaspase cleavage specificity, the C-terminal amino acid residue of the prodomain immediately preceding the prodomain cleavage site is D in both tobacco and rice phytaspases - Figure 1. Substitution of this Asp117 with Ala has impaired processing of the proenzyme (Chichkova et al., 2010). Of note, both of the phytaspase mutants with the processing defect, the Ser537Ala mutant and the Asp117Ala mutant, displayed impaired secretion into the apoplast, remaining mostly associated with the intracellular fraction.

Cumulatively, these data suggest that the phytaspase processing, as a part of its normal maturation process, involves the enzymatic activity of phytaspase itself and occurs on the route of the enzyme to cell exterior.

9. Retrograde trafficking of phytaspase during the programmed cell death

Phytaspase is involved in the accomplishment of PCD in plants triggered by abiotic stresses. Furthermore, phytaspase-mediated proteolysis of an intracellular target protein was reported to occur in the course of PCD (Chichkova et al., 2004). It was however not immediately clear how phytaspase, being localized in the apoplast, could get access to cell interior.

An important observation, shedding light on this problem, was made using a phytaspase-mRFP fusion protein transiently produced in *N. tabacum* leaves subject to abiotic stresses, such as treatment with MV or NaCl. It was found that phytaspase rapidly relocalizes from the apoplast to inside the cell in response to stress inducers (Chichkova et al., 2010) - Figure 2B. As the leaf treatment with cycloheximide, an inhibitor of protein biosynthesis, did not affect phytaspase relocalization, it appears likely that phytaspase accumulation within the dying cells originates from the redistribution of the presynthesized enzyme from the apoplast into the cytoplasm, rather than from the impairment of secretion of the newly synthesized protein.

The retrograde transport of the phytaspase to the cytoplasm appears to be specific, since another secreted protease, cathepsin B, under similar conditions retained its apoplastic localization.

A model describing phytaspase behaviour in healthy and dying plant cells has been suggested (Chichkova et al., 2010). According to this model (Figure 3), phytaspase is synthesized as a precursor protein which is constitutively secreted (due to the presence of a signal peptide) and autocatalytically processed on its route to the cell exterior. Mature

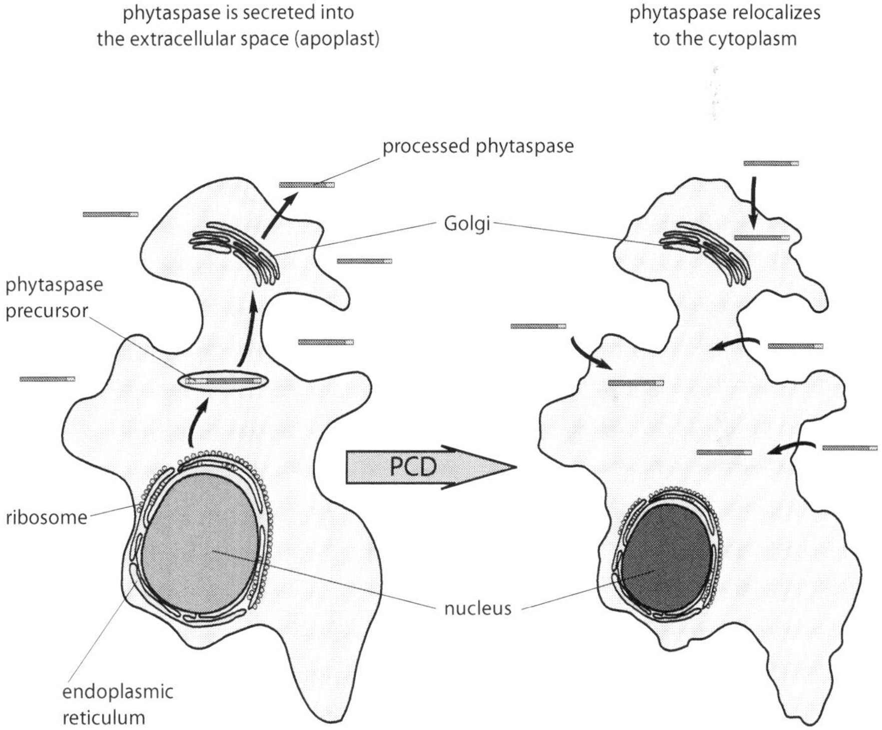

Fig. 3. Phytaspase trafficking in leaf tissue.

Left: Phytaspase is synthesized as an inactive precursor protein equipped with the N-terminal signal peptide and the prodomain. Within the secretory pathway, phytaspase undergoes autocatalytic processing, and the mature enzyme is accumulated in the apoplastic fluid.

Right: PCD-related abiotic stresses cause phytaspase relocalization from the apoplast to inside the cell.

phytaspase is stored within the apoplast without access to its intracellular targets, thus preventing unintended induction of cell death. Upon perception of PCD-inducing stimuli phytaspase becomes internalized via a yet undefined mechanism and participates in the accomplishment of PCD by introducing cleavages into target proteins.

The strategy employed by plant cells to control their death protease differs markedly from the one employed by animal cells. Animals store inactive (unprocessed) caspase precursors in the cytoplasm of healthy cells. Whenever an apoptogenic signal is received, caspases become activated via processing of their precursors which occurs in a form of a cascade. Caspase-mediated fragmentation of the target proteins ultimately causes cell death. The difference in the animal and plant strategies to manipulate with their PCD-related proteases may be indicative of additional role(s) played by phytaspase in the apoplast in the absence of cell death.

10. Role of phytaspase in biotic stress responses

In addition to its role in mediating plant PCD induced by abiotic stresses, phytaspase appears to be involved in protection of plants against pathogenic insults. One example is provided by the TMV infection of *N. tabacum* plants carrying the *N* gene. In this experimental system, hypersensitive response (HR, a form of plant PCD) is induced in infected plants due to the recognition of a viral protein by the *N* gene product (Erickson et al., 1999). This quick response serves to kill virus-infected cells and thus prevent virus multiplication and spread throughout the plant. It was demonstrated that the HR-associated cell death is markedly enhanced in phytaspase-overproducing plants and is suppressed in phytaspase-silenced plants (Chichkova et al., 2010). In accord with the protective role of phytaspase-mediated PCD, accumulation of TMV was suppressed in phytaspase overexpressors, whereas it was enhanced in phytaspase-silenced plants. Therefore, the PCD-promoting function of phytaspase confers resistance of tobacco plants to viral infection.

Another example of possible involvement of phytaspase in host-pathogen interactions comes from the initial observation that phytaspase is capable of introducing a cleavage into the C-terminal region of the VirD2 protein of *Agrobacterium tumefaciens* (Chichkova et al., 2004). This VirD2 region encompasses an NLS which allows VirD2 to direct the bacterial T-DNA complex into the plant cell nucleus to provide integration of bacterial DNA into the host genome and eventually plant cell transformation (Howard et al., 1992; Pitzschke & Hirt, 2010). Detachment of the NLS as a result of phytaspase-mediated VirD2 fragmentation is likely to interfere with the delivery of foreign DNA into the plant cell nucleus. In accordance with this scheme, an *Agrobacterium* strain encoding a mutant (phytaspase-resistant) VirD2 protein instead of the wild type one exhibited an enhanced capability to deliver and express foreign DNA in plant cell nucleus (Reavy et al., 2007) .

In general, apoplastic localization of phytaspase is consistent with a protective role which could be achieved by phytaspase-mediated fragmentation of the pathogen-encoded effector proteins. Examples of this kind are yet to be found.

11. Conclusions

PCD is an essential process which is frequently employed in animal and plant responses to abiotic stresses. Current data indicate that the role played by caspases in animal PCD is taken, at least in part, by plant subtilisin-like proteases, phytaspases. Caspases (Cys-dependent enzymes) and phytaspases (Ser-dependent proteases) are structurally very different, yet they share Asp cleavage specificity and a role in PCD. Although both types of the proteolytic enzymes are synthesized as inactive precursors, plants and animals employ distinct strategies to further deal with their death proteases. Caspases are stored as latent precursors within the animal cells and become activated/processed in response to PCD-inducing stimuli to accomplish fragmentation of their target proteins. Unlike this scenario, phytaspase precursors are constitutively and autocatalytically processed even in the absence of PCD. To avoid unintended proteolysis, phytaspases are secreted out of the plant cells to physically separate the enzyme from its intracellular targets. Re-entering of phytaspases into the cell occurs upon the induction of PCD and is accompanied by cleavage of target proteins.

Therefore, PCD-related responses to abiotic stresses in animals and plants display both common and distinct features. Further studies aimed on the unravelling of the cellular proteins fragmented by phytaspase, as well as elucidation of the mechanism underlying the retrograde phytaspase trafficking in the course of PCD should provide important insights into molecular mechanisms of plant responses to stresses.

12. Acknowledgments

Work in the Vartapetian lab is supported by grants from the Russian Foundation for Basic Research (NVCh and ABV) and Royal Society (ABV), and by Russian Ministry of Education and Science (contract numbers P334 and 14.740.11.0168 (ABV)).

13. References

Aravind, L., Dixit V.M. & Koonin E.V. (1999). The domains of death: Evolution of the apoptosis machinery. *Trends Biochem Sci,* Vol. 24, No. 2, pp. 47-53, 0968-0004 (Print) 0968-0004 (Linking)

Beers, E.P., Woffenden B.J. & Zhao C. (2000). Plant proteolytic enzymes: Possible roles during programmed cell death. *Plant Mol Biol,* Vol. 44, No. 3, pp. 399-415, 0167-4412 (Print) 0167-4412 (Linking)

Bonneau, L., Ge Y., Drury G.E. & Gallois P. (2008). What happened to plant caspases? *J Exp Bot,* Vol. 59, No. 3, pp. 491-499, 1460-2431 (Electronic) 0022-0957 (Linking)

Brady, G.F. & Duckett C.S. (2009). A caspase homolog keeps CED-3 in check. *Trends Biochem Sci,* Vol. 34, No. 3, pp. 104-107, 0968-0004 (Print) 0968-0004 (Linking)

Bryan, P., Pantoliano M.W., Quill S.G., Hsiao H.Y. & Poulos T. (1986). Site-directed mutagenesis and the role of the oxyanion hole in subtilisin. *Proc Natl Acad Sci U S A,* Vol. 83, No. 11, pp. 3743-3745, 0027-8424 (Print) 0027-8424 (Linking)

Chang, H.Y. & Yang X. (2000). Proteases for cell suicide: Functions and regulation of caspases. *Microbiol Mol Biol Rev,* Vol. 64, No. 4, pp. 821-846, 1092-2172 (Print) 1092-2172 (Linking)

Chichkova, N.V., Kim S.H., Titova E.S., Kalkum M., Morozov V.S., Rubtsov Y.P., Kalinina N.O., Taliansky M.E. & Vartapetian A.B. (2004). A plant caspase-like protease activated during the hypersensitive response. *Plant Cell,* Vol. 16, No. 1, pp. 157-171, 1040-4651 (Print) 1040-4651 (Linking)

Chichkova, N.V., Galiullina R.A., Taliansky M.E. & Vartapetian A.B. (2008). Tissue disruption activates a plant caspase-like protease with TATD cleavage specificity. *Plant Stress,* Vol. 2, No. pp. 89–95, ISSN: 1749-0359

Chichkova, N.V., Shaw J., Galiullina R.A., Drury G.E., Tuzhikov A.I., Kim S.H., Kalkum M., Hong T.B., Gorshkova E.N., Torrance L., Vartapetian A.B. & Taliansky M. (2010). Phytaspase, a relocalisable cell death promoting plant protease with caspase specificity. *EMBO J*, Vol. 29, No. 6, pp. 1149-1161, 1460-2075 (Electronic) 0261-4189 (Linking)

Coffeen, W.C. & Wolpert T.J. (2004). Purification and characterization of serine proteases that exhibit caspase-like activity and are associated with programmed cell death in Avena sativa. *Plant Cell*, Vol. 16, No. 4, pp. 857-873, 1040-4651 (Print) 040-4651 (Linking)

Cohen, G.M. (1997). Caspases: The executioners of apoptosis. *Biochem J*, Vol. 326 (Pt 1), No. pp. 1-16, 0264-6021 (Print) 0264-6021 (Linking)

Danon, A., Rotari, V.I., Gordon, A., Mailhac, N. & Gallois, P. (2004) Ultraviolet-C overexposure induces programmed cell death in Arabidopsis, which is mediated by caspase-like activities and which can be suppressed by caspase inhibitors, p35 and Defender against Apoptotic Death. *J Biol Chem*, Vol. 279, No. 1, pp.779-787.

del Pozo, O. & Lam E. (1998). Caspases and programmed cell death in the hypersensitive response of plants to pathogens. *Curr Biol*, Vol. 8, No. 20, pp. 1129-1132, 0960-9822 (Print) 0960-9822 (Linking)

del Pozo, O. & Lam E. (2003). Expression of the baculovirus p35 protein in tobacco affects cell death progression and compromises N gene-mediated disease resistance response to tobacco mosaic virus. *Mol Plant Microbe Interact*, Vol. 16, No. 6, pp. 485-494, 0894-0282 (Print) 0894-0282 (Linking)

Dickman, M.B., Park Y.K., Oltersdorf T., Li W., Clemente T. & French R. (2001). Abrogation of disease development in plants expressing animal antiapoptotic genes. *Proc Natl Acad Sci U S A*, Vol. 98, No. 12, pp. 6957-6962, 0027-8424 (Print) 0027-8424 (Linking)

Drew, M.C., He C.J. & Morgan P.W. (2000). Programmed cell death and aerenchyma formation in roots. *Trends Plant Sci*, Vol. 5, No. 3, pp. 123-127, 1360-1385 (Print) 1360-1385 (Linking)

Erickson, F.L., Dinesh-Kumar S.P., Holzberg S., Ustach C.V., Dutton M., Handley V., Corr C. & Baker B.J. (1999). Interactions between tobacco mosaic virus and the tobacco N gene. *Philos Trans R Soc Lond B Biol Sci*, Vol. 354, No. 1383, pp. 653-658, 0962-8436 (Print) 0962-8436 (Linking)

Evstafieva, A.G., Belov G.A., Kalkum M., Chichkova N.V., Bogdanov A.A., Agol V.I. & Vartapetian A.B. (2000). Prothymosin alpha fragmentation in apoptosis. *FEBS Lett*, Vol. 467, No. 2-3, pp. 150-154, 0014-5793 (Print) 0014-5793 (Linking)

Evstafieva, A.G., Belov G.A., Rubtsov Y.P., Kalkum M., Joseph B., Chichkova N.V., Sukhacheva E.A., Bogdanov A.A., Pettersson R.F., Agol V.I. & Vartapetian A.B. (2003). Apoptosis-related fragmentation, translocation, and properties of human prothymosin alpha. *Exp Cell Res*, Vol. 284, No. 2, pp. 211-223, 0014-4827 (Print) 0014-4827 (Linking)

Felle, H.H., Herrmann A., Huckelhoven R. & Kogel K.H. (2005). Root-to-shoot signalling: Apoplastic alkalinization, a general stress response and defence factor in barley (Hordeum vulgare). *Protoplasma*, Vol. 227, No. 1, pp. 17-24, 0033-183X (Print) 0033-183X (Linking)

Fujibe, T., Saji H., Arakawa K., Yabe N., Takeuchi Y. & Yamamoto K.T. (2004). A methyl viologen-resistant mutant of Arabidopsis, which is allelic to ozone-sensitive rcd1, is tolerant to supplemental ultraviolet-B irradiation. *Plant Physiol,* Vol. 134, No. 1, pp. 275-285, 0032-0889 (Print) 0032-0889 (Linking)

Hatsugai, N., Kuroyanagi M., Yamada K., Meshi T., Tsuda S., Kondo M., Nishimura M. & Hara-Nishimura I. (2004). A plant vacuolar protease, VPE, mediates virus-induced hypersensitive cell death. *Science,* Vol. 305, No. 5685, pp. 855-858, 1095-9203 (Electronic) 0036-8075 (Linking)

Hatsugai, N., Iwasaki S., Tamura K., Kondo M., Fuji K., Ogasawara K., Nishimura M. & Hara-Nishimura I. (2009). A novel membrane fusion-mediated plant immunity against bacterial pathogens. *Genes Dev,* Vol. 23, No. 21, pp. 2496-2506, 1549-5477 (Electronic) 0890-9369 (Linking)

Howard, E.A., Zupan J.R., Citovsky V. & Zambryski P.C. (1992). The VirD2 protein of A. tumefaciens contains a C-terminal bipartite nuclear localization signal: Implications for nuclear uptake of DNA in plant cells. *Cell,* Vol. 68, No. 1, pp. 109-118, 0092-8674 (Print) 0092-8674 (Linking)

Huh, G.H., Damsz B., Matsumoto T.K., Reddy M.P., Rus A.M., Ibeas J.I., Narasimhan M.L., Bressan R.A. & Hasegawa P.M. (2002). Salt causes ion disequilibrium-induced programmed cell death in yeast and plants. *Plant J,* Vol. 29, No. 5, pp. 649-659, 0960-7412 (Print) 0960-7412 (Linking)

Jackson, M.B. & Armstrong W. (1999). Formation of aerenchyma and the processes of plant ventilation in relation to soil flooding and submergence. *Plant Biology,* Vol. 1, No. 3, pp. 274-287, 1438-8677

Karapetian, R.N., Evstafieva A.G., Abaeva I.S., Chichkova N.V., Filonov G.S., Rubtsov Y.P., Sukhacheva E.A., Melnikov S.V., Schneider U., Wanker E.E. & Vartapetian A.B. (2005). Nuclear oncoprotein prothymosin alpha is a partner of Keap1: Implications for expression of oxidative stress-protecting genes. *Mol Cell Biol,* Vol. 25, No. 3, pp. 1089-1099, 0270-7306 (Print) 0270-7306 (Linking)

Kobayashi, T., Wang T., Maezawa M., Kobayashi M., Ohnishi S., Hatanaka K., Hige S., Shimizu Y., Kato M., Asaka M., Tanaka J., Imamura M., Hasegawa K., Tanaka Y. & Brachmann R.K. (2006). Overexpression of the oncoprotein prothymosin alpha triggers a p53 response that involves p53 acetylation. *Cancer Res,* Vol. 66, No. 6, pp. 3137-3144, 0008-5472 (Print) 0008-5472 (Linking)

Kraus, E. & Femfert U. (1976). Proteinase K from the mold Tritirachium album Limber. Specificity and mode of action. *Hoppe Seylers Z Physiol Chem,* Vol. 357, No. 7, pp. 937-947, 0018-4888 (Print) 0018-4888 (Linking)

Lam, E. & del Pozo O. (2000). Caspase-like protease involvement in the control of plant cell death. *Plant Mol Biol,* Vol. 44, No. 3, pp. 417-428, 0167-4412 (Print) 0167-4412 (Linking)

Levine, A., Pennell R.I., Alvarez M.E., Palmer R. & Lamb C. (1996). Calcium-mediated apoptosis in a plant hypersensitive disease resistance response. *Curr Biol,* Vol. 6, No. 4, pp. 427-437, 0960-9822 (Print) 0960-9822 (Linking)

Magneschi, L. & Perata P. (2009). Rice germination and seedling growth in the absence of oxygen. *Ann Bot,* Vol. 103, No. 2, pp. 181-196, 1095-8290 (Electronic) 0305-7364 (Linking)

Matsubayashi, Y. (2003). Ligand-receptor pairs in plant peptide signaling. *J Cell Sci,* Vol. 116, No. Pt 19, pp. 3863-3870, 0021-9533 (Print) 0021-9533 (Linking)

Pitzschke, A. & Hirt H. (2010). New insights into an old story: Agrobacterium-induced tumour formation in plants by plant transformation. *EMBO J*, Vol. 29, No. 6, pp. 1021-1032, 1460-2075 (Electronic) 0261-4189 (Linking)

Reavy, B., Bagirova S., Chichkova N.V., Fedoseeva S.V., Kim S.H., Vartapetian A.B. & Taliansky M.E. (2007). Caspase-resistant VirD2 protein provides enhanced gene delivery and expression in plants. *Plant Cell Rep*, Vol. 26, No. 8, pp. 1215-1219, 0721-7714 (Print) 0721-7714 (Linking)

Rojo, E., Martin R., Carter C., Zouhar J., Pan S., Plotnikova J., Jin H., Paneque M., Sanchez-Serrano J.J., Baker B., Ausubel F.M. & Raikhel N.V. (2004). VPEgamma exhibits a caspase-like activity that contributes to defense against pathogens. *Curr Biol*, Vol. 14, No. 21, pp. 1897-1906, 0960-9822 (Print) 0960-9822 (Linking)

Rubtsov, Y.P., Zolotukhin A.S., Vorobjev I.A., Chichkova N.V., Pavlov N.A., Karger E.M., Evstafieva A.G., Felber B.K. & Vartapetian A.B. (1997). Mutational analysis of human prothymosin alpha reveals a bipartite nuclear localization signal. *FEBS Lett*, Vol. 413, No. 1, pp. 135-141, 0014-5793 (Print) 0014-5793 (Linking)

Ryan, C.A., Pearce G., Scheer J. & Moura D.S. (2002). Polypeptide hormones. *Plant Cell*, Vol. 14 Suppl, No. pp. S251-264, 1040-4651 (Print) 1040-4651 (Linking)

Ryerson, D.E. & Heath M.C. (1996). Cleavage of nuclear DNA into oligonucleosomal fragments during cell death induced by fungal infection or by abiotic treatments. *Plant Cell*, Vol. 8, No. 3, pp. 393-402, 1532-298X (Electronic) 1040-4651 (Linking)

Schaller, A. (2004). A cut above the rest: The regulatory function of plant proteases. *Planta*, Vol. 220, No. 2, pp. 183-197, 0032-0935 (Print) 0032-0935 (Linking)

Seidah, N.G. & Chretien M. (1999). Proprotein and prohormone convertases: A family of subtilases generating diverse bioactive polypeptides. *Brain Res*, Vol. 848, No. 1-2, pp. 45-62, 0006-8993 (Print) 0006-8993 (Linking)

Thompson, C.B. (1995). Apoptosis in the pathogenesis and treatment of disease. *Science*, Vol. 267, No. 5203, pp. 1456-1462, 0036-8075 (Print) 0036-8075 (Linking)

Tripathi, L.P. & Sowdhamini R. (2006). Cross genome comparisons of serine proteases in Arabidopsis and rice. *BMC Genomics*, Vol. 7, No. pp. 200, 1471-2164 (Electronic) 1471-2164 (Linking)

Vartapetian, A.B., Tuzhikov A.I., Chichkova N.V., Taliansky M. & Wolpert T.J. (2011). A plant alternative to animal caspases: Subtilisin-like proteases. *Cell Death Differ.*, doi:10.1038/cdd.2011.49

Vartapetian, B.B., Sachs M.M. & Fagerstedt (2008). Plant anaerobic stress: Strategy of avoidance of anaerobiosis and other aspects of plant life under hypoxia and anoxia *Plant Stress*, Vol. 2, No. 1, pp. 1-19 ISSN: 1749-0359

Vercammen, D., Belenghi B., van de Cotte B., Beunens T., Gavigan J.A., De Rycke R., Brackenier A., Inze D., Harris J.L. & Van Breusegem F. (2006). Serpin1 of Arabidopsis thaliana is a suicide inhibitor for metacaspase 9. *J Mol Biol*, Vol. 364, No. 4, pp. 625-636, 0022-2836 (Print) 0022-2836 (Linking)

Watanabe, N. & Lam E. (2005). Two Arabidopsis metacaspases AtMCP1b and AtMCP2b are arginine/lysine-specific cysteine proteases and activate apoptosis-like cell death in yeast. *J Biol Chem*, Vol. 280, No. 15, pp. 14691-14699, 0021-9258 (Print) 0021-9258 (Linking)

Zakharova, N., Sokolov V., Roudko V., Melnikov S., Vartapetian A. & Evstafieva A. (2008). Effect of prothymosin alpha and its mutants on the activity of the p53 tumour suppressor. *Molecular Biology*, Vol. 42, No. 4, pp. 598-608, 0026-8933

8

Plant Plasma Membrane H^+-ATPase in Adaptation of Plants to Abiotic Stresses

Małgorzata Janicka-Russak
Wrocław University, Institute of Plant Biology, Department of Plant Physiology
Poland

1. Introduction

H^+-ATPase is a major enzyme protein of the plant plasma membrane (PM). This protein belongs to a large superfamily of pumps termed P-type ATPases. A characteristic feature of P-ATPases is the formation of a phosphorylated intermediate during the catalytic cycle. P-type ATPase genes can be divided into ten phylogenetic branches. Six of them are found in plants (Axelsen & Palmgren, 2001). The plasma membrane proton pump belongs to the P_3-type ATPase family. The H^+-ATPase branch is subdivided into five subfamilies (Arango et al., 2003). Plasma membrane H^+-ATPase is a tightly bound and integral transmembrane protein. The enzyme is a single polypeptide of ~100 kDa. By the use of the chemical energy of ATP, plasma membrane connected ATPases extrude protons from cells of plants to generate electrochemical proton gradients. The generation of this gradient has a major role in providing the energy for secondary active transport across the plasma membrane. The plant plasma membrane H^+-ATPase is a proton pump which plays a central role in physiological functions such as nutrient uptake, intracellular pH regulation, stomatal opening and cell growth. Besides regulation of physiological processes, the plasma membrane proton pump also plays a role in adaptation of plants to changing conditions, especially stress conditions. Thus, H^+-ATPase can be a mutual element for resistance mechanisms that are activated in various stress conditions. Many studies have shown the changes of gene expression of the plasma membrane H^+-ATPase in response to a variety of environmental factors. Moreover, besides the genetic regulation of the proton pump, its activity may undergo fast post-translational modulation.

2. H^+-ATPase structure and H^+ transport

The enzyme is a functional single polypeptide chain with mass of about 100 kDa. The protein can oligomerize to form dimeric and hexameric complexes (Kanczewska et al., 2005). The H^+-ATPase has N- and C-terminal segments, which emerge into the cytoplasm (Duby & Boutry, 2009). The structure of plasma membrane H^+-ATPase consists of domains A, M, P, N, and R. The A-domain (actuator) consists of the N-terminal segment and small loop. The M-domain (membrane) corresponds to a transmembrane domain with ten helices, M1 to M10. The P-domain (phosphorylation) is located in the large loop. The N-domain (nucleotide binding) is located between two parts of the sequence forming the P-domain. The R-domain (regulatory) consists of the C-terminal fragment of the protein which acts as an autoinhibitory domain (Pedersen et al., 2007; Duby & Boutry, 2009). The plant plasma membrane H^+-ATPase is kept at a low activity level by its C-terminal domain, the inhibitory function of which is thought to be mediated by two regions (regions I and II) interacting with cytoplasmic domains essential for the catalytic cycle (Speth et al., 2010).

The catalytic cycle of H^+-ATPase is described by two main conformational states, E_1 and E_2. In the E_1 conformation, the transmembrane binding site has high affinity for the proton and for ATP, whereas in E_2 the same site has low affinity for both ligands. The states E_1 and E_2 alternate during transport (Buch-Pedersen et al., 2009). The cytoplasmically positioned N, P and A domains are in charge of ATP hydrolysis. Conformational changes in these domains during catalysis lead to simultaneous movements in the membrane-embedded part that directs the proton transport. The transporting unit of plasma membrane proton pump is defined by a centrally located proton acceptor/donor - a single protonable aspartic acid residue (Asp684), an asparagine residue (Asn106), an arginine amino acid residue (Arg655), and a large central cavity likely to be filled with water (Buch-Pedersen et al., 2009). The Asp684 is in close contact with Asn106. In the E_1P structure protonation of Asp684 is believed to facilitate hydrogen bond formation between these two. Phosphorylation leads from E_1P to E_2P conformational changes and proton unloading from Asp684. Arg655 is suggested to play an important role in proton release and proton pumping against high membrane potentials. The positive charge of Arg655 approaching Asp684 will favour proton release from Asp684 and inhibit reprotonation of Asp684 with an extracellular proton (Buch-Pedersen et al., 2009). In plasma membrane H^+-ATPases, a single proton is believed to be transported per hydrolysed ATP (Palmgren, 2001). However, partial uncoupling between ATP hydrolysis and proton transport has been suggested (Buch-Pedersen et al., 2006). The plant plasma membrane H^+-ATPase is stimulated by potassium (Palmgren, 2001). K^+ is bound to the proton pump in the cytoplasmic phosphorylation domain (P-domain). Binding of K^+ promotes dephosphorylation of the phosphorylated E_1P reaction cycle and it controls the H^+/ATP coupling ratio. It was suggested that potassium acts as an intrinsic uncoupler of the plasma membrane H^+-ATPase (Buch-Pedersen et al., 2006).

3. Transcriptional regulation of plant plasma membrane H^+-ATPase

The molecular study of plant H^+-ATPase has shown that this enzyme is encoded by a multigene family. The genes have been identified in: *Arabidopsis thaliana* 12 genes (Palmgren, 2001), *Lycopersicon esculentum* 10-12 genes (Kalampanayil & Wimmers, 2001), *Oryza sativa* 10 genes (Baxter et al., 2003), *Cucumis sativus* 10 genes (Wdowikowska, data unpublished), *Nicotiana plumbaginifolia* 9 genes (Oufattole et al., 2000), *Vicia faba* 5 genes (Nakajima et al., 1995), and *Zea mays* 4 genes (Santi et al., 2003). Phylogenetic and gene structure analysis of plant H^+-ATPases divided them into five subfamilies (Arango et al., 2003). Expression of H^+-ATPase subfamilies I and II is not restricted to particular organs. These subfamilies are highly expressed in many cell types. In *A. thaliana* two genes, *AHA1* and *AHA2*, are expressed in all tissues and organs. *AHA1* is predominantly expressed in shoots, *AHA2* in roots. In *N. plumbaginifolia*, *pma4* was highly expressed in several cell types, including meristematic tissues, root epidermis, hairs, phloem, and guard cells (Moriau et al., 1999). In maize, *MHA2* was expressed in the same set of cells as *PMA4* (Frias et al., 1996). Conversely, expression of genes belonging to subfamilies III, IV, and V is limited to specific organs or cell types (Arango et al., 2003). In *A. thaliana*, *AHA6* and *AHA9* transcripts are present mainly in anthers, *AHA10* in the endothelium of the developing seed coat, and *AHA7* and *AHA8* in pollen (Gaxiola et al., 2007). It has been shown that various genes are expressed in the same organ. Moreover, even within the same cell type at the same developmental stage, at least two H^+-ATPase genes are expressed (Harms et al., 1994; Hentzen et al., 1996; Moriau et al., 1999). In *N. plumbaginifolia* two different plasma membrane H^+-ATPase genes, *PMA2* and *PMA4*, are expressed in guard cells (Moriau et al., 1999). This observation suggests that isoforms with distinct kinetics might co-exist in the same cell.

In addition to tissue-specific expression, the plasma membrane H^+-ATPases are differentially expressed according to environmental factors. Several studies have indicated that the H^+-ATPase genes might be activated by various abiotic and biotic stresses. With such a phenomenon the amount of H^+-ATPase might be increased under conditions requiring greater transport activity. The external signals result in changes in plant plasma membrane H^+-ATPase gene expression, include salt (Niu et al., 1993; Binzel, 1995; Janicka-Russak & Kłobus, 2007), low temperature (Ahn et al., 1999, 2000), heavy metals (Janicka-Russak et al., 2008), dehydration (Surowy et al., 1991), light conditions (Harms et al., 1994), mechanical stress (Oufattole et al., 2000) and externally applied hormones (Frías et al., 1996).

4. Post-translational regulation of plant plasma membrane H^+-ATPase

Plant plasma membrane H^+-ATPase is involved in many different physiological roles and what is more its activity is changed by a large number of physiological factors. Therefore, it can be assumed that there are multiple regulatory features that integrate signals from the environment. As to post-translational regulation, the best known mechanism described to date involves the autoinhibitory action of the C-terminal domain (approximately 100 amino acids) of the enzyme protein. Deletion of the carboxyl terminus by trypsin treatment or genetic engineering results in constitutively activated enzyme. Although it is clear that the carboxyl terminus is the main regulatory domain involved in activation of the H^+-ATPase, recent results suggest that the N-terminus may also play a role in modification of plasma membrane proton pump activity.

4.1 Regulation by phosphorylation

Phosphorylation and dephosphorylation of proteins is a very common example of post-translational modification that has the potential to alter protein activity. The activity of the enzyme is well known to be regulated by 14-3-3 proteins, the association of which requires phosphorylation of the penultimate H^+-ATPase residues of Thr 947 (Svennelid et al., 1999). The binding of 14-3-3 regulatory protein displaces the inhibitory R-domain, activating the enzyme. R-domains contains two regions (I and II), of about 20 residues each, which contribute to keeping the enzyme at a low activity level. Mutagenesis of these regions abolishes the inhibitory effect of the C-terminus (Morssome et al., 1998; Axelsen et al., 1999;). 14-3-3 binding to H^+-ATPase is stabilized by the fungal toxin fusicoccin, which decreases the dissociation rate. One 14-3-3 protein dimer binds two C-terminal polypeptides simultaneously, so a high activity state of H^+-ATPase could involve formation of dimers or multimeric complexes. An analysis with cryo-electron microscopy showed that PMA2-14-3-3 complex is a wheel-like structure with 6-fold symmetry, suggesting that the activated complex consists of six H^+-ATPase molecules and six 14-3-3 molecules (Kanczewska et al., 2005).

Activation of plasma membrane H^+-ATPase involves protein kinase mediated phosphorylation of penultimate Thr in the C-terminus (region II) of the enzyme protein. Which specific kinases catalyse the reaction is an unresolved question. The protein kinase responsible for this phosphorylation has not yet been identified. Recently, it was reported that calcium-dependent protein kinase (CDPK) stimulated by abscisic acid leads to phosphorylation-dependent activation of H^+-ATPase (Yu et al., 2006). In rice, a 55 kDa calcium-dependent protein kinase was shown to phosphorylate two Thr residues from the autoregulatory domain (region I) of the plasma membrane proton pump (Ookura et al., 2005).

Many reports indicate that the phosphorylation of amino acid residues at the C-terminus of the protein leads not only to raised enzyme activity but also to its inhibition (Vera- Estrella et al., 1994; Xing et al., 1996; Lino et al., 1998; Schaller & Oecking, 1999). A number of conserved serine and threonine residues found in the C-terminal regulatory domain serve as candidate residues for negative phosphorylation (Nühse et al., 2003). Some phospho-residues are close to each other and might influence the phosphorylation status of their neighbours (Duby & Boutry, 2009). In *Nicotiana tabacum* two new phosphorylation sites, Thr931 and Ser938, in the C-terminus of H^+-ATPase, were identified. Their mutation suggests that phosphorylation of Thr931 or Ser938 prevents 14-3-3 protein binding, although the penultimate Thr955 was still phosphorylated, and prevented full activation of the enzyme (Duby et al., 2009). PKS5, a Ser/Thr protein kinase, is a negative regulator of the plasma membrane proton pump. In *A. thaliana* this kinase phosphorylates the enzyme (AHA2) at Ser931 in the C-terminus. Phosphorylation at this site inhibits interaction between H^+-ATPase and 14-3-3 protein (Fuglsang et al., 2007). PKS5 is a salt overly sensitive (SOS) 2-like protein and interacts with the calcium-binding protein ScaBP1 (Fuglsang et al., 2007). During salt stress in *A. thaliana,* the calcium sensor Salt Overly Sensitive 3 (SOS3) binds to and activates the Ser/Thr protein kinase SOS2. The Ca^{2+}-SOS3-SOS2 complex phosphorylates and activates the Na^+/H^+ antiporter SOS1, resulting in regulation of Na^+ homeostasis and salt tolerance (Zhu, 2002). In *Arabidopsis*, a chaperone, J3 (DnaJ homolog 3, heat shock protein 40-like) activates plasma membrane H^+-ATPase activity by physically interacting with PKS5 kinase (Yang et al., 2010).

Two extensively expressed plasma membrane H^+-ATPase isoforms of *Nicotiana tabacum* (PMA2 and PMA4) are differentially regulated by phosphorylation of their penultimate threonine. Cold stress reduced the Thr phosphorylation of PMA2, whereas no significant changes in Thr phosphorylation of PMA4 were observed (Bobik et al., 2010a).

A phosphorylation event requires action of a protein phosphatases to make regulation reversible. Phosphatase 2A (PP2A) competes with binding of 14-3-3 protein to the C-terminus of H^+-ATPase (Fuglsang et al., 2006). Inhibition of phosphatase activity by administration of okadaic acid (inhibitor of phosphatase PP2A) to plants (maize roots) increased the bound level of 14-3-3 proteins and activity of H^+-ATPase (Camoni et al., 2000).

A novel interactor of the plant plasma membrane proton pump was identified. This protein is named PPI1 (proton pump interactor, isoform 1). This interactor is able to modulate the plasma membrane H^+-ATPase activity by binding to a site different from the 14-3-3 binding site (Morandini et al., 2002). PPI1 can only hyper-activate H^+-ATPase molecules whose C-terminus has been displaced by other factor such as 14-3-3 protein (Viotti et al., 2005).

Recent studies have shown that the N-terminus of the plant plasma membrane H^+-ATPase directly participates in pump regulation (Morsomme et al., 1998; Ekberg et al., 2010). It was suggested that transformation from low to high activity state of enzyme protein involves a structural rearrangement of both the C- and the N-terminus.

4.2 Regulation by membrane environment

The plant plasma membrane H^+-ATPase requires lipids for activity. This lipid dependency suggests a possible mode of regulation of the plasma membrane proton pump via modification of its lipid environment (Kasamo, 2003). Abiotic stresses lead to changes in the plasma membrane lipid composition altering the fluidity of the membrane. The modulation of the phospholipid environment of the plasma membrane regulates the activity of H^+-ATPase. This enzyme protein activity was abolished upon the removal of membrane lipids by detergents, but it was restored by exogenous addition of phospholipids (Kasamo & Nouchi, 1987, Kasamo 1990). The activation of H^+-ATPase is dependent on the degree of saturation or unsaturation of the fatty acyl chain and its length. The activity decreased with an increase in the length of the fatty acyl chain and in the degree of unsaturation of fatty acid (Hernandez et al., 2002; Kasamo, 2003, Martz et al., 2006).

Lysophosphatidylcholine (a natural detergent produced from phosphatidylcholine by phospholipase A_2) increased the plasma membrane H^+-ATPase activity (Pedechenko et al., 1990, Regenberg et al., 1995). It was found that lysophosphatidylcholine binds to the C-terminal region of the protein and by displacing the autoinhibitory domain leads to increase of ATPase activity.

5. Physiological roles

All living plant cells are thought to express a plasma membrane H^+-ATPase. However, it has been found that the abundance of this protein varies in different cell types and tissues. The main role of roots is uptake of nutrients from soil and translocation of those nutrients to the shoots of plants. In roots high amounts of immunodetectable H^+-ATPase antibodies were observed in epidermal cells, endodermis and phloem (Parets-Soler et al., 1990; Jahn et al., 1998). Because the plasma membrane H^+-ATPase is responsible for establishing the proton gradient involved in the membrane energization used for solute transport, this enzyme controls the major transport processes in the plant: root nutrient uptake, xylem and phloem loading.

In addition to its fundamental role in nutrient uptake, H^+-ATPase plays a major role in cell growth. The so-called "acid growth" theory suggests that protons extruded by an activated H^+-ATPase decrease the apoplastic pH and activate enzymes involved in cell wall loosening (Hager, 2003). It is well established that auxin activates the proton pump, resulting in loosening of the cell wall (Hager, 2003). In plants, an auxin increases membrane flow from the endoplasmic reticulum to the plasma membrane. Therefore it was assumed that this membrane flow could carry newly synthesized H^+-ATPase molecules to the PM. It was reported that IAA increased the amount of antibody-detectable H^+-ATPases in the PM (Hager et al., 1991). Besides, treatment of maize coleoptile segments with auxin resulted in increase of mRNA levels of plasma membrane H^+-ATPase, *MHA2* (Frias, et al., 1996). Recent evidence has shown that the PM H^+-ATPase appears to be a target of NO-mediated auxin action (Kolbert et al., 2008; Zandonadi et al., 2010).

This pump is also involved in intracellular pH regulation. Acidification of the cytosol activates the plasma membrane proton pump and enhances the extrusion of protons from the cytosol to the apoplast. This contributes to alkalinization of the cytosol (Sanders et al., 1981). In *Nicotiana tabacum* acidification of the cytosol increased PMA2 phosphorylation at the penultimate Thr in the C-terminus (Bobik et al., 2010b).

Moreover, the plasma membrane proton pump is involved in other important physiological functions, such as stomatal aperture. Cell turgor changes promote modification of the stomatal aperture. Activation of H^+-ATPase leads to plasma membrane hyperpolarization and subsequent opening of K^+ channels. K^+ and anion influx lead to water uptake, turgor increase, and cell swelling. On the other hand, an elevation of cytosolic Ca^{2+}, inhibition of plasma membrane H^+-ATPase, an increase in pH, a reduction in K^+, Cl^-, and organic solute contents in both guard cells surrounding the stomatal pore, are downstream elements of ABA-induced stomatal closure (Zhang et al., 2001). The plant hormone abscisic acid (ABA) is a key signal molecule, mediating responses to various environmental stresses, and has been demonstrated to induce stomatal closure, thereby preventing water loss (Assmann & Shimazaki, 1999). The opening of stomata is mediated by an accumulation of K^+ in guard cells, and K+ accumulation is driven by an inside-negative electrical potential across the plasma membrane. The electrical potential is created by the plasma membrane H^+-ATPase in response to blue light (Schroeder et al., 2001). Blue light activates the H^+-ATPase through the phosphorylation of Thr residues in the C-terminus. Next, phosphorylation induces the binding of 14-3-3 to the penultimate residue of Thr, which acts as a positive regulator for the H^+-ATPase (Kinoshita & Shimazaki, 2002). Blue light receptors that mediate activation of the plasma membrane proton pump in stomatal guard cells are phototropins (Kinoshita et al., 2001). Phototropins are autophosphorylating Ser/Thr protein kinases (Brigg & Christie, 2002). The activated phototropins transmit the signal to plasma membrane H^+-ATPase for its activation (Christie, 2007). In addition, it was demonstrated that phosphatase 1 (PP1, Ser/Thr protein phosphatase) mediates the signal between phototropins and H^+-ATPase in guard cells (Takemiya et al., 2006). It is known that ABA inhibits blue light dependent proton pumping by plasma membrane H^+-ATPase (Zhang et al., 2004). Biochemical and genetic studies have demonstrated that H_2O_2 mediates inhibition of the plasma membrane proton pump by ABA (Zhang et al., 2001). In guard cells, ABA induces production of phosphatidic acid (PA). Because PA also interacts with PP1 and decreases its phosphatase activity, it seems that PA suppresses the blue light signalling of guard cells (Takemiya & Shimazaki, 2010). Studies have demonstrated that ABA induces NO synthesis through H_2O_2 (Bright et al., 2006) and that NO causes PA production in guard cells (Distéfano et al., 2008).

6. H^+-ATPase in abiotic stress

The plant plasma membrane H^+-ATPase is an important functional protein, which plays a central role in plant physiology. PM H^+-ATPase couples ATP hydrolysis to proton transport out of the cell, and so establishes an electrochemical gradient across the plasma membrane, which is dissipated by secondary transporters using protons in symport or antiport. This enzyme controls the major transport processes in the plant, such as root nutrient uptake and xylem or phloem loading. Moreover, this pump has been proposed to be involved in other important physiological functions, such as stomata aperture, cell elongation, or cellular pH regulation. Generation of an electrochemical gradient across the membrane results in a proton-motive force that is used by secondary transport for assimilation of various nutrients, and also for releasing ions and toxic substances from cells. The plasma membrane proton pump is an enzyme whose activity is altered significantly in response to a number of factors, such as light, temperature, hormones and presence of salt or heavy metals in the environment. Besides regulation of growth and development processes, the plasma membrane proton pump also plays a role in plastic adaptation of plants to changing conditions, especially conditions of stress. Adaptation is a complex process. Some of the modifications in plants subjected to abiotic stress are indicated to be adaptive. Physiological modifications caused by environmental stress and allowing continued plant functions are ascribed by plant physiologists as being adaptive.

6.1 Salt stress

Salt stress is a complex abiotic stress in which both ionic and osmotic components are involved (Alvarez et al., 2003). Under conditions of elevated NaCl levels outside the cell, sodium passively enters into the roots through non-selective cation channels and the low-affinity sodium transporter HKT1. Salinity tolerance of plants is a complex trait involving adaptation at the level of cells, organs and the whole plant. The key factor of salinity tolerance, beside osmotic adjustment, is the control of intracellular ion homeostasis (Niu et al., 1993). The excess of Cl^- is passively effluxed across the membrane down the $\Delta\mu H^+$ generated by plasma membrane H^+-ATPase. To prevent accumulation of toxic Na^+ amounts in the cytosol, active sodium efflux into the apoplast and its compartmentalization inside the vacuole occur. Since Na^+ pumps responsible for sodium extrusion in animals and microorganisms are absent in higher plant cells, secondary sodium/proton antiporters in both the plasma membrane and the tonoplast are needed to translocate sodium ions against their electrochemical gradients (Apse and Blumwald 2007). Molecular analyses in *Arabidopsis* led to the identification of a plasma membrane SOS1 (Salt Overly Sensitive 1) and vacuolar NHX1, Na^+/H^+ antiporters upregulated at the genetic level in response to NaCl (Apse et al. 1999; Gaxiola et al. 1999; Shi et al. 2000, 2003). These proton-coupled sodium transporters use the proton-motive force created by the specific proton pumps. The only pump which generates an electrochemical proton gradient across the plasma membrane is H^+-ATPase (Palmgren, 2001). For this reason, it is believed that plant plasma membrane H^+-ATPase plays a major role in salt stress tolerance. The importance of plant plasma membrane H^+-ATPase in salt tolerance is demonstrated by observations such as increase in its activity in halophytes (Braun et al., 1986; Niu et al., 1993; Vera-Estrella et al., 1994, Sahu & Shaw, 2009). In halophytes and salt-tolerant cultivars significantly greater activity of the enzyme is observed in normal conditions, without NaCl (Sahu & Shaw, 2009). Moreover, it has been well documented that salt treatment of plants induces the activities of the plasma membrane proton pumps both in halophytes and glycophytes (Niu et al., 1993; Perez-Prat et al., 1994; Binzel, 1995; Kłobus and Janicka-Russak, 2004, Sahu & Shaw, 2009; Lopez-Perez et al., 2009; Shen et al., 2011). However, there are few reports about the inhibition or no effect of NaCl on the plasma membrane H^+-ATPase in leaves (Chelysheva et al., 2001; Zörb et al., 2005; Pitann et al., 2009; Wakeel et al., 2010). The authors observed an increase of apoplastic pH in salt stress conditions. The salinity would thus limit leaf elongation.

The aforementioned salt-dependent activation of the plasma membrane proton pump encompasses the transcriptional as well as post-translational level. Accumulation of mRNAs of PM H^+-ATPase under NaCl stress and the positive correlation with salt tolerance are well documented (Niu et al., 1993; Perez-Prat et al., 1994; Janicka-Russak & Kłobus 2007; Sahu and Shaw, 2009). In *Suaeda maritima* (a natural halophyte) and *Oryza sativa* salt-tolerant cultivar accumulation of PM H^+-ATPase gene transcript was greater than that in a non-tolerant cultivar of rice treated with NaCl (Sahu & Shaw, 2009). As mentioned earlier (subsection 11.3), plant plasma membrane ATPase is encoded by a large gene family. The existence of multiple isoforms of the enzyme creates the opportunity of their role in abiotic stress tolerance, particularly salt stress tolerance. In rice a new isoform of the enzyme (finding maximum homology with *OSA7*) in response to salt treatment was discovered. In the halophyte *Suaeda maritime* the related gene (*SM2*, which has 89% homology with the new isoform from rice) does not require the presence of salt for its expression. So the salt-inducible isoform of the plasma membrane H+-ATPase gene in rice remains constitutively expressed in the halophyte *S. maritime* (Sahu & Shaw, 2009). The isoform of PM H^+-ATPase from tomato *LHA8* is most closely related to *Nicotiana plumbaginifolia* gene *PMA6*. Both of these isoforms are specially induced by stress: *LHA8* by salt stress, *PMA6* by mechanical stress (Oufattole et al., 2000; Kalampanayil & Wimmers, 2001).

A direct role of H^+-ATPase in salt tolerance was confirmed by studies with transgenic tobacco, using a PMA4 mutant, lacking the autoinhibitory domain (ΔPMA4). In the mutant a constitutively activated PMA4 H^+-ATPase isoform was present. The ΔPMA4 plant roots showed better growth in saline conditions than those of untransformed plants (Gévaudant et al., 2007).

In rice phospholipase Dα (PLDα) is involved in salt tolerance by the mediation of H^+-ATPase activity and transcription (Shen et al., 2011). When rice suspension-cultured cells were treated with 100 mM NaCl, PLDα activity increased. The knockdown of OsPLDα1 prevented NaCl-induced increase in transcript levels of OSA2 (which encodes PM H^+-ATPase) as well as ATPase activity.

Recently evidence has been presented that NaCl also causes a rapid modulation of proton pumps, which is due to the reversible phosphorylation of enzyme proteins (Kerkeb et al., 2002; Kłobus and Janicka-Russak, 2004). In many plant tissues a salt-inducible shift in the cytoplasmic calcium level was observed (Heterington and Quatrano, 1991; Rock and Quatrano, 1995; Danielsson et al., 1996; Knight et al., 1997; Blumwald et al., 2000; Netting, 2000; Xiong et al., 2002), suggesting its involvement in the signalling pathway target under NaCl stress conditions. An SOS (salt overly sensitive) network exists in plants exposed to salt stress. SOS1, antiporter Na^+/H^+, enables Na^+ efflux across the plasma membrane and controls long-distance Na^+ transport between roots and leaves through loading and unloading of Na^+ in the xylem (Shi et al. 2002). Sodium efflux through SOS1 is mediated by SOS3-SOS2 complex (Qiu et al. 2002). Perception of salt stress induces a cytosolic calcium signal activating the calcium sensor, myristoylated protein SOS3. After binding with Ca^{2+}, SOS3 changes its conformation and interacts with the FISL motif of SOS2, a Ser/Thr protein kinase, and activates its substrate phosphorylation. The activated SOS3-SOS2 complex then phosphorylates SOS1 and activates its antiporter activity (Chinnusamy et al. 2005).

Evidence has also been presented that in the plasma membrane ATPase phosphorylation, as a target of activation by NaCl, a calcium/calmodulin-dependent protein kinase sensitive to staurosporine, is involved (Kłobus and Janicka-Russak, 2004).

Transient increases in cytosolic Ca^{2+} can induce the phosphorylation of different proteins in cells, improving the salt tolerance (Hasegawa et al. 2000). Evidence suggests that the major role in coupling the calcium signal to specific protein phosphorylation cascade(s) is played by the Ca^{2+}/calmodulin-dependent protein kinases (CDPKs) and the SOS3 family of Ca^{2+} sensors (Xiong et al. 2002, Zhu 2002). The results of Urao et al. (1994) and Saijo et al. (2000)

demonstrated that NaCl rapidly induced CDPK in different plant tissues. Furthermore, evidence has been presented that Ca^{2+}/calmodulin-dependent protein kinases are responsible for the phosphorylation of the plasma membrane H^+-ATPase protein (Van der Hoeven et al. 1996, Camoni et al., 2000).

Ca^{2+} has been identified as a possible mediator of ABA-induced stimulus-response coupling (Netting, 2000). Abscisic acid is known as a stress hormone, which mediates responses to a variety of stresses, including water and salt stress (Skriver and Mundy, 1990; Tan et al., 1994; Jia et al., 2001). The endogenous level of ABA increases when plants are stressed with drought or NaCl, and application of ABA to unstressed plants results in the induction of numerous water-deficit-related activities (La Rosa et al., 1985; 1987; Skriver and Mundy, 1990; Kefu et al., 1991; Cowan et al., 1997; Barkla, 1999). They include triggering of stomatal closure to reduce transpirational water loss by post-translational modulation of ion channels in guard cells (Grabov and Blatt, 1998), and alterations in gene expression through induction of ABA-responsive genes coding for structural, metabolic or transport proteins (Ingram and Bartels, 1996; Shinozaki and Yamaguchi-Shinozaki, 1996; Barkla et al., 1999).

It was reported that ABA treatment of cucumber plants, as well as NaCl treatment, increased activity of plasma membrane H^+-ATPases. ABA treatment of seedlings elevated the level of plasma membrane H^+-ATPase transcript (Janicka-Russak & Kłobus, 2007). Thus activation of this proton pump by salt and abscisic acid seems to involve the genetic level.

Nitric oxide (NO) and hydrogen peroxide (H_2O_2) function as signal molecules in plants subjected to abiotic stresses. In plants, the generation of H_2O_2 seems to be mediated by a plasma membrane bound NADPH oxidase complex (Yang et al., 2007; Lherminier et al., 2009). NO is synthesized via nitric oxide synthase (NOS) and nitrate reductase (Yamasaki & Sakihama, 2000). NO and H_2O_2 interact as signalling molecules in plants (Zhang et al., 2007). Under salt stress, both NO and H_2O_2 are produced. Treatment of plants with NO increased the activity of plasma membrane NADPH oxidase and thus the generation of hydrogen peroxide. Both these molecules stimulated the activity, as well as the expression of plasma membrane H^+-ATPase in *Populus euphratica* (Zhang et al., 2007). However, the increase of proton pump activity is induced by NO possibly via the regulation of PM NADPH oxidase. Additionally, activity of PM H^+-ATPase may depend on the membrane redox state. Plasma membrane oxidoreductase can modify activity of the PM H^+-ATPase proton pump (Kłobus and Buczek 1995). PM oxidoreductase, by transporting electrons across the PM, simultaneously acidifies the cytoplasm (Lüthje et al. 1997). Low pH of the cytoplasm and membrane depolarization stimulate the activity of PM H^+-ATPase (Hager and Moser 1985; Rubinstein and Stern 1986; Kłobus 1995).

Plants can increase salinity tolerance by modifying the biosynthesis of polyamines. In cucumber the level of polyamines decreased when the concentration of Na^+ in the cytosol increased, so the action of polyamines contributes to ionic equilibrium (Janicka-Russak et al., 2010). Moreover, it was found that polyamines decreased activity of the plasma membrane proton pump in cucumber. So a decrease in their biosynthesis under salinity seems to be beneficial in stress tolerance.

Fatty acids, as the main component of membrane lipids, are considered to be important in salt tolerance of plants. Salt-tolerant plants showed an increase of unsaturated fatty acids (Lin & Wu, 1996). In broccoli plants a high degree of unsaturation in the plasma membrane of roots was observed (López-Pérez et al., 2009). The activity of plant plasma membrane H^+-ATPase increased with an increase in the degree of unsaturation of fatty acid (Kasamo, 2003,

Martz et al., 2006). Non-tolerant plants subjected to salt stress commonly show decreased levels of 18:3 in their membranes (Upchurch, 2008). A study with transgenic tobacco showed that overexpression of ω-3 desaturases, which increases 18:3, elevated tolerance to salt stress (Zhang et al., 2005).

In potato the mRNA level of *StPPI1*, a homologue of the *A. thaliana* PPI1 (proton pump interactor isoform 1), increased under salt and cold stress conditions. StPP1 increases PM H^+-ATPase activity by hyperactivation of H^+-ATPase whose C-terminus has been displaced earlier by 14-3-3 protein (García et al., 2011).

6.2 Heavy metals

An increase in permeability related to membrane damage is observed in plants that have been subjected to heavy metal stress. It is well known that metal ions are easily bound to both the sulfhydryl groups of proteins and hydroxyl groups of phospholipids (Devi and Prasad, 1999). They can also replace the calcium ions at essential sites of cell membranes (Breckle and Kahle, 1991). All these events result in disruption of membrane integrity and ionic homeostasis of cells. Maintaining ionic balance and replenishing the loss of essential substances in repair processes is an important issue under such conditions. Support of active transport of ions and organic compounds through the plasma membrane requires increased generation of a proton gradient by PM H^+-ATPase. Generation of an electrochemical gradient across the membrane results in a proton-motive force that is used by passive transport for assimilation of various nutrients, as well as for releasing ions and toxic substances from cells.

To date, data concerning the action of heavy metals on plasma membrane H^+-ATPase are limited. A few observations have indicated that enzyme activity was changed under heavy metal (Cd, Cu, Ni, Al) stresses (Lindberg & Wingstrand, 1985; Kennedy and Gonsalves, 1989; Ros et al., 1992 a,b; Fodor et al., 1995; Demidchik et al., 1997; Astolfi et al., 2003; Burzyński & Kolano, 2003; Astolfi et al., 2005; Shen et al., 2005; Janicka-Russak et al., 2008; Kabała et al., 2008). The effect of metals on plasma membrane H^+-ATPase activity depends on time of exposure of plants to heavy metals, kind and concentration of heavy metal or plant species. In cucumber seedling roots, brief treatment of plants with Cd and Cu leads to inhibition of plasma membrane H^+-ATPase activity (Janicka-Russak et al., 2008, Kabała et al., 2008). The inhibition of the enzyme was partially diminished in the presence of cantharidin, a specific inhibitor of PP2A and PP1 phosphatases. Moreover, Western blot analysis with an antibody against phosphothreonine confirmed that decreased activity of plasma membrane H^+-ATPase in a short time (2 hours, 10 and 100 μM Cd or Cu) resulted from dephosphorylation of the enzyme protein (Janicka-Russak et al., 2008). However, longer time of treatment of cucumber roots with those heavy metals (6 days, 10 μM Cd or Cu) leads to increased activity of the enzyme measured both as hydrolysis of ATP and proton transport across the plasma membrane (Janicka-Russak, data unpublished). The same effect was observed in cucumber plants treated for 18 hours with 100 μM Cd and in maize with 100 μM both Cd and Cu (Burzyński & Kolano, 2003). In rice treated for 5 or 10 days with 100 and 500 μM Cd increased proton pump activity was observed (Ros et al., 1992a). On the other hand, in oat roots treated long term (7 and 21 days) with 100 μM Cd plasma membrane proton pump activity was inhibited (Astolfi et al., 2003). The same effect was observed in maize subjected for 4 days to cadmium (Astolfi et al., 2005). When plants were treated for 5 or 10 days with nickel, stimulation of H^+-ATPase in rice shoots' plasma membrane was observed both in 100 and 500 μM concentration. But in rice roots only 100 μM concentrations of Ni lead to stimulation of the enzyme protein (Ros et al., 1992a). The increase of ATPase activity in conditions of Ni and Cd metal stress was observed simultaneously with decrease in the degree of unsaturation and length of the phospholipid fatty acyl chain (Ros et al., 1992b). In contrast, Hernandez & Cook (1997) observed an increased degree of unsaturation of phospholipid-associated fatty acid in pea root plasma membrane as a result of 10-day cadmium treatment of plants. A similar effect was observed in the case of Cu in tomato membrane lipids as a result of 7 days' stress (Ouariti et al., 1997). It is known that PM H^+-ATPase is stimulated when the degree of unsaturation of fatty acids is increased (Hernandez et al., 2002). In soybean root, under aluminium stress, up-regulation of transcript, translation and threonine-oriented phosphorylation of plasma membrane H^+-ATPase was observed (Shen et al., 2005).

6.3 Low temperature

An increase in membrane permeability and a change in its viscosity and fluidity result in a decrease of cell turgor due to low temperature (Guy 1990). Membrane changes at low temperature concern positional redistribution of saturated and unsaturated fatty acids (Hughes and Dunn 1996). Probably the change in membrane fluidity in cold-sensitive species initiates a signal chain reaction that leads to acclimation to cold stress by increasing the expression of certain genes, for example genes encoding desaturases (Xiong and Zhu 2001; Chinnusamy et al. 2006). An increase in permeability related to membrane damage is observed in plants that have been subjected to low temperature. Maintaining ionic balance and replenishing the loss of essential substances in repair processes is an important issue under such conditions. Support of active transport of ions and organic compounds through the PM requires increased generation of a proton gradient by PM H^+-ATPase. Generation of an electrochemical gradient across the membrane results in a proton-motive force that is used by active transport for assimilation of various nutrients. Published data indicate that activity of PM H^+-ATPase is modulated under low temperature by changes in lipids associated with the PM proton pump (Lindberg et al. 2005; Martz et al. 2006). Plants capable of cold acclimation accumulate polyunsaturates during cold stress (Vega et al., 2004; Upchurch et al., 2008). In *Pinus resinosa* the activity of plasma membrane H^+-ATPase increased more than twofold following cold acclimation (Martz et al., 2006). In winter hardy tree species seasonal changes in PM H+-ATPase activity and fatty acid composition occur during cold acclimation and de-acclimation under natural conditions. Fatty acid-regulated plasma membrane proton pump activity is involved in the cellular response underlying cold acclimation and de-acclimation (Martz et al., 2006). Similarly, in rye roots H^+-ATPase activity increases during acclimation to low temperature, following increases in fatty acid unsaturation, particularly linoleic acid (White et al., 1990). An increase in plasma membrane H+-ATPase at low temperature (5 °C) was also observed in cells of winter wheat seedlings which were hardened earlier. In non-hardened tissues such an increase of enzyme activity was not observed (Ling-Cheng et al., 1982). In *Oryza sativa* plasma membrane H^+-ATPase activity increased at low temperature in chilling-insensitive plants whereas in chilling-sensitive plants a slight decrease in enzyme activity was observed (Kasamo, 1988). The length of plants' exposure to the low temperature affects activity of the PM proton pump in various ways. Brief (1 or 3 days) exposure of cucumber seedlings to low temperature inhibits hydrolytic and transport activity of H^+-ATPase (Lee et al., 2004, Janicka-Russak, data unpublished). However, stimulation of activity was observed after treating plants with a low temperature (10 °C) for 6 days (Janicka-Russak, data unpublished). On the other hand, Ahn and coworkers (2000), despite using the same plant species, cucumber, observed different results. The increase in H^+-ATPase activity after 1 day of low temperature treatment was reversed and gradually diminished as root temperatures of 10 °C continued for the next 6 days.

It was shown that regulation of the plasma membrane proton pump may be based on the interaction of H^+-ATPase and 14-3-3 protein. Rapid cooling of protoplasts derived from sugar beet cells results in activation of the proton pump. Moreover, cytoplasmic 14-3-3 protein associated with plasma membrane and thus the amount of ATPase-14-3-3 complexes increased (Chelsyheva et al., 1999).

The activity changes of PM H^+-ATPase may partly result from changes in the pattern of expression of PM H^+-ATPase genes (Ahn et al., 1999, 2000; Janicka-Russak, data unpublished).

7. Conclusion

According to the above information, we can conclude that plant plasma membrane H^+-ATPase is an important functional protein, which plays a central role in plant physiology in normal growth conditions and under abiotic stresses. In plants exposed to different abiotic stresses an increase in permeability related to membrane damage is observed. Maintaining ionic balance and replenishing the loss of essential substances in repair processes is an important issue under such conditions. Support of active transport of ions and organic compounds through the plasma membrane requires increased generation of a proton gradient by PM H^+-ATPase. In addition, stress conditions such as salinity or heavy metals leads to accumulation of a toxic excess of certain ions. Generation of an electrochemical gradient across the membrane results in a proton-motive force that is used by active transport for assimilation of various nutrients, as well as for releasing ions and toxic substances from cells. Thus, plasma membrane H^+-ATPase can be a mutual element for resistance mechanisms that are activated in various stress conditions.

8. References

Ahn, S.; Im, Y.; Chung, G. & Cho, B. (1999). Inducible expression of plasma membrane H^+-ATPase in the roots of figleaf gourd plants under chilling root temperature. *Physiologia Plantarum*, Vol.106, pp.35-40

Ahn, S.; Im, Y.; Chung, G.; Seong, K. & Cho, B. (2000). Sensitivity of plasma membrane H^+-ATPase of cucumber root system in response to low root temperature. *Plant Cell Reports*, Vol.19, pp. 831-835

Alvarez, I.; Tomaro, M. & Benavides M. (2003). Changes in polyamines, proline and ethylene in sunflower calluses treated with NaCl. I. *Plant Cell Tissue Organ Cult*, Vol.74, No.1, pp. 51-59

Apse, M.; Aharon, G.; Snedden, W. & Blumwald, E. (1999). Overexpression of a vacuolar Na^+/H^+ antiport confers salt tolerance in *Arabidopsis*. *Science*, Vol.285, pp. 1256-1258

Apse, M. & Blumwald, E. (2007). Na^+ transport in plants. FEBS Letters, Vol.581, pp. 2247-2254

Arango, M. ; Gevaudant, F. ; Oufattole, M. & Boutry, M.(2003). The plasma membrane proton pump ATPase : the significance of gene subfamilies. *Planta*, Vol.216, pp. 355-365

Assmann, S. & Schimazaki, K. (1999). The multisensory guard cell : stomatal responses to blue light and abscisic acid. *Plant Physiology*, 119, pp. 809-815

Astolfi, S. ; Zuchi, S. ; Chiani, A. & Passera, C. (2003). *In vivo* and *in vitro* effects of cadmium on H+-ATPase activity of plasma membrane vesicles from oat (*Avena sativa* L.) roots. *Journal of Plant Physiology*, Vol.160, pp. 387-393

Astolfi, S. ; Zuchi, S. Passera, C. (2005). Effect of cadmium on H^+-ATPase activity of plasma memebrane vesicles isolated from roots of different S-supplied maize (*Zea mays* L.) plants. *Plant Science*, Vol.169, pp. 361-368

Axelsen, K. & Palmgren, M. (2001). Inventory of the superfamily of P-type ion pumps in Arabidopsis. *Plant Physiology*, vol. 126, (June 2001), pp. 696-706

Barkla, B.; Vera-Estrella, R.; Maldonado-Gama, M. & Pantoja, O. (1999). Abscisic acid induction of vacuolar H^+-ATPase activity in *Mesembryanthemum crystallinum* is developmentally regulated. Plant Physiology, Vol.120, pp. 811-819

Baxter, I.; Tchieu, J.; Sussman, M.; Boutry, M.; Palmgren, M.; Gribskov, M.; Harper, J. & Axelsen, K. (2003). Genomic comparison of P-type ATPase ion pumps in Arabidopsis and rice. *Plant Physiology*, Vol.132, pp. 618-828

Binzel, M. (1995). NaCl induced accumulation of tonoplast and plasma membrane H^+-ATPase message in tomato. *Physiol Plantarum*, Vol.94, pp.722-728

Bobik, K.; Duby, G.; Nizet, Y.; Vandermeeren, C.; Stiernet, P., Kanczewska, J. & Boutry, M. (2010a). Two wdely espressed plasma membrane H+-ATPase isoforms of *Nicotiana tabacum* are differentially regulated by phosphorylation of their penultimate threonine. *The Plant Journal*, Vol.62, pp. 291-301

Bobik, K.; Boutry, M. & Duby, G. (2010b). Activation of the plasma membrane H^+-ATPase by acid stress. *Plant Signaling & Behavior*, Vol.5, No.6, (June) pp. 681-683

Braun, Y.; Hassidini, N.; Lerner, H. & Reinhold, L. (1986). Studies on H^+ translocating ATPases in plants of varying resistance to salinity. *Plant Physiology*, Vol.81, pp. 1050-1056.

Briggs,, W. & Christie, J. (2002). Phototropins 1 and 2: versatile plant blue-light receptors. *Trends in Plant Science*, Vol. 7, pp. 204-2010

Bright, J.; Desikaan, R.; Hancock, J.; Weir, I. & Neill, S. (2006). ABA-induced NO generation and stomatal closure in Arabidopsis are dependent on H_2O_2 synthesis. *Plant Journal*, Vol.45, pp. 113-122

Buch-Pedersen, M. ; Rudashevskaya, E. ; Berner, T. ; Venema, K. & Palmgren, M. (2006). Potassium as an intrinsic uncoupler of the plasma membrane H^+-ATPase. *The Journal of Biological Chemistry*, Vol.281, No.50, (December 15), pp. 38285-38292

Buch-Pedersen, M. ; Pedersen, P. ; Veierskov, B. ; Nissen, P. & Palmgren, M. (2009). Protons how they are transported by proton pumps. *Pflugers Archiv. European Journal of Physiology*, Vol.457, pp. 573-579

Burzyński, M. & Kolano, E. (2003). *In vivo* and *in vitro* effects of copper and cadmium on the plasma membrane H^+-ATPase from cucumber (*Cucumis sativus* L.) and maize (*Zea mays* L.) roots. *Acta Physiologiae Plantarum*, Vol.25, pp. 39-45

Camoni, L. ; Iori, V. ; Marra, M. & Aducci, P. (2000). Phosphorylation-dependent interaction between plant plasma membrane H^+-ATPase and 14-3-3 proteins. *Journal of Biological Chemistry*, Vol. 275, pp. 9919-9923

Chelysheva, V.; Smolenskaya, I.; Trofimova, M.; Babakov, A. & Muromtsev, G. (1999). Role of the 14-3-3 proteins in the regulation of H^+-ATPase activity in the plasma membrane of suspension cultured sugar beet cells under cold stress. *FEBS Letters*, Vol.456, pp. 22-26

Chelysheva, V. ; Zorinyants, S. ; Smolenskaya, I. & Babakov, A. (2001). Regulation of H+ pumping By plants plasmalemma under osmotic stress : The role of 14-3-3 proteins. *Russian Journal of Plant Physiology*, Vol.48, pp. 271-180

Chinnusamy, V.; Jagendorf, A. & Zhu, J. (2005). Understanding and improving salt tolerance in plants. *Crop Science*, Vol. 45, pp.437-448

Chinnusamy, V.; Zhu, J. & Zhu, J. (2006). Gene regulation during cold acclimation in plants. *Physiologia Plantarum*, Vol.126, pp. 52-61

Christie, J. (2007). Phototropin blue-light receptors. *Annual Review of Plant Biology*, Vol.58, pp. 21-45

Cowan, A.; Richardson, G. & Maurel, J. (1997). Stress-induced abscisic acid transients and stimulus-response- coupling. *Physiologia Plantarum*, Vol.100, pp. 491-499

Danielsson, A.; Larsson, C.; Larsson, K.; Gustafsson, L. & Adler, L. (1996) A genetic analysis of the role of calcineurin and calmodulin in Ca^{2+} dependent improvement of NaCl tolerance of *Saccharomyces cerevisiae*. *Current Genetics*, Vol. 30, pp.476-484

Demidchik, V.; Sokolik, A. & Yurin, V. (1997). The effect of Cu^{2+}on ion transport systems of the plant cell plasmalemma. *Plant Physiology*, Vol.114, pp. 1313-1325

Devi, S. & Prasad, M. (1999). Membrane lipid alterations in heavy metal exposed plants. In: Prasad MNV, Hagemeyer J, eds. *Heavy metal stress in plants. From molecules to ecosystems*. Berlin: Springer, pp. 99-116

Diestéfano, A. ; Garcia-Mata, C. ; Lamattina, L. & Laxalt, A. (20080. Nitric oxide- induced phosphatidic acid accumulation : a role for phospholipases C and D in stomatal closure. *Plant Cell Environment*, Vol.31, pp. 187-194

Duby, G. & Boutry, M. (2009). The plant plasma membrane proton pump ATPase: a highly regulated P-type ATPase with multiple physiological roles. *Pflugers Archiv. European Journal of Physiology*, Vol.457, pp. 645-655

Duby, G. ; Poręba, W. ; Piotrowiak, D. ; Bobik, K. ; Gerua, R. ; Waelkens, E. Boutry, M. (2009). Activation of plant plasma mmebrane H_-ATPase by 14-3-3 proteins is negatively controlled by two phosphorylation sites within the H+-ATPase C-terminal region. *The Journal of Biological Chemistry*, Vol.284, No. 7,(February 13), pp. 4213-4221

Ekberg, K. ; Palmgren, M. ; Veierskov, B. & Buch-Pedersen, M (2010). A novel mechanism of P-type ATPase autoinhibition involving both termini of the protein. *The journal of Biological Chemistry*, Vol.285, No.10, (March 5) pp. 7344-7350

Fodor, E. ; Szabo-Nagy, A. & Erdei L. (1995). The effects of cadmium on the fluidity and H^+-ATPase activity of plasma membrane from sunflower and wheat roots. *Journal of Plant Physiology*, Vol.147, pp. 87-92

Frías, I.; Caldeira, M.; Pérez-Castiiieira, J.; Navarro-Aviiió, J.; Culiaiiez-Macia, F. ; Kuppinger, O.; Stransky, H. ; Pagés, M. ; Hager, A. & Serrano, R. (1996). A Major isoform of the Maize Plasma embrane H^+-ATPase: Characterization and induction by Auxin in Coleoptiles. *The Plant Cell,* Vol.8, pp. 1533-1544

Fugslang, A,; Tulinius, G.; Ciu, N. & Palmgren, M. (2006). Protein phosphatases 2A scaffolding subunit A interacts with plasma membrane H^+-ATPase C-terminus in the same region as 14-3-3 protein. *Physiologia Plantarum,* Vol. 128, pp. 334-340

Fuglsang, A.; Guo, Y.; Cuin, T.; Qiu, Q.; Song, C.; Kristiansen, K.; Bych, K.; Schulz, A.; Shabala, S.; Schumaker, K.; Palmgren, M. & Zhu, M. (2007). *Arabidopsis* protein kinase PKS5 inhibits the plasma membrane H+-ATPase by preventing interaction with 14-3-3 protein. *The Plant Cell,* Vol.19, (May), pp. 17-17-1634

García, M.; País, S.; Téllez-Iñón, M. & Capiati, D. (2011). Characterization od StPPI1, a proton pump interactor from *Solanum tuberosum* L. that is up-regulated during tuber development and by abiotic stress. *Planta* , Vol.233, pp. 661-674

Gaxiola, R.; Rao, R.; Sherman, A.; Grisafi, P.; Alper, S. & Fink, G. (1999). The *Arabidopsis thaliana* proton transporters, AtNhx1 and Avp1, can function in cation detoxification in yeast. *PNAS,* Vol.96, pp. 1480-1485

Gaxiola, R.; Palmgren, M. & Schumacher, K. (2007). Plant proton pumps. *FEBS Letters,* Vol.581, pp. 2204-2214

Géwaudant, F.; Duby, G.; Stedingk, E. Zhao, R. (2007). Expression of a constitutively activated plasma membrane H^+-ATPase alters plant development and increases salt tolerance. *Plant Physiology,* Vol.144, (August), pp. 1763-1776

Grabov, A. & Blatt, M. (1998). Co-ordination of signaling elements in guard cell ion channel control. *Journal of Experimental Botany,* Vol.49, pp. 351-360

Guy, C. (1990). Cold acclimation and freezing stress tolerance: role of protein metabolism. *Annual Review of Plant Physiology and Plant Molecular Biology,* Vol.41, pp.187-223

Hager, A. (2003). Role of the plasma membrane H^+-ATPase in auxin induced elongation growth: historical and new aspects. *Journal of Plant Research,* Vol.116, pp. 483-505

Hager, A. & Moser, I. (1985). Acetic acid esters and permeable weak acids induce active proton extrusion and extension growth of coleoptile segments by lowering the cytoplasmic pH. *Planta,* Vol.163, pp. 391-400

Hager, A.; Debus, G.; Edel, H.; Stransky, H. & Serrano, R. (1991). Auxin induced exocytosis and the rapid synthesis of a high-turnover pool of plasma membrane H+-ATPase. *Planta,* Vol. 185, pp. 527-537

Harms, K.; Wohner, R.; Schulz, B. & Frommer, W. (1994). Isolation and characterization of P-type H^+-ATPase genes from potato. *Plant Molecular Biology,* Vol.26, pp. 978-988

Hasegawa, P.; Bressan, R.; Zhu, J. & Bohnert, H. (2000). Plant cellular and molecular responses to high salinity. *Annual Review of Plant Physiology and Plant Molecular Biology,* Vol. 51, pp.463-499

Hentzen, A.; Smart, L.; Wimmers, L.; Fang, H.; Schroeder, J. & Bennett, A. (1996). Two plasma membrane H^+-ATPase genes expressed in guard cells of *Vicia faba* are also expressed throughout the plant. *Plant and Cell Physiology,* Vol. 37, pp. 650-659

Hernandez, L. & Cook, D. (1997). Modification of the root plasma membrane lipid composition of cadmium treated *Pisum sativim. Journal of Experimental Botany,* Vol.48, pp. 1375-1381

Hernandez, A.; Cooke, D. & Clarkson, D. (2002). *In vivo* activation of plasma membrane H+-ATPase hydrolytic activity by complex lipid-bound unsaturated fatty acids in *Ustilago maydis. European Journal of Biochemistry,* Vol.269, pp. 1006-1011

Hetherington, A. & Quatrano, R. (1991). Mechanisms of action of abscisic acid at the cellular level. *New Phytologist*, Vol.119, pp. 9-32

Hughes, M. & Dunn, M. (1996). The molecular biology of plant acclimation to low temperature. *Journal of Experimental Botany*, Vol.47, pp. 291-305

Ingram, J. & Bartels, D. (1996). The molecular basis of dehydration tolerance in plants. *Annual Review of Plant Physiology and Molecular Biology*, Vol.47, pp. 377-403

Jahn, T.; Baluska, F.; Michalke, W.; Harper, J. & Volkmann, D. (1998). Plasma membrane H+-ATPase in the root apex: evidence for strong expression in xylem parenchyma and asymetric localization within cortical and epidermal cells.*Physiologia Plantarum*, Vol.104, pp. 311-316

Jang, Y.; Qin, Y.; Xie,, C.; Zhao, F.; Zhao, J.; Liu, D.; Chen, S.; Fuglsang, A.; Palmgren, M.; Schumaker, K.; Deng,X. & Guo, Y. (2010). The *Arabidopsis* chaperone J3 regulates the plasma membrane H^+-ATPase through interaction with the PKS5 kinase. *The Plant Cell*, Vol.22, pp.1313-1332

Janicka-Russak, M. & Kłobus, G. (2007). Modification of plasma membrane and vacuolar H^+-ATPase in response to NaCl and ABA. *Journal of Plant Physiology*, Vol.164, pp. 295-302

Janicka-Russak, M.; Kabała, K.; Burzyński, M. & Kłobus, G. (2008). Response of plasma membrane H^+-ATPase to heavy metal stress in *Cucumis sativus* roots. *Journal of Experimental Botany*, Vol.59, pp. 3721-3728

Janicka-Russak, M.; Kabała, K.; Młodzińska, E. & Kłobus, G. (2010). The role of polyamines in the regulation of the plasma membrane and the tonoplast proton pumps under salt stress. *Journal of Plant Physiology*, Vol.167, pp. 261-269

Jia, W.; Zhang, J. & Liang J. (2001). Initiation and regulation of water deficit-induced abscisic acid accumulaiton in maize leaves and roots: cellular volume and water relations. *Journal of Experimental Botany*, Vol.52, pp. 295–300

Kabała, K.; Janicka-Russak, M.; Burzyński, M. & Kłobus, G. (2008). Comparison of heavy metal effect on the proton pumps of plasma membrane and tonoplast in cucumber root cells. *Journal of Plant Physiology*, Vol.165, pp. 278-288

Kalampanayil, B. & Wimmers, L. (2001). Identification and characterization of a salt-stressed-induced plasma membrane H^+ATPase in tomato. *Plant Cell Environment*, Vol.24, pp. 999-1005

Kanczewska, J. ; Marco, S. ; Vandermeeren, C. ; Maudoux, O. ; Rigaud, J. & Boutry, M. (2005). Activation of the plant plasma membrane H^+-ATPase by phosphorylation and binding of 14-3-3 proteins convert a dimer into a hexamer. *PNAS*, Vol.102, No.33, (August 16), pp. 11675-11680

Kasamo, K. (1988). Response of tonoplast and plasma membrane ATPase in chilling sensitive and insensitive rice (Oryza sativa L.) culture cells to low temperature. Plant and Cell Physiology, Vol. 29, No.7, pp. 1085-1094.

Kasamo, K. (1990). Mechanism for activation of plasma membrane H+-ATPase from rice (*Oryza sativa*) culture cells by molecular species of a phospholipid. *Plant Physiology*, Vol.93, pp. 1049-1052

Kasamo, K. (2003). Regulation of plasma membrane H^+-ATPase activity by the membrane environment. *Journal of Plant Research*, Vol.116, pp. 517-523

Kasamo, K. & Nouchi, I. (1987). The role of phospholipids in plasma membrane ATPase activity in *Vagina radiata* L. (Mung bean) roots and hypocotyls. *Plant Physiology*, Vol.83, pp. 323-328

Kennedy, C. & Gonsalves, F. (1989). The action of divalent Zn, Cd, Hg, Cu and Pb ions on the ATPase activity of plasma membrane fraction isolated from roots of *Zea mays*. *Plant and Soil*, Vol.117, pp. 167-175

Kerkeb, L. ; Venema, K. ; Donaire, J. & Rodriguez-Rozales, M. (2002). Enhanced H^+/ATP coupling ratio of H^+-ATPase and increased 14-3-3 protein content in plasma membrane of tomato cells upon osmotic shock. *Physiologia Plantarum*, Vol.16, pp. 37-41

Kinoshita, T. & Shimazaki, K. (2001). Analysis of the phosphorylation level in guard-cell plasma membrane H+-ATPase in response to fusicoccin. *Plant Cell Physiology*, Vol.42, pp. 424-432

Kinoshita, T. & Shimazaki, K. (2002). Biochemical evidence for the requirement of 14-3-3 protein binding in activation of the guard-cell plasma membrane H+-ATPase by blue light. *Plant Cell Physiology*, Vol.43, pp. 1359-1365

Kłobus, G. & Janicka-Russak, M. (2004). Modulation by cytosolic components of proton pump activities in plasma membrane and tonoplast from *Cucumis sativus* roots during salt stress. *Physiologia Plantarum*, Vol.121, pp. 84-92

Kłobus, G. & Buczek, J. (1995). The role of plasma membrane oxidoreductase activity in proton transport. *Plant Physiology*, Vol.146, pp. 103-107

Knight, H.; Trewavas, A. & Knight, M. (1997). Calcium signalling in *Arabidopsis thaliana* responding to drought and salinity. *Plant Journal*, Vol.12, pp. 1067- 1078

Kolbert, Z. ; Bartha, B. & Erdei, L. (2008). Osmotic stress- and indolile-3-butyric acid-induced NO generation are partially distinct processes in root growth and development in *Pisum sativum*. *Physiologia Plantarum*, Vol.133, pp. 406-416

La Rosa, P.; Handa, A.; Hasegawa, P. & Bressan, R. (1985). Abscisic acid accelerates adaptation of cultured tobacco cells to salt. *Plant Physiology*, Vol.79, pp. 138-142.

La Rosa, P.; Hasegawa, P.; Rhodes, D. & Clithero, J. (1987). Abscisic acid stimulated osmotic adjustment and its involvement in adaptation of tobacco cells to NaCl. *Plant Physiology*, Vol. 85, pp. 174-185

Lee, S.; Singh, A.; Chung, G.; Ahn, S.; Noh, E. & Steudle, E. (2004) Exposure of roots of cucumber (Cucumis sativus) to low temperature severely reduces root pressure, hydraulic conductivity and active transport nutrients. *Physiologia Plantarum*, Vol.120, pp. 413-420

Lherminier, J.; Elmayan, T.; Fromentin, J.; Ekaraqui, K.; Vesa, S.; Morel, J.; Verrier, J.; Cailleteau, B.; Blein, J. & Simon-Plas, F. (2009). NADPH oxidase mediated reactive oxygen species production: subcellular localization and reassessment of its role in plant defence. *Molecular Plant Microbe Interaction*, Vol.22, No.7, pp. 868-881

Lin, H. & Wu, L. (1996). Effects of salt stress on root plasma memebrane characteristics of salt-tolerant and salt-sensitive buffalograss clones. *Environmental and Experimental Botany*, Vol.36, pp. 239-254

Linndberg, S. & Wingstrand, G. (1985). Mechanism for Cd^{2+} inhibition of (K^+ + Mg^{2+}) ATPase activity and K^+ ($^{86}Rb^+$) uptake join roots of sugar beet (*Beta vulgaris*). *Physiologia Plantarum*, Vol.63, pp. 181-186

Lindberg, S.; Banaś, A. & Stymne, S. (2005). Effects of different cultivation temperatures on plasma membrane ATPase activity and lipid composition of sugar beet roots. *Plant Physiology and Biochemistry*, Vol.43, pp. 261-268

Ling-Cheng, J.; Sun, L. & Dong, H. (1982). Adaptive changes in ATPase activity in the cells of winter wheat seedlings during cold hardening. *Plant Physiology*, Vol.70, pp. 127-131

Lino, B. ; Baizabal-Aguirre, V. & González de la Vara, L. (1998). The plasma H+-ATPase from beet root is inhibited by calcium-dependent phosphorylation. *Planta*, Vol.204, pp. 352-359

López-Pérez, L.; Martínez Ballesta, M.; Maurel, C. & Carvajal, M. (2009). Changes in plasma membrane lipids, aquaporins and proton pump of broccoli roots, as an adaptation mechanism to salinity. *Phytochemistry*, Vol.70, pp. 492-500

Lüthje, S.; Döring, O.; Heuer, S.; Lüthen, H. & Böttger, M. (1997). Oxidoreductases in plant plasma membranes. *Biochimica et Biophysica Acta*, Vol.1331, pp. 81-102

Martz, F.; Sutinen, M.; Kiviniemi, S. & Palta, J. (2006). Changes in freezing tolerance, plasma membrane H^+-ATPase activity and fatty acid composition in *Pinus resinosa* needles during cold acclimation and de-acclimation. *Tree Physiology*, Vol. 26, pp. 783-790

Morandini, P. ; Valera, M. ; Albumi, C. ; Bonza, M. ; Giacometti, S. ; Ravera, G. ; Murgia, I. ; Soave, C. & De Michelis, M. (2002). A nivel interaction partner for C-terminus of *Arabidopsis thaliana* plasma memebrane H+-ATPase (AHA1 isoform) : site and mechanism of action on H+-ATPase activity differ from those of 14-3-3 proteins. *The Plant Journal*, VOL.31, No.4, pp. 487-497

Moriau, L.; Michelet, B.; Bogaerts, P.; Lambert, L.; Michel, A.; Oufattole, M. & Boutry, M. (1999). Expression analysis of two gene subfamilies encoding the plasma membrane H^+-ATPase in *Nicotiana plumbaginifolia* reveals the major transport functions of this enzyme. *Plant Journal*, Vol.19, pp. 31–41

Morssome, P.; Dambly, S.; Maudoux, O. & Boutry, M. (1998). Single point mutations distributed in 10 soluble and mambrane regions of the Nicotiana plumbaginifolia plasma membrane PMA2 H^+-ATPase activate the enzyme and modify the structure of the C-terminal region. *Journal of Biological Chemistry*, Vol.273, pp. 34837-34842

Nakajima, N.; Saji, H.; Aono, M. & Kondo, N. (1996). Isolation of cDNA for a plasma membrane H^+-ATPase from guard cells of *Vicia faba* L. *Plant Cell Physiology*, Vol.36, pp. 919-924

Netting, A. (2000). pH, abscisic acid and the integration of metabolism in plants under stressed and non-stressed conditions: cellular responses to stress and their implication for plant water relations. *Journal of Experimental Botany*,Vol. 343, pp. 147-158

Niu, X.; Narasimhan, M. & Salzman, R. (1993). NaCl regulation of plasma membrane H+ATPase gene expression in *Glycophyte* and *Halophyte*. *Plant Physiology*, Vol.103, pp. 712-718

Nüchse, T.; Stensballe, A.; Jensen O. & Peck, S. (2003). Large-scale analysis of *in vivo* phosphotrylated membrane proteins by immobilized metal ion affinity. *Mollecular and Cellular Proteomics*, Vol. 1, pp. 1234-1243

Ookura, T.; Komatsu, S.; Kawamura, Y. & Kasamo, K. (2005). A 55-kDa calcium dependent protein kinase phosphorylated Thr residues from the auto-regulatory domain of plasma membrane H^+-ATPase in rice. *JARQ*, Vol.39, No2, pp. 99-104

Ouffatole, M.; Arango, M. & Boutry, M. (2000). Identification and expression of tree new *Nicotianaplumbaginifolia* genes which encode isoforms of a plasma-membrane H^+-ATPase, and one of which is induced by mechanical stress. *Planta*, Vol.210, pp. 715-722

Ouriti, O.; Boussama, N.; Zarrouk, M.; Cherif, A. & Ghorbal, M. (1997). Cadmium- and copper-induced changes in tomato membrane lipids. *Phytochemistry*, Vol.45, No.7, pp. 1343-1350

Qiu, Q.; Guo, Y.; Dietrich, M.; Schumaker, K. & Zhu, J. (2002). Regulation of SOS1, a plasma membrane Na^+/H^+ exchanger in *Arabidopsis thaliana*, by SOS2 and SOS3. *PNAS*, Vol.99, pp. 8436-8441

Palmgren, M. (2001). Plant plasma membrane H^+-ATPase : powerhous for nutrient uptake. *Annual Review of Plant Physiology and Molecular Biology*, Vol.52, pp. 817-845

Pedechenko, V. ; Nasirova, G. & Palladina, T. (1990). Lysophosphatidylcholinespecifically stimulates plasma membrane H+-ATPase from corn roots. *FEBS Letters*, Vol.275, pp. 205-208

Perez-Prat, E. ; Narasimhan, M. ; Niu, X. ; Botella, M. ; Bressan, R. ; Valupesta, V. ; Hasegawa, P. & Binzel, M. (1994). Growth cycle stage dependent NaCl induction of plasma membrane H^+-ATPase mRNA accumulation in de-adapted tobacco cells. *Plant, Cell and Environment*,Vol.17, pp. 327-333

Peters-Soler, A. ; Prado, J. & Serrano, R. (1990). Immunocytolocalization of plasma membrane H^+-ATPase. *Plant Physiology*, Vol.93, pp. 1654-1658

Pitan , B. ; Schubert, S. & Mühling, K. (2009). Decline in leaf growth under salt stress is due to an inhibition of H^+ pumping activity and increase in apoplastic pH of maize leaves. *Journal of Plant Nutrition and Soil Science*, Vol.172, pp. 535-43

Regenberg, B. ; Villalba, J. ; Lanfermeijer, F. & Palmgren, M. (1995). C-terminal deletion analysis of plant plasma membrane H+-ATPase : yeast as a model sysytem for solute transport across the plant plasma membrane. *Plant Cell*, Vol.7, pp. 1655-1666

Ros, R. ; Morales, A. ; Segura, J. & Picazo, I. (1992a). In vivo and in vitro effects of nickel and adcmium on the pasmalemma ATPase from rice (*Oryza sativa* L.) shoots and roots. *Plant Science*, Vol.83, pp. 1-6

Ros, R.; Cooke ,D.; Martinez-Cortina, C. & Picazo, I. (1992b). Nickel and cadmium related changes in growth, plasma membrane lipid composition, ATPase hydrolytic activity and proton-pumping of rice (*Oryza sativa* L. cv. Bahia) shoots. *Journal of Experimental Botany*, Vol.43, pp. 1475-1481

Rubinstein, B. & Stern, A. (1986). Relationship of transplasmalemma redox activity to proton and solute transport of Zea mays. *Plant Physiology*, Vol.80, pp. 805-811

Sahu, B. & Shaw, B. (2009). Salt-inducible isoform of plasma membrane H^+-ATPase gene in rice remains costitutively ezpressed in natural halophyte, *Suaeda maritima.Journal of Plant Physiology*, Vol. 166, pp. 1077-1089

Santi, S.; Locci, G.; Monte, R.; Pinton, R. & Varanini, Z. (2003). Induction of nitrate uptake in maize roots: expression of a putative high-affinity nitrate transporter and plasma membrane H^+-ATPase isoforms. *Journal of Experimental Botany*, Vol.54, pp. 1851-1864

Schaller, A. & Oecking, C. (1999). Modulation of plasma membrane H^+-ATPase activity differentially activates wound and pathogen defence responses in tomato plants. *Plant Cell*, Vol. 11, pp. 263-272

Schroeder, J.; Allen, G.; Hugouvieux, V.; Kwak, J. & Waner, D. (2001). Guard cell signal transduction. *Annual Review of Plant Physiology and Plant Molecular Biology*, Vol.52, pp. 627-658

Shen, H.; He, L.; Sasaki, T.; Yamamoto, Y.; Zheng, S.; Ligaba, Z.; Yan, X.; Ahn, S.; Yamaguchi, M.; Sasakawa, H. & Matsumoto, H. (2005). Citrate secretion coupled with the modulation of soybean root tip under aluminium stress. Up-regulation of transcription, translation, and threonine-oriented phosphorylation of plasma membrane H+-ATPase. *Plant Physiology*, Vol.138, (May), pp. 287-296

Shen, P.; Wang, R. Zhang, W. (2011). Rice phospholipase Dα is involved in salt tolerance by the mediation of H+-ATPase activity and transcription. Journal of Integrative Plant Biology, Vol. 53, No.4, (April), pp. 289-299

Shi, H.; Ishitani, M.; Kim, C. & Zhu, J. (2000). The *Arabidopsis thaliana* salt tolerance gene *SOS1* encodes a putative Na^+/H^+ antiporter. *PNAS*, Vol.97, pp. 6896-6901

Shi, H.; Quintero, F.; Pardo, J. & Zhu, J. (2002). The putative plasma membrane Na^+/H^+ antyporter SOS1 controls long-distance Na^+ transport in plants. *Plant Cell*, Vol.14, pp. 465-477

Shi, H.; Lee, B.; Wu, S. & Zhu, J (2003). Overexpression of a plasma membrane Na^+/H^+ antiporter gene improves salt tolerance in *Arabidopsis thaliana*. *Nature Biotechnology*, Vol.21, pp. 81-85

Shinozaki, K. & Yamaguchi-Shinozaki, K. (1996). Molecular responses to drought and cold stress. *Current Opinion Biotechnology*, Vol. 7, pp. 161-167

Skriver, K. & Mundy, J. (1990). Gene expression in response to abscisic acid and osmotic stress. *Plant Cell*, Vol. 2, pp. 503-512

Speth, C. ; Jaspert, N. ; Marcon, C. & Oecking, C. (2010). Regulation of the plant plasma membrane H+-ATPase by its C-terminal domain : what we know for sure? *European Journal of Cell Biology*, Vol.89, pp. 145-151

Surowy, T. & Boyer J. (1991). Low water potentials affect expression of genes encoding vegetative storage proteins and plasma membrane proton ATPase in soybean. *Plant Molecular Biology*, Vol.16, pp. 252–262

Svennelid, F.; Olsson, A.; Piotrowski, M.; Rosenquist, M.; Ottman, C.; Larsson, C.; Oecking, C. & Sommarin, M. (1999). Phosphorylation of Thr-948 at the C-terminus of the plasma membrane H^+-ATPase creates a binding site for regulatory 14-3-3 protein. *Plant Cell*, Vol.11, pp. 2379-2391

Takemiya, A. & Schimazaki,, K. (2010). Phosphatidic acid inhibits blue light-induced stomatal opening via inhibition of protein phosphatases 1. *Plant Physiology*, Vol.153, (August) pp. 1555-1562

Takemiya, A.; Kinoshita, T, Asanuma, M. & Shimazaki, K. (2006). Protein phosphatases 1 positively regulates stomatal opening in response to blue light in *Vicia faba*. *Proceedings of the National Academy of Sciences USA*, Vol.103, pp. 13549-13554

Tan, B.; Schwartz, S.; Zeevaart, J. & Mc Carty, R. (1994). Genetic control of abscisic acid biosynthesis in maize. *Proceedings of the National Academy of Sciences USA*, Vol.94, pp. 12235-12240

Upchurch, R. (2008). Fatty acid unsaturation, mobilization, and regulation in the response of plants to stress. *Biotechnological Letters*, Vol.30, pp. 967-977

Van der Hoeven, P.; Siderius, M.; Korthout, H.; Drabkin, A. & De Boer, A. (1996). A calcium and free fatty acid modulated protein kinase as putative effector of the fussicoccin 14-3-3 receptor. *Plant Physiology*, Vol.111, pp. 857-865

Vega, S.: Del Rio, A.; Bamberg, J. & Palta J. (2004). Evidence for the up-regulation of stearoyl-ACP (D9) desaturase gene expression during cold acclimation. *American Journal of Potato Research*, Vol.81, pp. 125–135

Vera-Estrella, R.; Barkla, B. Higgins, V. & Blumwald E. (1994). Plant defense response to fungal pathogens (activation of host-plasma membrane H+-ATPase by elicitor-induced enzyme dephosphorylation). *Plant Physiology*, Vol.104, pp. 209-215

Viotti, C; Luoni, L.; Morandini, P. & De Michelis, M. (2005). Characterization of the interaction between the plasma membrane H+-ATPase of *Arabidopsis thaliana* and a novel interactor (PP1). *FEBS Journal*, Vol.272, pp. 5864-5871

Wakeel, A. Hanstein, S.; Pitann, B. & Schubert, S. (2010). Hydrolytic and pumping activity of H+-ATPase from leaves of sugar beet (*Beata vulgaris* L.) as affected by salt stress. *Journal of Plant Physiology*, Vol.167, pp. 725-731

White, F.; Cooke, D.; Earnshaw, M.; Clarkson, D. & Burden, R. (1990). Does plant growth temperature modulate the membrane composition and ATPase activities of tonoplast and plasma membranes fractions from rye roots? *Phytochemistry*, Vol.29, pp. 3385-3393

Xing, T.; Higgins, V. & Blumwald E. (1996). Regulation of plant defense response to fungal pathogens: Two types of protein kinases in the reversible phosphorylation of the host plasma membrane H^+-ATPase. *Plant Cell*, Vol.8, pp. 555-564

Xiong, L. & Zhu, J. (2001). Abiotic stress signal transduction in plants: Molecular and genetic perspectives. *Physiologia Plantarum*, Vol.112, pp. 152-166

Xiong, L.; Schumaker, K. & Zhu, J. (2002). Cell Signaling during Cold, Drought, and Salt Stress. *Plant Cell*,Vol.14, pp.165-183

Yamasaki, H. & Sakihama, Y. (2000). Simultaneous production of nitric oxide and peroxynitriteby plant nitrate reductase: in vitro evidence for the NR-dependent formation of active nitrogen species. *FEBS Letters*, Vol.468, pp. 98-92

Yang, Y.; Xu, S.; An, L. & Chen, N. (2007). NADPH oxidase dependent hydrogen peroxide production, induced by salinity stress, may be involved in the regulation of total calcium in roots of wheat. *Journal of Plant Physiology*, Vol.164, pp. 1429-1435

Yu, X.; Li, M.; Gao, G.; Feng, H.; Geng, X.; Peng, C.; Zhu, S.; Wang, X.; Shen, Y. & Zhang, D. (2006). Abscisic acid stimulates a calcium-dependent protein kinase in grape berry. *Plant Physiology*, Vol.140, pp.558-579

Zandonadi, D.; Santos, M.; Dobbss, L.; Olivares, F.; Canellas, LL.;, Binzel, M.; Okorokova-Façanha, A. & Façanha, A. (2010). Nitric oxide mediates humic acids-induced root development and plasma membrane H^+-ATPase activation. *Planta*, (2010). Vol.231, pp. 1025-1036

Zhang, X.; Zhang, L.; Dong, F.; Gao, J.; Galbraith, D. & Song, C. (2001). Hydrogen peroxide id involved in abscisic acid-induced stomatal closure in *Vicia faba*. *Plant Physiology*, Vol.126, (August), pp. 1438-1448

Zhang, X.; Wang, H.; Takemiya, A.; Song, C.; Kinoshita, T. & Shimazaki, K. (2004). Inhibition of blue light-dependent H^+ pumping by Abscisic acid through hydrogen peroxide-induced dephosphorylation of the plasma membrane H^+-ATPase in guard cell protoplasts. *Plant Physiology* , Vol.136, (December), pp. 4150-4158

Zhang , m.; Barg, R.; Yin, M.; Gueta-Dahan, Y.; Leikin-Frenkel, A.; Salts, Y.; Shabtai, S. & Ben-Hayyim, G. (2005). Modulated fatty acid desturation via over-expression of two distinct ω-3 desaturases differentially alters tolerance to various abiotic stresses in transgenic tobacco cells and plants. *Plant Journal*, Vol.44, pp. 361-371

Zhang, F.; Wang, Y. & Wang, D. (2007). Role of nitric oxide and hydrogen peroxide during the salt resistance response. *Plant Signaling & Behavior*, Vol.2, No.6, (November/December), pp. 473-474

Zhu, J. (2002). Salt and drought stress signal transduction in plants. *Annual Review of Plant Biology*, Vol.53, pp. 247-273

Zörb, C.; Stracke, B.; Tramnitz, B.; Denter, D.; Sümer, A.; et al. (2005). Does H^+ pumping by plasmalemma ATPase limit leaf growth of corn (*Zea mays* L.) during the first phase of salt stress? *Journal of Plant Nutrition and Soil Science*, Vol.168, pp. 550-557

Part 3

Genes and Genomes

9

Plant Abiotic Stress: Insights from the Genomics Era

Erik R. Rowley and Todd C. Mockler
Oregon State University
United States

1. Introduction

Agricultural crop plants make up a large proportion of the world's economy and in many countries constitute the main sustenance for humans. Therefore maximizing crop yield is of extreme importance and interest. There are many factors that can limit the yield of a crop; however the main causes of crop failure are abiotic stresses such as salinity, drought, extremes in temperature, intense light, and oxidative stress caused by reactive oxygen species. Plants have evolved mechanisms and pathways allowing them to cope with the environment by modifying their physiological and cellular states. For example, plants living in colder regions undergo a phenomenon known as cold acclimation, resulting in cell membrane composition and protein concentration changes to reduce intracellular ice crystal formation and dehydration due to freezing (Thomashow, 1998).

Abiotic stresses alter gene networks and signaling cascades in an effort to restore cellular homeostasis. It has been demonstrated (Reddy, 2007, Palusa et al., 2007) that abiotic stress conditions alter the alternative splicing of a number of genes. Alternative pre-mRNA splicing in higher eukaryotes is a highly regulated mechanism, often allowing for many proteins (isoforms) to be derived from a single gene, thereby increasing overall proteome diversity. These alternative transcripts often result in functionally and structurally distinct proteins (Biamonti, 2009) with their own functions in development, cellular localization, and responses to the environment (Tanabe et al., 2006).

Recently genome and transcriptome-wide surveys (Rensink et al., 2005, Kreps et al., 2002, Egawa et al., 2006, Filichkin et al., 2010) have offered glimpses into transcript abundance profiles under abiotic stresses, demonstrating dramatic shifts in alternative splicing patterns, as well as upregulation of key transcription factors controlling stress-induced signaling cascades. This research offers the potential for discovery of candidate genes that, through genetic engineering, may confer increased tolerance to abiotic stresses, several examples of which will be discussed in this chapter.

Recent research has demonstrated the possibility of pre-disposing plants to stress tolerance by overexpressing a gene known to be upregulated in response to a certain stress (for example: Haake et al., 2002, Forment et al., 2002, Kim et al., 2010) often acting upstream or in conjunction with a longer signaling cascade, such as the mitogen-activated protein (MAP) kinase cascade (Xiong et al. 2003), salt overly sensitive (SOS) pathway (Liu and Zhu, 1998; Ishitani et al., 2000, Qiu et al., 2004), or C-repeat-binding factor/dehydration-responsive element (CBF/DREB) pathway (Thomashow, 2010).

A high degree of crosstalk exists between these pathways, as often the plant's physiological and cellular responses to different abiotic stresses are similar **(Figure 1)**. In drought and cold stress for example, two types of molecular responses occur simultaneously: those protecting cells from acute dehydration and those protein factors involved in further regulation of gene expression and signal transduction functioning in overall stress response (Shinozaki et al., 2000 and 2007). Other examples include crosstalk between cold and osmotic signaling pathways, as well as cold and abscisic acid (Ishitani et al., 1997). Abscisic acid regulates stomatal aperture changes and is a crucial signaling molecule in stress plant responses along with changes in metabolite concentrations.

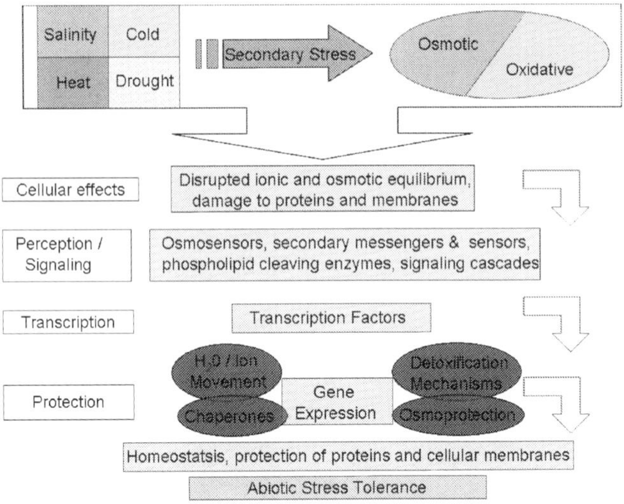

Fig. 1. Depiction of the overlapping and complex cellular responses resulting from abiotic stressors. These responses are a recurring theme throughout the chapter, as responses to many stress overlap, sharing regulons and transcriptional "hubs". Primary stresses (cold, drought, heat, and cold) result in secondary stresses, such as altered osmotic or ionic potential. The cell senses initial stress signals, such as osmotic and ionic effects, temperature, membrane fluidity changes, and triggers downstream signaling cascades, and amplifications of the signal. The final layer is the transcription factors, which initiate stress-responsive mechanisms in order to protect the cell, and restore homeostasis (Figure adapted from review by Wang et al., 2003).

Understanding the genes and gene networks that underlie abiotic stress resistance is integral not only in improving the world's long term food production and security, but also in piecing together the web of abiotic stress induced global gene regulation including post-transcriptional regulation such as alternative splicing and regulation by miRNAs. Recent insights into genes conferring abiotic stress tolerance, particularly salt tolerance, have come from the study of plants naturally adapted for growth in extreme conditions such as the halophytes, which thrive in areas of elevated salt levels. Differential gene expression in seedlings of the salt marsh halophyte *Kosteletzkya virginica* was recently analyzed (Guo et al., 2009), identifying genes necessary for re-establishing ion homeostasis and protecting the plant from stress damage, as well as those involved in metabolism and plant development under salt stress. Also demonstrated was the ability of *K. virginica* seedlings to sequester sodium, primarily in the roots. In another example, a dehydration and abscisic acid (ABA) induced transcription factor was functionally characterized in *Craterostigma plantagineum*, a plant possessing the ability to survive complete dehydration.

The knowledge gained from high-throughput sequencing (HTS) technologies and comparative studies of halophytes, coupled with our continually expanding knowledge of metabolites, and the molecular and physiological responses to abiotic stresses that are profiled within this chapter, will allow a plethora of opportunities for directed genetic engineering and breeding strategies that will allow us to meet the worlds demand for food despite a growing population. This chapter aims to offer insights from the past decade of plant abiotic stress research, and give an overview of the wealth of knowledge generated by the genomics era, such as advances from global gene expression surveys and differential gene expression between stresses.

2. Salt

Increasing salinity of soil leads to reduction of crop yields, and while soil salinity is not uncommon (Flowers et al., 1997), secondary effects such as irrigation-induced salination affects about 20% of the world's cultivated fields (Yeo, 1998) with 50% of lands predicted to be affected in the coming decades. These irrigated lands currently produce an estimated one-third of the world's food (Munns 2002). Irrigation water typically contains a variety of ions, such as Ca+, Mg+, as well as Na+ in the form of NaCl. It is when the water evaporates and the Ca+ and Mg+ precipitate that the Na+ ions begin to become dominate in the soil (Serrano et al., 1999). Plants do not have specific mechanisms for the uptake of Na+ ions; however several pathways exist for passive entry. For example, root cells uptake Na+ ions via cation channels, of which there are two main classifications. Voltage dependent channels, namely the K+ transporter *HKT1*, import Na+ ions into root cells (Na+ leakage), ultimately leading to higher concentrations of intracellular sodium. Excess salt in the soil now begins to present an issue due to this osmotic gradient, as elevated Na+ levels in the soil begins to drive water out of the cell. Initial plant response to salt stress, the osmotic stress component, share metabolic similarities with drought however long-term exposure introduces the ion toxicity component, the displacement of K+ ions with Na+ and Cl- ions. Plants differ in their methods of coping with Na+ entry: some prevent or minimize entry altogether (at the roots), while others reduce the cytoplasmic Na+ concentration by compartmentalization in the vacuoles, thus avoiding toxic effects on photosynthesis and other key metabolic processes (Chaves et al., 2009).

Na+ transport from roots to stem is quite rapid due to the transpiration stream in the xylem, and can only be returned to the roots via the phloem. The roots are able to regulate Na+ levels by export to either the stem of back to the soil; however once in the xylem the Na+ ions will accumulate as the leaves age and the water evaporates (Tester et al., 2003). This rapid accumulation of sodium ions has several detrimental metabolic effects to the plant cell.

Turgor pressure is necessary in order to stretch the cells walls during growth. When faced with the initial sudden influx of Na+ ions, the plant cell is able to sequester these ions in the vacuole, in effect reducing the osmotic potential in an attempt to restore homeostasis and equally importantly reducing degradation of cytosolic enzymes. Research (Carden et al.,

2003) comparing ion concentrations in the roots of two barley (*Hordeum vulgare*) varieties differing in NaCl tolerance indicated the cytosolic Na+ concentration requirements to be quite low; around 10-30mM. Na+ concentration within the vacuoles however, may be much higher.

During the initial osmotic phase of salt stress response, the expansion rate of growing leaves is reduced, along with stomatal aperture in response to leaf turgor decline, leading to decreased emergence of new leaves and therefore fewer branches. Among the cereals, barley is the most tolerant with rice (*Oryza sativa*) being the least tolerant.

Na+ ions also compete with K+ ions for binding sites, often required for crucial cellular and metabolic processes. Protein stability is also coupled with K+ concentration, as it is a cofactor for many enzymes, and the tRNA binding to ribosomes is also dependent on high K+ concentration (Zhu, 2002, Tester et al., 2003). Nutrient uptake from the roots is affected due to K+ ion channels being disrupted, and Na+ induced osmotic changes inhibit root growth.

Another way plants reduce osmotic stress is by the accumulation of cytoplasmic osmolytes such as proline and glyceinebetaine, shown to stabilize the chloroplasts and cellular membranes, as well as play a role in maintaining cell volume and fluid balance (Bohnert et al., 1996). These osmolytes also serve to protect proteins from degradation by reacting oxygen species (ROS). Salt stress (among other abiotic stresses) leads to the accumulation of high levels of ROS. When present at low levels, ROS may act to signal upregulation of the defense-responsive genes. Typically however, excessive production and accumulation of ROS such as hydrogen peroxide (H_2O_2), superoxide (O_2-) and hydroxyl radicals (OH-) can perturb the cellular redox homeostasis leading to oxidative injuries. There is also growing evidence that the cell's downstream ability to repair damage, and scavenge damaging reactive oxygen species (ROS) is equally as important as Na+ uptake and vacuolar sequestration itself. Expression of ROS defense transcripts was found to be elevated in *Arabidopsis* plants constitutively expressing the zinc finger protein Zat10 (Mittler et al., 2006). The plants displayed enhanced tolerance of salinity, heat and osmotic stress. Surprisingly, knockout and RNAi mutants of Zat10 were also more tolerant to osmotic and salinity stress suggesting that Zat10 plays a key role as both a positive and a negative regulator of plant defenses. Readers are directed toward a recent review (Miller et al., 2008) for a discussion on how ROS integrate cellular signals generated from abiotic stress.

Studies of salt tolerant clones of *Eucalyptus camaldulensis,* an important crop in Australia due to the use of its oils, have demonstrated a significant increase in shoot proline levels when exposed to 100mM NaCl (Woodward et al., 2005). Proline accumulation is known to be mediated by both ABA-dependent and ABA-independent signaling pathways (Hare et al., 1999). Assays on *in vitro* shoot cultures of *Populus euphratica* suggest accumulated proline and sugars promote both osmotic and salt tolerance (Watanabe et al., 2000). Another study using sugar beet cultivars (Ghoulam et al., 2001) report a positive trend with proline levels increasing with relation to salt tolerance; however the quantitative contribution of proline to osmotic adjustment in a salt tolerant variety was weak. It was determined the elevation of proline in *Arabidopsis,* acting as an osmoprotectant during salt stress adaptation, led to the enhancement of the enzymes scavenging reactive oxygen species (Abraham et al., 2003). Recently proline and glycine betaine were shown to improve salt stress in cultured tobacco cells via scavenging of hydrogen peroxide and methylglyoxal (Banu et al., 2010). Glycine betaine was shown to be in induced in the burning bush, *Kochia scoparia,* (Kern et al., 2004), and the Mediterranean shrub *Atriplex halimus* (Martinex et al., 2004). The cyanobacterium *Synechococcus* also displays increased tolerance to both salt and cold stress following induction of glycine betaine (Ohnishi et al., 2006).

The second major constraint, besides the osmotic stress of Na+ surrounding the roots, is intracellular Na+ toxicity. Because of the similarity in physicochemical properties between Na+ and K+ (i.e. ionic radius and ion hydration energy), the former competes with K+ for major binding sites in key metabolic processes in the cytoplasm, such as enzymatic reactions, protein synthesis and ribosome functions (Shabala et al., 2008). Increased concentrations of Na+ ions in the soil reduce the activity of many essential nutrients (including K^+), making them less available. Secondly, Na^+ competes with K^+ for uptake sites at the plasma membrane. Recent evidence indicates the K+/Na+ intracellular ratio is a key determinant of salt tolerance. The optimal cytosolic K+/Na+ ratio can be maintained by either restricting Na+ accumulation in plant tissues or by preventing K+ loss from the cell. At the cellular level, restricted Na+ uptake, active Na+ exclusion back to the soil solution (via the plasma membrane salt overly sensitive (SOS1) Na+/H+ antiporter; (Zhu, 2002) and compartmentalization of excessive Na+ in the vacuole by the tonoplast Na+/H1+exchanger (Zhang and Blumwald 2001) are considered central to salt tolerance.

The past decade of research into the SOS (Salt Overly Sensitive) pathway, utilizing *Arabidopsis* knock-out mutants and the plants basal tolerance to NaCl stress as a background concentration for screening, has elucidated key steps in the salt stress signaling pathway (Liu and Zhu, 1998; Ishitani et al., 2000, Qiu et al., 2004). A transient Ca2+ signal, an important secondary messenger for many cellular processes, is ultimately propagated by the secondary messenger IP3 and is the crucial first step in restoring cellular homeostasis. This process involves the sensing of the Ca+ ion by SOS3 (also known as AtCBL4: calcineurin B-like protein) followed by interaction with SOS2, a serine/threonine protein kinase, resulting in its activation (Halfter et al., 2000). These work in conjunction to phosphorylate and activate the transport activity of the plasma membrane Na+/H+ antiporter SOS1. SOS1 also has a large cytoplasmic domain predicted to act as potential novel Na+ sensor (Zhu 2002), which may act in feedback regulation. Recent research has indicated the C-terminal region of SOS1 interacts with RCD1 under salt and oxidative stresses (Katiyar-Agarwat et al., 2006). Typically a nuclear protein, RCD1 is found both in the nucleus and in the cytoplasm near the cell periphery during salt and oxidative stresses and demonstrated similar expression and tissue localization as SOS1, perhaps regulating transport of ROS across the cell membrane and oxidative-stress signaling.

There are likely more components to the SOS pathway, the function of which are the focus of current research. For example, there is a family of 9 SO3-like Ca+ binding proteins (the SCaBP's) present in *Arabidopsis* and 24 SOS2-like protein kinases. One of the SCaBP's, the putative calcium sensor *SCABP8*/CBL10 was shown to interact with the protein kinase SOS2 to protect *Arabidopsis* shoots from salt stress (Xie et al., 2009). Further screens under more stringent (100mM NaCl) conditions for salt-hypersensitive mutants have yielded more members of the SOS pathway. SOS4 encodes a pyridoxal kinase that is involved in the biosynthesis of pyridoxal-5-phosphate, an active form of vitamin B6, which is often found in roots and necessary for growth (Mahajan et al., 2005). Knock-out mutants for SOS4 are defective in root hair formation and root tip growth, perhaps acting as an integral upstream regulator of root hair development (Zhu et al. 2002). Another component, SOS5, has been shown to be a putative cell surface (Shi et al. 2003) adhesion protein that is required for normal cell. For a detailed overview into current insights into the SOS pathway, the reader is directed to the 2008 review by Mahajan.

The SOS pathway is far from a unique response in *Arabidopsis*, or for that matter glycophytes in general, as conservation of SOS pathway components have been identified in halophytes as well as cereals and also woody periennals. For example in halophytes, salt treatment of *Thellungiella halophila* led to increased expression of an AtSOS1 homologue in the plasma membrane and increased H+ transport and hydrolytic activity of the H+-ATPase was observed in both the plasma membrane as well the tonoplast (Vera-Estrella et al., 2005). *Chenopodium quinoa*, a halophyte native to the Andes Mountains, was found to contain 2 AtSOS1 homologs (Maughan et al., 2009), with future work to include complementation of a mutant sos1 *Arabidopsis* line with the homologues from *C. quinoa*. Homologues of AtSOS1 have also been identified for multiple glycophyte plant species such as rice (*Oryza sativa*), the seagrass *Cymodocea nodosa*, and *Populus trichocarpa*, the woody perennial poplar tree (Martínez-Atienza et al., 2007, Garciadeblás et al., 2007, Tang et al., 2007, respectively).

Not surprisingly, also conserved are genes controlling sodium entry, such as the previously mentioned K+ channel HKT1 and also genes controlling vacuole compartmentalization, of which the following discussion will focus primarily on AtNHX1, a gene encoding a vacuolar Na+/H+ exchanger. Shi et al., 2002 demonstrated AtNHX1 transcript up-regulation following treatment with NaCl, KCl or ABA, as well as detecting strong expression in guard cells and root hairs, suggesting AtNHX1 plays a role in pH regulation and/ K+ homeostasis along with storing Na+ in the enlarged vacuoles in root hair cells, respectively. As previously mentioned, one way to remove harmful Na+ ions from the cytosol and maintain osmotic balance within the cells is by compartmentalization in the vacuoles, and this aspect of Na+ tolerance has been the focus of much current research, with very encouraging results.

In 1999 Apse et al. demonstrated increased Na+ tolerance from overexpression of the AtNHX1 Na+/H+ antiporter in *Arabidopsis*, and also that salinity tolerance was correlated with higher-than-normal levels of *AtNHX1*transcripts, protein, and vacuolar Na^+/H^+(sodium/proton) antiport activity. In 2007 tomato (*Lycopersicon esculentum* cv. Moneymaker) was successfully transformed with an overexpressed AtHKT1, demonstrating not only the ability to grow in 200mM Na+ concentrations, but an accumulation of sodium ions in the leaves rather than the fruit (Zhang et al., 2001). This discovery was quite exciting in two ways: firstly this yielded the potential for agriculturally relevant crop, as the fruit quality was not adversely affected, and secondly it demonstrated an increased resistance to salt tolerance in an agriculturally important crop plant resulting from the modification of a single trait. Then an AtNHX homologue from a monocot halophyte, *Aeluropus littoralis*, was identified (AlNHX) and cloned (Zhang et al. 2008). This gene was then transformed into tobacco, which displayed the ability to grow in MS media containing 250mM NaCl, and survived 400mM NaCl in pots for one month. Na+ ions were found to be sequestered primarily in the roots rather than stem tissue, with the leaves maintaining a higher K+ level than the WT control plants. Notably the results indicate the halophyte AlNHX may play a role in root rather than shoot Na+ levels, which was different than observations in overexpressed OsNHX1 in transgenic rice (Fukuda et al., 2004).

In fact there have been many examples of increased salt tolerance resulting from overexpression of the NHX family of Na+/H+ antiporters from various plant species, selected examples being: perennial ryegrass transformed with OsNHX1 (Wu et al., 2005), wheat (Xue et al., 2004), *Petunia hybrida* with AtNHX1 (Xu et al., 2009) demonstrated increased salt and drought tolerance. Recently the SsNHX1 gene from the halophyte *Salsola soda* (Li et al., 2010) conferred salt tolerance when overexpressed in transgenic alfalfa (*Medicago sativa*). These transgenic alfalfa plants had the ability to grow normally for 50 days under Na+ treatment, with no apparent difference in growth detectable between transgenic plants and wild-type plants under normal conditions, likely due to the use of the stress inducible promoter rd29A rather than the typical constitutively active Ca35S promoter. Clearly the NHX family of Na+/H+ of antiporters are able to confer increased Na+ tolerance across a wide range of plant species, and aside from being a single trait, may be even more relevant from a genetic engineering standpoint as sixth-generation soybean plants expressing AtNHX1 proved to be just as resistant (Li et al., 2010) to salt stress as the first generation transgenic plants, indicating this single trait change in heritable.

Although much research has been conducted into the NHX family of Na+/H+ of antiporters, several others have also shown promise for genetic engineering. The plasma membrane Na"/H+ antiporter SOS1, activated it response to salt stresses by the SOS pathway reviewed above, has been shown to be critical for Na+ partitioning in plant organs as well as the ability for the plants to partition Na+ in the stems, preventing the ions to reach photosynthetic tissues (Olias et al., 2009). Ca+ antiporters, such as *Arabidopsis* H+/Ca2+ Antiporter CAX1 were found to confer increased Ca+ transport and salt tolerance (Chen et al 2004). Both salt and drought tolerance can be significantly increased in *Arabidopsis* plants by overexpressing AtAVP1, the gene encoding a vacuolar pyrophosphatase which acts as a vacuolar membrane proton pump (Gaxiola et al., 2001), moving more H+ into the vacuoles to create a higher electrochemical gradient. In addition to *Arabidopsis*, overexpression of AtAVP1 in tomato also enhances drought tolerance (Park et al., 2005), due the increased osmotic adjustment ability conferred by the increased vacuolar H+ concentration.

Gene expression studies in halophytes have yielded fascinating candidate genes for future study; root and leaf tissue collected from *Kosteletzkya virginica* seedlings (Guo et al., 2008) identified 34 differentially expressed gene fragments homologous to known genes from other species and 4 of novel function. The differentially expressed genes were classified into four groups: those necessary for re-establishing ion homeostasis those involved in metabolism or energy and resuming plant growth and development under salt stress, those involved in regulation of gene expression, and those responsible for signal transduction (Guo et al., 2008).

The halophyte *Craterostigma plantagineum,* known as the resurrection plant, has the ability to survive complete dehydration. In an attempt to further understand desiccation tolerance in this plant, the CpMYB10 transcription factor gene was functionally characterized (Villalobos et al., 2004) and found to be rapidly induced by dehydration and abscisic ABA treatments in leaves and roots, with no expression detected in fully hydrated tissues. Its subsequent overexpression in Arabidopsis also leads to salt tolerance of the transgenic lines. However, it also was found that plants overexpressing CpMYB10 also exhibited glucose-insensitive and ABA hypersensitive phenotypes. This finding exemplifies an issue in studies with model organisms in short-term laboratory settings: is there overlap between the molecular mechanisms to cope with stress in *Arabidopsis*, crops plants, and halophytes? Are there overlaps between gene regulation, transcriptional activators, and their tissue-specificity? These distinctions are essential in order for genetic engineering to beneficially be used in crop species for trait selection. One useful tool for candidate gene discovery is genome-wide profiling of both stress-induced expression and post-transcriptional events occurring as a result of stress exposure.

Recent microarray studies have provided sets of candidate genes for further investigation in order to define the transcriptome profile under salt stress. Tomato (Zhou et al., 2007) gene expression was profiled under salt stress, discovering several key enzyme genes in the metabolic pathways of carbohydrates, amino acids, and fatty acids to be initiated. Also higher transcript levels were detected for antioxidant enzymes, ion transporters, and genes known to be involved with numerous signal transduction pathways. The Euphrat poplar tree (*Populus euphratica*) that thrives in a saline and arid environment is expanding our understanding of stress induced gene networks in trees, which spend a much greater amount of time in soils due to their longer life and therefore must possess robust systems for dealing with abiotic stresses. Acclimation to increasing levels of Na+ requires adjustment to the osmotic pressure of leaves, achieved by accumulation of Na+ and compensatory decreases in Ca+ and soluble carbohydrates. The primary strategy of *P. euphratica* to protect the cytosol against sodium toxicity is apoplastic, instead of vacuolar, salt accumulation, suggesting that Na+ adaptation requires suppression of Ca+ related signaling pathways. Evidence also points to shifts in carbohydrate metabolism and suppression of reactive oxygen species in mitochondria under salt stress (Ottow et al., 2005). Overexpression of a single Ca+ dependent protein kinase in rice increases salt tolerance (Saijo et al., 2000), with levels of tolerance correlation to levels of protein.

A recent microarray study of *P. euphratica* by Brinker et al. (2010) noted three distinct transcriptome phase changes associated with salt stress, with the duration and intensity of these phases differing between the leaf and root tissues sampled. Key factors initially involved with salinity-stress are molecular chaperones, namely the dehydrins and osmotin, which assist with protein stabilization. Leaves initially suffered from dehydration stress, resulting in transcript level shifts of mitochondrial and photosynthetic genes, indicating adjustment of energy metabolism. Initially a decrease in known stress-associated genes occurs, with induction occurring later, after excessive sodium concentrations accumulate in the leaves. In roots a decrease in aquaporins occurs, potentially reducing water loss. Roots and leaves perceive physiologically different stress situations, and therefore activate unique stress responses; however sucrose synthase and chaperones from leaves were also found upregulated in roots as the only overlapping salt-responsive genes in roots and leaves. To identify the stress-specific genes within the poplar salt-stress responsive transcriptome Brinker et al., used *in silico* analyses with *Arabidopsis* orthologs to reduce the number of candidate genes for functional analysis. Ultimately two genes, a lipocalin-like gene and a gene encoding a protein with previously unknown functions were identified and shown to display salt-sensitive phenotypes in *Arabidopsis* knockout mutants, suggesting these genes play roles in salt tolerance. These results are quite exciting, since they demonstrate salt-susceptible plants harbor genes important for salt tolerance that cannot be identified by conventional salt screens relying on differential gene expression (Brinker et al. 2010).

Foxtail millet (*Setaria italica*) is a food and fodder grain crop grown in arid and semi-arid regions (Puranik et al., 2011) and is a self-pollinating, diploid, C4 grass. Comparative transcriptome analyses between two cultivars differing in response to short-term salinity stress identified 81 differentially expressed novel transcripts. These transcripts represent an "untapped genetic resource" (Puranik et al., 2011), in a model crop with natural increased resistance to abiotic stress.

High-throughput Illumina based RNA-seq experiments are allowing for genome-wide glimpses into transcript abundance and transcriptional regulation, having the benefit of not requiring previously annotated genes or being limited to specific probes present on a microarray. Genome-wide mapping of alternative splicing in Arabidopsis under abiotic stresses (Filichkin et al., 2010) have identified different types of stress differentially regulating known genes implemented in various pathways and cellular responses. For example, a splicing factor in the SR (serine/arginine rich) family, SRP30/SR30 (At1g09140), displays upregulation of the reference isoform under salt stress. This makes SR30 a candidate for further study in order to elucidate salt-stress responses from a splicing factor, rather than a transcriptional angle.

Another Illumina-based RNA-seq experiment using rice (*Oryza sativa* L. 'Nipponbare) cDNAs focused on the identification of salt-responsive unannotated transcripts derived from root and shoot mRNAs in rice and those transcripts encoding putative functional proteins (Mizuno et al., 2010). 7-day old rice seedlings were transferred to either 150mM NaCl solution or water (control) for 1hr. Of the total unannotated transcripts discovered, 1,525 in shoot and 1,659 in root were novel transcripts. Of these transcripts, 213 (shoot) and 436 (root) were differentially expressed in response to salinity stress. The predicted encoded proteins were associated with amino acid metabolism in response to abiotic stresses, and mechanosensitive ion channel function. These responses are gated directly by physical stimuli such as osmotic shock and known to transduce these stimuli into electrical signals. Also captured were previously identified genes involved in salinity tolerance; those associated with trehalose synthesis, dehydrin, ABA synthesis sugar transport, glycerol transferase, and transcription factors similar to those of the DREB family (Mizuno et al., 2010). The DREB transcription activators are involved in ABA-independent and abiotic stress response, binding to the consensus dehydration-responsive element (DRE), present in promoter regions of genes induced by osmotic, saline, and cold stresses (Stockinger et al., 1997). As a substantial number of transcripts were exclusively upregulated only in the root, being directly exposed to 1 hour of salinity stress, it was hypothesized it may take longer exposure time to induce a greater network of genes (Mizuno et al., 2010).

3. Cold

Low temperatures, both sudden and for sustained periods, cause dramatic decreases in crop sustainability and yield by affecting the germination and reproductive rate of plants. Low temperature induced cold stress leads to reduced cell expansion and consequently reduced leaf growth, with the loss in turgor pressure causing severe wilting of leaves, ultimately leading to plant death.

Plants differ in their abilities to survive both freezing (temperatures below 0°C) and chilling (0°C to around 20°C) conditions by modifying their physiological and cellular states. The seeds of plants native to latitudes undergoing a freezing winter period such as the winter cereals (certain barley and wheat cultivars, rye, and oats, among others) require a period of cold temperature, called vernalization, prior to germination. This epigenetic response alters the chromatin structure of a flowering repressor gene, in effect allowing the seedlings to "remember" the period of cold preceding the warmth of the growing season (Sung & Amasino, 2009). The vernalization period is necessary to prevent premature transition to the reproductive phase before the winter freezing threat has ended, however this does not continue past onset of the vegetative phase (Chinnusamy et al., 2007). During the warm growing season, these temperate region plants have little ability to withstand freezing, however as the temperatures gradually fall in the time preceding winter, they are able to increase their freezing tolerance by undergoing a phenomenon known as cold acclimation (Thomashow 1999). This results in cell membrane composition and protein concentration changes to reduce intracellular ice crystal formation and dehydration due to freezing (Thomashow, 1998). Plants that do not undergo this gradual acclimation phase have drastically reduced tolerance to freezing. Temperatures of -5°C kill non-acclimated rye yet after a period of gradual exposure to low nonfreezing temperatures the plant is able to survive freezing down to -30°C (Thomashow 1999). Plants native to warmer regions such as the tropics are much more sensitive to chilling and generally lack the ability to acclimatize. Several of these plants are agriculturally important; such as rice, tomato, soybean, grapes, and maize. To this end, efforts have been made to increase freezing tolerance of these plants by combinations of transgenic and conventional breeding approaches, which will be discussed in more detail later.

Much research in the past decade has been directed towards dissecting the mechanisms by which plants initially sense low temperature to subsequently activate the cold-acclimation response, along with regulation by transcription factors, post-transcriptional modifications, secondary messengers, and cross-talk with other stress responses at stress response "nodes". Research has focused on the identification of freezing-tolerance genes through microarray, high-throughput sequencing, and genetic approaches such as comparative studies of freezing-tolerant cultivars. Much information has been yielded thus far, however the story is far from complete. The following section of this chapter will initially provide an overview of the physiology and mechanisms causing freezing injury to the plant, work through our current understanding of the subsequent response pathway(s) and the players involved, before concluding with examples of genetic engineering for improved freezing tolerance and how the genomics era will continue to yield further insight into this multifaceted field. Our understanding of the cold response pathway and the roles of the genes involved is continually improving, and ultimately this will allow for directed single gene modification at multiple steps of the pathway, allowing for enhanced crop improvement.

There are two types of physiological changes a plant must confront upon the onset of cold temperatures: osmotic stress from low non-freezing temperatures and severe membrane damage from freezing. Chilling stress results in ratio changes between fatty acids and proteins as well as decreased membrane fluidity, due to fatty acid unsaturation in membrane lipids (Wang et al., 2006). Chilling also promotes dehydration due to the impairment of water uptake from the roots and a reduction in stomatal closure. Yet by far the largest cause of cold-associated crop loss is membrane damage as a result of freezing, along with the associated intracellular ice crystal formation leading to further dehydration.

Initially ice crystals form in the cell walls and intracellular spaces, decreasing the water potential outside the cell. The unfrozen water within the cell then travels the down the potential gradient, moving out of the cell and towards the intercellular spaces. This dehydration is what leads to the wilting phenotypes of leaf tissue after exposure to freezing temperatures, or in crops as a result of a "cold snap". Colder temperatures result in greater water loss: at -10°C 90% of the osmotically active water will move out of the cell into intercellular spaces (Thomashow, 1998). Freeze-induced cellular dehydration also results in a barrage of membrane damage: expansion-induced-lysis, lamellar to hexagonal - II phase transitions and fracture jump lesions (Uemura et al., 1995, Steponkus et al., 1993). Expansion-induced lysis occurs at temperatures around -2° to -4°C and is a result of the mechanical damage due to multiple freeze/thaw cycles, where the expansion and contraction of the plasma membrane leads to rupturing (lysing) of the cellular membrane.

Injury from fracture jump lesions is associated with the occurrence of localized deviations of the plasma membrane fracture plane to closely appressed lamellae (Webb et al., 1994). The cold acclimation process has been shown (Uemura et al., 1995, Steponkus et al., 1993) to prevent both expansion induced lysis and the formation of hexagonal II phase lipids in rye and other plants.

Multiple mechanisms are involved in the stabilization of the plant cell membrane. The content and composition of polar lipids and fatty acids in tomato (*Lycopersicon esculentum*) at 6°C suggests maintenance of high levels of chloroplast membrane lipids play an important role in the survival of cold-tolerant plants (Novitskaya et al., 2000). The *Arabidopsis d*SFR2 protein also compensates for changes in organelle volume and stabilizes the chloroplast membranes during freezing (Moellering et al., 2010).

The accumulation of sucrose and related simple sugars correspond with cold acclimation, quite likely contributing in part to the stabilization of plant plasma membranes. Investigation of sucrose metabolizing enzyme activity and sugar content during cold acclimation of perennial ryegrass (*Lolium perenne*) found that the sucrose metabolizing enzymes: phosphate synthase, sucrose synthase and sucrose phosphate synthase are similarly regulated by cold acclimation (Bhowmik et al., 2006). In *Arabidopsis*, sucrose was found to have a regulatory role in the acclimation of whole plants to cold, likely also playing an important role during diurnal dark periods (Rekarte-Cowie et al., 2008).

In addition there is emerging evidence that certain novel hydrophilic and LEA (late embryogenesis abundant) polypeptides also participate in the stabilization of membranes against freeze-induced injury, where disordered plant LEA proteins act as molecular chaperones (Kovacs et al., 2008). The level of expression of the winter barley LEA abscisic acid-regulated gene HVA1 accumulates upon cold acclimation, before disappearing 2 hours post exposure, with greater expression in the lesser of freezing-resistant cultivars (Sutton et al., 1992). Accumulation of chloroplast LEA proteins is correlated with the capacity of different wheat and rye cultivars to develop freezing tolerance (Dong et al., 2002). Transgenic *Arabidopsis* expressing a wheat LEA gene displays significant increases in freezing tolerance in cold-acclimated plants. *Arabidopsis* Cor15am is a late embryogenesis abundant (LEA) related protein shown to exhibit cryoprotective activity *in vitro*, likely by preventing protein aggregation (Nakayama et al., 2008). Global expression profiles of rice genes under abiotic stresses (Rabbani et al., 2003) using microarrays found an upregulation of LEA proteins post-stress. Genome-wide analysis of LEA proteins in *Arabidopsis* identified 51 LEA protein encoding genes in the having ABA and/or low temperature response elements in their promoters and, thus induced by ABA, cold, or drought (Hundertmark et al., 2008). The majority of LEA proteins were predicted to be highly hydrophilic and natively unstructured, but some were predicted to be folded. This comprehensive analysis will be an important starting point for future efforts to elucidate the functional role of these proteins (Hundertmark et al., 2008).

Plant cells initially sense cold stress resulting from the change in the fluidity of the cellular membrane. Cellular membranes are inherently dynamic, and cytoskeleton re-organization is an integral component in low-temperature signal transduction. The cold acclimation process is associated with gene expression requiring a transient influx of Ca+ from the cytosol. Under normal conditions the influx of Ca+ at 4C is nearly 15 times greater than at 25C, but when treated with chemical agents causing an increased Ca+ influx, cold acclimatization-specific genes are expressed at higher temperatures (Monroy & Dhindsa, 1995). When alfalfa (*Medicado sativa*) cells are treated with chemicals blocking this influx (Ovar et al., 2000), they are unable to cold-acclimatize. Furthermore Ovar et al. demonstrated the activation of cold-acclimation genes, Ca+ influx, and freezing tolerance at 4C to all be prevented by membrane stabilization, yet induced at 25C by the addition of an actin microfilament destabilizer, thereby linking the membrane rigidification process to the influx of Ca+ necessary to signal cold acclimation genes. Calcium sensing and sequestering proteins (Komatsu et al., 2007), and phosphoinositides also play roles as signaling molecules in the cold-stress pathway. Phosphoinositides are signaling molecules that regulate cellular events including vesicle targeting and interactions between membrane and cytoskeleton, and accumulate in salt, cold, and osmotically stressed plants (Williams et al., 2005). Mutations in the *Arabidopsis* phosphoinositide phosphatase gene SAC9 lead to overaccumulation of phosphoinositides and confer the characteristics of a constitutive stress response, including dwarfism, closed stomata. The mutations also upregulate stress-induced genes and overaccumulate ROS.

Accumulation of ROS (O_2-, H_2O_2, and HO) as secondary signals has strong impacts on plants ability to withstand cold. Once thought to only be an un-wanted byproduct of aerobic metabolism upregulated under biotic and abiotic stresses, ROS are now known to act as key regulators in numerous biological processes (Miller et al., 2008). An *Arabidopsis* mutant defective in the respiratory electron chain of mitochondria (*frostbite1)* constitutively produces ROS and displays reduced cold induction of stress-responsive genes such as *RD29A, KIN1, COR15A,* and *COR47.* The leaves also have a reduced capacity for cold acclimation, appear water-soaked, and leak electrolytes (Lee et al., 2002).

Hormones are also implicated in the response of plants to environmental stresses. The polyamine putrescine also progressively increases upon cold stress treatment and likely acts as regulator of hormone biosynthesis (Cuevas et al., 2008). Loss of function mutants and reverse complementation tests indicated that putrescine also modulates ABA biosynthesis at the transcriptional level in response to low temperature. Hormonal levels drive cell division and expansion and the plant hormone auxin is a key regulator of virtually every aspect of plant growth and development. Auxin plays a major role in cell expansion and growth, as well as being quite sensitive to temperature changes (Gray et al. 1998). Root growth and gravity response of *Arabidopsis* after cold stress suggests that cold stress affects auxin transport rather than auxin signaling (Shibasaki et al., 2009). Additionally, cold stress differentially affects various protein trafficking pathways, independently of cellular actin organization and membrane fluidity. Taken together, these results suggest that the effect of cold stress on auxin is linked to the inhibition of intracellular movement of auxin efflux carriers (Shibasaki et al 2009).

In 1991 Johnson-Flanagan et al. demonstrated increased freezing tolerance of *Brassica napus* suspension-cultured cells by the addition of the herbicide mefluidide or ABA to the culture medium. In 2000 Llorente et al. showed that ABA is required for full development of freezing tolerance in cold-acclimated *Arabidopsis,* and plays a role in mediating constitutive freezing tolerance. The *Arabidopsis* mutant *frs1* (freezing sensitive 1) is deficient in an allele of the ABA3 locus, displaying reduced constitutive freezing tolerance as well as tolerance post cold acclimation, producing the wilty phenotype corresponding with excessive water loss. Upon receiving an exogenous ABA treatment, *frs1* plants recover both their wild-type phenotype and capability to tolerate freezing temperatures and retain water. Gene expression in the *frs1* mutants was also altered in response to dehydration, suggesting dependence on ABA-regulated proteins allowing plants to cope with freeze-induced cellular dehydration (Llorente et al., 2000). Not all genes induced by low temperature are ABA-dependent, as evidenced by some of transcriptional regulators mentioned in the following section and indicative of the complexity and crosstalk of the regulatory network. Recent genome-wide profiling studies have begun to identify further downstream transcription factors and gene targets resulting from both pathways. In the ABA-dependent pathway, ABA likely activates the bZIP (basic leucine zipper) transcription factors, which regulate ABA dependent COR (COld Regulated) genes through ABA-responsive elements (ABRE) promoters. In the ABA-independent pathway, low temperature triggers the expression of the CBF family of transcription factors, which in turn activate downstream COR genes with other specific motifs in their promoters (Thomashow, 1999). There is also evidence (Knight et al., 2004, Talanova et al., 2008) of ABA initiating CBF expression although at lower levels than those resulting from cold acclimation. Both of these pathways confer or enhance freezing tolerance in plants and are described in more detail below.

The CBF cold response pathway plays a central role in cold acclimation and has been the focus of intense research for the past 2 decades. The *CBF/DREB* (C-repeat-binding factor/dehydration responsive element-binding factor) genes encode a small family of transcriptional activators that play an important role in freezing tolerance and cold acclimation (Thomashow 1999). In *Arabidopsis* there are three members CBF1-3 (also known as DREB1-B, C, and A, respectively), with transcripts beginning to accumulate within 15 minutes after exposure to cold temperatures. A microarray experiment to determine the core set of cold-induced genes in *Arabidopsis* (Vogel et al., 2005) found 302 genes to be upregulated upon cold stress, with 85% of these assigned to the CBF2 regulon and also induced upon CBF2 overexpression. The CBF proteins bind to the CRT/DRE motif (CCGAC) present in the promoters of a number of COR genes named the CBF regulon, which imparts freezing tolerance by activating the COR genes, with CBF induction occurring by ICE1 (Inducer of CBF Expression 1).

ICE1 was identified (Chinnusamy et al., 2003) as an upstream transcription factor regulating transcription of CBF genes in the cold. ICE1 encodes a MYC-like bHLH transcriptional activator that binds the CBF3 promoter. In *Arabidopsis* the *ice1* mutation blocks the expression of CBF3 as well as decreases the expression of genes downstream of CBFs, leading to a significant reduction in plant chilling and freezing tolerance. It is also constitutively expressed at low levels, and its overexpression in wild-type plants enhances the expression of the CBF regulon in the cold and improves freezing tolerance of the transgenic plants. ICE2, another bHLH transcription factor and homologue to ICE1, confers decreased levels of carbohydrate and increased levels of lipids when overexpressed in *Arabidopsis* (Fursova et al., 2008). CBF1 displayed differential expression in transgenic plants compared to wild-type control plants, suggesting a regulatory role provided by ICE2. HOS1 is negative regulator of ICE1, mediating its ubiquitination and subsequent degradation both *in vitro* and *in vivo* (Dong et al., 2006). Overexpression of HOS1 represses expression of the *CBF*s and their downstream genes, conferring increased sensitivity to freezing stress.

SIZ1, a SUMO E3 ligase, is a positive regulator of ICE1 and the sumoylation of ICE1 may activate and/or stabilize the protein, facilitating expression of CBF3/DREB1A and repression of MYB15, leading to low temperature tolerance (Miura et al., 2007). *Arabidopsis* knockouts *siz1-2* and *siz1-3* cause freezing and chilling sensitivities indicating that the SIZ1 is a controller of low temperature adaptation in plants. Interestingly a protein associated with stomatal differentiation, SCREAM, was shown to in fact be ICE1 (Kanaoka et al., 2008). This creates a potential link between cold acclimation and stomatal differentiation and a basis for future research.

All three CBF genes do not play the same roles in freezing tolerance. The function of CBF2 was not only demonstrated as having a distinct function from CBF1 and CBF3, but was shown to be a negative regulator of their activity. Reverse genetic approaches using an *Arabidopsis* knockout mutant for CBF2 displayed an increased capacity to tolerate freezing both before and after cold acclimation, and the plants displayed increased tolerance to dehydration and salt stresses (Novillo et al., 2004). The mutants also had stronger and more sustained expression of CBF/DREB1-regulated genes, resulting from increased expression of CBF1 and CBF3 in the *cbf2* plants, with the authors suggesting CBF1/CBF3 induction to precede CBF2. Indeed, DNA motifs for the calmodulin binding transcription activator (CAMTA) family of transcription factors have been identified in the promoters of *CBF2,* as well as the transcription factor *ZAT12,* conferring both negative and positive regulation. One of these binding sites (CAMTA), was shown to be a positive regulator of *CBF2* expression, with mutant plants impaired in freezing tolerance. CAMTA proteins may play a role in cold acclimation by linking Ca+ and calmodulin signaling with expression of COR genes (Doherty et al., 2009). Both ICE1 and CAMTA binding sites are found in the promoter of CBF2, potentially directly linking Ca+ signaling to cold response.

Low temperature induction of the *Arabidopsis* CBFs is also gated by the circadian clock (Fowler et al., 2005) with the highest and lowest levels of cold-induced CBF1-3 transcript occurring at 4 and 16 h after subjective dawn, respectively. Other transcription factors induced by cold in parallel with CBF1-3 are also gated by the circadian clock; however cycle in the opposite phase. This suggests nonidentical, though potentially overlapping, signaling pathways. Similar results in wheat (Badawi et al., 2007) and tomato (Pennycooke et al., 2008) suggest circadian regulation under homeostatic conditions concurring with dawn and dusk periods, maybe overlapping with stomatal aperture changes.

Light is also implicated in regulating the CBF pathway, for instance a low red to far-red ratio of light is sufficient to increase CBF gene expression and confer freezing tolerance at temperatures higher than those required for cold acclimation (Franklin et al., 2007), providing evidence for a second temperature-regulated step in this pathway. Phytochrome-Interacting Factor7 (PIF7) functions as a transcriptional repressor for DREB1C (CBF2) expression and its activity is regulated by components of the red light photoreceptor, and circadian oscillator (Kidokoro et al., 2009). DREB1/CBF expression may be important for avoiding plant growth retardation by the accumulation of DREB1/CBF proteins under unstressed conditions (Kidokoro et al., 2009). Downregulation occurs through a complex network of transcription factors, such as ZAT12 downregulating CBF2 (Vogel et al., 2005), and MYB15 interacting with ICE1, subsequently binding to the MYB recognition sequences in the CBF promoters (Agarwal et al., 2006)

The CBF pathway is not only present in dicots such as *Arabidopsis*, but is widespread through monocots and multiple plant genera, including those native to warm regions and not inherently cold tolerant, with variation in CBF gene copy numbers (Qin et al., 2004, Skinner et al., 2006, Badawi et al., 2007, Stockinger et al., 2007, Tamura et al., 2007, , Pennycooke et al., 2008, Knox et al., 2010). A recent paper by one of the pioneers of the field presents a detailed overview of the status of CBF research today (Thomashow, 2010)., and readers wishing for further detail are directed to this review

Alternative cold tolerance pathways also initiate transcription of cold-responsive genes, for example *Arabidopsis* SFR2 encodes a novel β-glycosidase, contributing to freezing tolerance and distinct from the CBF pathway (Thorlby et al., 2004). The null mutant (*sfr2-1*) causes freezing sensitivity in *Arabidopsis* possibly due to electrolyte leakage. Homologous genes are present and expressed in many terrestrial plants, including those unable to tolerate freezing. Each of these homologues however, has the ability to complement the freezing sensitivity of the *Arabidopsis sfr2* mutant (Fourrier et al., 2008). In *Arabidopsis* the SFR2 protein is localized to the chloroplast outer envelope membrane, with the chloroplasts of the *sfr2* mutant displaying rapid damage post-freezing.

MYBS3 is a single DNA-binding repeat MYB transcription factor previously shown to mediate sugar signaling in rice and also indicated to play a novel role in cold adaptation (Su et al., 2010). Transgenic rice constitutively overexpressing MYBS3 displayed no yield penalty in normal field conditions while tolerating temperatures of 4°C for at least 1 week. Su et al. demonstrated repression of CBF-dependent signaling by MYBS3 at the transcriptional level, with distinct pathways likely acting in parallel for short- and long-term cold stress in rice. This previously undiscovered cold adaptation pathway adds another layer to the complex web of plant responses to cold stress.

RNA processing and nuclear export / stabilization are critical mechanisms in a plants response to cold stress. Recent research has shown cold shock proteins (CSPSs) play roles in promoting cold tolerance, however much remains to be discovered in order to determine the mechanism in plants to promote cold tolerance. A protein from wheat with homology to an *E. coli* cold shock protein has been linked to the regulation of translation under low temperature; potentially by acting as a RNA chaperone to destabilize secondary structure (Nakaminami et al., 2006). Cold shock domain proteins and glycine-rich RNA-binding proteins from *Arabidopsis* have been shown to promote cold adaptation process *E. coli* (Kim et al., 2006). Excitingly, 2 novel cold shock domain proteins were cloned and characterized from rice, a plant unable to cold acclimatize. Nonetheless, *in vivo* functional analysis confirmed these OsCSPs complement a cold-sensitive bacterial strain that lacks four endogenous cold shock proteins (Chaikam et al., 2008). Transcripts were also shown to be upregulated during temperature decreases. Two structurally differed CSPs of *Arabidopsis* perform different functions in seed germination and growth under stress conditions, even rescuing cold tolerance from an RNA-binding protein null mutant (Park et al., 2009). Other RNA-binding proteins such as known splicing factors alter expression under cold stress, for example the serine-arginine (SR) rich splicing factor SRP34 (At1g02840), as recently reviewed (Filichkin et al., 2010), displays exon skipping under both drought and cold conditions, with several novel introns predicted through alternative splicing. Alternative splicing of another SR protein, SR1, was reported (Iida et al., 2004) under cold stress, as well as in response to hormones (Palusa *et al.*, 2007). Different isoforms of a splicing factor likely alter the binding preference and spicing of a host of downstream targets, presenting an exciting area for future research. The *Arabidopsis* STABILIZED1 gene encodes a U5 snRNP-associated splicing factor required for both pre-mRNA splicing and transcript turnover. Of interest, it is also upregulated by cold stress, and the *sta1-1* mutant plants are defective in the splicing of COR15A (Lee et al., 2006).

Thanks to the genomics revolution, the role of microRNAs (miRNAs) in abiotic stress regulation is being elucidated. Endogenous miRNA levels change as plants are exposed to abiotic stresses (Sunkar et al., 2004), and readers are pointed towards reviews (Jones-Rhoades et al., 2006, Sunkar et al., 2007) providing the backstory of miRNA research in plants with regards to stresses and classes of miRNA, A tiling array (Matsiu et al., 2008) global transcriptome analysis of *Arabidopsis* discovered 7,719 non-AGI (*Arabidopsis* Genome Initiative) transcriptional units (TUs) in the unannotated "intergenic" regions of *Arabidopsis* genome, and most of these are hypothetical non-protein-coding RNAs. Close to 80% of the previously un-annotated TUs belonged to pairs of the fully overlapping sense-antisense transcripts, suggesting stress or ABA induction of antisense TUs in the fully overlapping sense. These non-coding small RNAs exhibit stress-responsive expression patterns; however are also implicated in a very broad web of *in planta* regulation. For example, in wheat (*Triticum aestivum*) the *trans* targets of miRNAs include both transcription factors implicated in development and a plethora of genes involved multiple physiological processes (Yao et al., 2010). Further research will be necessary to pinpoint small RNA targets and the effects of their regulation with regards to cold stress, and other stresses.

Transcriptome profiling using microarrays is allowing for the identification of groups and networks of genes that respond to cold stress. A consensus among microarray studies (Fowler and Thomashow 2002, Rabbini et al., 2003, Lee et al. 2006, Oono et al. 2006) is that genes induced by abiotic stress fall into 2 categories: functional proteins such as the aforementioned LEA proteins, proteins playing roles in osmoprotection, and transporters. The second category is the regulatory proteins; the transcription factors, kinases, phosphatases, and other molecules dealing with signaling, directly or indirectly, such as those of the MAP kinase cascade. Such broad profiling allows for glimpses into regulatory networks and offers the potential for further research into specific up or down-regulated genes or gene families.

Initial work into engineering for cold tolerance focused on the CBF transcription factors. It was initially shown that overexpression of AtCBF3 in *Arabidopsis* confers freezing and drought tolerance, however also causes a dwarf phenotype (Liu et al., 1998). In comparison when AtCBF3 is overexpressed in rice, which is unable to cold acclimate, increased tolerance to drought and high salinity stress (and not low-temperature) without stunting growth results (Oh et al., 2005). Readers are directed towards a recent review (Thomashow, 2010) for further insights gained from the CBF pathway in genetic engineering.

Rice is a staple food in much of the world and the seedlings are particularly sensitive to chilling in high-elevation areas. Much research has been conducted into enhancing the cold tolerance of rice to allow for growth in different geographic regions to increase production. Recently (Hu et al., 2008) isolated SNAC2, a nuclear stress-responsive NAC gene from upland rice (*Oryza sativa* L. ssp japonica) characterized for its role in stress tolerance. Transgenic plants overexpressing SNAC2 had higher cell membrane stability than wild type during cold stress with over half of the transgenic plants, and none of the WT plants, surviving after 5 days at 4C (Hu et al., 2008). Another transcription factor, *OsMYB3R-2*, functions in both stress and developmental processes in rice, with transgenic rice plants overexpressing *OsMYB3R-2* shown to exhibit enhanced cold tolerance by regulating the progress of the cell cycle during chilling stress (Ma et al., 2009), suggesting cell cycle regulation as possible resistance mechanism to stress.

Molecules involved in the Ca+ signaling pathway, acting upstream of transcription factors, can also enhance cold tolerance. Over-expression of a calcium-dependent protein kinase and a calreticulin interacting protein has been shown to enhance cold tolerance in rice plants, emphasizing the importance of signaling components in the response to cold stress in rice (Komatsu et al., 2007). As mentioned in the salt section, overexpression of a single Ca+ dependent protein kinase in rice increases cold tolerance (Saijo et al., 2000), as well as salt tolerance, indicative of the crosstalk between abiotic stress response pathways.

Research into ethylene response factor (ERF) proteins is demonstrating their roles in plant stress responses via interaction with DRE/CBF genes, yet the regulatory mechanism is not well elucidated. Overexpressing TERF2/LeERF2 in tobacco not only activates expression of cold-related genes, but reduces electrolyte leakage (Zhang et al., 2010). The authors demonstrated RNAi knockdown TERF2/LeERF2 transgenic lines to have reduced freezing tolerance, rescued to normal levels with treatment of a precursor to ethylene. Overexpression of OsTERF2 in rice enhanced cold tolerance without affecting growth or agronomic traits (Tian et al., 2010). The transgenic lines displayed increased accumulation of osmotic substances and chlorophyll, as well as reduced ROS and decreased electrolyte leakage. The overexpression of OsTERF2 was shown to initiate expression of downstream cold regulated genes such as OsMyb, OsICE1, and OsCDPK7.

Our understanding of the cold response pathway has thus far allowed for single trait genetic engineering to improve or alter cold tolerance, often with very promising results however not without the caveats present in most abiotic stress research, such the differences between monocots and dicots. The traditional model system is the dicot *Arabidopsis,* and as evidenced by the OEX CBF growth phenotype differences between *Arabidopsis* and the monocot rice, homologous genes don't always confer conserved responses. The recent development of molecular, genetic, and genomic resources for the grasses *Brachypodium distachyon, Setaria italica, and Setaria viridis* provide model platforms for future studies of cold, and other abiotic stress research in general, in monocot systems.

4. Severe desiccation and water deficit: heat and drought stress

Plants face additive and interacting responses to drought and heat stress such as water loss through the evapotranspiration resulting from the opening of stomata for heat dissipation, and detrimental alterations to photosynthesis. There are subsets of genes that are induced by a combination of heat and drought stress that are not induced by each stress independently, as in a laboratory growth chamber. Under field growing conditions resulting in limited water supply, crop plants would be exposed to both stresses simultaneously. Therefore heat and drought stresses will be profiled in the same overarching section, and will have an additional portion at the end focusing on the overlap and further insights into the crosstalk present between these networks.

4.1 Heat stress

A transient elevation in temperature, 10-15C above ambient, is typically defined as heat stress (Wahid et al., 2007); however the effects vary with the duration and amount of temperature increase. Plants differ in their abilities to cope with rising temperatures; corn and rice are more thermotolerant than wheat, for example. As with all stresses, the onset of heat immediately changes the cellular state, alters membrane fluidity and lipid composition, and initiates the signaling cascades that ultimately lead to transcript accumulation for genes encoding protective and chaperone activities. The following section will profile the cellular changes that occur post-heat stress, with emphasis on how genetic engineering is utilizing these response mechanisms to both elucidate the stress response network and improve heat tolerance in agriculturally important crop species. For a detailed overview of plant heat tolerance, readers are directed to the aforementioned review by Wahid et al. (2007).

Gradual nonlethal heat treatment confers a phenomenon known as thermotolerance; an increase in heat resistance over non-acclimated plants similar in principle to the cold acclimation detailed in the previous section, however mechanistically different and fully elucidated. What is understood however, is genetic manipulations of some aspects of the response pathway are able to confer single trait heat tolerance to the resulting transgenic plants. Such advances and insights thus far are profiled in the following section.

Engineering for tolerance encompasses many facets of the cells natural defense mechanisms to stress, such as reducing the damaging effects of oxidative stress and subsequent buildup of ROS during heat (and other abiotic stresses). Transgenic potato plants were generated containing both the superoxide dismutase (SOD) and ascorbate peroxidase (APX) genes encoding two key chloroplast enzymes for ROS detoxification under the control of the chloroplast *SWPA2* oxidative stress-inducible promoter (Tang et al., 2006). Under high temperature treatment, the transgenic plants displayed a photosynthetic activity decrease of only 6%, whereas wild-type plants displayed a 29% decrease. These results suggest that manipulation of the antioxidative mechanism is likely a valuable tool for the creation of heat tolerant crop plants.

The osmolyte glycinebetaine (mentioned earlier in the section on salt) has been implicated in heat tolerance, although the exact mechanism though which tolerance is gained remains unknown. Tobacco (*Nicotiana tabacum)* lines transgenically accumulating glycinebetaine display higher thermotolerance than WT plants, especially when heat stress occurs under light, suggesting that the accumulation of glycinebetaine leads to increased tolerance to heat-enhanced photoinhibition. This tolerance is likely achieved by accelerating repair of photosystem II (PSII), possibly due to the reduced accumulation of ROS in the transgenic plants with elevated levels of glycinebetaine (Yang et al., 2007). Isoprene is a volatile compound emitted from leaves of many plant species, and has also has been implemented in heat tolerance. Recently the *Populus alba* isoprene synthase gene was introduced into *Arabidopsis* and shown to confer elevated heat tolerance in the transgenic lines over wild type (Sasaki et al., 2007).

As with chilling stress, it is becoming evident that heat stress promotes fatty acid unsaturation in membrane lipids, altering the ratio between membrane fatty acids and proteins and resulting in membrane fluidity changes. Protein transfer across membranes is mediated by protein machinery embedded in the membrane, with different lipid classes within a membrane is known to influence the efficiency of some protein translocation processes (Ma et al., 2006). To this end, membrane associated proteins involved in lipid metabolism have been successfully utilized to increase thermotolerance in both model and crop plants. Fatty acid omega-3 desaturase (FAD) is the key enzyme catalyzing the formation of trienoic fatty acids, the most common fatty acids in membrane lipids, comprising 70% of the membrane lipids in the chloroplast and implemented with defense response (Yaeno et al., 2004).

By investigating transgenic tobacco plants with reduced trienoic fatty acid content (Murakami et al., 2000) it was revealed that decreased contents of trienoic fatty acids play an important role in high-temperature tolerance. Transgenic rice plants in which the content of dienoic fatty acids was increased were more tolerant to high temperatures than WT, having increases in both chlorophyll content and growth. The maximum photochemical efficiency of PSII was also higher in transgenic plants plants upon high temperature stress (Sohn et al., 2007). Recently, antisense expression of tomato chloroplast omega-3 fatty acid desaturase gene (*LeFAD7*) was demonstrated to enhance high-temperature tolerance, again through reductions of trienoic fatty acids and increases of dienoic fatty acids (Liu X. et al., 2010).

Photosynthesis, the light driven carbon dioxide assimilation process and the primary means of energy production in plants, is extremely sensitive to elevated temperatures.

Heat stress inhibits photosynthesis in part by reducing the activation of Rubisco, due initially to the denaturation of Rubisco activase (Salvucci et al., 2001). Loss of activase activity during heat stress is caused by exceptional sensitivity of the protein to thermal denaturation and is responsible in part for deactivation of Rubisco itself. The effects of heat stress on *Arabidopsis* plants in which Rubisco activase or chloroplast thylakoid membrane fluidity had been altered demonstrated that a) plants having less polyunsaturation of thylakoid lipids display lower net photosynthetic rates than the WT and b) the rate of Rubisco deactivation affects the temperature dependence of photosynthesis (Kim et al., 2005). To test the hypothesis that a non-degraded Rubisco activase can improve photosynthesis under elevated temperatures, several thermostable *Arabidopsis* isoforms of Rubisco activase were introduced into a Rubisco activase null mutant line. The transgenics displayed higher photosynthetic rates, with increased biomass and increased seed yields, compared to wild-type activase, providing evidence for Rubisco activase as a limiting factor in photosynthesis elevated temperatures (Kurek et al., 2007). Rubisco activase is a potential target for future genetic manipulation in improving crop plants productivity under heat stress (Kurek et al., 2007). In addition, down-regulation of photosynthesis in temperature stressed plants is caused by reduced post-translational import into the chloroplast of plastidic proteins required for the replacement of impaired proteins coded by the nuclear genome (Dutta et al., 2010).

Heat stress also inhibits synthesis and promotes degradation of cytokinins, important hormones for regulation of growth and development processes, such as cell division, leaf senescence, and root growth (Xu et al., 2010), however the underlying mechanisms are poorly understood. Xu et al. used transgenic *Agrostis stolonifera,* a C3 perennial grass species, to survey protein changes in response to elevated temperatures. The gene controlling cytokinin synthesis was used to create 2 transgenic lines, each with different inducible promoters, and a null mutant line. Protein content changes in leaf and root tissue were found to primarily regulate energy metabolism, protein destination and storage. In the transgenic lines, 6 leaf proteins and 9 root proteins were found to be elevated or remain at steady state comparable WT levels, and among these was the small subunit of Rubisco, Hsp90, and glycolate oxidase, a suggesting a definite regulatory role for cytokinins in metabolic pathway regulation associated with heat tolerance in C3 perennial grass species (Xu et al., 2010).

Much research has been conducted on heat shock proteins (HSPs), the molecular chaperones regulating proper protein folding, localization, degradation, and stabilization of under homeostatic and stress conditions (Feder et al., 1999). There are several families of HSPs present in both plants and animals, named based on their respective molecular weights. There are 5 classes of HSPs in plants (for comprehensive reviews see Baniwal et al., 2004, Wang et al., 2004, Kotak et al., 2007); the Hsp70 class which prevents protein aggregation and assists with transcriptional activation and import, the Hsp60 chaperonin class which assists with folding and re-folding, the Hsp90 class which plays a role in assisting other signaling molecules, the Hsp100 class preventing unfolding, and finally the sHSP class which act to stabilize non-native proteins (**Figure 2**). Much of our current knowledge regarding HSPs contribution towards plant survival under heat stress is based off homology with other eukaryotes and extrapolation based molecular chaperoning activity and *in vitro,* with little specific *in vivo* information (Kotak et al., 2007).

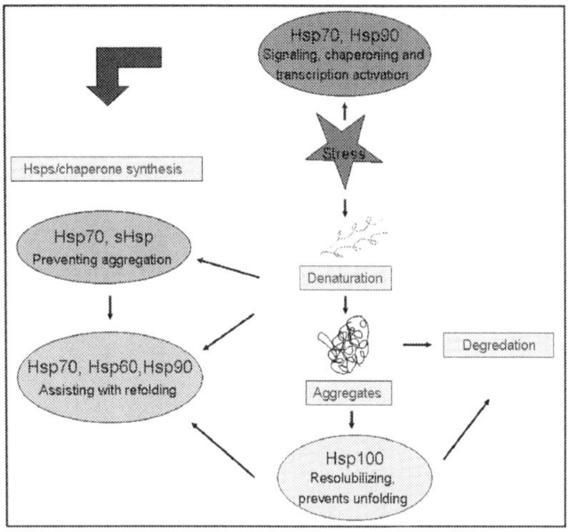

Fig. 2. Heat shock protein (Hsp) network during abiotic stress response, preventing protein degradation and assisting with maintaining cellular homeostasis. (Figure adapted from Wang et al., 2004)

In the absence of heat shock, cytosolic HSP90 appears to negatively regulate heat-inducible genes by actively suppressing Hsp function, however in is transiently inactivated following heat shock, leading to Hsf activation (Yamada et al., 2007). AtHsp101, when constitutively expressed in rice, enhances thermotolerance (Katiyar-Agarwal et al., 2003).

It is the heat stress transcription factor (Hsf) family of more than 20 members, which are the central regulation proteins of heat stress response and defense (Baniwal et al., 2004). These modular classes (A, B, and C) share motifs for DNA binding and transcriptional activation, and are defined by differences in the hydrophobic amino acid residues required for oligomerization. The B and C class Hsfs are believed to function in conjunction with class A Hsfs to amplify or regulate signals, rather than function on their own. The majority of our knowledge into the mechanisms into plant Hsfs has come from studies with 2 dicots: Arabidopsis and tomato (Figure 3).

Tomato has 17 members of the Hsf family, however despite this complexity; HsfA1 has a unique function as being the "master regulator" for induced thermotolerance and cannot be replaced with any of the other Hsf members (Mishra et al., 2002). This is not the case in Arabidopsis however, where sequencing of Arabidopsis genome revealed unique complexity of the Hsf family. Hsfs comprising 21 members were assigned to 3 classes and 14 groups based structural and phylogenetic comparison to homologues in other eukaryotes and plants (Nover et al 2001). No master regulator has been yet identified in Arabidopsis, where even double knockouts only affect a small subset of genes. While offering a beginning for homology comparisons, there is not complete overlap between the Hsfs in tomato and Arabidopsis however, for example while HsfA1a and HsfA1b are highly conserved between species. In Arabidopsis (unlike tomato) they have the capacity to functionally replace each other. Recent work implicates HsfA1a/1b in cooperation at a number of target gene promoters also regulated by HsfA2, possibly indicating a recruitment of HsfA2 and replacement of HsfA1a/A1b at the same target gene promoters (Li et al., 2010).

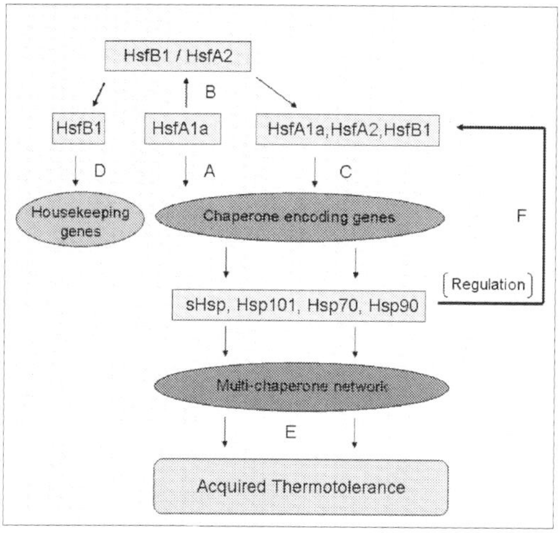

Fig. 3. Current model of the heat stress response pathway in tomato: the interaction of Hsfs and Hsps. A) HsfA1a is the master regulator responsible for induction of heat stress includes gene expression and encoding chaperones (Hsps), along with HsfA2 and HsfB1, next B) HsfA2 forms a hetero-oligomer with HsfA1a to activate Hsp gene expression while C) HsfB1 functions as coregulator of HsfA1a it an enhancesome-like complex by recruitment of the histone acetyl transferase HAC1. Finally, D) heat stress transcription factor (Hsf)B1, in conjunction with other transcription factors, controls expression of housekeeping genes to restore homeostasis. E) The large network of Hsps acting as molecular chaperones are an essential aspect of acquired thermotolerance, while F) the Hsp complexes themselves then modulate the heat shock response by interacting with the Hsfs. Figure adapted from a 2007 review by von Koskull-Doring et al. and references therein.

To date only a few of the Hsfs have been studied in depth, and the following brief summary of the Hsf network comes from an excellent review by von Koskull-Doring et al., from 2007, which readers are encouraged to read for a detailed summary of Hsf structure and function. HsfA2 is the dominant Hsf present in thermotolerant cells in both *Arabidopsis* and tomato, and may initiate transcription of a core subset of heat stress induced genes. Recent work has also implicated AtHsfA2 in anoxia tolerance in *Arabidopsis* (Banti et al., 2010), further demonstrating the overlapping and redundancy of this complex network. AtHsfA2 also plays an important role in linking heat shock with oxidative stress signals (Li et al., 2005). A recent study (Cohen-Peer et al., 2010) has demonstrated AtSUMO1 of AtHsfA2 to be involved with the plant's regulatory response to heat stress and acquired thermotolerance. Post-translational modification of target proteins by SUMO proteins (see cold section for background) regulates many cellular processes, and adds a further layer to this complex network In a recent study to identify potential regulatory components involved in thermotolerance, a reverse genetics approach was used by screening *Arabidopsis* T-DNA insertion mutants for lines displaying phenotypic decreased thermotolerance. The Hsf AtHsfA2 fell out as the only mutant line more sensitive to severe heat stress than WT following long recovery periods, and able to be complemented by the introduction of WT AtHsfA2. This depicts HsfA2 as a heat-inducible transactivator, sustaining expression of

Hsp genes and extending the duration of acquired thermotolerance in *Arabidopsis* (Charng et al., 2007), as well as being an attractive candidate for continued research in the orthologous genes of crop plants under field conditions.

LeHsfB1 interacts with HsfA1a in a synergistic fashion to form an "enhanceosome" complex to possibly regulate the expression of housekeeping genes during periods of heat stress. Both tomato and *Arabidopsis* HsfA5 acts as an inhibitor of the activator HsfA4, by initiating the formation of hetero-oligomer complexes. HsfA9 plays a role in seed development and maturation, likely working in conjunction with other networks during heat stress, and is shown to induce expression of small heat shock proteins (sHsps) and Hsp101 in *Arabidopsis* leaves under non-stressed conditions (Koskukll-Doring et al., 2007, and all references contained therein). HsfA3 is implicated is a crosstalk network with drought stress, with transcription in fact being induced DREB2A, in a cascade resulting in the transcription of genes encoding protective Hsps (Schramm et al., 2008, Yoshida et al., 2008, Chen et al., 2010).

The Hsp/Hsf network in plants response to heat stress is quite complex, and still being fully elucidated. Heat responses in monocots may increase the complexity of the network yet again. In contrast to tomato and *Arabidopsis* containing only one *HsfA2*, rice has five *HsfA2* genes (von Koskull-Döring et al., 2007). Expression profiles of 12 class A OsHsfAs suggest different regulatory networks between heat and non-heat stress (Liu et al., 2009). A population of *Arabidopsis* was transformed with a full-length rice cDNA library in order to isolate the rice genes responsible for high-temperature stress tolerance (Yokotani et al., 2008),. A thermotolerant line encoding the rice the heat stress transcription factor OsHsfA2 fell out of the analysis as highly expressing several classes of heat-shock proteins (Yokotani et al., 2008), and also displaying tolerance to high-salinity stress. A genome-wide analysis of rice, including *Oryza sativa* L. ssp japonica and *Oryza sativa* L. ssp indica (Wang et al., 2009), identified 25 rice *Hsf* genes. Promoter analysis identified a number of stress-related *cis*-elements in the promoter regions, however no correlation was found between heat-shock gene responses and their *cis*-elements. This study sets the foundation for future research into *OsHsf* function and tolerance. A recent study in the dicot grape (*Vitis vinifera)* identified four genes strongly upregulated by heat stress, whose overexpression resulted in the acquisition of thermotolerance in *Arabidopsis* (Kobayashi et al., 2010). Further *in vivo* studies in grape are underway to elucidate chaperone mechanisms, localization, and functions under stress conditions (Wang et al., 2009).

The complete mechanisms of Hsp mediated thermotolerance remains to be fully elucidated in plants; however work with transgenics has shown that altered levels of Hsps and Hsfs have dramatic effects on plants resistance to elevated temperatures. This offers a promising outlook for future research, utilizing non-model organisms and trials under realistic field conditions. Indeed, based on transcriptional response profiling, *Arabidopsis* Hsf and Hsp expression has been shown to be strongly induced by heat, cold, salt, (stresses sharing osmotic components), and upon wounding, suggesting an interaction point between multiple stress response pathways, warranting functional analysis under conditions apart from heat shock treatments, presenting another area for future research (Swindell et al., 2007).

4.2 Drought

A water deficit, along with freezing and increased Na+ concentration, all disturb the water content of the cell, thus altering membrane fluidity, protein stability, and water potential gradients. This osmotic stress leads to wilting associated with loss of turgor pressure, and ultimately complete desiccation. Cellular sensors initially perceive and respond to the drought induced signaling, triggering gene expression changes to synthesize additional signals such as ABA. Further signaling cascades are then initiated to signal new gene expression patterns that are proposed to play a role in cellular adaptation to water-deficit stress (Bray 2002). Drought stress shares many of the same response pathways as the other osmotic stress pathways profiled earlier on the sections on salt and cold.

As with these other osmotic stressors, there are two classes of proteins synthesized as a result of cellular perception of drought stress. First the regulatory proteins; kinases, components of signaling transduction and amplification pathways such as ABA, and the transcription factors activating genes encoding protective proteins. The second class of proteins in the functional proteins, those serving protecting and chaperoning roles such as LEA proteins, osmoprotectants such as proline, proteins regulating water channels for turgor pressure, and proteases. It is these classes of proteins that have been the focus of genetic engineering for drought tolerance, as well as the focus of the following section.

A plethora of recent reviews exist on the physiological responses of the plant to drought stress (Yordanov et al., 2000, Wang et al., 2003, Bartels et al., 2005, Umezawa et al., 2006, Barnabas et al., 2008), and methods to engineer tolerance to water deficit. Readers interested in detailed background on drought stress are encouraged to read these reviews; this section offers only a small summary of some of the most recent discoveries and genetic engineering.

Dehydrins are members of the LEA family of proteins and as mentioned previously aid in stabilizing proteins and other molecules during stresses, likely by replacing water to maintain homeostasis. Other hypotheses for roles played by LEA proteins are: compensating for the increasing ionic concentration by binding ions in dehydrated cells, and interaction with carbohydrates to prevent cellular damage during dehydration (Bartels 2005). Dehydrins are also present in fungi as well as plants; in the white truffle (*Tuber borchii*), a novel dehydrin-like coding gene displays increases transcript abundance during cellular dehydration (Abba et al., 2006). A promoter region from a dehydrin in coffee (*Coffea canephora)* has been cloned and implicated in osmotic stress-specific gene expression (Hinniger et al., 2005), and will be useful for studying control of gene expression during osmotic stress in coffee, an important crop.

Xerophytes are plants that are adapted to life in a low water environment, typically by employing altered root function. Watermelon (*Citrullus lanatus*) is one of these plants, and a recent study (Yoshimura et al., 2008) provides insights into the molecular mechanisms behind their adapted root system. In the early stages of drought stress watermelon displays enhanced root development, a drought avoidance mechanism for absorbing water deeper beneath the surface layer of soil. Proteome analysis revealed proteins induced in the early stage of drought stress to be involved in root morphogenesis and carbon/nitrogen metabolism, likely promoting rapid root development and growth. In later stages of drought stress however, the protein ratios shifted to lignin synthesis-related proteins and molecular chaperones, enhancing desiccation tolerance and protein stability. Developed root systems are not the only method plants use to survive in arid environments. Succulent xenophytes also show a greater abundance of free proline, up to 16 times greater than plants native to non-arid environments, as well as larger accumulations of Na^+ rather than K^+ for osmotic adjustment (Wang et al., 2004), perhaps acting as an effective strategy for their adaptation to arid environments. The tonoplast Na^+/H^+ antiporter (NHX) is involved in the compartmentalization of cytosolic Na^+ into vacuoles. *Zygophyllum xanthoxylum* is a succulent xerophyte with a recently characterized ZxNHX antiporter demonstrated to be most active in the leaves. The transcript abundance of *ZxNHX* under salt stress up to 8.4 times greater than unstressed plants and up to 4.4 times greater under drought conditions than unstressed controls (Wu et al., 2011), and may prove useful for future studies with crop species to predispose tolerance to both Na+ and drought.

Desiccation tolerance is an adaptation to extreme environmental conditions, perhaps leading to abundant expression of hydrophilic proteins as a survival mechanism, such as the "resurrection plant", *Craterostigma plantagineu* (Bartels 2005). *C. plantagineum* has limited genomic information available, yet it is becoming evident (and not surprisingly) that desiccation tolerance is a complex trait, and probably linked to the recognition of *cis*-regulatory sequences (Bartels 2005). However, it has been possible, based on homology inferences, to define four broad categories of *C. plantagineum* genes induced by dehydration: protective proteins including such as hydrophilins, regulatory proteins and RNA, carbohydrate metabolism enzymes, and proteins involved in water transport (Bartels 2005).

Other insights have come from naturally occurring desiccation tolerant plants, recently a 31-kDa putative dehydrin polypeptide was discovered in the desiccation-tolerant fern *Polypodium polypodioide,* found to be localized at the cell walls and present only during drying (Layton et al. 2010). The protein rapidly dissipated upon tissue rehydration, along with changing the hydrophilicity of leaf surfaces and enabling reversible cell wall deformation. This suggests this protein potentially plays a role in avoiding mechanical failure during drought, and another angle to pursue with genetic engineering.

Transcriptome analysis of sub-lethal drought stress conditions in *Arabidopsis* identified three distinct stages of plant responses: initially an early "priming and preconditioning" stage, with early accumulation of ABA and associated signaling genes, which with a decrease in stomatal conductance, an intermediate stage preparatory for acclimation, and a late stage of new homeostasis with reduced growth. This is accompanied by a peak in expression of genes involved in cell wall expansion, likely as a preparatory step toward drought acclimation by the adjustment of the cell wall (Harb et al., 2010).

A recent microarray investigating genes responsible for drought tolerance between genotypes of barley; two drought-insensitive and one drought-sensitive, identified 17 genes that may play a role in enhancing tolerance (Guo et al., 2009). These genes are likely constitutively expressed in the two drought-insensitive genotypes, with their encoded proteins playing a role in their tolerance. These genes include those controlling stomatal closure via carbon metabolism (NADP malic enzyme, NADP-ME), those synthesizing the osmoprotectant glycine-betaine, those generating protectants against ROS scavenging, and those stabilizing membranes and proteins. Also found were genes enhancing Ca+ signaling and molecular chaperoning. These findings allow a basis for selecting single genes conferring drought tolerance on cereals by transgenic means, and engineering for drought avoidance has taken advantage of the mechanisms for stomatal closure. For example, expression of the NADP-ME gene from maize has resulted in altered stomatal behavior and water relations when introduced into tobacco. The majority of water lost from plants occurs through stomata. When stomata are open, ions accumulate in order to increase the turgor pressure of the guard cells, which results in increased pore size (Laporte et al., 2002). Guard cells are present in pairs on the underside of leaves and surround the stomatal pores. These control both the CO_2 influx required for photosynthesis and the loss of water to the atmosphere due to transpiration. Drought induced ABA synthesis signals stomatal closing, thus reducing stomatal aperture and ultimately reducing water loss (Schroeder et al., 2001). The transgenic NADP-ME tobacco displayed reduced stomatal conductance, yet importantly remained similar to the wild type in their growth and rate of development (Laporte et al., 2002).

MAPK cascades are implicated various signaling pathways involved in plant development and stress responses. A tobacco MAPKKK (NPK1) was constitutively expressed in maize, improving drought tolerance by maintaining higher photosynthesis rates than WT without affecting yield, producing a kernel yield comparable to well-watered WT plants (Shou et al., 2004). A novel MAPKKK gene, DSM1, has recently been characterized in rice, which functions as an early signaling component in regulating responses to drought stress by regulating scavenging of ROS (Ning et al., 2010). Overexpression of DSM1 in rice increases resistance to dehydration stress at the seedling stage.

In a study exposing nodulated alfalfa plants to drought conditions, it was shown much of the loss of alfalfa performance was due to reduced photosynthesis (in the leaves) and nitrogenase activity (in the nodules). Proteomic profiling showed a marked increase in proline levels, likely as a result of the intracellular increase in ROS (Aranjuelo et al., 2010). High levels of the osmoprotectant proline also resulted from the introduction of AtDREB1A/CBF3, driven by the inducible rd29A promoter, into tall fescue (*Festuca arundinacea*) (Zhao et al., 2007), which displayed increased resistance to drought.

The NAC family of plant-specific transcription factors plays roles in plant organ development, division, and resistance to pathogen attack (Hu et al., 2006) and all references therin). Some members of the NAC family are stress responsive, for example SNAC1 (STRESS-RESPONSIVE NAC 1) in induced in guard cells by drought stress, and when overexpressed in rice confers enhanced drought and salt resistance, without phenotypic changes or yield penalty under field conditions (Hu et al., 2006). Another novel rice NAC gene, ONAC045, is induced by drought, salt, cold, and ABA treatment (Zheng et al., 2009). ONAC045 was shown to function as a transcriptional activator, and rice plants overexpressing ONAC045 displayed enhanced tolerance to both drought and salt. Of the 140 *OsNAC* genes predicted in rice, 18 have been identified as being induced by stress conditions (Jeong et al., 2010). Of these, a recent functional genomics approach identified a rice NAC-domain gene, *OsNAC10,* which when under the control root-specific promoter (rather that the constitutive promoter *GOS2*) *RCc3* displayed improved the drought tolerance and yield of transgenic rice plants grown under field drought conditions (Jeong et al., 2010). This represents another example of how spatial or temporal expressions of transgenes affect the growth habit and yield of the plant. Future research would benefit from utilizing promoters other that those offering constitutive expression when creating transgenics to both resist stress and maintain reproduction and yield.

As introduced in the salt section, promising results have been demonstrated by overexpressing vacuolar membrane H+ pumps (Gaxiola et al., 2001, Park et al., 2005), allowing for an increase in vacuolar solute content, allowing for enhanced osmotic adjustment capacity. Recently AtAVP1 was introduced into cotton (Pasapula et al., 2010), with the transgenic phenotype displaying an increased vacuolar proton gradient, resulting in solute accumulation and water retention. The AVP1-expressing cotton plants also displayed a 20% increase in fiber yield when grown under field conditions. These results suggest a promising role for AVP1 in both drought and salt tolerance, perhaps offering the ability to reclaim farmland in arid regions (Pasapula et al., 2010).

The C4 grass foxtail millet (*Setaria italica*) not only harbors novel genes for increased salt tolerance (Puranik et al., 2011), but is also resistant to dehydration stress. Comparative transcriptome analysis under early and late drought stress identified the major upregulated transcripts to be involved in metabolism, signaling, transcriptional regulation, and proteolysis (Lata et al., 2010). Five cultivars of varying drought sensitivity were also screened for their dehydration tolerance, with differentially expressed transcripts identified between them. Selected examples of upregulated transcripts include: DREB2 (with a 5 fold increase after 6 hours. and an 11-fold increase after 24 hours), Ca+ dependent kinases (likely due to enhanced Ca+ signaling), a member of the aquaporin superfamily, and the osmoprotectant thionin. Also shown to be upregulated in the more drought tolerant cultivars was the U2-snRNP, one of the 5 small ribonucleoprotein particles that make up the spliceosome, the regulator of both constitutive and alternative splicing in eukaryotes. This suggests an altered network of alternative splicing and gene regulation in foxtail millet under drought stress, offering an avenue for future research to pursue the gene targets and transcripts that undergo alternative splicing.

4.3 Combinations of abiotic stressors: Profiles of heat and drought

The combined physiological and molecular effects of heat and drought stress are quite complex, and it remains extremely difficult, if not impossible, to deduce these effects from observing the responses from one stress alone. For example, high leaf temperatures are a result of the combined effect because plants lose the ability for transpirational cooling when water availability is limited. When faced with high temperatures, plants will open their stomata in an effort to cool, however when drought is also introduced plants reduce their stomatal aperature in an effort to reduce water loss, which in turn increases temperatures within the leaf. This increase greatly perturbs cellular homeostasis and the activities of enzymes, membranes, and cellular homeostasis. A recent study in the perennial grass Leymus chinensis indicates high temperatures, combined with drought stress, reduces the function of PSII, weakens nitrogen anabolism, increases protein degradation, and provokes the peroxidation of lipids (Xu and Zhou, 2006).

The knowledge of the molecular effects of this combination in cereals remains rather limited, however a recent review (Barnabus et al., 2008) offers insights into the current physiological knowledge, and readers are directed to this detailed overview for more information. From initial development, to fertilization, to the development of reproductive organs and successful seed set, high light and drought stress put severe pressures on cereals. Of agricultural importance is the combination of these stresses on grain filling, the final stage of growth in cereals where fertilized ovaries develop into caryopses. This process is dependent on the remobilization of carbon from vegetative tissues to developing grain. Water stress during the grain-filling period induces early senescence, reduced photosynthesis, and shortens the grain-filling period; it increases the remobilization of nonstructural carbohydrates from the vegetative tissues to the grain (for a review see Yang et al., 2005). Under a combination of drought and heat stress, the amount of starch accumulation is greatly reduced, along with the activity of the enzymes responsible for starch synthesis, reducing grain weight. Around 65% of the dry weight of cereals can be accounted for by starch (Barnabus et al., 2008). ADP-glucose pyrophosphorylase (AGPase) is considered the rate limiting step in starch synthesis, and differs in thermostability between plants, for example the AGPases of cereal endosperms are heat labile, while those in potato (*Solanum tuberosum*) tubers, are heat stable (Linebarger et al., 2005). Recent research identified an N-terminal motif unique to heat-stable AGPases, and when inserted into corn (*Zea mays*) was shown to increase heat stability more than 300-fold. This thermostability stems from a cysteine residue within the motif, giving rise to small subunit homodimers not found in the wild-type maize enzyme (Linebarger et al., 2005).

Trancriptome analysis is shedding light on the extent of the crosstalk network that exists between these abiotic stresses, which often result in the transcription of both overlapping and unique gene sets. This is not surprising given that a) several abiotic stresses share an osmotic (water loss) component and b) outside of a laboratory setting plants are exposed to combinations of stresses simultaneously, such as drought/heat or salt/drought, with the combined effects initiating unique transcriptome responses. Arabidopsis plants subjected to drought and heat stress display partial overlap of the two stress defense pathways individually, as well as 454 transcripts found to be specifically expressed during a combination of drought and heat stress (Rizhsky et al., 2004). These transcripts are characterized by enhanced respiration, suppressed photosynthesis, a complex expression pattern of defense and metabolic transcripts, and the accumulation of sucrose and other sugars. Interestingly heat stress was found to ameliorate the toxicity of proline to cells, suggesting that during a combination of drought and heat stress sucrose replaces proline in plants as the major osmoprotectant (Rizhsky et al., 2004). Microarray analysis of transgenic overexpressing DREB2A discovered upregulation of known drought and salt responsive genes but also heat shock related genes (Sakuma et al., 2006), implicating its function in an ABA independent regulon (Nakashima and Yamaguchi-Shinozaki et al., 2005).

Significant research is required to tease apart the molecular basis for this additive effect, yet given the current state of the field and the advent of high-throughput sequencing technologies combined with molecular cloning and characterization, the future remains bright for engineering plants with one or more additive genes conferring tolerance to heat and drought stress.

5. High light

Increases in light intensity over and above that which a plant can utilize in photosynthetic reactions is considered high light stress, and is extremely detrimental to the plant. This is in part due to accumulation of ROS, as well as the severely detrimental effects on photosynthesis and carbon fixation, all of which lead to cellular perturbations and ultimately result in crop loss or yield reduction. This high light induced photoinhibition causes reductions in the photosystem II (PSII) complex, and reduced photosynthetic CO_2 fixation (Krause et al., 2005). The light driven PSII is found in the thylakoid membrane of chloroplasts, and also cyanobacteria, with the D1 and D2 protein complexes at its core. These sub-units act as the reaction center, binding chlorophyll, phenophytin, and plastoquinine co-factors involved in transmembrane induced charge separation (Nixon et al., 2010), with the redox state of the plastoquinine pool affecting signaling and chlorophyll fluorescence (Hohmann-Marriott et al., 2010). High light also synthesizes chloroplast antioxidant enzymes, with plastoquinol shown to be the main lipid-soluble antioxidant synthesized in *Arabidopsis* during the acclimation process (Szymańska et al., 2009).

Arabidopsis leaves respond to high light conditions by a gradual loss of chlorophyll; decreases to 79%, 78%, and 66% of the initial value after 24, 48, and 72 h of high light acclimation, respectively, have been observed (Zelisko et al., 2005). The 2010 review by Nixon et al., summarizes the past 30 years of research into the assembly and repair of PSII, and readers are directed to this manuscript in depth discussion and mechanisms. The D1 and D2 proteins are subject to photodamage under high light. In mature chloroplasts, expression of the genes encoding D1 and D2 are transcriptionally upregulated in response to light, maintaining high rates of synthesis of the reaction centers, and therefore PSII activity under high intensity light conditions (Onda et al., 2008 and references therein).

The light harvesting complex (LHC II) of PSII is located in the thylakoid membrane of the chloroplast, collecting energy from sunlight and transferring it to the PSII reaction centers (Zelisko et al., 2005), and although the main function of LHC II is energy collection and transfer, it also is involved in the distribution of excitation energy between PS II and PS I. LHC II also plays a role in preventing damage to photosynthetic machinery when there is an excess of light, necessary for high light acclimatization. Recent work into elucidating the regulatory network in of proteases responsible discovered a chloroplast-targeted protease, AtFtsH6, identified as being responsible for the degradation of LHC II in *Arabidopsis*, with an ortholog in *Populus trichocarpa*. It is likely that FtsH6 is a general LHC II protease and that FtsH6-dependent LHC II proteolysis is a feature of all higher plants (Zelisko et al., 2005), and may play a role in the high light acclimatization process.

Leaf anatomy changes during photosynthetic light acclimation, for example leaves under shade display a reduction in the mesophyll cell palisade layer, allowing a wider area for light harvesting tissues, while chloroplasts under sunlight display more active carbon fixation carriers (such as Rubisco) and reaction centers (Weston et al., 2000), with lower amounts of thylakoids per chloroplast area. *Arabidopsis* leaves have been shown to develop elongated palisade mesophyll cells and increase leaf thickness under exposure to increased fluence rates (Weston et al., 2000). In addition to their role as photosynthetic centers, the chloroplasts also produce fatty acids and amino acids that act as secondary messengers and the building blocks of protein synthesis (López-Juez 2007).

Microarray studies have offered insights into global gene expression changes in response to high light stress in *Arabidopsis*. Enzymes of the phenylpropanoid pathway, specifically those involved in lignin and anthocyanin synthesis, were shown to accumulate under long exposures to high light (Kimura et al., 2003), perhaps as acting cellular protection mechanisms. Upregulation of stress-specific sigma factors (the Sig family) are evident as well. AtSig5 (AiSigE) is upregulated in response to light stress, suggesting a regulatory role in chloroplast gene expression under high light, and has also been shown to be induced under blue-light (470 nm) illumination (Onda et al., 2008). AtSIG5 likely protects plants from stresses by assisting and increasing repair of the PSII reaction center (Nagashima et al., 2004), as well as play a crucial role in plant reproduction (Yao et al., 2003). There appears to be at least some homology between dicot and monocot systems; recently the nuclear genes *OsSIG5* and *OsSIG6* were identified and demonstrated to encode chloroplast localized sigma factors in rice, as the first example of Sig5 in crop plants (Kubota et al., 2007).

Interestingly, but not entirely surprising, was the fact the array demonstrated high light stress induces genes associated with other abiotic stresses such as LEA14, COR15a, KIN1, and RD29a, as well as fibrillins (suggesting a role for the lipid protein plastoglobulin in chloroplast protection) and lipid transfer proteins. This is not surprising given the amounts of crosstalk in the abiotic stresses, as these genes are responsible for the encoding of proteins involved in protection of chaperoning, membrane protection, and other cellular components. The transcription factor DREB2A again fell out as overlapping with drought and high light specifically, likely induced by increasing ROS levels in chloroplasts under high light conditions (Kimura et al., 2003). One of the genes demonstrated to be upregulated more than 3-fold by Kimura et al., was a member of the early light-inducible protein (ELIP) family, ELIP2. Photoinhibition by high light also induces ELIP transcription in thylakoid membranes, corresponding to the degree of photoinhibition (Adamska et al., 1992), with chloroplast ELIP levels paralleling the decrease in the amount of D1 protein subunit of PSII. The expression pattern of ELIPs suggests a role in protection of the photosynthetic apparatus against photooxidative damage. Since *Arabidopsis* carries two ELIP genes (ELIP1/2), a double null mutant was created, of which the sensibility to photoinhibition and ability to recover from light stress was not different from WT (Rossini et al., 2006), raising questions about the photoprotective function of these proteins. Constitutive expression of AtELIP2 in *Arabidopsis* leaves decreased chloroplast chlorophyll content and caused a decrease in all photosynthetic pigments, however did not alter the composition, organization, or functionality of the photosystems. This indicates ELIPs are likely not directly involved in the synthesis and assembly of specific photosynthetic complexes, but rather affect the biogenesis of all chlorophyll-binding complexes (Tzvetkova-Chevolleau et al., 2007). Continued study will be necessary to fully elucidate the photoprotective role of the ELIP family, perhaps suggesting they may not be the best candidates for genetic engineering to increase high light tolerance.

There has been little research using genetic engineering to increase the photosynthetic capability of agriculturally important crop species, yet given the insights we have gained from global transcriptome studies, and traditional genetic approaches have characterized genes enhancing cellular protection, altering photosynthesis, and involved with various aspects of high light acclimatization. The coming years now have the benefit of a wealth of genomic information, and identification of factors participating in signaling between the nucleus and chloroplast; allowing for directed studies into increased photochemical quenching, dissipation of excess light energy, and reduction of ROS.

6. Future of abiotic stress research: Incorporating the genomics revolution

The next decade of research into abiotic stress tolerance promises to be both an exiting and fruitful one. It has the advantage of an existing bank of knowledge in the form of public gene expression data from microarray and HTS experiments, new emerging monocot model systems closely related to the cereals, and the coupling of traditional breeding with genetic engineering. New insights into the gene regulatory networks regulating stress-relevant pathways are continuing to emerge, and natural variation between cultivars or accessions, when coupled with high-throughput sequencing and quantitative phenotyping for improved stress tolerance, can pinpoint candidate genes for future study.

Since the advent of genome-wide surveys of expression patterns and differential regulation under various conditions, huge datasets of stress-specific genes have begun to amass. These datasets are incorporated into public databases, and freely searchable by the research community. Now that we have begun to identify subsets of genes and gene families induced under stresses, it is time to utilize this knowledge towards high-throughput screens of transgenic plants expressing genes under stress induced or tissue specific promoters. It is through such large scale functional genomic approaches that genes or gene combinations will be identified that are capable of conferring tolerance to the abiotic stress of interest without detrimental effects to reproduction or yield. To date most such transgenic studies have relied most often on a candidate gene first identified in the dicot *Arabidopsis,* fused to a constitutive promoter, and grown under laboratory conditions for a short duration. While often resulting in a plant demonstrating improved stress tolerance, such studies are of limited value unless conducted in a crop species and under realistic field conditions, and can be likened to chipping away at an iceberg. Given the genome scale datasets available now it is plausible to directly identify novel genes or groups of genes in a crop itself or a closely related model system suited for laboratory study. Two relevant and recently emerging models are the grasses *Brachypodium* and *Setaria*. *Brachypodium* is member of the Pooideae subfamily of grasses and a well suited model system due to its relatively small fully sequenced and annotated genome, growing mutant collection, transcriptome sequence data, and other genetic resources (The International Brachypodium Initiative et al., 2010). *Setaria* is a C4 grass (Brutnell et al., 2010), as are corn, sugarcane, and sorghum, and therefore lends itself as a model for these agriculturally important crops. C4 plants have the ability to withstand higher light intensities and temperatures than C3 plants (wheat, barley, etc.) and information derived from *Setaria* may allow for improved viability of other crop species in new geographic regions.

One crucial issue that has only been touch upon briefly in this chapter, due to space constraints, is traditional breeding for increased stress tolerance. Abiotic stress tolerance is a complex trait, and it remains difficult to breed for tolerance without effecting yield or viability. There are many previous reviews (Bruce et al., 2001, Price et al., 2002, Withcombe et al., 2008, Ashraf et al., 2010) focusing on QTL and breeding cereals for stress tolerance, as well as genetic engineering coupled with breeding; and readers are directed to these reviews for further information. Future efforts will likely combine breeding and genetic engineering to maximize the benefits to both tolerance and yield. For example, a QTL involved in stress tolerance may bring undesirable closely linked traits, which may in turn be compensated by with complementary transgenes. The technology for such approaches is available now, and the challenge will be translating the laboratory discoveries into field studies and vice versa.

Research into improving stress tolerance has historically focused primarily on transcription factors. Transcription factors are master regulators of the response network, directly controlling either a single gene of multiple gene products. In addition, post-transcriptional regulation is mediated by splicing factors, specifically, by members the SR family of splicing factors, that are themselves alternatively spliced under abiotic stresses (Palusa et al., 2007, Filichkin et al., 2010). This layer of regulation of gene expression likely alters the splicing of a host of downstream genes in response to abiotic stresses, including transcription factors, and may simultaneously target multiple response mechanisms. Future research towards understanding the regulatory web of transcription factors, splicing factors, and their targets will be necessary in order to elucidate the foundations of abiotic stress tolerance in plants.

The next decade of abiotic stress research in plants has the potential to take great strides towards fully understanding stress response gene networks and translating this combined knowledge into increased crop yields. The knowledge gained from high-throughput and genome-scale technologies, coupled with the work of breeders, may allow us to meet the world's ever increasing demand for food, despite our growing population.

7. References

Abba, S., Ghignone, S., Bonfante, P. (2006). A dehydration-inducible gene in the truffle Tuber borchii identifies a novel group of dehydrins. *BMC Genomics*, Vol.7, No.39, (March 2006), ISSN 1471-2164

Abraham, E., Rigo, G., Szekely, G., Nagy, R., Koncz, C., Szabados, L. (2003). Light-dependent induction of proline biosynthesis by abscisic acid and salt stress is inhibited by brassinosteroid in Arabidopsis. *Plant Molecular Biology*, Vol.51, No.3, (Febuary 2003), pp. 363-372, ISSN 1677-0420

Adamska, I. (1997). ELIPs - Light-induced stress proteins. *Physiologia Plantarum* ,Vol.100, No.4, (August 1997), pp. 794-805, ISSN 0031-9317

Agarwal, M., Hao, Y., Kapoor, A., Dong, C.-H., Fujii, H., Zheng, X., et al. (2006). A R2R3 Type MYB Transcription Factor Is Involved in the Cold Regulation of CBF Genes and in Acquired Freezing Tolerance. *Journal of Biological Chemistry , Vol. 281, No.*49, (October 2006), pp. 37636-37645, ISSN 0021-9258

Apse, M. P., Aharon, G. S., Snedden, W. A., Blumwald, E. (1999). Salt Tolerance Conferred by Overexpression of a Vacuolar Na+/H+ Antiport in Arabidopsis. *Science,*Vol.285, No.5431, (August 1999), pp. 1256-1258, ISSN 0028-0836

Aranjuelo, I., Molero, G., Erice, G., Avice, J. C., Noques, S. (2010). Plant physiology and proteomics reveals the leaf response to drought in alfalfa (Medicago sativa L.). *Journal of Experimental Botany*, Vol.62, No.1, (August 2010), pp.1-13, ISSN 0022-0957

Ashraf, M. (2010). Inducing drought tolerance in plants: Recent advances. *Biotechnology Advances*, No. 28,Vol.1, (January 2010), pp. 169-183, ISSN 0734-9750

Badawi, M., Reddy, Y. V., Agharbaoui, Z., Tominaga, Y., Danyluk, J., Sarhan, F., Houde, M. (2008) Structure and functional analysis of wheat ICE (Inducer of CBF Expression) genes. *Plant and Cell Physiology*, Vol.48, No.9, (July 2008), pp. 1237-1249, ISSN 0032-0781

Baniwal, S., Bharti, K., Chan, K., Fauth, M., Ganguli, A., Kotak, S., Mishra, S., Nover, L., Port, M., Scharf, K.-D., Tripp, J., Weber, C., Zielinski, D., von Koskull-Doring, P. (2004). Heat stress response in plants, a complex game with chaperones and more than twenty heat stress transcription factors. *Journal of Biosciences*, Vol.29, No.4, (December 2004), pp. 471-487, ISSN 0250-5991

Banti, V., Mafessoni, F., Loreti, E., Alpi, A., Perata, P. (2010). The Heat-Inducible Transcription Factor HsfA2 Enhances Anoxia Tolerance in Arabidopsis. *C*, Vol.152, No.3, (March 2010), pp. 1471-1483, ISSN 0032-0889

Barnabás, B., Jäger, K., Fehér, A. (2008). The effect of drought and heat stress on reproductive processes in cereals. *Plant cell environment*, Vol.31, No.1, (January 2008), pp. 11-38, ISSN 0140-7791

Bartels, D. (2005). Desiccation Tolerance Studied in the Resurrection Plant Craterostigma plantagineum. *Integrative and Comparative Biology*, Vol.45, No.5, (December 2001), pp. 696-701, ISSN 1540-7063

Biamonti, G. and Caceres, J. F. (2009). Cellular stress and RNA splicing. *Trends in Biochemical Sciences*, Vol.34, No.3, (March 2009), pp.146-153, ISSN 0968-0004

Bohnert, H., J., Nelson, D. E. & Jensen, R. G. (1995) Adaptations to Environmental Stresses. *The Plant Cell Online*, Vol.7, No.7, (July 1005), pp. 1099-1111, ISSN 1040-4651

Bray, E. A. (2002). Abscisic acid regulation of gene expression during water-deficit stress in the era of the Arabidopsis genome. *Plant, Cell Environment*, Vol.25, No.2, (Febuary 2002), pp. 153-161, ISSN 0140-7791

Brinker, M., Brosch, M., Vinocur, B., Abo-Ogiala, A., Fayyaz, P., Janz, D. (2010). Linking the Salt Transcriptome with Physiological Responses of a Salt-Resistant Populus Species as a Strategy to Identify Genes Important for Stress Acclimation. *Plant Physiology*, Vol.154, No.4, (December 2010), pp. 1697-1709, ISSN 0032-0889

Bruce, W. B., Edmeades, G. O., Barker, T. C. (2002). Molecular and physiological approaches to maize improvement for drought tolerance. *Journal of Experimental Botany*, Vol.53, No.366, (Jamuary 2002), pp. 13-25, ISSN 0022-0957

Brutnell, T. P., Wang, L., Swartwood, K., Goldschmidt, A., Jackson, D., Zhu, X.-G., Kellogg, E. & Van Eck, J. (2010). Setaria viridis, A Model for C4 Photosynthesis.*The Plant Cell Online,* Vo.22, No.8, (August 2010), pp. 2537-2544, ISSN 1532-298X

Carden, D. E., Walker, D. J., Flowers, T. J., Miller, A. J. (2003). Single-Cell Measurements of the Contributions of Cytosolic Na+ and K+ to Salt Tolerance. *Plant Physiology,* Vol.131, No.2, (Febuary 2003), pp. 676-683, ISSN 0032-0889

Chaikam, V., Karlson, D. (2008). Functional characterization of two cold shock domain proteins from Oryza sativa. *Plant, Cell Environment,*Vol.31, No.7, (July 2008), pp. 995-1006, ISSN 0140-7791

Charng, Y.,Liu, H., Liu, N., Chi, W., Wang, C., Chang, S.(2007). A Heat-Inducible Transcription Factor, HsfA2, Is Required for Extension of Acquired Thermotolerance in Arabidopsis. *Plant Physiology,*Vol.143, No.1, (January 2007), pp. 251-262, ISSN 0032-0889

Chaves, M. M., Flexas, J., Pinheiro, C. (2009). Photosynthesis under drought and salt stress: regulation mechanisms from whole plant to cell. *Annals of Botany,* Vol.103, No.4, (Febuary 2009), pp. 551-560, ISSN 0305-7364

Chen, H., Hwang, J. E., Lim, C. J., Kim, D. Y., Lee, S. Y., Lim, C. O. (2010). Arabidopsis DREB2C functions as a transcriptional activator of HsfA3 during the heat stress response. *Biochemical and Biophysical Research Communications,*Vol. 401, No.2, (October 2010), pp.238-244, ISSN 1090-2104

Chen, W., J., Zhu, T. (2004). Networks of transcription factors with roles in environmental stress response. *Trends in Plant Science,* Vol.9, no.12, (December 2004), pp. 591-596, ISSN 1360-1385

Chinnusamy, V., Zhu, J., Zhu, J.-K. (2007). Cold stress regulation of gene expression in plants. *Trends in Plant Science,* Vol. 12, No.10, (October 2007), pp. 444-451, ISSN 1360-1385

Cohen-Peer, R., Schuster, S., Meiri, D., Breiman, A., Avni, A. (2010). Sumoylation of Arabidopsis heat shock factor A2 (HsfA2) modifies its activity during acquired thermotholerance. *Plant Molecular Biology,* Vol.74, No.1-2, (September 2010), pp. 33-45, ISSN 0167-4412

Craterostigma plantagineum. Integrative and Comparative Biology, Vol.45, No.5, (November 2005), pp. 696-701, ISSN 1540-7063

Cuevas, J. C., Lopez-Cobollo, R., Alcazar, R., Zarza, X., Koncz, C., Altabella, T., Salinas, J., Tiburcio, A. F., Ferrando, A. (2008). Putrescine is involved in Arabidopsis freezing tolerance and cold acclimation by regulating ABA levels in response to low temperature. *Plant Physiology,* Vol.148, No.2, (October 2008), pp. 1094–1105, ISSN 0032-0889

Doherty, C. J., Van, H. A., Myers, S. J., Thomashow, M. F. (2009). Roles for Arabidopsis CAMTA Transcription Factors in Cold-Regulated Gene Expression and Freezing Tolerance. *The Plant Cell,* Vol.21, No.3, (March 2009), pp. 972-984, ISSN 1040-4651

Dong, C., Danyluk, J., Wilson, K. E., Pocock, T., Huner, N. P., Sarhan, F. (2002). Cold-Regulated Cereal Chloroplast Late Embryogenesis Abundant-Like Proteins.

Molecular Characterization and Functional Analyses. *Plant Physiology,* Vol.129, No.3, (July 2002), pp.1368-1381, ISSN 0032-0889

Dong, C.-H., Agarwal, M., Zhang, Y., Xie, Q., Zhu, J.-K. (2006). The negative regulator of plant cold responses, HOS1, is a RING E3 ligase that mediates the ubiquitination and degradation of ICE1. *Proceedings of the National Academy of Sciences,* Vol.103, No.21, (May 2006), pp. 8281-8286, ISSN 0027-8424

Egawa, C., Kobayashi, F., Ishibashi, M., Nakamura, T., Nakamura, C., Takumi, S. (2006). Differential regulation of transcript accumulation and alternative splicing of a DREB2 homolog under abiotic stress conditions in common wheat. *Genes Genetic Systems,* Vol.81, No.2, (April 2006), pp. 77-91. ISSN 1341-7568

Feder, M. E., Hofmann, G. E. (1999). Heat-shock proteins, molecular chaperones, and the stress response: evolutionary and ecological physiology. *Annual Review of Physiology,*Vol.61, No.1, (March 1999), pp. 243-282, ISSN 0066-4278

Filichkin, S. A., Priest, H. D., Givan, S. A., Shen, R., Bryant, D. W., Fox, S. E., Wong, W. K., Mockler, T. C. (2009). Genome-wide mapping of alternative splicing in Arabidopsis thaliana. *Genome Research,* Vol.20, No.1, (January 2010), pp. 45-58, ISSN 1088-9051

Flowers, T., Garcia, A., Koyama, M. & Yeo, A. (1997) Breeding for salt tolerance in crop plants – the role of molecular biology *Acta Physiologiae Plantarum,* Vol.10, No.4, (1997), pp. 427-433, ISSN 1861-1664

Forment, J., Naranjo, M. Á., Roldán, M., Serrano, R., Vicente, O. (2002). Expression of Arabidopsis SR-like splicing proteins confers salt tolerance to yeast and transgenic plants. *The Plant Journal,* Vol.30, No.5, (June 2002), pp. 511-519, ISSN 0960-7412

Fourrier, N., Bédard, J., Lopez-Juez, E., Barbrook, A., Bowyer, J., Jarvis, P., Warren, G., Thorlby, G. (2008) A role for SENSITIVE TO FREEZING2 in protecting chloroplasts against freeze-induced damage in Arabidopsis. *The Plant Journal,* Vol.55, No.5, (September 2008), pp.734-745, ISSN 0960-7412

Fowler, S. G., Cook, D., Thomashow, M. F. (2005). Low Temperature Induction of Arabidopsis CBF1, 2, and 3 Is Gated by the Circadian Clock. *Plant Physiology,* Vol.137, No.3, (March 2005), pp. 961-968, ISSN 0032-0889

Franklin, K., Whitelam, G. (2007). Light-quality regulation of freezing tolerance in Arabidopsis thaliana. *Nature Genetics,* Vol.39, No.5, (November 2007), pp. 1410-1413, ISSN 1061-4036

Fukuda, A., Nakamura, A., Tagiri, A., Tanaka, H., Miyao, A., Hirochika, H., Tanaka, Y. (2004). Function, Intracellular Localization and the Importance in Salt Tolerance of a Vacuolar Na+/H+ Antiporter from Rice. *Plant and Cell Physiology,* Vol.45, No.2, (January 2004), pp. 146-159, ISSN 0032-0781

Fursova, O. V., Pogorelko, G. V., Tarasov, V. A. (2009). Identification of ICE2, a gene involved in cold acclimation which determines freezing tolerance in Arabidopsis thaliana. *Gene,* Vol.429, No.1-2, (January 2009), pp. 98-103, ISSN 0378-1119

Garciadeblás, B., Haro, R., Benito, B. (2007). Cloning of two SOS1 transporters from the seagrass Cymodocea nodosa, SOS1 transporters from Cymodocea and Arabidopsismediate potassium uptake in bacteria. *Plant Molecular Biology,* Vol.63, No.4, (March 2007), pp. 479-490, ISSN 0167-4412

Gaxiola, R. A., Li, J., Undurraga, S., Dang, L. M., Allen, G. J., Alper, S. L., Fink, G. R. (2001). Drought- and salt-tolerant plants result from overexpression of the AVP1 H+-

pump. *Proceedings of the National Academy of Sciences*, Vol.98, No.20, (September 2001), pp. 11444-11449, ISSN 0027-8424

Ghoulam, C., Foursy, A., Fares, K. (2002). Effects of salt stress on growth, inorganic ions and proline accumulation in relation to osmotic adjustment in five sugar beet cultivars. *Environmental and Experimental Botany,Vol.47, No.*1, (January 2002), pp. 39-50, ISSN 0098-8472

Gilmour, S. J., Zarka, D. G., Stockinger, E. J., Salazar, M. P., Houghton, J. M., Thomashow, M. F. (1998). Low temperature regulation of the Arabidopsis CBF family of AP2 transcriptional activators as an early step in cold-induced COR gene expression. *The Plant Journal,* Vol.16. No.4, (November 1998), pp. 433-442, ISSN 1365-313X

Gray, W. M., Ostin, A., Sandberg, G., Romano, C. P., Estelle, M. (1998). High temperature promotes auxin-mediated hypocotyl elongation in Arabidopsis. *Proceedings of the National Academy of Sciences,* Vol.95, No.12, (June 1998), pp. 7197-7202, ISSN 0027-8424

Guo, P., Baum, M., Grando, S., Ceccarelli, S., Bai, G., Li, R. (2009). Differentially expressed genes between drought-tolerant and drought-sensitive barley genotypes in response to drought stress during the reproductive stage. *Journal of Experimental Botany,* Vol.60, No.12, (June 2009) , pp. 3531-3544, ISSN 0022-0957

Guo, Y.-Q., Tian, Z.-Y., Qin, G.-Y., Yan, D.-L., Zhang, J., Zhou, W.-Z, Qin, P. (2009). Gene expression of halophyte Kosteletzkya virginica seedlings under salt stress at early stage. *Genetica,* Vol.*137*, No.2, (November 2009), pp. 189-199, ISSN 0016-6707

Haake, V., Cook, D., Riechmann, J., Pineda, O., Thomashow, M. F., Zhang, J. Z. (2002). Transcription Factor CBF4 Is a Regulator of Drought Adaptation in Arabidopsis. *Plant Physiology,* Vol.130, No.2, (October 2002), pp. 639-648, ISSN 0032-0889

Halfter, U., Ishitani, M., Zhu, J.-K. (2000). The Arabidopsis SOS2 protein kinase physically interacts with and is activated by the calcium-binding protein SOS3. *Proceedings of the National Academy of Sciences,* Vol.97, No.1, (March 2000), pp. 3735-3740, ISSN 0027-8424

Harb, A., Krishnan, A., Ambavaram, M. M., Pereira, A. (2010). Molecular and Physiological Analysis of Drought Stress in Arabidopsis Reveals Early Responses Leading to Acclimation in Plant Growth. *Plant Physiology,* Vol.154, No.3, (August 2010), pp.1254-1271, ISSN 0032-0889

Hare, P., Cress, W. & van Staden, J. (1999). Proline synthesis and degradation, a model system for elucidating stress-related signal transduction. *Journal of Experimental Botany,* Vol.50, No.333, (January 1999), pp.413-434, ISSN 0022-0957

Hinniger, C., Caillet, V., Michoux, F., BenAmor, M., Tanksley, S., Lin, C., Mccarthy, J. (2006). Isolation and Characterization of cDNA Encoding Three Dehydrins Expressed During Coffea canephora (Robusta) Grain Development. *Annals of Botany,* Vol.97, No.5, (May 2006), pp. 755-765, ISSN 0305-7364

Hohmann-Marriott, M. F., Takizawa, K., Eaton-Rye, J. J., Mets, L., Minagawa, J. (2010). The redox state of the plastoquinone pool directly modulates minimum chlorophyll fluorescence yield in Chlamydomonas reinhardtii. *FEBS Letters,* Vol.584, No.5, (March 2010), pp. 1021-1026, ISSN 0014-5793

Hu, H., Dai, M., Yao, J., Xiao, B., Li, X., Zhang, Q., Xiong, L. (2006). Overexpressing a NAM, ATAF, and CUC (NAC) transcription factor enhances drought resistance and salt tolerance in rice. *Proceedings of the National Academy of Sciences,* Vol.103, No.35, (June 2006), pp. 12987-12992, ISSN 0027-8424

Hu, H., You, J., Fang, Y., Zhu, X., Qi, Z., Xiong, L. (2008). Characterization of transcription factor gene SNAC2 conferring cold and salt tolerance in rice. *Plant Molecular Biology,*Vol.67, No.102, (May 2008), pp. 169-181, ISSN 0167-4412

Huang, J., Hirji, R., Adam, L., Rozwadowski, K. L., Hammerlindl, J. K., Keller, W. A., Selvaraj, G. (2000). Genetic Engineering of Glycinebetaine Production toward Enhancing Stress Tolerance in Plants: Metabolic Limitations. *Plant Physiology,* Vol.122, No.3, (March 2000), pp. 747-756, ISSN 0032-0889

Hundertmark, M., Hincha, D. (2008). LEA proteins and their encoding genes in Arabidopsis thaliana. *BMC Genomics,* Vol.9, No.1, (March 2008), pp. 118, ISSN 1471-2164

Iida, K., Seki, M., Sakurai, T., Satou, M., Akiyama, K., Toyoda, T., Konagaya, A., Shinozaki, K. (2004). Genome-wide analysis of alternative pre-mRNA splicing in Arabidopsis thaliana based on full-length cDNA sequences. *Nucleic Acids Research,*Vol. 32, No.17, (September 2004), pp. 5096-5103, ISSN 0305-1048

International Brachypodium Initiative. (2010). Genome sequencing and analysis of the model grass Brachypodium distachyon. *Nature,* Vol.463, No.7282, (February 2010), pp. 763-768, ISSN 0028-0836

Ishitani, M., Liu, J., Halfter, U., Kim, C., Shi, W., Zhu, J. (2000). SOS3 function in plant salt tolerance requires N-myristoylation and calcium binding. *Plant Cell,* Vol.12, No.9, (September 2000), pp. 1667-78, ISSN 1040-4651

Ishitani, M., Xiong, L., Stevenson, B., Zhu, J. K. (1997). Genetic Analysis of Osmotic and Cold Stress Signal Transduction in Arabidopsis: Interactions and Convergence of Abscisic Acid-Dependent and Abscisic Acid-Independent Pathways. *The Plant Cell Online , 9* (11), 1935-1949.

Jaspers, P., & Kangasjärvi, J. (2010) Reactive oxygen species in abiotic stress signaling. *Physiologia Plantarum,* Vol.138, No.4, (April 2010), pp. 405-413, ISSN 1399-3054

Jeong, J. S., Kim, Y. S., Baek, K. H., Jung, H., Ha, S.-H., Choi, Y. D., et al. (2010). Root-Specific Expression of OsNAC10 Improves Drought Tolerance and Grain Yield in Rice under Field Drought Conditions. *Plant Physiology,* Vol.153, No.1, (May 2010), pp. 185-197, ISSN 0032-0889

Johnson-Flanagan, A. M., Huiwen, Z., Thiagarajah, M. R., Saini, H. S. (1991). Role of Abscisic Acid in the Induction of Freezing Tolerance in Brassica napus Suspension-Cultured Cells. *Plant Physiology,* Vol.95, No.4, (April 1991), pp. 1044-1048, ISSN 0032-0889

Jones-Rhoades, M. W., Bartel, D. P., Bartel, B. (2006). MicroRNAs and their regulatory roles in plants. *Annual Review of Plant Biology,* Vol.57, (June 2006) pp. 19-53, ISSN 1543-5008

Kanaoka, M. M., Pillitteri, L. J., Fujii, H., Yoshida, Y., Bogenschutz, N. L., Takabayashi, J., Zhu, J.-K., Torii, K. U. (2008). SCREAM/ICE1 and SCREAM2 Specify Three Cell-State Transitional Steps Leading to Arabidopsis Stomatal Differentiation. *The Plant Cell,* Vol.20, No.7, (July 2008), pp.1775-1785, ISSN 1040-4651

Katiyar-Agarwal, S., Agarwal, M., Grover, A. (2003). Heat-tolerant basmati rice engineered by over-expression of hsp101. *Plant Molecular Biology,* Vol.51, No.5, (March 2003), pp. 677-686, ISSN 0167-4412

Katiyar-Agarwal, S., Zhu, J., Kim, K., Agarwal, M., Fu, X., Huang, A., Zhu, J. (2006). The plasma membrane Na+/H+ antiporter SOS1 interacts with RCD1 and functions in oxidative stress tolerance in Arabidopsis. *Proceedings of the National Academy of Sciences,* Vol. 103, No.49, (December 2006) pp. 18816-18821, ISSN 0027-8424

Kern, A., J., and Dyer, W. E. (2004). Glycine Betaine Biosynthesis Is Induced by Salt Stress but Repressed by Auxinic Herbicides in Kochia scoparia. *Journal of Plant Growth Regulation,* Vol.23, No.1, (March 2004), pp. 9-19, ISSN 0721-7595

Kidokoro, S., Maruyama, K., Nakashima, K., Imura, Y., Narusaka, Y., Shinwari, Z. K., Osakabe, Y., Fujita, Y., Mizoi, J., Shinozaki, K., Yamaguchi-Shinozaki, K. (2009). The Phytochrome-Interacting Factor PIF7 Negatively Regulates DREB1 Expression under Circadian Control in Arabidopsis. *Plant Physiology,* Vol.151, No.4, (December 2009), pp.2046-2057, ISSN 0032-0889

Kim, D.-Y., Jin, J.-Y., Alejandro, S., Martinoia, E., Lee, Y. (2010). Overexpression of AtABCG36 improves drought and salt stress resistance in Arabidopsis. *Physiologia Plantarum, Vol.139, No.*2, (June 2010), pp. 170-180, ISSN 0031-9317

Kim, J. S., Park, S. J., Kwak, K. J., Kim, Y. O., Kim, J. Y., Song, J., Boseung Jang, B., Jung, C., Kang, H. (2006). Cold shock domain proteins and glycine-rich RNA-binding proteins from Arabidopsis thaliana can promote the cold adaptation process in Escherichia coli. *Nucleic Acids Research,* Vol.35, No.2, (December 2006), pp. 506-516, ISSN 0305-1048

Kim, K., Portis, A. R. (2005). Temperature Dependence of Photosynthesis in Arabidopsis Plants with Modifications in Rubisco Activase and Membrane Fluidity. *Plant and Cell Physiology,* Vol.46, No.3, (Febuary 2005), pp. 522-530, ISSN 0032-0781

Kimura, M., Yamamoto, Y. Y., Seki, M., Sakurai, T., Sato, M., Abe, T., Yoshida, S., Manabe, K., Shinozaki, K., Matsui, M. (2003). Identification of Arabidopsis Genes Regulated by High Light–Stress Using cDNA Microarray. *Photochemistry and Photobiology,* Vol.77, No.2, (February 2003), pp.226-233, ISSN 0031-8655

Kishitani, S., Watanabe, K., Yasuada, S., Arakawa, K., Takabe, T. (1994). Accumulation of glycinebetaine during cold acclimation and freezing tolerance in leaves of winter and spring barley plants. *Plant, Cell & Environment,* Vol.17, No.1, (January 1994), pp. 89-95, ISSN 0140-7791

Knight, H., Zarka, D. G., Okamoto, H., Thomashow, M. F., Knight, M. R. (2004) Abscisic Acid Induces CBF Gene Transcription and Subsequent Induction of Cold-Regulated Genes via the CRT Promoter Element. *Plant Physiology,* Vol.135, No.3, (July 2004), pp. 1710-1717, ISSN 0032-0889

Knox, A., Dhillon, T., Cheng, H., Tondelli, A., Pecchioni, N., Stockinger, E. (2010). CBF gene copy number variation at Frost Resistance-2 is associated with levels of freezing tolerance in temperate-climate cereals. *Theoretical and Applied Genetics,* Vol.121, No.1, (June 2010), pp. 21-35, ISSN 0040-5752

Komatsu, S., Yang, G., Khan, M., Onodera, H., Toki, S., Yamaguchi, M. (2007). Over-expression of calcium-dependent protein kinase 13 and calreticulin interacting protein 1 confers cold tolerance on rice plants. *Molecular Genetics and Genomics,* Vol.277, No.6, (June 2007), pp. 713-723, ISSN 1617-4615

Kotak, S., Vierling, E., Bäumlein, H., von Koskull-Döring, P. (2007). A Novel Transcriptional Cascade Regulating Expression of Heat Stress Proteins during Seed Development of Arabidopsis. The Plant Cell, Vol.19, No.1, (January 2007), pp. 182-195, ISSN 1040-4651

Kovacs, D., Agoston, B., Tompa, P. (2008).Disordered plant LEA proteins as molecular chaperones. *Plant Signal Behavior*, Vol.3, No.9, (September 2008), pp.710-713, ISSN 1559-2316

Krause, G. H., Gallé, A., Virgo, A., García, M., Bucic, P., Jahns, P., Winter, K. (2006). High-Light Stress does not Impair Biomass Accumulation of Sun-Acclimated Tropical Tree Seedlings (Calophyllum longifolium Willd. and Tectona grandis L. f.). *Plant Biology*, Vol.8, No.1, (January 2006), pp. 31-41, ISSN 1435-8603

Kreps, J. A., Wu, Y., Chang, H.-S., Zhu, T., Wang, X. & Harper, J. F. (2002) Transcriptome Changes for Arabidopsis in Response to Salt, Osmotic, and Cold Stress. *Plant Physiology*, Vol.130, No.4, (December 2002), pp. 2129-2141, ISSN 1532-2548

Kubota, Y., Miyao, A., Hirochika, H., Tozawa, Y., Yasuda, H., Tsunoyama, Y., Niwa, Y., Imamura, S., Shirai, M., Asayama, M. (2007). Two Novel Nuclear Genes, OsSIG5 and OsSIG6, Encoding Potential Plastid Sigma Factors of RNA Polymerase in Rice: Tissue-Specific and Light-Responsive Gene Expression. *Plant and Cell Physiology*, Vol.48, No.1, (January 2007), pp. 186-192, ISSN 1471-9053

Kurek, I., Chang, T. K., Bertain, S. M., Madrigal, A., Liu, L., Lassner, M. W., Zhu, G. (2007). Enhanced Thermostability of Arabidopsis Rubisco Activase Improves Photosynthesis and Growth Rates under Moderate Heat Stress. *The Plant Cell*, Vol.19, No.10, (October 2007), pp. 3230-3241, ISSN 1040-4651

Laporte, M. M., Shen, B., Tarczynski, M. C. (2002). Engineering for drought avoidance: expression of maize NADP malic enzyme in tobacco results in altered stomatal function. *Journal of Experimental Botany*, Vol.53, No.369, (April 2002), pp. 699-705, ISSN 1460-2431

Lata, C., Sahu, P. P., Prasad, M. (2010). Comparative transcriptome analysis of differentially expressed genes in foxtail millet (Setaria italica L.) during dehydration stress. *Biochemical and Biophysical Research Communications*, Vol.393, No.4, (March 2010), pp. 720-727, ISSN 1090-2104

Layton, B. E., Boyd, M. B., Tripepi, M. S., Bitonti, B. M., Dollahon, M. N. R., Balsamo, R. A. (2010). Dehydration-induced expression of a 31-kDa dehydrin in Polypodium polypodioides (Polypodiaceae) may enable large, reversible deformation of cell walls. *American Journal of Botany*, Vol.97, No.4, (March 2010), pp. 535-544, ISSN 0002-9122

Lee, B., Kapoor, A., Zhu, J., Zhu, J.-K. (2006). STABILIZED1, a Stress-Upregulated Nuclear Protein, Is Required for Pre-mRNA Splicing, mRNA Turnover, and Stress Tolerance in Arabidopsis. *The Plant Cell*, Vol.18, No.7, (July 2006), pp. 1736-1749, ISSN 1040-4651

Lee, B.-h., Lee, H., Xiong, L., Zhu, J.-K. (2002). A Mitochondrial Complex I Defect Impairs Cold-Regulated Nuclear Gene Expression. *The Plant Cell*, Vol.14, No.6, (June 2002), pp. 1235-1251. ISSN 1040-4651

Li, C., Chen, Q., Gao, X., Qi, B., Chen, N., Xu, S., Chen, J., Wang, X. (2005). AtHsfA2 modulates expression of stress responsive genes and enhances tolerance to heat and oxidative stress in Arabidopsis. *Science in China Series C: Life Sciences*, Vol.46, No.8, (December 2005), pp.540-550, ISSN 1006-9305

Li, M., Berendzen, K., Schöffl, F. (2010). Promoter specificity and interactions between early and late Arabidopsis heat shock factors. *Plant Molecular Biology*, Vol.73, No.4-5 (July 2010), pp. 559-567, ISSN 0167-4412

Li, P., Brutnell, T. P. (2011). Setaria viridis and Setaria italica, model genetic systems for the Panicoid grasses. *Journal of Experimental Botany*, Epub ahead of print, (May 2011), ISSN 0022-0957

Li, T., Zhang, Y., Liu, H., Wu, Y., Li, W., Zhang, H. (2010). Stable expression of Arabidopsis vacuolar Na^+/H^+ antiporter gene AtNHX1 and salt tolerance in transgenic soybean for over six generations. *Chinese Science Bulletin*,Vol. 55, No.12, (April 2010), pp. 1127-1134, ISSN 1001-6538

Li, W., Wang, D., Jin, T., Chang, Q., Yin, D., Xu, S., Liu, B.,Liu, L. (2010). The Vacuolar Na+/H+ Antiporter Gene SsNHX1 from the Halophyte Salsola soda Confers Salt Tolerance in Transgenic Alfalfa (Medicago sativa L.). *Plant Molecular Biology Reporter*, Vol.29, No.2, (July 2010), pp. 278-290, ISSN 1572-9818

Linebarger, C. R. L., Boehlein, S. K., Sewell, A. K., Shaw, J., Hannah, L. C. (2005). Heat Stability of Maize Endosperm ADP-Glucose Pyrophosphorylase Is Enhanced by Insertion of a Cysteine in the N Terminus of the Small Subunit. *Plant Physiology*, Vol.139, No.4, (December 2005), pp. 625-1634, ISSN 0032-0889

Liu, A.-L., Zou, J., Zhang, X.-W., Zhou, X.-Y., Wang, W.-F., Xiong, X.-Y., Chen, L.-Y., Chen, X.-B. (2010). Expression Profiles of Class A Rice Heat Shock Transcription Factor Genes Under Abiotic Stresses. *Journal of Plant Biology*, Vol.53, No.2, (January 2010), pp. 142-149, ISSN 1226-9239

Liu, J., Zhu, J.-K. (1998). A Calcium Sensor Homolog Required for Plant Salt Tolerance. *Science*, Vol.280, No.5371, (June 1998), pp. 1943-1945, ISSN 0036-8075

Liu, Q., Kasuga, M., Sakuma, Y., Abe, H., Miura, S., Yamaguchi-Shinozaki, K., Shinozaki, K. (1998). Two Transcription Factors, DREB1 and DREB2, with an EREBP/AP2 DNA Binding Domain Separate Two Cellular Signal Transduction Pathways in Drought- and Low-Temperature-Responsive Gene Expression, Respectively, in Arabidopsis. *The Plant Cell*, Vol. 10, No.8, (August 1998), pp. 1391-1406, ISSN 1040-4651

Liu, X., Yang, J., Li, B., Yang, X., Meng, Q. (2010) Antisense expression of tomato chloroplast omega-3 fatty acid desaturase gene (LeFAD7) enhances the tomato high-temperature tolerance through reductions of trienoic fatty acids and alterations of physiological parameters. *Photosynthetica*, Vol.48, No.1, (January 2010) pp. 59-66, ISSN 0300-3604

Liu, Y., Xiong, Y., Bassham, D. C. (2009). Autophagy is required for tolerance of drought and salt stress in plants. *Autophagy*,Vol. 5, No.7, (October 2009), pp. 954-963, ISSN 1554-8627

Llorente, F., Oliveros, J. C., Martínez-Zapater, J. M., Salinas, J. (2000). A freezing-sensitive mutant of Arabidopsis frs1 is a new aba3 allele. *Planta*, Vol.211, No.5, (October 2000), pp. 648-655, ISSN 0032-0935

Lopez-Juez, E. (2007). Plastid biogenesis, between light and shadows. *Journal of Experimental Botany*,Vol.58, No.1, (January 2007), pp. 11-26, ISSN 0022-0957

Ma, S., Gong, Q., Bohnert, H. J. (2006). Dissecting salt stress pathways. *Journal of Experimental Botany*, Vol.57, No.5, (March 2006), pp. 1097-1107, ISSN 0022-0957

Mahajan, S., Tuteja, N. (2005). Cold, salinity and drought stresses: An overview. *Archives of Biochemistry and Biophysics,* Vol.444, No.2, (December 2005), pp. 139-158, ISSN 0003-9861

Martanez, J.-P., Lutts, S., Schanck, A., Bajji, M., Kinet, J.-M. (2004). Is osmotic adjustment required for water stress resistance in the Mediterranean shrub Atriplex halimus L? *Journal of Plant Physiology, Vol.161, No.9,* (September 2004), pp. 1041-1051, ISSN 0032-0889

Martínez-Atienza, J., Jiang, X., Garciadeblas, B., Mendoza, I., Zhu, J.-K., Pardo, J., M., Quintero, F., J. (2007). Conservation of the Salt Overly Sensitive Pathway in Rice. *Plant Physiology,* Vol.143, No.2, (February 2007), pp. 1001-1012, ISSN 0032-0889

Matsui, A., Ishida, J., Morosawa, T., Mochizuki, Y., Kaminuma, E., Endo, T. A., Okamoto., M., Nambara., E., Nakajima., M., Kawashima., M., Satou., M., Kim., J., Kobayashi., N., Toyoda., T., Shinozaki., K., Seki., M. (2008). Arabidopsis Transcriptome Analysis under Drought, Cold, High-Salinity and ABA Treatment Conditions using a Tiling Array. *Plant and Cell Physiology,* Vol.49, No.8, (July 2008), pp. 1135-1149, ISSN 0032-0781

Maughan, P.,J., Turner, T.,B., Coleman, C.,E., Elzinga, D.,B., Jellen, E.,N., Morales JA, Udall JA, Fairbanks DJ, Bonifacio A. (2009). Characterization of Salt Overly Sensitive 1 (SOS1) gene homoeologs in quinoa (Chenopodium quinoa Willd.). *Genome,* Vol.52, No.7, (July 2009), pp. 647-657, ISSN 0831-2796

Miller, G., Shulaev, V., Mittler, R. (2008). Reactive oxygen signaling and abiotic stress. *Physiologia Plantarum,* Vol.133, No.33, (July 2008), pp. 481-489, ISSN 0031-9317

Mishra, S. K., Tripp, J., Winkelhaus, S., Tschiersch, B., Theres, K., Nover, L., Scharf, K.-D. (2002). In the complex family of heat stress transcription factors, HsfA1 has a unique role as master regulator of thermotolerance in tomato. *Genes & Development,* Vol.16, No.12, (May 2002), pp. 1555-1567, ISSN 0890-9369

Mittler, R., Kim, Y., Song, L., Coutu, J., Coutu, A., Ciftci-Yilmaz, S., Lee, H., Stevenson, B., Zhu, J. (2006). Gain- and loss-of-function mutations in Zat10 enhance the tolerance of plants to abiotic stress. *FEBS Letters,* Vol.580, No.28-29, (December 2006), pp. 6537-6542, ISSN 0014-5793

Miura, K., Jin, J. B., Lee, J., Yoo, C. Y., Stirm, V., Miura, T., et al. (2007). SIZ1-Mediated Sumoylation of ICE1 Controls CBF3/DREB1A Expression and Freezing Tolerance in Arabidopsis. *The Plant Cell,* Vol.19, No.4, (April 2007), pp. 1403-1414, ISSN 1040-4651

Mizuno, H., Kawahara, Y., Sakai, H., Kanamori, H., Wakimoto, H., Yamagata, H., Oono, Y., Wu, J., Ikawa, H., Itoh, T., Matsumoto, T. (2010). Massive parallel sequencing of mRNA in identification of unannotated salinity stress-inducible transcripts in rice (Oryza sativa L.). *BMC Genomics,* Vol.11, No.1, (December 2010), pp. 683, ISSN 1471-2164

Moellering, E. R., Muthan, B., Benning, C. (2010). Freezing Tolerance in Plants Requires Lipid Remodeling at the Outer Chloroplast Membrane. *Science,* Vol.330, No.6001, (October 2010), pp. 226-228, ISSN 0036-8075

Mohanty, S. and Tripathy, B. (2010). Early and late plastid development in response to chill stress and heat stress in wheat seedlings. *Protoplasma,* Epub ahead of print, (November 2010), pp.1-12, ISSN 1615-6102

Monroy, A. F., Dhindsa, R. S. (1995). Low-Temperature Signal Transduction: Induction of Cold Acclimation-Specific Genes of Alfalfa by Calcium at 25C. *The Plant Cell,* Vol.7, No.3, (March 1995), pp. 321-331, ISSN 1040-4651

Munns, R. (2002). Comparative physiology of salt and water stress. *Plant, Cell & Environment,* Vol. 25, No.2, (February 2002), pp. 239-250, ISSN 0140-7791

Nagashima, A., Hanaoka, M., Shikanai, T., Fujiwara, M., Kanamaru, K., Takahashi, H., Tanaka, K. (2004). The Multiple-Stress Responsive Plastid Sigma Factor, SIG5, Directs Activation of the psbD Blue Light-Responsive Promoter (BLRP) in Arabidopsis thaliana. *Plant and Cell Physiology,* Vol.55, No.4, (April 2004), pp. 357-368, ISSN 0032-0781

Nakaminami, K., Karlson, D. T., Imai, R. (2006). Functional conservation of cold shock domains in bacteria and higher plants. *Proceedings of the National Academy of Sciences,* Vol.103, No.26, (June 2006), pp. 10122-10127, ISSN 0027-8424

Nakashima, K., and Yamaguchi-Shinozaki, K. (2005). Molecular Studies on Stress-Responsive Gene Expression in Arabidopsis and Improvement of Stress Tolerance in Crop Plants by Regulon Biotechnology. *Physiologia Plantarum,* Vol.126, No.1, (January 2006), pp. 62–71, ISSN 0021-3551

Nakayama, K., Okawa, K., Kakizaki, T., Inaba, T.(2008). Evaluation of the Protective Activities of a Late Embryogenesis Abundant (LEA) Related Protein, Cor15am, during Various Stresses in Vitro. *Bioscience, Biotechnology, and Biochemistry,* Vol.72, No.6, (June 2006), pp. 1642-1645, ISSN 0916-8451

Ning, J., Li, X., Hicks, L. M., Xiong, L. A. (2010) Raf-Like MAPKKK Gene DSM1 Mediates Drought Resistance through Reactive Oxygen Species Scavenging in Rice. *Plant Physiology,* Vol.152, No.2, (February 2010), pp. 876-890, ISSN 0032-0889

Nixon, P. J., Michoux, F., Yu, J., Boehm, M., Komenda, J. (2010) Recent advances in understanding the assembly and repair of photosystem II. *Annals of Botany,* Vol.106, No.1, (March 2010), pp. 1-16, ISSN 0305-7364

Nover, L., Bharti, K., Döring, P., Mishra, S., K., Ganguli, A., Scharf, K-D. (2001). *Arabidopsis* and the heat stress transcription factor world: how many heat stress transcription factors do we need? *Cell Stress Chaperones,* Vol.6, No.3, (April 2001), pp.177.189, ISSN 1355-8145

Novillo, F., Alonso, J. M., Ecker, J. R., Salinas, J. (2004). CBF2/DREB1C is a negative regulator of CBF1/DREB1B and CBF3/DREB1A expression and plays a central role in stress tolerance in Arabidopsis. *Proceedings of the National Academy of Sciences of the United States of America ,* Vol.101, No.11, (March 2004), pp. 3985-3990, ISSN 0027-8424

Novitskaya, G., V., Suvorova, T. A., Trunova, T. I. (2000). Lipid Composition of Tomato Leaves as Related to Plant Cold Tolerance. *Russian Journal of Plant Physiology,* Vol. 47, No.6, (January 2000), pp. 728-733, ISSN 1021-4437

Oh, S.-J., Song, S. I., Kim, Y. S., Jang, H.-J., Kim, S. Y., Kim, M., et al. (2005). Arabidopsis CBF3/DREB1A and ABF3 in Transgenic Rice Increased Tolerance to Abiotic Stress without Stunting Growth. *Plant Physiology,* Vol.138, No.1, (May 2005), pp. 341-351, ISSN *0032-0889*

Ohnishi, N. and Murata, N. (2006). Glycinebetaine Counteracts the Inhibitory Effects of Salt Stress on the Degradation and Synthesis of the D1 Protein during Photoinhibition in Synechococcus. *Plant Physiology*, Vol.141, No.2, (June 2006), pp. 758-765, ISSN 0032-0889

Olías, R., Eljakaoui, Z., Pardo, J. M., Belver, A. (2009). The Na+/H+ exchanger SOS1 controls extrusion and distribution of Na+ in tomato plants under salinity conditions. *Plant Signal Behavior*, Vol.4, No.10, (October 2009),pp.973–976,ISSN1559-2316

Onda, Y., Yagi, Y., Saito, Y., Takenaka, N., Toyoshima, Y. (2008). Light induction of Arabidopsis SIG1 and SIG5 transcripts in mature leaves: differential roles of cryptochrome 1 and cryptochrome 2 and dual function of SIG5 in the recognition of plastid promoters. *The Plant Journal*, Vol.55, No.6, (September 2008), pp. 968-978. ISSN 0960-7412

Ottow, E. A., Brinker, M., Teichmann, T., Fritz, E., Kaiser, W., Brosché, M., Kangasjärvi, J., Jiang, X., Polle, A. (2005). Populus euphratica Displays Apoplastic Sodium Accumulation, Osmotic Adjustment by Decreases in Calcium and Soluble Carbohydrates, and Develops Leaf Succulence under Salt Stress. *Plant Physiology*, Vol.139, No.4, (November 2005), pp. 1762-1772, ISSN 0032-0889

Palusa, S. G., Ali, G. S., Reddy, A. S. (2007). Alternative splicing of pre-mRNAs of Arabidopsis serine/arginine-rich proteins: regulation by hormones and stresses. *The Plant Journal*, Vol.49, No.6, (March 2007), pp. 1091-1107, ISSN 0960-7412

Park, S. J., Kwak, K. J., Oh, T. R., Kim, Y. O., Kang, H. (2009). Cold Shock Domain Proteins Affect Seed Germination and Growth of Arabidopsis thaliana Under Abiotic Stress Conditions. *Plant and Cell Physiology*, Vol.50, No.4, (April 2009), pp.869-878, ISSN 0032-0781

Park, S., Li, J., Pittman, J. K., Berkowitz, G. A., Yang, H., Undurraga, S., et al. (2005). Up-regulation of a H+-pyrophosphatase (H+-PPase) as a strategy to engineer drought-resistant crop plants. *Proceedings of the National Academy of Sciences of the United States of America*, Vol.102, No.52, (December 2005), pp. 18830-18835, ISSN 0027-8424

Pasapula, V., Shen, G., Kuppu, S., Paez-Valencia, J., Mendoza, M., Hou, P., Chen, J., Qiu, X., Zhu, L., Zhang, X., Auld, D., Blumwald, E., Zhang, H., Gaxiola, R., Payton, P. (2011). Expression of an Arabidopsis vacuolar H+-pyrophosphatase gene (AVP1) in cotton improves drought- and salt tolerance and increases fibre yield in the field conditions. *Plant Biotechnology Journal*, Vol.9, No.1, (January 2011), pp. 88-99, ISSN 1467-7644

Pennycooke, J., Cheng, H., Roberts, S., Yang, Q., Rhee, S., Stockinger, E. (2008). The low temperature-responsive, Solanum CBF1 genes maintain high identity in their upstream regions in a genomic environment undergoing gene duplications, deletions, and rearrangements. *Plant Molecular Biology*, Vol.67, No.5, (July 2008), pp. 483-497, ISSN 0167-4412

Price, A. H., Cairns, J. E., Horton, P., Jones, H. G., Griffiths, H. (2002). Linking drought-resistance mechanisms to drought avoidance in upland rice using a QTL approach: progress and new opportunities to integrate stomatal and mesophyll responses. *Journal of Experimental Botany*, Vol.53, No.371, (January 2002), pp. 989-1004, ISSN 0022-0957

Qin, F., Sakuma, Y., Li, J., Liu, Q., Li, Y.-Q., Shinozaki, K., Yamaguchi-Shinozaki, K. (2004). Cloning and Functional Analysis of a Novel DREB1/CBF Transcription Factor Involved in Cold-Responsive Gene Expression in Zea mays L. *Plant and Cell Physiology*, Vol.45, No.8, (May 2004), pp. 1042-1052, ISSN 0032-0781

Rabbani, M. A., Maruyama, K., Abe, H., Khan, M. A., Katsura, K., Ito, Y., Yoshiwara, K., Seki, M., Shinozaki, K., Yamaguchi-Shinozaki., K. (2003). Monitoring Expression Profiles of Rice Genes under Cold, Drought, and High-Salinity Stresses and Abscisic Acid Application Using cDNA Microarray and RNA Gel-Blot Analyses. *Plant Physiology*, Vol.133, No.4, (December 2003), pp. 1755-1767, ISSN 0032-0781

Reddy, A. S. (2007). Alternative Splicing of Pre-Messenger RNAs in Plants in the Genomic Era. *Annual Review of Plant Biology*, Vol.58, (June 2007), pp. 267-294, ISSN 1543-5008

Rekarte-Cowie, I., Ebshish, O. S., Mohamed, K. S., Pearce, R. S. (2008). Sucrose helps regulate cold acclimation of Arabidopsis thaliana. *Journal of Experimental Botany*, Vol.59, No.15, (November 2008), pp. 4205-4217, ISSN 0022-0957

Rensink, W., Hart, A, Liu, J., Ouyang, S., Zismann, V., Buel,l CR. (2005) Analyzing the potato abiotic stress transcriptome using expressed sequence tags. *Genome*, Vol.48,, No.4, (August 2005), pp. 598-605, ISSN 0225-7149

Rizhsky, L., Liang, H., Shuman, J., Shulaev, V., Davletova, S., Mittler, R. (2004). When Defense Pathways Collide. The Response of Arabidopsis to a Combination of Drought and Heat Stress. *Plant Physiology*, Vol.134, No.4, (April 2004), pp.1683-1696, ISSN 0032-0889

Rorat, T. (2006). Plant dehydrins – Tissue location, structure and function. *Cellular & Molecular Biology Letters*, Vol.*11*, No.4, (September 2006), pp. 536-556, ISSN 1425-8153

Rossini, S., Casazza, A. P., Engelmann, E. C., Havaux, M., Jennings, R. C., Soave, C. (2006). Suppression of Both ELIP1 and ELIP2 in Arabidopsis Does Not Affect Tolerance to Photoinhibition and Photooxidative Stress. *Plant Physiology*, Vol.141, No.4, (August 2006), pp. 1264-1273, ISSN 0032-0889

Saijo, Y., Hata, S., Kyozuka, J., Shimamoto, K., Izui, K. (2000). Over-expression of a single Ca2+-dependent protein kinase confers both cold and salt/drought tolerance on rice plants. *The Plant Journal*, Vol.23, No.3, (August 2000), pp. 319-327, ISSN 0960-7412

Sakamoto, A., Murata, N. (2000). Genetic engineering of glycinebetaine synthesis in plants: current status and implications for enhancement of stress tolerance. *Journal of Experimental Botany*,Vol.51, No.342, (January 2000), pp. 81-88, ISSN 0022-0957

Sakuma, Y., Maruyama, K., Qin, F., Osakabe, Y., Shinozaki, K., Yamaguchi-Shinozaki, K. (2006). Dual function of an Arabidopsis transcription factor DREB2A in water and heat-stress-responsive gene expression. *Proceedings of the National Academy of Sciences*, Vol.103, No.49, (December 2006), pp. 18822-18827, ISSN

Salvucci, M. E., Osteryoung, K. W., Crafts-Brandner, S. J., Vierling, E. (2001). Exceptional Sensitivity of Rubisco Activase to Thermal Denaturation in vitro and In vivo. *Plant Physiology*, Vol.127, No.3, (November 2001), pp. 1053-1064, ISSN 0032-0781

Sasaki, K., Saito, T., Lämsä, M., Oksman-Caldentey, K.-M., Suzuki, M., Ohyama, K., Muranaka, T., Ohara, K., Yazaki, K. (2007). Plants Utilize Isoprene Emission as a Thermotolerance Mechanism. *Plant and Cell Physiology*, Vol.48. No.9, (August 2007), pp. 1254-1262, ISSN 00320781

Schramm, F., Larkindale, J., Kiehlmann, E., Ganguli, A., Englich, G., Vierling, E., Von Koskull-Döring, P. (2008). A cascade of transcription factor DREB2A and heat stress transcription factor HsfA3 regulates the heat stress response of Arabidopsis. *The Plant Journal*, Vol.53, No.2, (January 2008), pp. 264-274, ISSN 0960-7412

Schroeder, J., Kwak, J., Allen, G. (2001). Guard cell abscisic acid signalling and engineering drought hardiness in plants. *Nature*, Vol.410, No.6826, (March 2001), pp. 327-30, ISSN 0028-0836

Serrano, R., Mulet, J. M., Rios, G., Marquez, J. A., F., I. i., Leube, M. P., Mendizabal, I., Pascual-Ahuir, A., Proft, M., Ros, R., Montesinos, C. (1999). A glimpse of the mechanisms of ion homeostasis during salt stress. *Journal of Experimental Botany*, Vol.50, No. Special Issue, (June 1999), pp. 1023-1036, ISSN 0022-0957

Shabala, S. and Cuin, T.A.(2008). Potassium transport and plant salt tolerance. *Physiologia Plantarum*, Vol.133,No.4,(August 2008), pp. 651-669, ISSN 0031-9317

Shi, H., Kim, Y., Guo, Y., Stevenson, B., Zhu, J.-K. (2002).The Arabidopsis SOS5 Locus Encodes a Putative Cell Surface Adhesion Protein and Is Required for Normal Cell Expansion. *The Plant Cell Online*, Vol.15, No.1, (January 2002), pp. 19-32, ISSN 1532-298X

Shibasaki, K., Uemura, M., Tsurumi, S. & Rahman, A. (2009). Auxin Response in Arabidopsis under Cold Stress, Underlying Molecular Mechanisms. *The Plant Cell*, Vol.21, No.12, (December 2009), pp. 3823-3838, ISSN 1040-4651

Shinozaki, K., Yamaguchi-Shinozaki, K. (2000). Molecular responses to dehydration and low temperature: differences and cross-talk between two stress signaling pathways. *Current Opinion in Plant Biology*, Vol.3, No.3, (June 2000), pp. 217-223, ISSN

Shinozaki, K., Yamaguchi-Shinozaki, K. (2007). Gene networks involved in drought stress response and tolerance. *Journal of Experimental Botany*, Vol.58, No.2, (January 2007), pp. 221-227, ISSN 0022-0957

Shou, H., Bordallo, P., Wang, K. (2004). Expression of the Nicotiana protein kinase (NPK1) enhanced drought tolerance in transgenic maize. *Journal of Experimental Botany*, Vol.75, No.5, (May 2005), pp. 1013-1019, ISSN 0022-0957

Skinner, J., Szucs, P., Zitzewitz, J. v., Marquez-Cedillo, L., Filichkin, T., Stockinger, E., Thomashow, M., F., Chen, T., H., H., Hates, P., M.. (2006). Mapping of barley homologs to genes that regulate low temperature tolerance in Arabidopsis. *TAG Theoretical and Applied Genetics*, Vol.*112*, No.5, (January 2006), pp.832-842, ISSN 0040-5752

Sohn, S., Back, K. (2007). Transgenic rice tolerant to high temperature with elevated contents of dienoic fatty acids. *Biologia Plantarum*, Vol.51, No.2, (June 2007), pp. 340-342, ISSN 0006-3134

Steponkus, P., L., Lynch, D., V. (1989). Freeze/thaw-induced destabilization of the plasma membrane and the effects of cold acclimation. *Journal of Bioenergetics and Biomembranes*, Vol.21, No.1, (February 1989), pp. 21-41, ISSN 0145-479X

Stockinger, E. J., Skinner, J. S., Gardner, K. G., Francia, E., Pecchioni, N. (2007) Expression levels of barley Cbf genes at the Frost resistance-H2 locus are dependent upon alleles at Fr-H1 and Fr-H2. *The Plant Journal*, Vol.51, No.12, (July 2007), pp.308-321, ISSN 0960-7412

Su, C.-F., Wang, Y.-C., Hsieh, T.-H., Lu, C.-A., Tseng, T.-H., Yu, S.-M. (2010). A Novel MYBS3-Dependent Pathway Confers Cold Tolerance in Rice. *Plant Physiology*, Vol.*153*, *No*.1, (May 2010), pp. 145-158, ISSN 0032-0889

Sung, S., Amasino, R. M. (2004). Vernalization and epigenetics: how plants remember winter. *Current Opinion in Plant Biology*, Vol.7, No.1, (Feb 2004), pp. 4-10, ISSN 1369-5266

Sunkar, R., Zhu, J.-K. (2004). Novel and Stress-Regulated MicroRNAs and Other Small RNAs from Arabidopsis. *The Plant Cell Online, Vol.16, No.*8, (August 2004), pp. 201-2019, ISSN 1040-4651

Sunkar, R., Chinnusamy, V., Zhu, J., Zhu, J.-K. (2007). Small RNAs as big players in plant abiotic stress responses and nutrient deprivation. *Trends in Plant Science*, Vol.12, No.7, (July 2007), pp. 301-309, ISSN 1360-1385

Sutton, F., Ding, X., Kenefick, D. G. (1992). Group 3 LEA Gene HVA1 Regulation by Cold Acclimation and Deacclimation in Two Barley Cultivars with Varying Freeze Resistance . *Plant Physiology*, Vol.99, No.1, (May 1992), pp. 338-340, ISSN 0032-0889

Swindell, W., Huebner, M., Weber, A. (2007). Transcriptional profiling of Arabidopsis heat shock proteins and transcription factors reveals extensive overlap between heat and non-heat stress response pathways. *BMC Genomics*,Vol.8, No.125 (March 2007), ISSN 1471-2164)

Szymanska, R. and Kruk, J. (2010). Plastoquinol is the Main Prenyllipid Synthesized During Acclimation to High Light Conditions in Arabidopsis and is Converted to Plastochromanol by Tocopherol Cyclase. *Plant and Cell Physiology*, Vol.51, No.4, (February 2010), pp. 537-545, ISSN 1471-9053

Takuhara, Y., Kobayashi, M., Suzuki, S. (2011) Low-temperature-induced transcription factors in grapevine enhance cold tolerance in transgenic Arabidopsis plants. *Journal of Plant Physiology*, Vol.168, No.9, (June 2011), pp. 967 - 975, ISSN 0176-1617

Tamura, K., Yamada, T. (2007). A perennial ryegrass CBF gene cluster is located in a region predicted by conserved synteny between Poaceae species. *TAG Theoretical and Applied Genetics, Vol.114*, No.2, (January 2007), pp. 273-283, ISSN 0040-5752

Tanabe, N., Yoshimura, K., Kimura, A., Yabuta, Y., Shigeoka, S. (2007). Differential Expression of Alternatively Spliced mRNAs of Arabidopsis SR Protein Homologs, atSR30 and atSR45a, in Response to Environmental Stress. *Plant and Cell Physiology*, Vol.48, No.7, (July 2007), pp. 1036-1049, ISSN 0032-0781

Tang, L., Kwon, S.-Y., Kim, S.-H., Kim, J.-S., Choi, J., Cho, K., Chang K. Sung, K., C., Kwak, S., Le, H. (2006). Enhanced tolerance of transgenic potato plants expressing both superoxide dismutase and ascorbate peroxidase in chloroplasts against oxidative stress and high temperature. *Plant Cell Reports*, Vol.25, No. 12, (July 2006), pp. 1380-1386 ISSN 0721-7714

Tester, M. and Davenport, R. (2003), Na+ Tolerance and Na+ Transport in Higher Plants. *Annals of Botany*, Vol.91, No.5, (April 2003), pp. 503-527, ISSN 0305-7364

Thomashow, M. (1999). Plant cold acclimation, Freezing tolerance genes and regulatory mechanisms. *Annual Review of Plant Physiology and Plant Molecular Biology*, Vol.50, (June 1999), pp. 571-599, ISSN 1040-2519

Thomashow, M. F. (1998). Role of Cold-Responsive Genes in Plant Freezing Tolerance. *Plant Physiology*, Vol.118, No.1, (September 1998), pp. 1-8, ISSN 0032-0889

Thomashow, M. F. (2001). So What's New in the Field of Plant Cold Acclimation? Lots! *Plant Physiology*, Vol.*125, No.*1, (January 2001), pp. 89-93, ISSN 0032-0889

Thomashow, M., F. (2010). Molecular basis of plant cold acclimation, insights gained from studying the CBF cold response pathway. *Plant Physiology*, Vol.154, No.2, (October 2010), pp. 571-577, ISSN 0032-0889

Thorlby, G., Fourrier, N., Warren, G. (2004). The SENSITIVE TO FREEZING2 Gene, Required for Freezing Tolerance in Arabidopsis thaliana, Encodes a beta-

Glucosidase. *The Plant Cell,* Vol. 16, No.8, (August 2004), pp. 2192-2203, ISSN 1040-4651

Tian, Y., Zhang, H., Pan, X., Chen, X., Zhang, Z., Lu, X., Huang, R. (2010). Overexpression of ethylene response factor TERF2 confers cold tolerance in rice seedlings. *Transgenic Research,* Epub ahead of print, (December 2010), ISSN 0962-8819

Tzvetkova-Chevolleau, T., Franck, F., Alawady, A. E., Dall'Osto, L., Carrière, F., Bassi, R., Grimm, B., Nussaume, L., Havaux, M. (2007). The light stress-induced protein ELIP2 is a regulator of chlorophyll synthesis in Arabidopsis thaliana. *The Plant Journal* , Vol. 50, No.5, (June 2007), pp. 795-809, ISSN 0960-7412

Uemura, M., Joseph, R., A., Steponkus, P.,l. (1995) Cold Acclimation of Arabidopsis thaliana (Effect on Plasma Membrane Lipid Composition and Freeze-Induced Lesions). *Plant Physiology,* Vol.109, No.1, (September 1995), pp. 15-30, ISSN 0032-0889

Umezawa, T., Fujita, M., Fujita, Y., Yamaguchi-Shinozaki, K., Shinozaki, K. (2006). Engineering drought tolerance in plants: discovering and tailoring genes to unlock the future. *Current Opinion in Biotechnology,* Vol.*17,* No.2, (April 2006), pp. 113-122, ISSN 0958-1669

Vera-Estrella, R., Barkla, B. J., García-Ramírez, L. and Pantoja, O. (2005). Salt Stress in Thellungiella halophila Activates Na+ Transport Mechanisms Required for Salinity Tolerance. *Plant Physiology,* Vol.139, No.3, (November 2005), pp. 1507-1517, ISSN 0032-0889

Villalobos, M. A., Bartels, D. & Iturriaga, G. (2004). Stress Tolerance and Glucose Insensitive Phenotypes in Arabidopsis Overexpressing the CpMYB10 Transcription Factor Gene. *Plant Physiology,*Vol.*135,* No.1, (May 2004), pp. 309-324 , ISSN 1532-2548

Vogel, J. T., Zarka, D. G., Van, H. A., Fowler, S. G., Thomashow, M. F. (2005). Roles of the CBF2 and ZAT12 transcription factors in configuring the low temperature transcriptome of Arabidopsis. *The Plant Journal,* Vol.41, No.2, (January 2005), pp. 195-211, ISSN 0960-7412

von Koskull-Döring, P., Scharf, K.-D., Nover, L. (2007). The diversity of plant heat stress transcription factors. *Trends in Plant Science,* Vol.12, No.10, (October 2007), pp. 452 - 457, ISSN 1360-1385

Wahid, A., Gelani, S., Ashraf, M., Foolad, M. (2007). Heat tolerance in plants, An overview. *Environmental and Experimental Botany,* Vol.61, No.3, (December 2007), pp. 199 - 223, ISSN 0098-8472

Wang, C., Zhang, Q., Shou, H.-x. (2009). Identification and expression analysis of OsHsfs, in rice. *Journal of Zhejiang University - Science B,* Vol.10, No.4, (April 2009), pp. 291-300, ISSN 1673-1581

Wang, S., Wan, C., Wang, Y., Chen, H., Zhou, Z., Fu, H., Sosebee, R., E. (2004). The characteristics of Na+, K+ and free proline distribution in several drought-resistant plants of the Alxa Desert, China. *Journal of Arid Environments,* Vol.56 No.3, (Febuary 2004), pp. 525-539, ISSN 0140-1963

Wang, W., Vinocur, B., Altman, A. (2003). Plant responses to drought, salinity and extreme temperatures: towards genetic engineering for stress tolerance. *Planta,* Vol.*218, No.*1, (November 2003), pp. 1-14, ISSN 0032-0935

Wang, W., Vinocur, B., Shoseyov, O., Altman, A. (2004). Role of plant heat-shock proteins and molecular chaperones in the abiotic stress response. *Trends in Plant Science,*Vol. 9, No.5, (May 2004), pp. 244-252, ISSN 1360-1385

Wang, W., Vinocur, B., Shoseyov, O., Altman, A. (2004). Role of plant heat-shock proteins and molecular chaperones in the abiotic stress response. *Trends in Plant Science*, Vol.9, No.5, (May 2004), pp. 244 - 252, ISSN 1360-1385

Wang, X., Li, W., Li, M., Welti, R. (2006). Profiling lipid changes in plant response to low temperatures. *Physiologia Plantarum*, Vol.126, No.1, (January 2006), pp. 90-96, ISSN 0031-9317

Watanabe, S., Kojima, K., Ide, Y., Sasaki, S. (2000). Effects of saline and osmotic stress on proline and sugar accumulation in Populus euphratica in vitro. *Plant Cell, Tissue and Organ Culture*, Vol.63, No. 3, (November 2000), pp. 199-206, ISSN 0167-6857

Webb, M. S., Uemura, M., Steponkus, P. L. (1994). A Comparison of Freezing Injury in Oat and Rye: Two Cereals at the Extremes of Freezing Tolerance. *Plant Physiology*, Vol.104, No.2, (Febuary 1994), pp. 467-478. ISSN 0032-0889

Webb, M. S., Uemura, M., Steponkus, P. L. (1994). A Comparison of Freezing Injury in Oat and Rye, Two Cereals at the Extremes of Freezing Tolerance. *Plant Physiology*, Vol.104, No.2, (February 1994), pp. 467–478, ISSN 0032-0889

Weston, E., Thorogood, K., Vinti, G., López-Juez, E. (2000). Light quantity controls leaf-cell and chloroplast development in Arabidopsis thaliana, wild type and blue-light-perception mutants. *Planta*, Vol.211, No.6, (November 2000) pp. 807-815, ISSN 0032-0935

Williams, M. E., Torabinejad, J., Cohick, E., Parker, K., Drake, E. J., Thompson, J. E., Hortter, M., DeWald, D., B. (2005). Mutations in the Arabidopsis Phosphoinositide Phosphatase Gene SAC9 Lead to Overaccumulation of PtdIns(4,5)P2 and Constitutive Expression of the Stress-Response Pathway. *Plant Physiology*, Vol.138, No.2, (June 2005), pp. 686-700, ISSN 0032-0889

Witcombe, J., Hollington, P., Howarth, C., Reader, S., Steele, K. (2008). Breeding for abiotic stresses for sustainable agriculture. *Philosophical Transactions of the Royal Society B: Biological Sciences*, Vol.363, No.1492, (Febuary 2008), pp. 703-716, ISSN 1471-2970

Woodward, A. J., Bennett, I. J. (2005). The effect of salt stress and abscisic acid on proline production, chlorophyll content and growth of in vitro propagated shoots of Eucalyptus camaldulensis. *Plant Cell, Tissue and Organ Culture*, Vol.82, No.2, (January 2005), pp. 189-200, ISSN 0167-6857

Wu, G.-Q., Xi, J.-J., Wang, Q., Bao, A.-K., Ma, Q., Zhang, J.-L., Wang, S-M. (2011). The ZxNHX gene encoding tonoplast Na+/H+ antiporter from the xerophyte Zygophyllum xanthoxylum plays important roles in response to salt and drought. *Journal of Plant Physiology*, Vol.168, No.8, (May 2011), pp. 758-767, ISSN 0032-0889

Wu, Y.-Y.,Chen, Q.-J., Chen, M., Chen, J., and Wang, X.-C. (2005). Salt-tolerant transgenic perennial ryegrass (Lolium perenne L.) obtained by Agrobacterium tumefaciens-mediated transformation of the vacuolar Na+/H+ antiporter gene. *Plant Science*, Vol.169, No.1, (July 2005), pp. 65-73, ISSN 0306-4484

Xie, C., G., Lin, H., Deng, H., W., Guo, Y. (2009). Roles of SCaBP8 in salt stress response. *Plant Signal Behavior*, Vol.4, No.10, (October 2009), pp. 956-958, ISSN1559-2316

Xiong, L. and Yang, Y. (2003). Disease Resistance and Abiotic Stress Tolerance in Rice Are Inversely Modulated by an Abscisic Acid Inducible Mitogen-Activated Protein Kinase. *The Plant Cell*, Vol.15, No.3, (March 2003), pp.45-759, ISSN 1040-4651

Xu, K., Hong, P., Luo, L., Xia, T. (2009). Overexpression of *AtNHX1*, a Vacuolar Na^+/H^+ Antiporter from *Arabidopsis thalina*, in *Petunia hybrida* Enhances Salt and Drought

Tolerance. *Journal of Plant Biology,* Vol.52, No.5, (August 2009), pp. 453-461, ISSN 1226-9239

Xu, Y., Gianfagna, T., Huang, B. (2010). Proteomic changes associated with expression of a gene (ipt) controlling cytokinin synthesis for improving heat tolerance in a perennial grass species. *Journal of Experimental Botany,*Vol.61, No.12, (June 2010), ISSN 1460-2431

Xu, Z. and Zhou, G. (2006). Combined effects of water stress and high temperature on photosynthesis, nitrogen metabolism and lipid peroxidation of a perennial grass Leymus chinensis. *Planta,* Vol.224, No.5, (October 2006), pp. 1080-1090, ISSN 0032-0935

Xue, Z.-Y., Zhi, D.-Y., Xue, G.-P., Zhang, H., Zhao, Y.-X., Xia, G.-M. (2004). Enhanced salt tolerance of transgenic wheat (Tritivum aestivum L.) expressing a vacuolar Na+/H+ antiporter gene with improved grain yields in saline soils in the field and a reduced level of leaf Na+. *Plant Science,* Vol.167, No.4, (October 2004), pp. 849 - 859, ISSN 0306-4484

Yaeno, T., Matsuda, O., Iba, K. (2004). Role of chloroplast trienoic fatty acids in plant disease defense responses. *The Plant Journal,* Vol.40, No.6, (September 2004), pp. 931-941, ISSN 0960-7412

Yamada, K., Fukao, Y., Hayashi, M., Fukazawa, M., Suzuki, I., Nishimura, M. (2007). Cytosolic HSP90 Regulates the Heat Shock Response That Is Responsible for Heat Acclimation in Arabidopsis thaliana. *Journal of Biological Chemistry,* Vol.282, No.52, (December 2007), pp. 37794-37804, ISSN 0021-9258

Yang, J. and Zhang, J. (2006). Grain filling of cereals under soil drying. *New Phytologist,* Vol.169, No.2, (January 2006), pp. 223-236, ISSN 0028-646X

Yang, X., Wen, X., Gong, H., Lu, Q., Yang, Z., Tang, Y., et al. (2007). Genetic engineering of the biosynthesis of glycinebetaine enhances thermotolerance of photosystem II in tobacco plants. *Planta,* Vol.225, No.3 (September 2007), pp. 719-733, ISSN 0032-0935

Yao, J., Roy-Chowdhury, S., Allison, L. A. (2003). AtSig5 Is an Essential Nucleus-Encoded Arabidopsis s-Like Factor. *Plant Physiology,* Vol.132, No.2, (June2003), pp. 739-747, ISSN 0032-0889

Yao, Y., Ni, Z., Peng, H., Sun, F., Xin, M., Sunkar, R., et al. (2010). Non-coding small RNAs responsive to abiotic stress in wheat (Triticum aestivum). *Functional &; Integrative Genomics,Vol.10,* No.2, (May 2010), pp. 187-190, ISSN 1438-7948

Yeo, A. (1998). Predicting the interaction between the effects of salinity and climate change on crop plants. *Scientia Horticulturae,* Vol.78, No.1-4, (November 1998), pp. 159-174, ISSN 0304-4238

Yokotani, N., Ichikawa, T., Kondou, Y., Matsui, M., Hirochika, H., Iwabuchi, M., et al. (2008). Expression of rice heat stress transcription factor OsHsfA2e enhances tolerance to environmental stresses in transgenic Arabidopsis. *Planta,* Vol.227, No.5, (April 2008), pp. 957-967, ISSN 0032-0935

Yordanov, I., Velikova, V., Tsonev, T. (2000). Plant responses to drought and stress tolerance. *Photosynthetica.* Vol.30, No.2, (July 2000), pp.187-206, ISSN 03003604

Yoshida, T., Sakuma, Y., Todaka, D., Maruyama, K., Qin, F., Mizoi, J., et al. (2008). Functional analysis of an Arabidopsis heat-shock transcription factor HsfA3 in the transcriptional cascade downstream of the DREB2A stress-regulatory system.

Biochemical and Biophysical Research Communications, Vol.368, No.3, (April 2008), pp.515-521., ISSN 1090-2104

Yoshimura, K., Masuda, A., Kuwano, M., Yokota, A., Akashi, K. (2008). Programmed Proteome Response for Drought Avoidance/Tolerance in the Root of a C3 Xerophyte (Wild Watermelon) Under Water Deficits. *Plant and Cell Physiology*, Vol. 49. No.2, (January 2008), pp. 226-241, ISSN 0032-0781

Zelisko, A., García-Lorenzo, M., Jackowski, G., Jansson, S., Funk, C. (2005). AtFtsH6 is involved in the degradation of the light-harvesting complex II during high-light acclimation and senescence. *Proceedings of the National Academy of Sciences of the United States of America*, Vol.102, No.38, (September 2005), pp. 13699-13704, ISSN 0027-8424

Zhang, G.-H., Su, Q., An, L.-J., Wu, S. (2008). Characterization and expression of a vacuolar Na+/H+ antiporter gene from the monocot halophyte Aeluropus littoralis. *Plant Physiology and Biochemistry*, Vol.46, No.2, (February 2008), pp. 117-126, ISSN 0981-9428

Zhang, H. & Blumwald, E. (2001). Transgenic salt-tolerant tomato plants accumulate salt in foliage but not in fruit. *Nature Biotechnology*, Vol.19, No.8, (August 2001), pp. 756-758, ISSN 1087-0156

Zhang, H., Irving, L. J., McGill, C., Matthew, C., Zhou, D., Kemp, P. (2010). The effects of salinity and osmotic stress on barley germination rate: sodium as an osmotic regulator. *Annals of Botany*,Vol.106, No.6, (December 2010), pp. 1027-1035, ISSN 03057364

Zhang, J., Jia, W., Yang, J., Ismail, A. M. (2006). Role of ABA in integrating plant responses to drought and salt stresses. *Field Crops Research*,Vol.97, No.1, (May 2006), pp. 111-119, ISSN 0378-4290

Zhang, J.-L., Flowers, T., Wang, S.-M. (2010). Mechanisms of sodium uptake by roots of higher plants. *Plant and Soil*. Vol.326, No.1-2, (July 2010), pp. 45-60, ISSN 1573-5036

Zhang, Z. and Huang, R. (2010). Enhanced tolerance to freezing in tobacco and tomato overexpressing transcription factor TERF2/LeERF is modulated by ethylene biosynthesis. *Plant Molecular Biology*, Vol.73, No.3, (June 2010), pp. 241-249, ISSN 0167-4412

Zhao, J., Ren, W., Zhi, D., Wang, L., Xia, G. (2007). Arabidopsis DREB1A/CBF3 bestowed transgenic tall fescue increased tolerance to drought stress. *Plant Cell Reports*, Vol.26, No.9, (September 2010), pp. 1521-1528, ISSN 0721-7714

Zheng, X., Chen, B., Lu, G., Han, B. (2009). Overexpression of a NAC transcription factor enhances rice drought and salt tolerance. *Biochemical and Biophysical Research Communications*, Vol.379, No.4, (Febuary 2009), pp. 985-989, ISSN 0006-291X

Zhou, S., Wei, S., Boone, B., Levy, S. (2007). Microarray analysis of genes affected by salt stress in tomato. *African Journal of Environmental Science and Technology*, Vol.1, No.2, (September 2007), pp.014-026, ISSN 1996-0786

Zhu, J.-K. (2002). Salt and Drought Stress Signal Transduction In Plants. *Annual Review of Plant Biology*, Vol.53, No.1, (June 2002), pp. 247-273, ISSN 1 543-5008

10

Role of Plant Transcription Factors in Abiotic Stress Tolerance

Charu Lata, Amita Yadav and Manoj Prasad
National Institute of Plant Genome Research, Aruna Asaf Ali Marg, New Delhi, India

1. Introduction

Plants are constantly exposed to a wide range of environmental stresses such as drought, high salt, heat and extremes of temperature. Growth constraints due to these abiotic stresses result in reduced productivity and significant crop losses globally. Drought and salinity affect more than 10% of arable land, which results in more than 50% decline in the average yields of important crops worldwide (Bray et al., 2000). Tolerance or susceptibility to these stresses is also a very intricate event as stress may affect multiple stages of plant development and often several stresses concurrently affect the plants (Chinnusamy et al., 2004). Therefore, the basic mechanisms of abiotic stress tolerance and adaptation have been the area of comprehensive research.

Plants counter adverse environmental conditions in a complex, integrated way depending on the timing and length that allows them to respond and adapt to the existing constraints present at a given time. Plant stress tolerance involves changes at whole-plant, tissue, cellular, physiological and molecular levels. Exhibition of a distinct or a combination of intrinsic changes ascertains the capacity of a plant to sustain itself under unfavorable environmental conditions (Farooq et al., 2009). This comprises a range of physiological and biochemical adjustments in plants including leaf wilting, leaf area reduction, leaf abscission, root growth stimulation, alterations in relative water content (RWC), electrolytic leakage (EL), production of reactive oxygen species (ROS) and accumulation of free radicals which disturb cellular homeostasis ensuing lipid peroxidation, membrane damage, and inactivation of enzymes thus influencing cell viability (Bartels and Sunkar, 2005). Other than these, abscissa acid (ABA), a plant stress hormone, induces leaf stomata closure, thus reducing transpirational water loss and photosynthetic rate which improves the water-use efficiency (WUE) of the plant. Molecular responses to abiotic stress on the other hand include perception, signal transduction, gene expression and ultimately metabolic changes in the plant thus providing stress tolerance (Agarwal et al., 2006).

Several genes are activated in response to abiotic stresses at the transcriptional level, and their products are contemplated to provide stress tolerance by the production of vital metabolic proteins and also in regulating the downstream genes (Kavar et al., 2007). Transcript profiling can be a significant tool for the characterization of stress-responsive genes. Extensive transcriptome analyses have divulged that these gene products can largely be classified into two groups (Bohnert et al., 2001; Seki et al., 2002; Fowler and Thomashow, 2002). First group comprises of genes that encode for proteins that defend the cells from the effects of water-deficit. These genes mainly include those that regulate the accumulation of compatible solutes (enzymes for osmolyte biosynthesis like proline, betaine, sugars, etc.); passive and active transport systems across membranes (water channel proteins and membrane transporters); and protection and stabilization of cell structures from damage by ROS (the detoxification enzymes such as glutathione *S*-transferase, catalase, superoxide dismutase , ascorbate peroxidase, etc.); fatty acid metabolism enzymes, proteinase inhibitors, ferritin and lipid-transfer proteins; and other proteins for the protection of macromolecules (LEA protein, osmotin, chaperons, etc.). Another group of genes stimulated by abiotic stresses includes regulatory proteins that further regulate the stress signal transduction and alter gene expression and hence possibly function in stress response. They comprise several transcription factors (TFs) emphasizing the role of various transcriptional regulatory mechanisms in the stress signal transduction pathways; protein kinases (MAP kinase, CDP kinase, receptor protein kinase, etc.); protein phosphatases and proteinases implicated in the regulation of stress signaling and gene expression (Seki et al., 2003; Shinozaki and Yamaguchi- Shinozaki, 2007).

2. Role of transcription factors in abiotic stress responses

Transcription factors (TFs) are proteins that act together with other transcriptional regulators, including chromatin remodeling/modifying proteins, to employ or obstruct RNA polymerases to the DNA template (Udvardi et al., 2007). Plant genomes assign approximately 7% of their coding sequence to TFs, which proves the complexity of transcriptional regulation (Udvardi et al., 2007). The TFs interact with *cis*-elements in the promoter regions of several stress-related genes and thus up-regulate the expression of many downstream genes resulting in imparting abiotic stress tolerance (Agarwal and Jha, 2010). In *Arabidopsis thaliana* genome about 1500 TFs are described which are considered to be involved in stress responsive gene expression (Riechmann et al., 2000). Transcriptome data in *Arabidopsis* and in numerous other plants suggest that there are several pathways that independently respond to environmental stresses (in both ABA dependent- and independent- manner), suggesting that stress tolerance or susceptibility is controlled at the transcriptional level by an extremely intricate gene regulatory network. (Fig.1) (Fowler and Thomashow, 2002; Umezawa et al., 2006).

The phytohormone ABA is the central regulator of abiotic stress particularly drought resistance in plants, and coordinates a complex gene regulatory network enabling plants to cope with decreased water availability (Cutler et al., 2010; Kim et al., 2010). ABA-dependent signaling systems have been illustrated as pathways that mediate stress adaptation by induction of atleast two separate regulons (a group of genes controlled by a certain TF): (1) the AREB/ABF (ABA-responsive element-binding protein/ ABA-binding factor) regulon; and (2) the MYC (myelocytomatosis oncogene)/MYB (myeloblastosis oncogene) regulon (Abe et al., 1997; Busk and Pagés, 1998; Saibo et al., 2009). While ABA-independent regulons are: (1) the CBF/DREB regulon; and (2) the NAC (NAM, ATAF and CUC) and ZF-HD (zinc-finger homeodomain) regulon (Nakashima et al. 2009; Saibo et al., 2009). However in addition, several studies have identified the existence of both ABA-dependent and -independent pathways of stress response that function through AP2/EREBP (ERF) family members (Yamaguchi-Shinozaki and Shinozaki, 1994; Kizis and Pagés, 2002). In addition to these well-known regulons, a large number of other TFs are also involved in abiotic stress responses, thereby playing a crucial role in imparting stress endurance to plants. Although

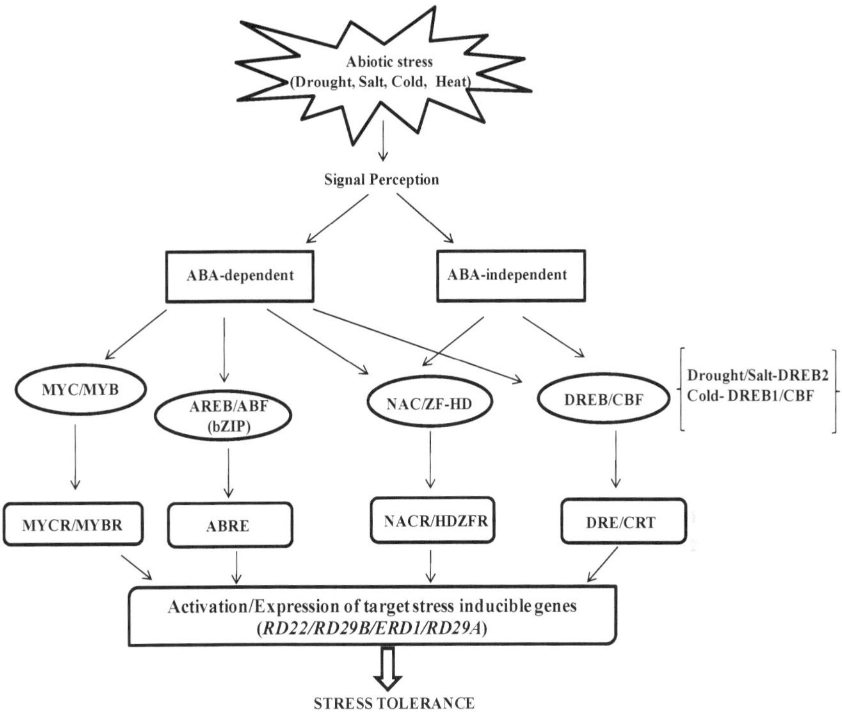

Fig. 1. A schematic representation of transcriptional regulatory networks of *cis*-acting elements and transcription factors involved in abiotic-stress-responses. Transcription factors are shown in ellipses; *cis*-acting elements are shown in boxes; and target stress inducible genes are shown in long rectangular box at the bottom.

these different stress responsive TFs usually function independently, it is undoubtedly possible that some level of cross-talk exists between them.

This chapter focuses on these TFs and their role in regulating abiotic stress responses in plants (Table 1) as well as their utility in engineering stress tolerance for crop improvement programs (Table 2).

Family	Gene	ABA response	Inducible by	Species	cis-element	References
bZIP	ABF1	Yes	Cold	*Arabidopsis thaliana*	ABREs (CACGTGGC/CGCGTGGC)	Choi et al., 2000
	ABF2	Yes	Salt, Drought	*Arabidopsis thaliana*	ABREs	Choi et al., 2000
	ABF3	Yes	Salt	*Arabidopsis thaliana*	ABREs	Choi et al., 2000
	ABF4	Yes	Drought, Salt, Cold	*Arabidopsis thaliana*	ABREs	Choi et al., 2000
	GmbZIP44	Yes	Drought, Salt	*Glycine max*	GLM (GTGAGTCAT)	Liao et al., 2008a
	GmbZIP62	Yes	Cold, Drought, Salt	*Glycine max*	ABRE (CCACGTGG)	Liao et al., 2008a
	GmbZIP78	Yes	Drought, Salt	*Glycine max*	PB-like (TGAAAA)	Liao et al., 2008a
	GmbZIP132	Yes	Cold, Drought, Salt	*Glycine max*	GCN4-like motif (GTGAGTCAT)	Liao et al., 2008b
	Wlip19	Yes	Cold, Drought	*Triticum aestivum*	NA	Kobayashi et al., 2008
	OsABI5	Yes	Salt	*Oryza sativa*	G-box element (CACGTG)	Zou et al., 2008

Family	Gene	ABA response	Inducible by	Species	cis-element	References
	ZmbZIP17	Yes	Drought, Heat, Salt	*Zea mays*	NA	Jia et al., 2009c
	OsbZIP23	Yes	Drought, Salt, PEG	*Oryza sativa*	NA	Xiang et al., 2008
	OsAREB1	Yes	Drought, Heat	*Oryza sativa*	ABRE cis-element (ACGTGCC)	Jin et al. 2010
MYC	AtMYC2	Yes	Drought, Salt, Cold	*Arabidopsis thaliana*	CACATG	Abe et al., 2003
MYB	AtMYB2	Yes	Drought, Salt	*Arabidopsis thaliana*	TGGTTAG	Abe et al., 2003
	AtMYB4	Yes	Salt, Ethylene, JA, SA	*Arabidopsis thaliana*	NA	Yanhui et al., 2006
	AtMYB6	Yes	Salt, Ethylene, JA, SA	*Arabidopsis thaliana*	NA	Yanhui et al., 2006
	AtMYB7	Yes	Salt, Ethylene, JA, SA	*Arabidopsis thaliana*	NA	Yanhui et al., 2006
	AtMYB44	Yes	Salt, Ethylene, JA, SA	*Arabidopsis thaliana*	NA	Yanhui et al., 2006
	AtMYB73	Yes	Salt, Ethylene, JA, SA	*Arabidopsis thaliana*	NA	Yanhui et al., 2006
	MYB15	Yes	Drought, Salt, Cold	*Arabidopsis thaliana*	Type IIG Myb recognition sites	Agarwal et al., 2006; Ding et al., 2009
	GmMYB76	No	Salt	*Glycine max*	MBSI (TATAACGGTTTTTT)	Liao et al., 2008c
	GmMYB92	No	Cold, Salt	*Glycine max*	MBSI, MRE4 (TCTCACCTACC)	Liao et al., 2008c
	GmMYB177	No	Drought, Salt	*Glycine max*	MBSI	Liao et al., 2008c
	OsMYB3R-2	NA	Drought, Salt, Cold	*Oryza sativa*	NA	Dai et al., 2007
CBF/DREB	DREB1A	No	Cold	*Arabidopsis thaliana*	DRE sequence (TACCGACAT)	Liu et al., 1998
	DREB2A	No	Drought, Salt	*Arabidopsis thaliana*	DRE sequence	Liu et al., 1998
	DREB2C	No	Salt, Mannitol, Cold	*Arabidopsis thaliana*	C-repeat/DRE	Lee et al., 2010
	CBF1	No	Cold	*Arabidopsis thaliana*	DRE/CRT	Gilmour et al., 1998
	CBF2	NA	Cold	*Arabidopsis thaliana*	DRE/CRT	Gilmour et al., 1998
	CBF3	NA	Cold	*Arabidopsis thaliana*	DRE/CRT	Gilmour et al., 1998
	CBF4	Yes	Drought	*Arabidopsis thaliana*	NA	Haake et al., 2002
	OsDREB1A	No	Cold, Salt, Wounding	*Oryza sativa*	GCCGAC	Dubouzet et al., 2003
	OsDREB1B	No	Cold	*Oryza sativa*	DRE/CRT	Dubouzet et al., 2003
	OsDREB1C	Yes	Drought, Salt, Cold, Wound	*Oryza sativa*	DRE/CRT	Dubouzet et al., 2003
	OsDREB1D	No	None	*Oryza sativa*	DRE/CRT	Dubouzet et al., 2003
	OsDREB1F	No	Drought, Salt, Cold	*Oryza sativa*	DRE/CRT	Wang et al., 2008
	OsDREB2A	No	Drought, Salt, faintly to Cold	*Oryza sativa*	DRE/CRT	Dubouzet et al., 2003
	OsDREB2B	No	Heat, Cold	*Oryza sativa*	DRE/CRT	Matsukura et al., 2010
	OsDREB2C	No	None	*Oryza sativa*	DRE/CRT	Matsukura et al., 2010
	OsDREB2E	No	None	*Oryza sativa*	DRE/CRT	Matsukura et al., 2010
	TaDREB1	No	Cold, Drought	*Triticum aestivum*	DRE sequence (TACCGACAT)	Shen et al., 2003a
	WDREB2	Yes	Drought, Salt, Cold	*Triticum aestivum*	NA	Egawa et al., 2006
	HvDRF1	Yes	Drought, Salt	*Hordeum vulgare*	T(T/A)ACCGCCTT)	Xue and Loveridge, 2004
	HvDREB1	No	Drought, Salt, Cold	*Hordeum vulgare*	DRE/CRT elements	Xu et al., 2009
	ZmDREB2A	No	Drought, Salt, Cold, Heat	*Zea mays*	DRE sequence	Qin et al., 2007
	PgDREB2A	No	Drought, Salt, Cold	*Pennisetum glaucum*	(DRE) ACCGAC	Agarwal et al., 2007
	SbDREB2	NA	Drought	*Sorghum bicolor*	NA	Bihani et al., 2011
	SiDREB2	No	Drought, Salt	*Setaria italica*	NA	Lata et al., 2011
	CaDREB-LP1	No	Drought, Salt, Wounding	*Capsicum annum*	NA	Hong and Kim, 2005
	AhDREB1	NA	Salt	*Artiplex hortensis*	DRE sequence	Shen et al., 2003b
	GmDREBa	No	Cold, Drought, Salt	*Glycine max*	DRE sequence	Li et al., 2005
	GmDREBb	Yes	Cold, Drought, Salt	*Glycine max*	DRE sequence	Li et al., 2005
	GmDREBc	No	Drought, Salt	*Glycine max*	DRE sequence	Li et al., 2005
	GmDREB2	No	Drought, Salt	*Glycine max*	DRE sequence	Chen et al., 2007
	DmDREBa	Yes	Cold	*Dendronthema x moriforlium*	DRE sequence	Yang et al., 2009

Family	Gene	ABA response	Inducible by	Species	cis-element	References
	DmDREBb	Yes	Cold	*Dendronthema x moriforlium*	DRE sequence	Yang et al., 2009
	PNDREB1	No	Drought, Cold	*Arachis hypogea*	DRE sequence	Mei et al., 2009
	CAP2	Yes	Drought, Salt, Auxin	*Cicer arietinum*	C-repeat/DRE (TACCGACAT)	Shukla et al., 2006
	DvDREB2A	Yes	Drought, Heat, Cold	*Dendrathema*	DRE sequence (TACCGACAT)	Liu et al., 2008
	SbDREB2A	NA	Drought, Salt, Heat	*Salicornia brachiata*	DREs (ACCGAC and GCCGAC)	Gupta et al., 2010
NAC	ATAF1		Drought	*Arabidopsis thaliana*	NA	Lu et al., 2007
	AtNAC2		Salt	*Arabidopsis thaliana*	NA	He et al., 2005
	AtNAC019		Drought, Salt	*Arabidopsis thaliana*	CATGTG motif	Tran et al., 2004
	AtNAC055		Drought, Salt	*Arabidopsis thaliana*	CATGTG motif	Tran et al., 2004
	AtNAC072		Drought, Salt	*Arabidopsis thaliana*	CATGTG motif	Tran et al., 2004
	OsNAC6		Cold, Drought, Salt	*Oryza sativa*	NA	Nakahsima et al., 2007
	SNAC1		Cold, Drought, Salt	*Oryza sativa*	NAC recognition sequence	Hu et al., 2006
	SNAC2		Cold, Drought, Salt	*Oryza sativa*	NAC recognition sequence	Hu et al., 2008
	CarNAC5		Drought, Heat	*Cicer arietinum*	NA	Peng et al., 2009
	TaNAC4		Cold, Salt, Wounding, Ethylene, MeJA	*Triticum aestivum*	NA	Xia et al., 2010
	GhNAC2		Cold, Drought	*Gossypium hirsutum*	NA	Meng et al., 2009
	GhNAC3		Cold	*Gossypium hirsutum*	NA	Meng et al., 2009
	GhNAC4		Cold, Drought, Salt	*Gossypium hirsutum*	NA	Meng et al., 2009
	GhNAC5		Cold, Drought	*Gossypium hirsutum*	NA	Meng et al., 2009
	GhNAC6		Cold, Drought, Salt	*Gossypium hirsutum*	NA	Meng et al., 2009
	SiNAC		Drought, Salt, Ethephone, MeJA	*Setaria italica*	NA	Puranik et al., 2011
WRKY	OsWRKY45			*Oryza sativa*	NA	Qiu and Yu, 2009
	GmWRKY21			*Glycine max*	W-box (TTGAC)	Zhou et al., 2008
	GmWRKY54			*Glycine max*	W-box	Zhou et al., 2008
	GmWRKY13			*Glycine max*	W-box	Zhou et al., 2008
	NbWRKY			*Nicotiana benthamiana*	NA	Archana et al., 2009
ZFP	ZPT2-3			*Petunia sp.*	NA	Sugano et al., 2003
	OSISAP1			*Oryza sativa*	NA	Mukhopadhyay et al., 2004
	CaZF			*Cicer arietinum*	EP1S (TGACAGTGTCA)	Jain et al., 2009

Table 1. Response of transcription factors to various stresses.

Family	Gene	Transgenic Plants	Stress Tolerance	References
bZIP	ABF2	Arabidopsis	Drought	Kim et al., 2004
	ABF3	Arabidopsis	Drought	Kang et al., 2002
	ABF4	Arabidopsis	Drought	Kang et al., 2002
	GmbZIP44	Arabidopsis	Salinity, Freezing	Liao et al., 2008a
	GmbZIP62	Arabidopsis	Salinity, Freezing	Liao et al., 2008a
	GmbZIP78	Arabidopsis	Salinity, Freezing	Liao et al., 2008a
	GmbZIP132	Arabidopsis	Salinity	Liao et al., 2008b
	AtbZIP60	Arabidopsis	Salinity	Fujita et al., 2007
	Wlip19	Tobacco	Freezing	Kobayashi et al., 2008
	OsABI5	Rice	Salinity	Zou et al., 2008

Family	Gene	Transgenic Plants	Stress Tolerance	References
	OsbZIP23	Rice	Drought, Salinity	Xiang et al., 2008
	OsAREB1	Rice	Drought, Heat	Jin et al. 2010
bHLH	AtMYC2	Arabidopsis	Osmotic stress	Abe et al., 2003
MYB	AtMYB2	Arabidopsis	Osmotic stress	Abe et al., 2003
	MYB15	Arabidopsis	Drought, Salinity	Ding et al., 2009
	OsMYB3R-2	Arabidopsis	Drought, Salinity, Cold	Dai et al., 2007
	OsMYB4	Arabidopsis	Freezing	Vannini et al., 2004
	OsMYB4	Arabidopsis	Drought	Mattana et al. 2005
	OsMYB4	Tomato	Drought	Vannini et al., 2007
CBF/DREB	AtDREB1A	Arabidopsis	Drought	Liu et al., 1998
	AtDREB1A	Tobacco	Freezing, Drought	Kasuga et al., 2004
	AtDREB1A	Wheat	Drought	Pellegrineschi et al., 2004
	AtDREB1A	Rice	Drought, Salinity	Oh et al., 2005
	AtDREB1A	Potato	Salinity	Behnam et al., 2006
	AtDREB1A	Peanut	Drought	Bhatnagar-Mathur et al., 2006
	AtDREB2A-CA	Arabidopsis	Drought	Sakuma et al., 2006
	AtCBF1	Arabidopsis	Salinity	Jaglo-Ottosen et al.,1998
	AtCBF2	Tomato	Freezing	Hseih et al., 2002
	AtCBF3	Arabidopsis	Freezing	Gilmour et al., 2000
	AtCBF4	Arabidopsis	Freezing	Haake et al., 2002
	BNCBF5	Brassica napus	Freezing	Savitch et al., 2005
	BNCBF17	Brassica napus	Freezing	Savitch et al., 2005
	AtDREB2C	Arabidopsis	Thermotolerance	Lim et al., 2007
	OsDREB1A	Arabidopsis	Drought, Salinity, Freezing	Dubouzet et al., 2003
	OsDREB2B	Arabidopsis	Drought, Thermotolerance	Matsukura et al., 2010
	OsDREB1F	Rice, Arabidopsis	Drought, Salinity, Freezing	Wang et al., 2008
	OsDREB1G	Rice	Drought	Chen et al., 2008
	ZmDREB2A	Arabidopsis	Drought, Thermotolerance	Qin et al., 2007
	PgDREB2A	Tobacco	Hyperionic, Hyperosmotic	Agarwal et al., 2010
	AhDREB1	Tobacco	Drought, Salinity	Shen et al., 2003b
	LeCBF1	Arabidopsis	Freezing	Zhang et al., 2004
	GhDREB1	Tobacco	Freezing	Shan et al., 2007
	CAP2	Tobacco	Drought, Salinity	Shukla et al., 2006

Family	Gene	Transgenic Plants	Stress Tolerance	References
	PeDREB2	Tobacco	Salinity	Chen et al., 2009
	HvDREB1	Arabidopsis	Salinity	Xu et al., 2009
NAC	AtNAC2	Arabidopsis	Drought	Tran et al., 2004
	AtNAC019	Arabidopsis	Drought	Tran et al., 2004
	AtNAC055	Arabidopsis	Drought	Tran et al., 2004
	SNAC1	Rice	Drought, Salinity	Hu et al., 2006
	OsNAC6	Rice	Drought, Salinity	Nakashima et al., 2007
	ONAC045	Rice	Drought, Salinity	Zheng et al., 2009
ERF	SodERF3	Tobacco	Drought, Salinity	Trujillo et al., 2008
	GmERF3	Tobacco	Drought, Salinity	Zhang et al., 2009
WRKY	OsWRKY89	Rice	UV irradiation	Wang et al., 2007
	OsWRKY45	Arabidopsis	Drought, Salinity	Qiu and Yu, 2009
	GmWRKY21	Arabidopsis	Freezing	Zhou et al., 2008
	GmWRKY54	Arabidopsis	Drought, Salinity	Zhou et al., 2008
ZFP	Alfin1	Alfalfa	Salinity	Winicov and Bastola, 1999
	SCOF-1	Tobacco	Freezing	Kim et al., 2001
	ZPT2-3	Petunia	Drought	Sugano et al. 2003
	OSISAP1	Tobacco	Drought, Salinity, Freezing	Mukhopadhyay et al., 2004
	OSISAP2	Onion	Salinity	Xu and Que, 2007
	Zat12	Arabidopsis	Oxidative, Light Stress	Davletova et al., 2005
	Zat7	Arabidopsis	Salinity	Ciftci-Yilmaz et al., 2007
	CaZF	Tobacco	Salinity	Jain et al., 2009
Others	HARDY	Rice	Drought, Salinity	Karaba et al., 2007

Table 2. Stress response of overexpressing transcription factors in transgenic plants.

3. The AREB/ABF regulon

A conserved *cis*-element named as ABA-responsive element (ABRE; PyACGTGG/TC) was identified from the promoters of ABA-inducible genes (Bray, 1994; Giraudat et al., 1994; Busk and Page`s, 1998). Subsequently it was revealed that ABA-responsive gene expression needs multiple ABREs or the combination of an ABRE with a coupling element (CE) as a functional promoter (Yoshida et al., 2010). For example, ABRE and coupling elements, including coupling element 1 (CE1) and coupling element 3 (CE3), constitute an ABA-responsive complex in the regulation of wheat *HVA1* and *HVA22* genes (Shen et al., 1996). For the expression of *RD29B* in seeds and vegetative tissues of *Arabidopsis*, two ABRE *cis*-acting elements are required (Uno et al., 2000; Nakashima and Yamaguchi-Shinozaki, 2006). The AREB or ABFs are bZIP (basic leucine zipper) TFs that bind to the ABRE motif and activate ABA-dependent gene expression were first isolated in a yeast one-hybrid screening (Choi et al., 2000; Uno et al., 2000). It was reported that in the ABA-deficient *aba2* and ABA-

insensitive *abi1* mutants, the AREB/ABF proteins have less activity while they show an enhanced activity in the ABA hypersensitive *era1* mutant of Arabidopsis suggesting that these TFs require an ABA-mediated signal for their activation (Uno et al., 2000). The reason possibly may be an ABA-dependent phosphorylation of the AREB/ABF proteins (Shinozaki and Yamaguchi-Shinozaki, 2007). The 75 AtbZIPs have been divided into 11 groups, and the ABFs/AREBs are classified to group A (Jakoby et al., 2002) which usually act in ABA signaling during seed maturation or stress conditions. Several studies have suggested that ABFs function in different stress response pathways; i.e. ABF1 in cold; ABF2 in salt, drought, heat and glucose; ABF3 in salt; ABF4 in cold, salt, and drought signaling pathways (Kim et al., 2004; Fujita et al., 2005). AREB/ABFs are phosphorylated by ABA-responsive 42-kDa kinases which suggest that ABA-dependent phosphorylation may be involved in activation of AREB subfamily proteins (Uno et al., 2000). These kinases (SnRK2-type) such as OST1/SRK2E in Arabidopsis phosphorylate Ser/Thr residues of R-X-X-S/T sites in the conserved regions of AREB1 (Mustilli et al., 2002; Yoshida et al., 2002; Furihata et al., 2006).
AREB/ABF genes are mostly redundant in tissue-specific expression either in vegetative tissues or seeds (Choi et al., 2000; Uno et al., 2000). *AREB1/ABF2*, *AREB2/ABF4*, and *ABF3* were mainly expressed in vegetative tissues, whereas *ABI5* and *EEL* were expressed during seed maturation and/or germination (Choi et al., 2000; Uno et al., 2000; Bensmihen et al., 2002; Fujita et al., 2005; Nakashima and Yamaguchi-Shinozaki, 2006). Rice homolog *TRAB1* and barley homolog *HvABI5* activated ABA-responsive gene expression in seeds (Hobo et al., 1999; Casaretto and Ho, 2003). Expression of *OsABI5* was stimulated by ABA and high salinity, but was down-regulated by drought and cold stress in seedlings, and its overexpression also improved salinity tolerance in rice (Zou et al., 2008; Nakashima at al., 2009). *ZmbZIP17* was up-regulated by drought, heat, ABA and NaCl stress in maize seedlings (Jia et al., 2009).
Overexpression of *ABF3* and *ABF4* resulted in reduced transpiration and improved drought tolerance (Kang et al., 2002). *AREB1/ABF2* was found to be a crucial component of glucose signaling, and its over-expression improved drought stress tolerance (Kim et al., 2004). Overexpressing *OsbZIP23*, a member of AREB/ABF subfamily can also significantly improve drought and high salinity resistance of transgenic rice at the reproductive stage (Xiang et al., 2008). Enhanced tolerance to drought and heat was also observed in *35S-OsAREB1* transgenic Arabidopsis plants (Jin et al. 2010). The over-expression of the constitutively active form of AREB1 in transgenic *Arabidopsis* plants showed ABA hypersensitivity and enhanced drought tolerance, and LEA-class genes and ABA- and dehydration-stress-inducible regulatory genes such as linker histone H1 and AAA ATPase were upregulated. Over-expressing *SRK2C* caused hypersensivity to ABA, improved drought tolerance and lowered transpiration rate (Umezawa et al., 2004). Overexpression of *AtbZIP60* led to improved salt tolerance (Fujita et al., 2007).

4. The MYC /MYB regulon

The MYC/MYB families of proteins are universally found in both plants and animals and known to have varied functions. Both MYC/MYB TFs participate in the ABA-dependent pathway of stress signaling for the upregulation of the abiotic stress responsive genes. The first MYB gene identified was the v-MYB gene of avian myeloblastosis virus (AMV) (Klempnauer et al., 1982). The first plant MYB gene, C1, was identified in *Zea mays*. It encodes a c-MYB-like TF that is involved in anthocyanin biosynthesis (Paz-Ares et al., 1987).

Wide existence of MYB genes indicates that these are very ancient evolutionarily. A MYB domain is usually composed of one to three imperfect repeats, each with about 52 amino acid residues that adopt a helix-turn-helix conformation intercalating in the major groove of the DNA (Yanhui et al. 2006). Plant MYB proteins are categorized into three major groups: (i) R2R3-MYB having two adjacent repeats; (ii) R1R2R3-MYB having three adjacent repeats; and (iii) MYB-related proteins, usually containing a single MYB repeat (Rosinski and Atchley, 1998; Jin and Martin, 1999; Stracke et al., 2001). The R2R3 family contains the largest number of MYB genes. Yanhui et al. (2006) have reported that there are 198 and 183 MYB genes in the Arabidopsis and rice genomes, respectively.

MYB TFs play important roles in many physiological processes under normal or unfavorable growth conditions (Jin and Martin, 1999; Chen et al., 2006; Yanhui et al., 2006) and also in secondary metabolism (Paz-Ares et al., 1987), cell morphogenesis (Higginson et al., 2003), meristem formation and floral and seed development (Kirik et al., 1998), cell cycle control (Araki et al., 2004), defense and stress responses (Abe et al., 2003), and hormone signaling (Newman et al., 2004). MYC and MYB TFs accumulate only after ABA accumulation. *AtMYB4* (At1g22640), *AtMYB6* (At4g09460), *AtMYB7* (At2g16720), *AtMYB44* (At5g67300), *AtMYB73* (At4g37260), *AtMYB77* (At3g50060), and *AtMYBCDC5* (At1g09770) were found to be constitutively expressed in all organs and during all stress treatments (Yanhui et al., 2006). *AtMYB2* and *AtMYC2* function cooperatively as transcriptional activators in the dehydration- and ABA-inducible *rd22* expression (Urao et al., 1993; Abe et al., 2003). According to Denekamp and Smeekens (2003), *AtMYB102* integrates dehydration, osmotic, or salinity stress, ABA application, and wound-signaling pathways. *AtMYB60* and *AtMYB61*are involved in light-induced opening of stomata (Cominelli et al., 2005) and dark-induced closure of stomata, respectively (Liang et al., 2005). *AtMYB44, AtMYB73,* and *AtMYB77* are activated by wounding (Cheong et al., 2002), white-light (Ma et al., 2005), cold stress (Fowler and Thomashow, 2002), and salt stress (Kamei et al., 2005). *AtMYB44* and *AtMYB77* expression is reduced in *fus3* (fusca3), *lec1* (leafy cotyledon1), and *abi3* (ABA-insensitive3) mutants that are defective in development of dormancy and drought tolerance during late embryogenesis and seed maturation (Kirik et al., 1998). *AtMYB44* TF confers abiotic stress tolerance through enhancing stomatal closure in an ABA-independent manner (Jung et al., 2008). Recent studies have shown that *AtMYB15* expression is detectable in both vegetative and reproductive organs and is up-regulated by cold and salt stresses (Agarwal et al., 2006). *AtMYB15* has been found to negatively regulate freezing tolerance in *Arabidopsis* with its ability to repress the expression levels of *CBF* genes (Agarwal et al., 2006). *AtMyb41* from Arabidopsis is transcriptionally regulated in response to salinity, drought, cold, and ABA (Lippold et al., 2009). Liao et al. (2008c) identified 156 *GmMYB* genes of which the expression of 43 genes changed on treatment with ABA, salt, drought and/or cold stress.

Overexpression of *MYB15* results in improved drought and salt tolerance in *Arabidopsis* (Ding et al., 2009). Increased expression levels of *AtMYB2, AtMYC2* or both enhance ABA sensitivity and improve osmotic tolerance (Abe et al., 2003). Overexpression of *35S:AtMYC2* and *35S:AtMYB2* and *35S:AtMYC2+AtMYB2* in *Arabidopsis* induced ABA responsive stress genes and showed an ABA-hypersensitive phenotype with increased osmotic stress tolerance (Abe et al., 2003). Transgenic plants overexpressing *AtMyb41* showed dwarf phenotype due to alterations of cell expansion and cuticle integrity and enhanced drought sensitivity (Cominelli et al., 2008). Overexpression of *AtMyb75* and *AtMyb90* led to increased anthocyanin levels (Borevitz et al., 2000; Xie et al., 2006), while Met-derived glucosinolate content of Arabidopsis increased with overexpression of *AtMyb28* (Gigolashvili et al., 2007). In contrast, *OsMYB3R-2* transgenic plants showed enhanced tolerance to freezing, drought and salt stress and decreased sensitivity to ABA (Dai et al., 2007). Different level of tolerance was imparted by overexpression of *OsMYB4* depending on the nature of the host plants. *Arabidopsis* transgenic plants overexpressing *OsMYB4* showed increased chilling and freezing tolerance with a dwarf phenotype (Vannini et al., 2004), the tomato transgenic showed higher tolerance to drought stress (Vannini et al., 2007), whereas increased drought and cold tolerance was observed in the apple transgenic (Pasquali et al., 2008). Overexpression of a *StMYB1R-1* transgene in potato plants improved plant tolerance to drought stress while having no significant effects on other agricultural traits (Shin et al. 2011).

5. The CBF/DREB regulon

The dehydration responsive element binding proteins (DREBs) are important AP2/ERF plant TFs that induce a set of abiotic stress-related genes, thus imparting stress tolerance to plants. These play an important role in the ABA-independent pathways that activates stress responsive genes. The first isolated cDNAs encoding DRE binding proteins, CBF1 (CRT binding factor1), DREB1A and DREB2A were identified through yeast one-hybrid screening from *Arabidopsis* (Stockinger et al., 1997; Liu et al., 1998). Since then, many DREBs have been isolated from various plants. These proteins specifically bind to and activate the expression of genes regulated by the DRE sequence (5'-TACCGACAT-3') and were first identified in the promoter of the drought-responsive gene *rd29A* (Yamaguchi-Shinozaki and Shinozaki 1993). DREB1 and DREB2 are two main subgroups of DREB subfamily, involved in two different signal transduction pathways under cold and dehydration respectively. *DREB1B/CBF1*, *DREB1A/CBF3* and *DREB1C/CBF2* genes are positioned in consonance on chromosome 4 of *Arabidopsis* (Gilmour et al., 1998; Liu et al., 1998). *Arabidopsis* also contains major DREB2 proteins namely, DREB2A and DREB2B (Liu et al., 1998). DREB1/DREB2-homologous genes have also been identified in various cereals and millet crops (Nakashima et al., 2009; Lata et al., 2011).

The DREB TFs contain an extremely conserved AP2/ERF DNA-binding domain throughout plant kingdom. The domain consists of a three-stranded β-sheet and one α -helix running almost parallel to it that contacts DNA via Arg and Trp residues located in the β-sheet (Magnani et al., 2004). Two conserved functional amino acids (valine and glutamic acid) at 14th and 19th residues respectively, exist in the DNA binding domain, which are crucial sites for the binding of DREBs and DRE core sequences (Liu et al., 1998). An alkaline N-terminal amino acid region that serve as a nuclear localization signal (NLS) and a conserved Ser/Thr-rich region responsible for phosphorylation near the AP2/ERF DNA binding domain are also mostly present (Liu et al., 1998; Agarwal et al., 2006).The proteins contain an acidic C-terminal region which might be functional in *trans*-activation activity (Stockinger et al., 1997).

The activation of these transcripts is organ-specific and comparative to the extent of the stress given. When exposed to salt stress, *AhDREB1* was highly expressed in roots but less significantly in stems and leaves (Shen et al., 2003b). It was observed that *OsDREB1F* was constitutively expressed throughout the plant with highest expression in panicles and callus than in the other tissues (Wang et al., 2008). *AtDREB2A* accumulated in roots, stems and leaves under control conditions (Liu et al., 1998). *DREB2C* expressed in mature embryo and the cotyledons of germinating seedlings (Lee et al., 2010). Almoguera et al. (2009) reported that sunflower *HaDREB2* expresses in all vegetative tissues. Chrysanthemum *DvDREB2A* was expressed in all organs under normal conditions (Liu et al., 2008). *SiDREB2*, a DREB2 gene accumulated in leaves, roots, young and mature spikelets of foxtail millet indicating its function in developmental pathways also (Lata et al., 2011).

AtDREB1 was induced within 10 min at 4 ^{0}C (Liu et al., 1998). The transcript of *CBF* genes was detectable after 30 min at 4^{0}C with highest accumulation at 1 h (Medina et al., 1999). HvDREB1 gene in barley leaves significantly accumulated on salt, drought, and low-temperature treatments (Xu et al., 2009). *OsDREB1A* and *Os-DREB1B* were induced early (within 40 min) after cold exposure but not on ABA treatment. *OsDREB1A* was induced within 5 h of salinity stress whereas *OsDREB1C* showed constitutive expression (Dubouzet et al., 2003). *PNDREB1* strongly responded to low temperature and dehydration (Mei et al., 2009). However, hot pepper *Ca-DREBLP1* was quickly activated by dehydration, high salinity and mechanical wounding but not at all by cold stress (Hong and Kim, 2005). The expression of *Arabidopsis DREB2A* and its homolog *DREB2B* were stimulated by dehydration and high salinity, but not by cold and ABA (Liu et al., 1998; Nakashima et al., 2000). Similarly, ABA, mannitol and cold treatments had minimal effect on DREB2C expression (Lee et al., 2010). A detailed study of all five rice *OsDREB2s* showed that *OsDREB2A* expressed to the highest levels under the control condition, and its expression was increased to some extent by high temperature, drought and high salinity, but not by low temperature treatments. Expression of *OsDREB2B* was markedly increased after 20 min of high and 24 h of low temperature stress. While the transcript levels of *OsDREB2C*, *OsDREB2E* and *OsABI4* were low under the control condition and were transiently induced by the abiotic stresses (Matsukura et al., 2010). Wheat *TaDREB1*and *WDREB2*, maize *ZmDREB2A*, and pearl millet *PgDREB2* are responsive to cold stress while foxtail millet *SiDREB2* was not (Shen et al., 2003a; Egawa et al., 2006; Agarwal et al., 2007; Qin et al., 2007; Lata et al., 2011). Expression of chickpea *CAP2* was induced by dehydration, NaCl, ABA and auxin treatments but not by low temperature, salicylic acid and jasmonic acid (Shukla et al., 2006). The transcript expression of *Salicornia brachiata SbDREB2A* was stimulated by NaCl, drought and heat stress (Gupta et al., 2010).

Transgenic *Arabidopsis* plants over-expressing *DREB1B/CBF1* or *DREB1A/CBF3* show strong tolerance to freezing, drought, and high salinity stresses implying that DREBs/CBFs affect multiple genes (Jaglo-Ottosen et al., 1998; Liu et al., 1998; Kasuga et al., 1999). DREB1A/CBF3 overexpressing transgenics accumulated proline and various sugars under non-stress conditions (Gilmour et al., 2000). Transgenic *Arabidopsis* and rice plants overexpressing *OsDREB1A* too displayed tolerance to low temperatures, high salinity and drought (Dubouzet et al., 2003; Ito et al., 2006). The *rd29A:DREB1A/CBF3* wheat transgeneic showed improved drought stress tolerance (Pellegrineschi et al., 2004). Likewise, the constitutively overexpressing *CBF3/DREB1A* and *ABF3* transgenic rice showed better drought and salinity tolerance without any growth inhibition or phenotypic anomalies (Oh et al., 2005). The overexpression of *AhDREB1* accumulated putative downstream target genes and also conferred improved survival rate to transgenic tobacco plants under salt stress as compared to the wild-type plants (Shen et al., 2003b). The over-expression of *OsDREB1F* greatly enhanced tolerance of plants to high salinity, drought, and low-temperature both in rice and *Arabidopsis*, thus playing a significant role in plant stress signal transduction (Wang et al., 2008). Microarray analysis of transgenic *Arabidopsis* plants

suggested that over-expression of DREB2A-CA induced drought-, salt-responsive and heat-shock (HS)-related genes. These transgenic plants also exhibited enhanced thermotolerance which was significantly decreased in DREB2A knockout plants (Sakuma et al., 2006). Overexpression of *DREB2C* was also found to activate the expression of many HS responsive genes (Lim et al., 2007). Transgenic *Arabidopsis* plants overexpressing maize *ZmDREB2A* were dwarf and also displayed improved drought and heat stress tolerance. Transgenic *Arabidopsis* plants overexpressing *OsDREB2B* showed enhanced expression of DREB2A target genes and improved drought and heat-shock stress tolerance (Matsukura et al., 2010). Transgenic tobacco plants overexpressing *PgDREB2A* showed better tolerance to both hyperionic and hyperosmotic stresses (Agarwal et al. 2010). Transgenic tobacco plants overexpressing *CAP2* showed improved growth and development, and tolerance to dehydration and salt stress (Shukla et al., 2006). While its expression in yeast (*Saccharomyces cerevisae*) enhanced heat tolerance, with increased expression of heat shock factor 1 (Hsf1) and its target yeast heat shock protein 104 (Hsp 104) suggesting strong evolutionary conservation of the stress response mechanisms (Shukla et al., 2009). In another remarkable study it was described that the recombinant *E. coli* cells expressing *SbDREB2A* exhibited better growth in basal LB medium as well as if supplemented with NaCl, PEG and mannitol (Gupta et al., 2010).
These studies indicate that the DREB proteins are important TFs in regulating abiotic stress-related genes and play a critical role in imparting stress endurance to plants.

6. The NAC (NAM, ATAF and CUC) and ZF-HD (zinc-finger homeodomain) regulon

The NAC family of plant-specific TFs is one of the largest in the plant genome, with 106 and 149 members in Arabidopsis and rice, respectively (Gong et al., 2004, Xiong et al., 2005). NAC family TFs contains a highly conserved N-terminal DNA-binding domain and a diversified C-terminal domain (Hu et al., 2008). NAC was derived from the names of the first three described TFs containing NAC domain, namely NAM (no apical meristem), ATAF1-2 and CUC2 (cup-shaped cotyledon) (Souer et al., 1996; Aida et al., 1997). The *cis*-element of NAC TF [NAC recognized sequence (NACRS)] was also identified in *Arabidopsis* (Tran et al., 2004).
Numerous studies have examined the involvement of several types of NAC TFs in plant developmental programs (Sablowski and Meyerowitz 1998; Xie et al. 2000; Weir et al. 2004), and disease resistance (Collinge and Boller, 2001; Oh et al., 2005; Nakashima et al., 2007). A few *NAC* genes were found to be involved in response to various environmental stresses also such as *ANAC019*, *ANAC055*, and *ANAC072* from *Arabidopsis* (Tran et al., 2004), and BnNAC from *Brassica* (Hegedus et al., 2003). *SNAC1* is activated mainly in guard cells under dehydration (Hu et al., 2006). *AtNAP* and its homologs play an important role in leaf senescence in Arabidopsis (Guo and Gan et al., 2006). *ERD1* promoter analysis showed that TFs belonging to the NAC family and ZF-HD are important for the activation of the ERD1 (early responsive to dehydration stress 1) gene (Tran et al., 2007). *XND1* is expressed in xylem and associated with stress, ABA response and leaf senescence in *Arabidopsis* (Zhao et al., 2008). In soybean 101 NAC domain containing proteins, identified as functionally non-redundant were involved in response to abiotic stresses and in cell death events whereas *GmNAC2*, *GmNAC3* and *GmNAC4* were strongly induced by osmotic stress (Pinheiro et al., 2009). Soybean NACs *GmNAC3* and*GmNAC4* were also induced by ABA, JA and salinity

but differed in their response to cold. *GmNAC1*, *GmNAC5* and *GmNAC6* transiently expressed in tobacco leaves, resulting in cell death and enhanced expression of senescence markers. Flavonoid biosynthesis is regulated by *ANAC078* under high-light (Morishita et al., 2009). A rice NAC gene, *ONAC045* was induced by drought, high salt, low temperature, and ABA treatment in leaves and roots (Zheng et al., 2009). The transcription level of *CaNAC1* could be elevated by exogenous SA, ET, and MeJA treatment (Oh et al., 2005). A novel wheat NAC TF, *TaNAC4* was found to be induced in response to cold, salt, wounding, ABA, ethylene and MeJA, suggesting a significant cross-talk between abiotic and biotic stress conditions (Xia et al., 2010). Kim et al. (2008) reported that a salt-inducible *NTL8* (membrane associated NAC) regulates gibberellic acid -mediated salt signaling in seed germination. Very recently a membrane associated NAC TF from foxtail millet was found to be up-regulated in drought, salinity, ethephone and MeJA treatments (Puranik et al., 2011).

Several target genes of the ANAC019, ANAC055, and ANAC072 transcriptional activators were identified in the Arabidopsis transgenic plants using cDNA microarray. These transgenic plants also exhibited improved drought tolerance (Tran et al., 2004). The *SNAC1*-overexpressing transgenic rice seedlings showed significantly higher survival rate than wild type under drought treatment and significantly enhanced salinity tolerance as well (Hu et al., 2006). A rice *R2R3-MYB* gene (*UGS5*) containing putative NACRS in the promoter region was also induced in the *SNAC1*-overexpressing plants (Hu et al., 2006). Many abiotic and biotic stress responsive genes were upregulated in the *OsNAC6* transgenic plants, and the transgenics were tolerant to dehydration, high salt stresses (Nakashima et al., 2007). *ONAC045* overexpressing rice plants showed enhanced tolerance to drought and salt treatments (Zheng et al., 2009). *XND1*overexpressing showed severe stunting, premature death, and repression of TE differentiation (Zhao et al., 2008). Hence NAC TFs play an indispensable role in physiological adaptation for successful plant propagation under abiotic stress conditions.

7. Other TFs in abiotic stress response and tolerance

There are a number of TFs which are involved in abiotic stress responses other than the TFs belonging to the well known regulons described above. A new class of homeodomain TF known as HIGHER EXPRESSION OF OSMOTICALLY RESPONSIVE GENE 9 (HOS 9) and a R2R3-type MYB protein HOS 10 have been identified recently which are found to be associated with cold stress (Zhu et al., 2004, 2005). *hos9* and *hos 10* mutants show freezing hypersensitivity but at the same time enhance expression of *RD29A* and other cold responsive genes without changes in the CBF/DREB1 regulon implicating their role as negative regulators of cold stress-responsive genes. Another homeodomain TF, *HDG11* which codes for HD-START TF plays a significant role in drought tolerance by enhancing the water homeostasis of the plants (Yu et al., 2008).

HARDY (HRD), an AP2-EREBP IIIc TF gene is expressed in inflorescence tissue to protect it from desiccation (Nakano et al., 2006). Rice plants overexpressing *HRD* exhibited drought and salinity tolerance as well as improved WUE (Karaba et al., 2007). ERFs (ethylene responsive factors) also belong to the AP2-EREBP TF and have been found to be involved in growth, development, metabolic regulation and biotic and abiotic stress responses (Hussain et al., 2011). Transgenic tobacco plants expressing *SodERF3* exhibited extremely improved drought and salt tolerance (Trujillo et al., 2008). Zhang et al., (2009) reported that transgenic tobacco plants overexpressing soybean *GmERF3* exhibited tolerance not only to high salinity and drought stresses but also to various pathogens, suggesting its crucial role in both abiotic and biotic stresses.

WRKYs are another important class of plant TFs which have shown to possess multiple functions in plants including abiotic stress responses. *OsWRKY45* in rice was up-regulated by dehydration, cold, heat and salt (Qiu and Yu, 2009). *Arabidopsis* overexpressing *OsWRKY45* also showed improved drought tolerance. They have suggested that *OsWRKY45* may be involved in ABA synthesis that induces a signaling cascade resulting in lowered transpiration and enhanced tolerance to drought. Overexpression of the *OsWRKY89* in rice led to growth inhibition at early stages of plant development, but showed increased tolerance to UV irradiation and fungal infection (Wang et al., 2007). *GmWRKY13*, *GmWRKY21* and *GmWRKY54* were found to be differentially expressed under abiotic stresses (Zhou et al., 2008). Transgenic Arabidopsis plants overexpressing *GmWRKY21*were tolerant to cold stress, whereas *GmWRKY54* conferred salt and drought tolerance, possibly through the regulation of *DREB2A* and *STZ/Zat10*. However, transgenic plants over-expressing *GmWRKY13* showed increased sensitivity to salt and mannitol stress, decreased sensitivity to ABA, and an increase in lateral roots. Archana et al., (2009) reported that down-regulation of *NbWRKY*; an abiotic stress related WRKY TF, by virus-induced gene silencing produced chlorosis and senescing phenotype in tobacco plants.

Zinc finger proteins (ZFPs) are one of the important TFs found abundantly in plants and animals. They contain sequence motifs in which cysteines and/or histidines coordinate zinc atom(s) forming local peptide structures required for their specific functions (Singh et al., 2010). Cys2/His2 (C2H2)-type ZFPs containing the EAR transcriptional repressor domain, play a key role in regulating the defense responses of plants to biotic and abiotic stress conditions (Singh et al., 2010). Over-expression of *Alfin1*, a novel member of the ZFP family confers salt tolerance to the transgenic Alfalfa plants (Winicov and Bastola, 1999). The constitutive over-expression of soybean SCOF-1 induced cold-regulated (COR) gene expression and transgenic *Arabidopsis* and tobacco plants (Kim et al., 2001). *ZPT2-3*, a C2H2-type Petunia ZFP, when constitutively over-expressed in petunia, resulted in dehydration tolerance of transgenic plants (Sugano et al. 2003). *OSISAP1* from rice was inducible by cold, desiccation, salt, submergence, heavy metals and wounding, and its overexpression in tobacco exhibited cold, dehydration and salt tolerance at the seed germination/seedling stages (Mukhopadhyay et al., 2004). Constitutive expression of *Zat12* in Arabidopsis resulted in the increased expression of oxidative- and light stress responsive genes (Davletova et al., 2005). Transgenic Arabidopsis plants constitutively expressing the *Zat7* exhibited suppressed growth and were more tolerant to salinity stress (Ciftci-Yilmaz et al., 2007). *CaZF*, a C2H2 ZFP provided salinity-tolerance in transgenic tobacco (Jain et al., 2009). Interestingly, heterologous expression of *CaZF* provided osmotolerance in *S. cerevisiae* through Hog1p and calcineurin dependent as well as independent pathways (Jain et al., 2009).

8. Conclusion and future perspectives

In response to abiotic stresses such as, drought, salinity, heat, cold and mechanical wounding many genes are regulated, and their gene products function in providing stress tolerance to plants. Understanding the molecular mechanisms of plant responses to abiotic stresses is very important as it facilitates in exploiting them to improve stress tolerance and productivity. This review summarizes the role of important plant TFs namely; ABRE,

MYC/MYB, CBF/DREBs and NAC that regulate various stress responsive gene expression. They play a crucial role in providing tolerance to multiple stresses generally in both ABA-dependent and -independent manner and through respective *cis*-elements and DNA binding domains. These TFs can be genetically engineered to produce transgenics with higher tolerance to drought, salinity, heat and cold using different promoters. Functional analysis of these TFs will thus provide more information on the intricate regulatory networks involved in abiotic stress responses and the cross-talk between different signaling pathways during stress adaptation. Further, considering TFs as candidate genes in breeding and other crop improvement programs will give us a clear understanding of abiotic stress related signal transduction events and eventually will lead us to develop crop varieties superior in stress tolerance by genetic manipulation.

9. Acknowledgements

This study was supported by the Department of Biotechnology, Govt. of India, New Delhi and core grant from the National Institute of Plant Genome Research (NIPGR). Ms Charu Lata acknowledges the award of SRF from the UGC, New Delhi. We are thankful to Prof. A. K. Tyagi (Director, NIPGR) for support and encouragement.

10. References

Abe H, Yamaguchi-Shinozaki K, Urao T, Iwasaki T, Hosokawa D, Shinozaki K (1997) Role of *Arabidopsis* MYC and MYB homologs in drought- and abscisic acid-regulated gene expression. The Plant Cell 9: 1859-1868.

Abe H, Urao T, Ito T, Seki M, Shinozaki K, Yamaguchi-Shinozaki K (2003) Arabidopsis AtMYC2 (bHLH) and AtMYB2 (MYB) function as transcriptional activators in abscisic acid signaling. The Plant Cell 15: 63–78.

Agarwal M, Hao Y, Kapoor A, Dong CH, Fuji H, Zheng X, Zhu JK (2006) A R2R3 type MYB transcription factor is involved in the cold regulation of CBF genes and in acquired freezing tolerance. Journal of Biological Chemistry 281: 37636-37645.

Agarwal PK, Agarwal P, Reddy MK, Sopory SK (2006) Role of DREB transcription factors in abiotic and biotic stress tolerance in plants. Plant Cell Reports 25: 1263-1274.

Agarwal P, Agarwal PK, Nair S, Sopory SK, Reddy MK (2007) Stress inducible DREB2A transcription factor from *Pennisetum glaucum* is a phosphoprotein and its phosphorylation negatively regulates its DNA binding activity. Molecular Genetics and Genomics 277: 189–198.

Agarwal P, Agarwal PK, Joshi AJ, Sopory SK, Reddy MK (2010) Overexpression of PgDREB2A transcription factor enhances abiotic stress tolerance and activates downstream stress-responsive genes. Molecular Biology Reports 37: 1125–1135.

Agarwal PK, Jha B (2010) Transcription factors in plants and ABA dependent and independent abiotic stress signaling. Biologia Plantarum 54: 201-212.

Aida M, Ishida T, Fukaki H, Fujisawa H, Tasaka M (1997) Genes involved in organ separation in *Arabidopsis*: an analysis of the cup-shaped cotyledon mutant. The Plant Cell 9: 841-857.

Almoguera C, Prieto-Dapena P, Díaz-Martín J, Espinosa JM, Carranco R, Jordano J (2009) The HaDREB2 transcription factor enhances basal thermotolerance and longevity of seeds through functional interaction with HaHSFA9. BMC Plant Biology 9:75 doi: 10.1186/1471-2229-9-75

Araki S, Ito M, Soyano T, Nishihama R, Machida Y (2004) Mitotic cyclins stimulate the activity of c-Myb-like factors for transactivation of G2/M phase-specific genes in tobacco. Journal of Biological Chemistry 279: 32979–88.

Archana K, Rama N, Mamrutha HM, Nataraja KN (2009) Down-regulation of an abiotic stress regulated *Nicotiana benthamiana* WRKY transcription factor induces physiological abnormalities. Indian Journal of Biotechnology 8: 53-60.

Bartels D, Sunkar R (2005) Drought and salt tolerance in plants. Critical Reviews in Plant Science 21:1–36.

Behnam B, Kikuchi A, Celebi-Toprak F, Yamanaka S, Kasuga M, Yamaguchi-Shinozaki K, Watanabe KN (2006) The Arabidopsis DREB1A gene driven by the stress-inducible rd29A promoter increases salt-stress tolerance in proportion to its copy number in tetrasomic tetraploid potato (*Solanum tuberosum*). Plant Biotechnology 23: 169–177.

Bensmihen S, Rippa S, Lambert G, Jublot D, Pautot V, Granier F, Giraudat J, Parcy F (2002) The homologous ABI5 and EEL transcription factors function antagonistically to fine-tune gene expression during late embryogenesis. Plant Cell 14: 1391–1403.

Bhatnagar-Mathur P, Devi M J, Reddy DS, Vadez V, Yamaguchi-Shinozaki K, Sharma KK (2006) Overexpression of *Arabidopsis thaliana* DREB1A in transgenic peanut (*Arachis hypogaea* L.) for improving tolerance to drought stress (poster presentation). In: Arthur M. Sackler Colloquia on "From Functional Genomics of Model Organisms to Crop Plants for Global Health", April 3-5, 2006. National Academy of Sciences, Washington, DC.

Bihani P, Char B, Bhargava S (2011) Transgenic expression of sorghum *DREB2* in rice improves tolerance and yield under water limitation. The Journal of Agricultural Science 149: 95-101.

Bohnert HJ, Ayoubi P, Borchert C, Bressan RA, Burnap RL, Cushman JC, Cushman MA, Deyholos M, Fisher R, Galbraith DW, Hasegawa PM, Jenks M, Kawasaki S, Koiwa H, Kore-eda S, Lee B-H, Michalowski CB, Misawa E, Nomura M, Ozturk N, Postier B, Prade R, Song C-P, Tanaka Y, Wang H, Zhu JK (2001). A genomic approach towards salt stress tolerance. Plant Physiology and Biochemistry 39:295–311.

Borevitz JO, Xia YJ, Blount J, Dixon RA, Lamb C (2000) Activation tagging identifies a conserved MYB regulator of phenylpropanoid biosynthesis. The Plant Cell 12: 2383–2393.

Bray EA, Bailey-Serres J, and Weretilnyk E (2000) Responses to abiotic stresses. In Biochemistry and Molecular Biology of Plants, Buchanan BB, Gruissem W, and Jones RL, eds (Rockville, MD: American Society of Plant Biologists), pp. 1158–1203.

Bray EA (1994) Alterations in gene expression in response to water deficit. In Stress-Induced Gene Expression in Plants (Basra, A.S., ed.). Amsterdam: Harwood Academic, pp. 1–23.

Busk PK, Pagès M (1998) Regulation of abscisic acid-induced transcription. Plant Molecular Biology 37:425-435.

Casaretto J, Ho TH (2003) The transcription factors HvABI5 and HvVP1 are required for the abscisic acid induction of gene expression in barley aleurone cells. Plant Cell 15:271–284.

Chen J, Xia X, Yin W (2009) Expression profiling and functional characterization of a DREB2-type gene from *Populus euphratica*. Biochemical and Biophysical Research Communications 378: 483–487.

Chen JQ, Meng XP, Zhang Y, Xia M, Wang XP (2008) Over-expression of OsDREB genes lead to enhanced drought tolerance in rice. Biotechnology Letter 30: 2191-2198.

Chen M, Wang QY, Cheng XG, Xu ZS, Li LC, Ye XG, Xia LQ, Ma YZ (2007) *GmDREB2*, a soybean DRE-binding transcription factor, conferred drought and high-salt tolerance in transgenic plants. Biochemical and Biophysical Research Communications 353: 299-305.

Chen Y, Yang X, He K, Liu M, Li J, Gao Z, Lin Z, Zhang Y, Wang X, Qiu X, Shen Y, Li Z, Deng X, Luo J, Deng X-W, Chen Z, Gu H, Qu L-J (2006) The MYB transcription factor superfamily of *Arabidopsis*: Expression analysis and phylogenetic comparison with the rice MYB family. Plant Molecular Biology 60: 107-124.

Cheong YH, Chang HS, Gupta R, Wang X, Zhu T, Luan S (2002) Transcriptional profiling reveals novel interactions between wounding, pathogen, abiotic stress, and hormonal responses in Arabidopsis. Plant Physiology 129: 661–677.

Chinnusamy V, Schumaker K, Zhu J (2004) Molecular genetic perspectives on cross-talk and specificity in abiotic stress signaling in plants. Journal of Experimental Botany 55: 225-236.

Choi H, Hong J, Ha J, Kang J, Kim SY (2000) ABFs, a family of ABA responsive element binding factors. Journal of Biological Chemistry 275: 1723–1730.

Ciftci-Yilmaz S, Morsy MR, Song L, Coutu A, Krizek BA, Lewis MW, Warren D, Cushman J, Connolly EL, Mittler R (2007) The EAR-motif of the Cys2/His2-type zinc finger protein Zat7 plays a key role in the defense response of *Arabidopsis* to salinity stress. Journal of Biological Chemistry 282: 9260–9268.

Collinge M, Boller T (2001) Differential induction of two potato genes, Stprx2 and StNAC, in response to infection by *Phytophthora infestans* and to wounding. Plant Molecular Biology 46: 521-529.

Cominelli E, Galbiati M, Vavasseur A, Conti L, Sala T, Vuylsteke M, Leonhardt N, Dellaporta SL, Tonelli C (2005) A guard-cell-specific MYB transcription factor regulates stomatal movements and plant drought tolerance. Current Biology 15: 1196–1200.

Cominelli E, Sala T, Calvi D, Gusmaroli G, Tonelli C (2008) Overexpression of the Arabidopsis AtMYB41 gene alters cell expansion and leaf surface permeability. The Plant Journal 53: 53–64.

Cutler SR, Rodriguez PL, Finkelstein RR, Abrams SR (2010) Abscisic acid: Emergence of a core signaling network. Annual Reviews in Plant Biology 61: 651-679.

Dai X, Xu Y, Ma Q, Xu W, Wang T, Xue Y, Chong K (2007) Overexpression of an R1R2R3 MYB gene, *OsMYB3R-2*, increases tolerance to freezing, drought, and salt stress in transgenic *Arabidopsis*. Plant Physiology 143: 1739-1751.

Davletova S, Schlauch K, Coutu J, Mittler R (2005) The zinc-finger protein Zat12 plays a central role in reactive oxygen and abiotic stress signaling in *Arabidopsis*. Plant Physiology 139: 847–856.

Denekamp M, Smeekens SC (2003) Integration of wounding and osmotic stress signals determines the expression of the AtMYB102 transcription factor gene. Plant Physiology 132: 1415–1423.

Ding Z, Li S, An X, Liu X, Qin H, Wang D (2009) Transgenic expression of MYB15 confers enhanced sensitivity to abscisic acid and improved drought tolerance in *Arabidopsis thaliana*. Journal of Genetics & Genomics 36: 17–29.

Dubouzet JG, Sakuma Y, Ito Y, Kasuga M, Dubouzet EG, Miura S, Seki M, Shinozaki K, Yamaguchi-Shinozaki K (2003) *OsDREB* genes in rice, *Oryza sativa* L., encode transcription activators that function in drought-, high-salt- and cold-responsive gene expression. The Plant Journal 33: 751–763.

Egawa C, Kobayashi F, Ishibashi M, Nakamura T, Nakamura C, Takumi S (2006) Differential regulation of transcript accumulation and alternative splicing of a DREB2 homolog under abiotic stress conditions in common wheat. Genes & Genetic Systems 81: 77–91.

Farooq M, Wahid A, Kobayashi N, Fujita D, Basra SMA (2009) Plant drought stress: effects, mechanisms and management. Agronomy of Sustainable Development 29: 185–212.

Fowler S, Thomashow MF (2002) Arabidopsis transcriptome profiling indicates that multiple regulatory pathways are activated during cold acclimation in addition to the CBF cold response pathway. The Plant Cell 14: 1675–1690.

Fujita M, Mizukado S, Fujita Y, Ichikawa T, Nakazawa M, Seki M, Matsui M, Yamaguch-Shinozaki K, Shinozaki K (2007) Identification of stress-tolerance-related transcription-factor genes via mini-scale full length cDNA over-eXpressor (FOX) gene hunting system. Biochemical and Biophysical Research Communications 364: 250-257.

Fujita Y, Fujita M, Satoh R, Maruyama K, Parvez MM, Seki M, Hiratsu K, Masaru OT, Kazuo S, Kazuko YS (2005) AREB1 is a transcription activator of novel ABRE-dependent ABA signaling that enhances drought stress tolerance in Arabidopsis. The Plant Cell 17: 3470-3488.

Furihata T, Maruyama K, Fujita Y, Umezawa T, Yoshida R, Shinozaki K, Yamaguchi-Shinozaki K (2006) Abscisic acid dependent multisite phosphorylation regulates the activity of a transcription activator AREB1. Proceedings of the National Academy of Sciences USA 103: 1988–1993.

Gigolashvili T, Yatusevich R, Berger B, Muller C, Flugge UI (2007) The R2R3-MYB transcription factor HAG1/MYB28 is a regulator of methionine derived glucosinolate biosynthesis in Arabidopsis thaliana. The Plant Journal 51: 247–261.

Gilmour SJ, Sebolt AM, Salazar MP, Everard JD, Thomashow MF (2000) Overexpression of Arabidopsis CBF3 transcriptional activator mimics multiple biochemical changes associated with cold acclimation. Plant Physiology 124: 1854–1865.

Gilmour SJ, Zarka DG, Stockinger EJ, Salazar MP, Houghton JM, Thomashow MF (1998) Low temperature regulation of the Arabidopsis CBF family of AP2 transcriptional activators as an early step in cold-induced COR gene expression. The Plant Journal 16: 433–442.

Giraudat J, Parcy F, Bertauche N, Gosti F, Leung J, Morris P, Bouvier-Durand M, Vartanian N (1994) Current advances in abscisic acid action and signaling. Plant Molecular Biology 26: 1557–1577.

Gong W, Shen YP, Ma LG, Pan Y, Du YL, Wang DH, Yang JY, LD Hu, Liu XF, Dong CX, Ma L, Chen YH, Yang XY, Gao Y, Zhu D, Tan X, Mu JY, Zhang DB, Liu YL, Dinesh-Kumar SP, Li Y, Wang XP, Gu HY, Qu LJ, Bai SN, Lu YT, Li JY, Zhao JD, Zuo

J, Huang H, Deng XW, Zhu YX (2004) Genome-wide ORFeome cloning and analysis of Arabidopsis transcription factor genes . Plant Physiology 135: 773-782.

Guo Y, Gan S (2006) AtNAP, a NAC family transcription factor, has an important role in leaf senescence The Plant Journal 46: 601-612.

Gupta K, Agarwal PK, Reddy MK, Jha B (2010) SbDREB2A, an A-2 type DREB transcription factor from extreme halophyte *Salicornia brachiata* confers abiotic stress tolerance in *Escherichia coli*. Plant Cell Reports 29: 1131-1137.

Haake V, Cook D, Riechmann JL, Pineda O, Thomashow MF, Zhang JZ (2002) Transcription factor CBF4 is a regulator of drought adaptation in *Arabidopsis*. Plant Physiology 130: 639-648.

He X-J, Mu R-L, Cao W-H, Zhang Z-G, Zhang J-S, Chen S-Y(2005) AtNAC2, a transcription factor downstream of ethylene and auxin signaling pathways, is involved in salt stress response and lateral root development. The Plant Journal 44: 903-916.

Hegedus, D, Yu M, Baldwin D, Gruber M, Sharpe A, Parkin I, Whitwill S, Lydiate, D (2003) Molecular characterization of *Brassica napus* NAC domain transcriptional activators induced in response to biotic and abiotic stress. Plant Molecular Biology 53: 383-397.

Hsieh TH, Lee JT, Yang PT, Chiu LH, Charng YY, Wang YC, Chan MT (2002) Heterology expression of the Arabidopsis C-repeat/dehydration response element binding factor 1 gene confers elevated tolerance to chilling and oxidative stresses in transgenic tomato. Plant Physiology 129: 1086–1094.

Higginson T, Li SF, Parish RW (2003) AtMYB103 regulates tapetum and trichome development in Arabidopsis thaliana. The Plant Journal 35: 177–192.

Hobo T, Kowyama Y, Hattori T (1999) A bZIP factor, TRAB1, interacts with VP1 and mediates abscisic acid-induced transcription. Proceedings of the National Academy of Sciences USA 96: 15348–15353.

Hong JP, Kim WT (2005) Isolation and functional characterization of the *Ca-DREBLP1*gene encoding a dehydration-responsive element binding-factor-like protein 1 in hot pepper (*Capsicum annuum* L. cv Pukang). Planta 220:875–888.

Hu H, Dai M, Yao J, Xiao B, Li Xianghua, Zhang Q, Xiong L (2006) Overexpressing a NAM, ATAF, and CUC (NAC) transcription factor enhances drought resistance and salt tolerance in rice. Proceedings of the National Academy of Sciences, USA 103: 12987–12992.

Hu H, You J, Fang Y, Zhu X, Qi Z, Xiong L (2008) Characterization of transcription factor gene SNAC2 conferring cold and salt tolerance in rice. Plant Molecular Biology 67: 169-181.

Hussain SS, Kayani MA, Amjad M (2011) Transcription factors as tools to engineer enhanced drought tolerance in plants. American Institute of Chemical Engineers DOI: 10.1002/btpr.514.

Ito Y, Katsura K, Maruyama K, Taji T, Kobayashi M, Shinozaki K, Yamaguchi-Shinozaki K (2006) Functional analysis of rice DREB1/CBF-type transcription factors involved in cold-responsive gene expression in transgenic rice. Plant Cell Physiology 47: 141–153.

Jaglo-Ottosen KR, Gilmour SJ, Zarka DG, Schabenberger O, Thomashow MF (1998) *Arabidopsis* CBF1 overexpression induces *cor* genes and enhances freezing tolerance. Science 280: 104–106.

Jain D, Roy N, Chattopadhyay D (2009) CaZF, a plant transcription factor functions through and parallel to HOG and calcineurin pathways in *Saccharomyces cerevisiae* to provide osmotolerance. PLoS One 4: e5154.

Jakoby M, Weisshaar B, Dröge-Laser W, Vicente-Carbajosa J, Tiedemann J, Kroj T, Parcy F (2002) bZIP transcription factors in Arabidopsis. Trends in Plant Science 7: 106-111.

Jia Z, Lian Y, Zhu Y, He J, Cao Z, Wang G (2009) Cloning and characterization of a putative transcription factor induced by abiotic stress in *Zea mays*. African Journal of Biotechnology 8: 6764-6771.

Jin H, Martin C (1999) Multifunctionality and diversity within the plant MYB-gene family. Plant Molecular Biology 41: 577–585.

Jin X-F, Jiong A-S, Peng R-H, Liu J-G, Gao F, Chen J-M, Yao Q-H (2010) OsAREB1, an ABRE-binding protein responding to ABA and glucose, has multiple functions in Arabidopsis. BMB Reports 43: 34-39.

Jung C, Seo JS, Han SW, Koo YJ, Kim CH, Song SI, Nahm BH, Choi YD, Cheong J-J (2008) Overexpression of AtMYB44 enhances stomatal closure to confer abiotic stress tolerance in transgenic Arabidopsis. Plant Physiology 146: 623–635.

Kamei A, Seki M, Umezawa T, Ishida J, Satou M, Akiyama K, Zhu JK, Shinozaki K (2005) Analysis of gene expression profiles in Arabidopsis salt overly sensitive mutants sos2-1 and sos3-1. Plant, Cell & Environment 28: 1267–1275

Kang JY, Choi HI, Im MY, Kim SY (2002) Arabidopsis basic leucine zipper proteins that mediate stress-responsive abscisic acid signaling. The Plant Cell 14: 343–357.

Karaba A, Dixit S, Greco R, Aharoni A, Trijatmiko KR, Marsch-Martinez N, Krishnan A, Nataraja KA, Udayakumar M, Pereira A (2007). Improvement of water use efficiency in rice by expression of HARDY, an Arabidopsis drought and salt tolerance gene. Proceedings of the National Academy of Sciences USA 104: 15270–15275.

Kasuga M, Liu Q, Miura S, Yamaguchi-Shinozaki K, Shinozaki K (1999) Improving plant drought, salt, and freezing tolerance by gene transfer of a single stress-inducible transcription factor. Nature Biotechnology 17: 287–291.

Kasuga M, Miura S, Shinozaki K, Yamaguchi-Shinozaki K (2004) A combination of the *Arabidopsis* DREB1A gene and stress-inducible *rd29A* promoter improved drought- and low-temperature stress tolerance in tobacco by gene transfer. Plant Cell Physiology 45: 346–350.

Kavar T, Maras M, Kidric M, Sustar-Vozlic J, Meglic V (2007) Identification of genes involved in the response of leaves of *Phaseolus vulgaris* to drought stress. Molecular Breeding 21: 159–172.

Kim JC, Lee SH, Cheong YH, Yoo CM, Lee SI, Chun HJ, Yun DJ, Hong JC, Lee SY, Lim CO, Cho MJ (2001) A novel cold-inducible zinc finger protein from soybean, SCOF-1, enhances cold tolerance in transgenic plants. The Plant Journal 25: 247–259.

Kim S, Kang JY, Cho DI, Park JH, Kim SY (2004) ABF2, an ABRE-binding bZIP factor, is an essential component of glucose signaling and its overexpression affects multiple stress tolerance. Plant Journal 40: 75-87.

Kim SG, Lee AK, Yoon HK, Park CM (2008) A membrane bound NAC transcription factor NTL8 regulates gibberellic acid-mediated salt signaling in *Arabidopsis* seed germination. The Plant Journal 55: 77-88.

Kim T-H, Bo¨hmer M, Hu H, Nishimura N, Schroeder JI (2010) Guard cell signal transduction network: Advances in understanding abscisic acid, CO2, and Ca2+ signaling. Annual Reviews in Plant Biology 61: 561–591.

Kirik V, Ko¨lle K, Mise´ra S, Ba¨umlein H (1998) Two novel MYB homologues with changed expression in late embryogenesis-defective Arabidopsis mutants. Plant Molecular Biology 37: 819–827.

Kizis D, Pagès M (2002) Maize DRE-binding proteins DBF1 and DBF2 are involved in rab17 regulation through the drought-responsive element in an ABA-dependent pathway. The Plant Journal 30: 679-689.

Klempnauer K.-H, TJ Gonda, JM Bishop (1982) Nucleo- tide sequence of the retroviral leukemia gene *v-myb* and its cellular progenitor *c-myb:* The architecture of a transduced oncogene. The Cell 31: 453-463.

Kobayashi F, Maeta E, Terashima A, Kawaura K, Ogihara Y, Takumi S (2008) Development of abiotic stress tolerance via bZIP-type transcription factor LIP19 in common wheat. Journal of Experimental Botany 59: 891-905.

Lata C, Bhutty S, Bahadur RP, Majee M, Prasad M (2011) Association of a SNP in a novel DREB2-like gene *SiDREB2* with stress tolerance in foxtail millet [*Setaria italica* (L.)]. Journal of Experimental Botany DOI: 10.1093/jxb/err016.

Lee SJ, Kang JY, Park HJ, Kim MD, Bae MS, Choi HI, Kim SY (2010) DREB2C interacts with ABF2, a bZIP protein regulating abscisic acid-responsive gene expression, and its overexpression affects abscisic acid sensitivity. Plant Physiology 153: 716–727.

Li XP, Tian AG, Luo GZ, Gong ZZ, Zhang JS, Chen SY (2005) Soybean DRE-binding transcription factors that are responsive to abiotic stresses. Theoretical and Applied Genetics 110: 1355–1362.

Liang YK, Dubos C, Dodd IC, Holroyd GH, Hetherington AM, Campbell MM (2005) AtMYB61, an R2R3-MYB transcription factor controlling stomatal aperture in Arabidopsis thaliana. Current Biology 15: 1201–1206.

Liao Y, Zhang J-S, Chen S-Y, Zhang, W-K (2008b) Role of soybean GmbZip132 under abscisic acid and salt stresses. Journal of Integrative Plant Biology 50: 221-230.

Liao Y, Zou H, Wei W, Hao YJ, Tian AG, Huang J, Liu Y-F, Zhang J-S, Chen S-Y (2008a) Soybean *GmbZIP44, GmbZIP62* and *GmbZIP78* genes function as negative regulator of ABA signaling and confer salt and freezing tolerance in transgenic *Arabidopsis*. Planta 228: 225-240.

Liao Y, Zou H-F, Wang H-W, Zhang W-K, Ma B, Zhang J-S (2008c) Soybean *GmMYB76, GmMYB92*, and *GmMYB177* genes confer stress tolerance in transgenic *Arabidopsis* plants. Cell Research 18: 1047-1060.

Lim CJ, Hwang JE, Chen H, Hong JK, Yang KA, ChoiMS, Lee KO, Chung WS, Lee SY, Lim CO (2007) Over-expression of the Arabidopsis DRE/CRT-binding transcription factor DREB2C enhances thermotolerance. Biochemical and Biophysical Research Communications 362: 431–436.

Lippold F, Sanchez DH, Musialak M, Schlereth A, Scheible WR, Hincha DK, Udvardi MK (2009) AtMyb41 Regulates Transcriptional and Metabolic Responses to Osmotic Stress in Arabidopsis[W][OA]. Plant Physiology 149: 1761-1772.

Diego H. Magdalena Musialak, Armin Schlereth, Wolf-Ruediger Scheible, Dirk K. Hincha and Michael K. Udvardi

Liu Q, Kasuga M, Sakuma Y, Abe H, Miura S, Goda H, Shimada Y, Yoshida S, Shinozaki K, Yamaguchi-Shinozaki K (1998) Two transcription factors, DREB1 and DREB2, with an EREBP/AP2 DNA binding domain separate two cellular signal transduction pathways in drought- and low-temperature-responsive gene expression, respectively, in *Arabidopsis*. The Plant Cell 10: 391–406.

Liu L, Zhu K, Yang Y, Wu J, Chen F, Yu D (2008) Molecular cloning, expression profiling and trans-activation property studies of a DREB2-like gene from chrysanthemum (*Dendranthema vestitum*). Journal of Plant Research 121: 215–226.

Lu P-L, Chen N-Z, An R, Su Z, Qi B-S, Ren F, Chen J, Wg X-C (2007) A novel drought-inducible gene, ATAF1, encodes a NAC family protein that negatively regulates the expression of stress-responsive genes in *Arabidopsis*. Plant Molecular Biology 63: 289-305.

Ma L, Sun N, Liu X, Jiao Y, Zhao ·H, Deng XW (2005) Organ-specific expression of Arabidopsis genome during development. Plant Physiology 138: 80–91.

Magnani E, Sjölander K, Hake S (2004) From endonucleases to transcription factors: evolution of the AP2 DNA binding domain in plants. The Plant Cell 16: 2265–2277.

Matsukura S, Mizoi J, Yoshida T, Todaka D, Ito Y, Maruyama K, Shinozaki K, Yamaguch-Shinozaki K (2010) Comprehensive analysis of rice *DREB2*-type genes that encode transcription factors involved in the expression of abiotic stress-responsive genes. Molecular Genetics and Genomics 283: 185–196.

Mattana M, Biazzi E, Consonni R, Locatelli F, Vannini C, Provera S, Coraggio I (2005) Overexpression of *Osmyb4* Enhances Compatible Solute Accumulation and Increases Stress Tolerance of *Arabidopsis thaliana*. Physiologia Plantarum 125: 212-223.

Medina J, Bargues M, Terol J, P´erez-Alonso M, Salinas J (1999) The *Arabidopsis CBF* gene family is composed of three genes encoding AP2 domain-containing proteins whose expression is regulated by low temperature but not by abscisic acid or dehydration. Plant Physiology 119: 463–469.

Mei Z, Wei L, Yu-Ping B, Zi-Zhang W (2009) Isolation and Identification of *PNDREB1*: A New DREB Transcription Factor from Peanut (*Arachis hypogaea* L.) Acta Agronomica Sinica 35: 1973-1980.

Morishita T, Kojima Y, Maruta T, Nishizawa-Yokoi A, Yabuta Y, Shigeoka S (2009) Arabidopsis NAC transcription factor, ANAC078, regulates flavonoid biosynthesis under high-light. Plant Cell Physiology 50: 2210-2222.

Mukhopadhyay A, Vij S, Tyagi AK (2004) Overexpression of a zinc-finger protein gene from rice confers tolerance to cold, dehydration, and salt stress in transgenic tobacco. Proceedings of National Academy Sciences USA 101: 6309–6314.

Mustilli AC, Merlot S, Vavasseur A, Fenzi F, Giraudat J (2002) Arabidopsis OST1 protein kinase mediates the regulation of stomatal aperture by abscisic acid and acts upstream of reactive oxygen species production. The Plant Cell 14: 3089–3099.

Nakano T, Suzuki K, Fujimura T, Shinshi H (2006) Genome-Wide Analysis of the ERF Gene Family in Arabidopsis and Rice. Plant Physiology 140: 411–432.

Nakashima K, Ito Y, Yamaguchi-Shinozaki K (2009) Transcriptional Regulatory Networks in Response to Abiotic Stresses in Arabidopsis and Grasses. Plant Physiology 149: 88-95.

Nakashima K, Shinwar ZK, Sakuma Y, Seki M, Miura S, Shinozaki K, Yamaguchi-Shinozaki K (2000) Organization and expression of two *Arabidopsis DREB2*genes encoding DRE-binding proteins involved in dehydration- and high salinity-responsive gene expression. Plant Molecular Biology 42: 657–665.

Nakashima K, Tran L-SP, Nguyen DV, Fujita M, Maruyama K, Todaka D, Ito Y, Hayashi N, Shinozaki K, Yamaguchi-Shinozaki K (2007) Functional analysis of a NAC-type transcription factor OsNAC6 involved in abiotic and biotic stress-responsive gene expression in rice. The Plant Journal 51: 617-630.

Nakashima K, Yamaguchi-Shinozaki K (2006) Regulons involved in osmotic stress-responsive and cold stress responsive gene expression in plants. Physiologia Plantarum 126: 62–71.

Newman LJ, Perazza DE, Juda L, Campbell MM (2004) Involvement of the R2R3-MYB, AtMYB61, in the ectopic lignification and dark-photomorphogenic components of the det3 mutant phenotype. The Plant Journal 37: 239–50.

Oh SJ, Song SI, Kim YS, Jang HJ, Kim SY, Kim M, Kim YK, Nahm BH, Kim JK (2005) *Arabidopsis* CBF3/DREB1A and ABF3 in transgenic rice increased tolerance to abiotic stress without stunting growth. Plant Physiology 138: 341–351.

Oh S-K, Lee S, Yu SH, Choi D (2005) Expression of a novel NAC domain containing transcription factor (CaNAC1) is preferentially associated with incompatible interactions between chili pepper and pathogens. Planta 222: 876–887.

Pasquali G, Biricolti S, Locatelli F, Baldoni E, Mattana M (2008) *Osmyb4* expression improves adaptive responses to drought and cold stress in transgenic apples. Plant Cell Reports 27: 1677-1686.

Paz-Ares J, Ghosal D, Wienand U, Peterson PA, Saedler H (1987) The regulatory *cl* locus of *Zea mays* encodes a protein with homology to *myb* proto-oncogene products and with structural similarities to transcriptional activators. EMBO f. 6: 3553-3558.

Pellegrineschi A, Reynolds M, Pacheco M, Brito RM, Almeraya R, Yamaguchi-Shinozaki K, Hoisington D (2004) Stress-induced expression in wheat of the *Arabidopsis thaliana DREB1A* gene delays water stress symptoms under greenhouse conditions. Genome 47: 493–500.

Peng H, Cheng HY, Yu XW, Shi QH, Zhang H, Li JG, Ma H (2009) Characterization of a chickpea (*Cicer arietinum* L.) NAC family gene, CarNAC5, which is both developmentally- and stress-regulated. Plant Physiology & Biochemistry 47: 1037-45.

Pinheiro GL, Marques CS, Costa MDBL, Reis PAB, Alves MS, Carvalho CM, Fietto LG, Fontes EPB (2009) Complete inventory of soybean NAC transcription factors: Sequence conservation and expression analysis uncover their distinct roles in stress response. Gene 444: 10-23.

Puranik S, Bahadur RP, Srivastava PS, and Prasad M (2011) Molecular cloning and characterization of a membrane associated NAC family gene, SiNAC from foxtail millet [*Setaria italica* (L.) P. Beauv.]. Molecular Biotechnology DOI: 10.1007/s12033-011-9385-7.

Qin F, Kakimoto M, Sakuma Y, Maruyama K, Osakabe Y, Tran LS, Shinozaki K, Yamaguchi-Shinozaki K (2007) Regulation and functional analysis of ZmDREB2A in response to drought and heat stresses in *Zea mays* L. The Plant Journal 50: 54–69.

Qiu Y, Yu D (2009) Over-expression of the stress-induced OsWRKY45 enhances disease resistance and drought tolerance in Arabidopsis. Environmental and Experimental Botany 65: 35-47.

Riechmann JL, Heard J, Martin G, Reuber L, Jiang C, Keddie J, Adam L, Pineda O, Ratcliffe OJ, Samaha RR, Creelman R, PilgrimM, Broun P, Zhang JZ, Ghandehari D, Sherman BK, Yu G (2000) *Arabidopsis* transcription factor: genome wide comparative analysis among eukaryotes. Science 290: 2105–2110.

Rosinski JA, Atchley WR (1998) Molecular evolution of the Myb family of transcription factors: evidence for polyphyletic origin. Journal of Molecular Evolution 46: 74–83.

Sablowski RW, Meyerowitz EM (1998) A homolog of NO APICAL MERISTEM is an immediate target of the floral homeotic genes APETALA3/PISTILLATA. The Cell 92: 93-103.

Saibo NJM, Lourenc¸o T, Oliveira MM (2009) Transcription factors and regulation of photosynthetic and related metabolism under environmental stresses. Annals of Botany 103: 609-623.

Sakuma Y, Maruyama K, Qin F, Osakabe Y, Shinozaki K, Yamaguchi-Shinozaki K (2006) Dual function of an Arabidopsis transcription factor DREB2A in water-stress-responsive and heat-stress-responsive gene expression. Proceedings of National Academy of Sciences USA 103: 18822–18827.

Savitch LV, Allard G, Seki M, Robert LS, Tinker NA, Huner NPA, Shinozaki K, Singh J (2005) The effect of overexpression of two *Brassica CBF/DREB1*-like transcription factors on photosynthetic capacity and freezing tolerance in *Brassica napus*. Plant Cell Physiology 46: 1525-1539.

Seki M, Narusaka M, Ishida J, Nanjo T, Fujita M, Oono Y, Kamiya A, Nakajima M, Enju A, Sakurai T, Satou M, Akiyama K, Taji T, Yamaguchi-Shinozaki K, Carninci P, Kawai J, Hayashizaki Y, Shinozaki K (2002) Monitoring the expression profiles of 7000 *Arabidopsis* genes under drought, cold and high salinity stresses using a full-length cDNA microarray. The Plant Journal 31: 279–292.

Seki M, Kamei A, Satou M, Sakurai T, Fujita M, Oono Y, Yamaguch-Shinozaki K, Shinozaki K (2003) Transcriptome analysis in abiotic stress conditions in higher plants. Topics Curr Genet 4: 271–295.

Shan D-P, Huang J-G, Yang Y-T, Guo Y-H, Wu C-A, Yang G-D, Gao Z, Zheng C-C (2007) Cotton GhDREB1 increases plant tolerance to low temperature and is negatively regulated by gibberellic acid. New Phytologist 176: 70–81.

Shen, Q., Zhang, P. and Ho, T. (1996) Modular nature of abscisic acid (ABA) response complexes: composite promoter units that are necessary and sufficient for ABA induction of gene expression in barley. Plant Cell 8: 1107–1119.

Shen YG, Zhang WK, He SJ, Zhang JS, Liu Q, Chen SY (2003a) An EREBP/AP2-type protein in *Triticum aestivum* was a DRE-binding transcription factor induced by cold, dehydration and ABA stress. Theoretical and Applied Genetics 106: 923–930.

Shen YG, Zhang WK, Yan DQ, Du BX, Zhang JS, Liu Q, Chen SY (2003b) Characterization of a DRE-binding transcription factor from a halophyte *Atriplex hortensis*. Theoretical and Applied Genetics 107: 155–161.

Shin D, Moon S-J, Han S, Kim B-G, Park SR, Lee S-K, Yoon H-J, Lee H-E, Kwon H-B, Baek D, Yi BY, Byun M-O (2011) Expression of *StMYB1R-1*, a novel potato single MYB-like domain transcription factor, increases drought tolerance. Plant Physiology 155: 421-432.

Shinozaki K. and Yamaguchi-Shinozaki K (2007) Gene networks involved in drought stress tolerance and response. Journal of Experimental Botany 58: 221-227.

Shukla RK, Raha S, Tripathi V, Chattopadhyay D (2006) Expression of CAP2, an APETALA2-family transcription factor from chickpea, enhances growth and tolerance to dehydration and salt stress in transgenic tobacco. Plant Physiology 142: 113-123.

Shukla RK, Tripathi V, Jain D, Yadav RK, Chattopadhyay D (2009) CAP2 enhances germination of transgenic tobacco seeds at high temperature and promotes heat stress tolerance in yeast. FEBS Journal 276: 5252-5262.

Singh AK, Sopory SK, Wu R, Singla-Pareek SL (2010) Transgenic Approaches. In: Abiotic Stress Adaptation in Plants: Physiological, Molecular and Genomic Foundation, A Pareek, SK Sopory, HJ Bohnert and Govindjee (eds.), pp.417–450.

Souer E, Van Houwelingen A, Kloos D, Mol J, Koes R (1996) The no apical meristem gene of *Petunia* is required for pattern formation in embryos and flowers and is expressed at meristem and primordial boundaries. The Cell 85: 159-170.

Stockinger EJ, Gilmour SJ, Thomashow MF (1997) Arabidopsis thaliana CBF1 encodes an AP2 domain-containing transcriptional activator that binds to the C-repeat/DRE, a cis-acting DNA regulatory element that stimulates transcription in response to low temperature and water deficit. Proceedings of the National Academy of Sciences USA 94: 1035–1040.

Stracke R, Werber M, Weisshaar B (2001) The R2R3-MYB gene family in Arabidopsis thaliana. Current Opinion in Plant Biology 4: 447–456.

Sugano S, Kaminaka H, Rybka Z, Catala R, Salinas J, Matsui K, Ohme-Takagi M, Takatsuji H (2003) Stress-responsive zinc finger gene ZPT2-3 plays a role in drought tolerance in petunia. The Plant Journal 36:830–841.

Tran L-SP, Nakashima K, Sakuma Y, Simpson SD, Fujita Y, Maruyama K, Fujita M, Seki M, Shinozaki K, Yamaguchi-Shinozaki K (2004) Isolation and functional analysis of *Arabidopsis* stress-inducible NAC transcription factors that bind to a drought-responsive *cis*-element in the *early responsive to dehydration stress 1* promoter. The Plant Cell 16: 2481-2498.

Tran LS, Nakashima K, Sakuma Y, Osakabe Y, Qin F, Simpson SD, Maruyama K, Fujita Y, Shinozaki K, Yamaguchi-Shinozaki K (2007) Co-expression of the stress-inducible zinc finger homeodomain ZFHD1 and NAC transcription factors enhances expression of the ERD1 gene in Arabidopsis. The Plant Journal 49: 46–63.

Trujillo LE, Sotolongo M, Menendez C, Ochogava ME, Coll Y, Hernandez I, Borras-Hidalgo O, Thomma BPHJ, Vera P, Hernandez L (2008) SodERF3, a novel sugarcane ethylene responsive factor (ERF), enhances salt and drought tolerance when over-expressed in tobacco plants. Plant Cell Physiology 49: 512–515.

Udvardi MK, Kakar K, Wandrey M, Montanri O, Murray J, Andraiankaja A, Zhang J-Y, Benedito V, Hofer JMI, Cheng F, Town CD (2007) Legume transcription factors: global regulators of plant development and response to the environment. Plant Physiology 144: 538-549.

Umezawa T, Fujita M, Fujita Y, Yamaguchi-Shinozaki K, and Shinozaki K (2006) Engineering drought tolerance in plants: discovering and tailoring genes to unlock the future. Current Opinion in Biotechnology 17: 113-122.

Umezawa T, Yoshida R, Maruyama K, Yamaguchi-Shinozaki K, Shinozaki K (2004) SRK2C. A SNF1-related protein kinase 2, improves drought tolerance by controlling stress-responsive gene expression in Arabidopsis thaliana. Proceedings of the National Academy of Sciences, USA 101: 17306–17311.

Uno Y, Furihata T, Abe H, Yoshida R, Shinozaki K, Yamaguchi-Shinozaki K (2000) Arabidopsis basic leucine zipper transcription factors involved in an abscisic acid-dependent signal transduction pathway under drought and high-salinity conditions. Proceedings of National Academy of Sciences USA 97: 11632–11637.

Urao T, Yamaguchi-Shinozaki K, Urao S, Shinozaki K (1993) An Arabidopsis myb homolog is induced by dehydration stress and its gene product binds to the conserved MYB recognition sequence. The Plant Cell 5: 1529–1539.

Vannini M, Campa M, Iriti M, Genga A, Faoro F, Carravieri S, Rotino GL, Rossoni M, Spinardi A, Bracale M (2007) Evaluation of transgenic tomato plants ectopically expressing the rice *Osmyb4* gene. Plant Science 173: 231-239.

Vannini C, Locatelli F, Bracale M, Magnani E, Marsoni M, Osnato M, Mattana M, Baldoni E, Coraggio I (2004) Overexpression of the rice Osmyb4 gene increases chilling and freezing tolerance of Arabidopsis thaliana plants. The Plant Journal 37: 115–127.

Wang H, Hao J, Chen X, (2007) Overexpression of rice WRKY89 enhances ultraviolet B tolerance and disease resistance in rice plants. Plant Molecular Biology 65: 799-815.

Wang Q, Guan Y, Wu Y, Chen H, Chen F, Chu C (2008) Overexpression of a rice OsDREB1F gene increases salt, drought, and low temperature tolerance in both Arabidopsis and rice. Plant Molecular Biology 67: 589–602.

Weir I, Lu J, Cook H, Causier B, Schwarz-Sommer Z, Davies B (2004) CUPULIFORMIS establishes lateral organ boundaries in Antirrhinum. Development 131: 915-922.

Winicov I, Bastola DR (1999) Transgenic overexpression of the transcription factor Alfin1 enhances expression of the endogenous MsPRP2 gene in alfalfa and improves salinity tolerance of the plants. Plant Physiology 120: 473–480.

Xia N, Zhang G, Liù X-Y, Deng L, Cai G-L, Zhang Y, Wang X-J, Zhao J, Huang L-L, Kang Z-S (2010) Characterization of a novel wheat NAC transcription factor gene involved in defense response against stripe rust pathogen infection and abiotic stresses. Molecular Biology Reports 37: 3703-3712.

Xiang Y, Tang N, Du H, Ye H, Xiong L (2008) Charaterization of Osb-ZIP23 as a key player of the basic leucine zipper transcription factor family for conferring abscisic acid sensitivity and salinity and drought tolerance in rice. Plant Physiology 148: 1938–1952.

Xie Q, Frugis G, Colgan D, Chua NH (2000) Arabidopsis NAC1 transduces auxin signal downstream of TIR1 to promote lateral root development. Genes & Development 14: 3024-3036.

Xie DY, Sharma SB, Wright E, Wang ZY, Dixon RA (2006) Metabolic engineering of proanthocyanidins through co-expression of anthocyanidin reductase and the PAP1 MYB transcription factor. The Plant Journal 45: 895–907.

Xiong Y, Liu T, Tian C, Sun S, Li J, Chen M (2005) Transcription factors in rice: a genome-wide comparative analysis between monocots and eudicots . Plant Molecular Biology 59: 191-203.

Xu QJ, Cui CR (2007) Genetic transformation of OSISAP1 gene to onion (*Allium cepa* L.) mediated by amicroprojectile bombardment. Journal of Plant Physiology and Molecular Biology (Article in Chinese) 33: 188-96.

Xu ZS, Ni ZY, Li ZY, Li LC, Chen M, Gao DY, Yu XD, Liu P, Ma YZ (2009) Isolation and functional characterization of HvDREB1-a gene encoding a dehydration-responsive element binding protein in *Hordeum vulgare.* Journal of Plant Research 122:121-130.

Xue GP, Loveridge CW (2004) HvDRF1 is involved in abscisic acid mediated gene regulation in barley and produces two forms of AP2 transcriptional activators, interacting preferably with a CT-rich element. The Plant Journal 37: 326-339.

Yamaguchi-Shinozaki K, Shinozaki K (1994) A novel *cis*-acting element in an *Arabidopsis* gene is involved in responsiveness to drought, low-temperature, or high-salt stress. The Plant Cell 6: 251-264.

Yamaguchi-Shinozaki K, Shinozaki K (1993) The plant hormone abscisic acid mediates the drought-induced expression but not the seed-specific expression of *rd22,* a gene responsive to dehydration-stress in *Arabidopsis thaliana.* Molecular and General Genetics 238: 17-25.

Yang Y, Wu J, Zhu K, Liu L, Chen F, Yu D (2009) Identification and characterization of two chrysanthemum (*Dendronthema × morifolium*) DREB genes, belonging to the AP2/EREBP family. Molecular Biology Reports 36: 71-81.

Yanhui C, Xiaoyuan Y, Kun H, Meihua L, Jigang L, Zhaofeng G, Zhiqiang L, Yunfei Z, Xiaoxiao W, Xiaoming Q, Yunping S, Xiaohui D, Jingchu L, Xing-Wang D, Zhangliang C, Hongya G, Li-Jia Q (2006) The MYB transcription factor superfamily of Arabidopsis: expression analysis and phylogenetic comparison with the rice MYB family. Plant Molecular Biology 60: 107-124.

Yoshida T, Fujita Y, Sayama H, Kidokoro S, Maruyama K, Mizoi J, Shinozaki K, Yamaguchi-Shinozaki K (2010) AREB1, AREB2, and ABF3 are master transcription factors that cooperatively regulate ABRE-dependent ABA signaling involved in drought stress tolerance and require ABA for full activation. The Plant Journal 61: 672-685.

Yoshida R, Hobo T, Ichimura K, Mizoguchi T, Takahashi F, Alonso J, Ecker JR, Shinozaki K (2002) ABA-activated SnRK2 protein kinase is required for dehydration stress signaling in Arabidopsis. Plant Cell Physiology 43: 1473-1483.

Yu H, Chen X, Hong Y-Y, Wang Y, Xu P, Ke S-D, Liu H-Y, Zhu J-K, Oliver DJ, Xiang C-B (2008) Activated expression of an Arabidopsis HD-START protein confers drought tolerance with improved root system and reduced stomatal density. The Plant Cell 20: 1134-1151.

Zhang G, Chen M, Li L, Xu Z, Chen X, Guo J, Ma Y (2009) Overexpression of the soybean GmERF3 gene, an AP2/ERF type transcription factor for increased tolerances to salt, drought and diseases in transgenic tobacco. Journal of Experiment Botany 60: 3781-3796.

Zhang JZ, Creelman RA, Zhu JK (2004) From laboratory to field. Using information from Arabidopsis to engineer salt, cold, and drought tolerance in crops. Plant Physiology 135: 615-621.

Zhao C, Avci U, Grant EH, Haigler CH, Beers EP (2008) XND1, a member of the NAC domain family in *Arabidopsis thaliana*, negatively regulates lignocellulose synthesis and programmed cell death in xylem. The Plant Journal 53: 425-436.

Zheng X, Chen B, Lu G, Han B (2009) Overexpression of a NAC transcription factor enhances rice drought and salt tolerance. Biochemical and Biophysical Research Communications 379: 985-989.

Zhou QY, Tian AG, Zou HF, Xie ZM, Lei G, Huang J, Wang CM, Wang HW, Zhang JS, Chen SY (2008) Soybean WRKY-type transcription factor genes, GmWRKY13, GmWRKY21, and GmWRKY54, confer differential tolerance to abiotic stresses in transgenic Arabidopsis plants. Plant Biotechnology Journal 6: 486-503.

Zhu J, Shi H, Lee BH, Damsz B, Cheng S, Stirm V, Zhu J-K, Hasegawa PM, Bressan RA (2004) An Arabidopsis homeodomain transcription factor gene, HOS9, mediates cold tolerance through a CBF-independent pathway. Proceedings of the National Academy of Sciences, USA 101: 9873–9878.

Zhu J, Verslues PE, Zheng X, Lee B-H, Zhan X, Manabe Y, Sokolchik I, Zhu Y, Dong C-H, Zhu J-K, Hasegawa PM, Bressan RA (2005) HOS10 encodes an R2R3-type MYB transcription factor essential for cold acclimation in plants. Proceedings of the National Academy of Sciences, USA 102: 9966–9971.

Zou M, Guan Y, Ren H, Zhang F, Chen F (2008) A bZIP transcription factor, OsABI5, is involved in rice fertility and stress tolerance. Plant Molecular Biology 66: 675–683.

11

The Roles of Germin Gene Products in Plants Under Salt Stress

Mahmut Caliskan
Mustafa Kemal University, Faculty of Arts & Sciences
Biology Department, Hatay
Turkey

1. Introduction

The members of plants response various internal and external signals differently. The responses of plants to biotic and abiotic stress factors involve biochemical, physiological, morphological and developmental changes. Among the various circumstances salt stress is particularly under extensive studies due to low salt tolerance of crop plants. Germin and germin-like gene products were previously announced to be involved in various aspects of plant development such as plant defence, embryonic development and they are responsive to biotic and abiotic stress including salt. The responses of germin and germin like genes to salt stress are found to be various in different plants.

The salinity of soil is an important problem in agriculture, particularly since the majority of crop plants have low salt tolerance. The response of plants to salt stress is a complex phenomenon that involves biochemical and physiological processes as well as morphological and developmental changes (Flowers et al., 1977; Greenway & Munns, 1980). The identification of genes whose expression enables plants to adapt to or tolerate to salt stress is essential for breeding programs, but little is known about the genetic mechanisms for salt tolerance. One approach in clarifying the molecular mechanisms involved in salt stress is to identify the genes whose levels change as a result of salt stress. In this aspect, Hurkman et al. (1989) reported that in barley, gene regulation is altered by salt stress and the levels of translatable mRNAs change with salt treatment. Among the salt stress responsive gene products, germin and germin-like proteins (GLP) were identified (Caliskan, 1997; Hurkman et al., 1989). Cereal germin protein is a homopentameric apoplastic glycoprotein whose synthesis is associated with the onset of growth in germinating wheat embryos (Lane, 1991). Germin genes and their proteins were first detected in germinating cereals (Grzelczak et al., 1985), but subsequently, germin-like proteins were also identified in dicotyledonous angiosperms (Michalowski and Bohnerd, 1992), gymnosperms (Domon et al., 1995) and mosses (Yamahara et al., 1999). Germins are suggested to be a member of "superfamily" which comprises various growth-related genes (Dunwell et al., 2000).

Cereal germin proteins display strong oxalate oxidase activity (Lane et al., 1993), an activity that generates one mole of H_2O_2 and two moles of CO_2 from the degradation of oxalic acid. It is reported that H_2O_2 might act as a signaling molecule at low concentration (Luthell, 1993) or a component of cell wall modifications at high concentrations (Showalter, 1993). Another germin-like protein isolated from the cells of a moss, *Barbula unguiculata*, was shown to have manganese superoxide dismutase activity (Yamahara et al., 1999). Germin genes and proteins have been shown to be associated with various aspects of plant development (Caliskan, 2000; Lane, 2002) such as defense system (Berna and Bernier, 1999; Donaldson et al. 2001), embryonic development (Caliskan and Cuming, 2001), photoperiodic oscillations (Ono et al. 1996), and hormonal stimuli (Berna & Bernier, 1997). The accumulation of germin gene products in wheat and barley seedlings in the presence of NaCl was analyzed previously, but little is known about the possible role of germin gene products during salt stress (Hurkman et al., 1991, 1994; Hurkman & Tanaka, 1996a; Berna & Bernier, 1999).

In this paper, we have brought together the results of our studies and the results of other researchers who are working in the field of germin, germin-like proteins and their enzymatic activities.

2. The stress factors and the early plant development

Plants are a distinct kingdom of organisms that possess unique properties of reproduction, development, physiology and metabolism. The early development of plants is also quite different from that of animals in various ways. The main features that distinguish embryogenesis in angiosperms from the related pathways in animal systems are (1) the process of double fertilization and the subsequent interaction between embryo and endosperm, (2) the totipotency of certain plant cells by contrast with the uniqueness of the zygote in animals, (3) the activation of large numbers of zygotic genes during very early stages of embryogenesis, (4) the formation of apical meristems which produce the basic body plan of the "adult" plant, (5) the absence of a germ line established early in development, (6) the absence of cellular migration during embryo development. Due to the above discrepancies, studies of animal systems (*e.g. Drosophila*) may not be directly applicable to plants.

Embryogenesis is a developmental stage which covers the time period beginning with the formation of the zygote and ending with the formation of a mature embryo within a seed. Embryogenesis has a central role in the life cycle of flowering plants since it results in the production of a structurally and functionally organized miniature adult plant which is called the embryo. The conversion of an ovule into a seed is triggered by fertilization. The cells in male (anther) and female (ovary) organs of the flower undergo meiotic and mitotic divisions which give rise to a male gametophyte (pollen grain) and a female gametophyte (embryo sac). The pollen grain contains two sperm cells, whereas the embryo sac contains only a single egg cell in association with a number of accessory haploid cells. Upon penetration of the embryo sac by the pollen tube, one of the sperm nuclei fertilizes the egg cell, while the other unites with the two additional haploid nuclei of the embryo sac to form the triploid endosperm. Therefore, unlike animal fertilization, plants undergo a double fertilization process. As the ovule develops into a seed after fertilization, the zygote enters a pathway of cell division and differentiation to produce an embryo. At the end of this pathway, the embryo is composed of an embryonic axis that bears the root meristem at one end, the cotyledon(s) and the shoot meristem at the other end. Based on the cotyledon number, angiosperms are defined as monocotyledonous or dicotyledonous plants. The endosperm tissue may comprise the bulk of the mature seed, as a storage tissue, as in the case of cereal grains, or it may be reduced in size and importance in seeds where the principal storage reserves are kept within the cotyledons, in dicotyledonous seeds.

In seeds of dicotyledonous plants, the embryo is typically comprised of an axial region (root and shoot) and two large cotyledons. The other organs are rudimentary, at best. By contrast, in monocotyledonous seeds the embryonic organs are usually highly differentiated at maturity. Primary root, numerous leaf primordia, shoot apex and coleoptile are well developed in cereal embryonic axes. Moreover, in monocots the cotyledon is often modified to form a relatively diminutive scutellum; whilst the endosperm is a large structure at maturity and is the main storage tissue. In wheat embryogenesis, five characteristic morphological stages have been identified by Rogers & Quatrano (1983). Fertilization and the subsequent stages of cell proliferation and differentiation end in the production of a functionally mature embryo. This embryo can germinate immediately, but normally it remains in a state of suspended growth (dormancy). The early stages of plant embryo development are characterized by cell division and morphogenesis. For instance, during wheat embryogenesis (~7 weeks), the zygote (1 cell) gives rise to a mature embryo (~10^5 cells) and the DNA content of the organism increases correspondingly (10^5-fold) (Lane, 1988). This is followed by a period of cell specialization and embryonic maturation in preparation for coming dormancy and germination periods.

Dormancy is not a period of metabolic inactivity because this period is characterized by the massive accumulation of storage materials in the appropriate tissues or organs, causing an increase in size and weight, the maturation stage. There are some proteins which characterize this stage of embryo development such as Em "Embryo **m**aturation" and LEA "**L**ate **E**mbryogenesis **A**bundant" proteins. LEA polypeptides are non-storage proteins and synthesized during stage 3 of embryogenesis. It has been suggested that non-storage LEA proteins might be involved in the acquisition of desiccation tolerance, during stage 4, which is a characteristic feature of the development of most seeds (Galau et al., 1986). The most abundant cytosolic wheat embryo protein is the Em protein in mature embryos (Grzelczak et al., 1982). Levels of the Em polypeptide increase rapidly in the cytosol and continue throughout desiccation of the embryo. Upon subsequent imbibition, both the mRNA and its polypeptide product are rapidly degraded, and completely disappear by 24 hours post-imbibition (Cuming, 1984). It was suggested that Em could have a limited storage function but was more likely to be associated with the acquisition of desiccation tolerance of embryo during embryogenesis and it is considered to be an example of the "group-1" LEA polypeptides (Butler & Cuming, 1993; Galau et al. 1986). The dormant condition is relieved only by the final stage in seed development, dehydration. Dehydration is an essential part of embryo development, being both necessary to maintain seeds in a viable form for long time periods and to effect a switch in their pattern of development. Seed desiccation also effects a permanent change in metabolism, in that, upon rehydration, synthesis of proteins associated with development ceases, and that associated with germination and seedling establishment commences. During the maturation period, the embryo accumulates reserve compounds in an effectively anabolic lifestyle. As a result of dehydration, the embryo is potentiated to germinate immediately upon the rehydration, its metabolic activity switched towards the massive catabolism of storage compounds to support growth.

During the early cell division stage of development, very little storage material is synthesized. Most of the synthesis of storage proteins, carbohydrates and lipids takes place during subsequent cell expansion. The final stage of seed development is characterized by the loss of water during maturation drying, when reserve synthesis stops and the seed becomes metabolically inactive. Drying of the seed enables the embryo to be a dispersal structure that is resistant to environmental disturbances and that remains quiescent until conditions are suitable for germination and growth. Germination of the seed starts with the imbibition of water, resulting in the rapid resumption of protein synthesis, using those components of the synthetic complex conserved within the dry seed. Upon emergence of the radicle from the seed, germination is complete and the seedling becomes established, with the rapid elaboration of the clearly defined root and shoot regions. During the early development (2 days) of a mature wheat embryo (<1 mg) into a seedling (> 10mg), the rapid and striking (> 10 fold) increase in mass is accompanied by limited cell division (Lane, 1988), and must therefore be a consequence of cell expansion driven by water uptake. This change in the size of the cells must necessarily involve significant adjustments in the structure of the extra cellular matrix (ECM) (apoplast, cell walls) in the growing seedling.

The role of plant growth regulators in the germination process is uncertain although it is known that exogenous application of abscisic acid inhibits germination in many species, while gibberellic acid promotes germination. However, the relationship between these experimental phenomena and the embryo's endogenous hormonal metabolism remains obscure (Bryant & Cuming, 1993). Gibberellic acid has a well-established role in the mobilization of the seed's nutrient reserves in the cereals during germination, and the mobilization of reserves is clearly essential for successful seedling development in the longer term, but for the first 48 hours of germination, very little of the endosperm reserves is mobilized, and the nutrients utilized must therefore derive from the stores held within the embryo (principally lipids and storage proteins). When a seed is provided with water and oxygen at an appropriate temperature, water is taken up by imbibition in order to initiate the germination. Protein synthesis must play an important role in germination and early plant development. It is important for the growth of the embryonic axis and in the synthesis of hydrolytic enzymes as well as for the other cellular processes involved in the mobilization of food reserves. Germination can therefore be recognized as a catabolic stage because of this breakdown of reserves. Dry seeds contain residual mRNAs which were synthesized during embryogenesis. The mRNAs translated during early germination are soon replaced as germination progresses by the synthesis of new mRNA and of additional proteins. Not surprisingly, investigations of these new proteins have characterized many of them to be hydrolytic enzymes including amylases, proteases and lipases (for the mobilization of starch, protein and oil respectively) (Fincher, 1989; Bewley & Black, 1985).

A close examination of the *in vitro* translation products from embryo mRNA sets isolated at different times of development has led to the identification of classes of genes expressed with characteristic patterns. The majority of genes are described to be produced constitutively, such as those encoding actins and tubulins (Dure, 1985), the remaining sets are expressed in a stage specific manner (Goldberg et al., 1989). Jendrisak (1980) reported that *de novo* mRNA synthesis was needed for resumption of growth in germinating wheat embryos. He suggested that the importance and significance of stored mRNA in the dry seed and its role in germination was minimal, since stored mRNA was by itself insufficient to allow for the resumption of germinative growth. As an imbibing seed undergoes the rapid transition from quiescence to vigorous metabolism many biochemical and physiological processes are activated. An obvious expectation is an increase in the expression of many different genes whose products provide for the higher levels of steady-state metabolic rate reached during this period (Caliskan et al., 2003). Whilst some of the biochemical events might be peculiar to this period, the majority would be expected to be related to normal growth-maintenance. Undoubtedly there are genes coding for proteins that control the activation of genes whose products provide for the higher steady-state metabolic rates seen. Moreover, there are probably a number of genes that code for proteins involved in the more complex developmental aspects of germination and seedling establishment. Seedling establishment needs the mobilization of stored reserves, involving such supporting tissues as the cotyledons and endosperm in dicotyledons seeds and the aleurone layer and endosperm in seeds of monocots. These tissues will undergo increases in expression of growth-maintenance genes, but there will be major activation of growth regulatory genes in accord with the role of these tissues in supporting germination and growth, *e.g.* the postgerminative production of hydrolytic enzymes by storage tissues.

An early event upon rehydration of seed tissues is an increase in protein synthesis and the proliferation of cellular organelles. The first consequence of imbibition may be an increase in ATP to an appropriate threshold level allowing this basic metabolic process to commence (Moreland, 1974). Axis germination would then proceed through an initial phase during which substrates and enzymes already present in the dry seed provide the required biochemical activity, to a final phase in which the primary interactants are mobilized substrates and proteins synthesized from newly transcribed mRNAs. After germination of both wheat and barley, the cells of the aleurone layer (the secretory cells that surround the nonliving starchy endosperm) synthesize several hydrolytic enzymes, including, characteristically, α-amylase which increases its activity 100-fold after germination. It was shown that activation of α-amylase is under control of GA (gibberellic acid). In addition, a number of other hydrolytic activities are induced in the barley aleurone by GA, including nucleases, β-glucanases, and proteases (Bewley & Marcus, 1990).

Although the induction of hydrolytic enzyme activities is essential during germination, plants need to synthesize some other proteins and enzymes which are required for a successful seedling development and autotrophic life. Typically, these are those associated with the establishment of photosynthetically competent tissues, and include all those structural proteins and enzymes found within the chloroplasts, in aeriel parts of the plant. However, although these gene products are needed for sustained growth, they can not be considered as germination-specific. A number of uncharacterized polypeptides have been shown to be uniquely associated with the onset of germination. One such protein has been isolated in 4 day old pea seedlings and it is called "C3 protein". This displays a shoot-specific pattern of expression (de Vries et al., 1983). The absence of its mRNA in any other organs and also the insensitivity of its accumulation to illumination have indicated that the protein was not related to photosynthesis but rather is associated with shoot elongation. The gene for phytochrome is also stimulated in expression in germinated pea axes (Konomi et al., 1987). The amount of phytochrome mRNA and that of the mRNA for glutamine synthetase (glutamate-ammonia lyase EC 6.3.1.2) (clone 2A2; Datta et al., 1987), peaks shortly after the onset of increased embryo fresh weight, suggesting that these mRNAs can be regarded as functioning primarily in the maintenance of steady-state growth. Glutamine synthetase gene expression is also abundant in soybean seedling root tips (where the predominant activity is cell division and early elongation) and in rapidly growing soybean cell cultures (very little or no differentiation occurs), a conclusion again consistent with the notion that this gene is particularly involved in growth-maintenance, presumably through its role in nitrogen assimilation. By contrast, two other soybean seedling mRNAs have been identified (clones 4D7 and 2E2; Datta et al., 1987) which are not expressed in seedling root tips, nor in growing cells in culture, and thus were presumed to be involved in more subtle aspects of seedling development. Another defined mRNA, whose expression commences with the onset of germination, occurs in the wheat seedling. The product of this mRNA was called germin to indicate that it was associated in germinative growth (Grzelczak & Lane, 1984). Germin (G) was not found in immature embryos or in mature, dry embryos before their hydration, nor was it found in mature wheat organs (Lane et al., 1992).

2.1 The possible roles of germin gene products

Germination is a critical period in plant development in which the rapid growth of the embryo is driven by water uptake. The water content of a mature wheat embryo in an ungerminated grain is less than 5%; upon germination it rises to about 60% in less than 1 hour. Between 1 and 5 hours of imbibition there is no further in fresh mass or water content. A "secondary water uptake" phase then raises the water content from 60% to 85% by 24 hours postimbibition (Marcus, 1969). Biochemical analysis of wheat embryo germination indicates that there is only a limited accumulation of new gene products during germination (Thompson & Lane, 1980) and until recently, just one had been observed to signal the onset of early plant development (Lane & Tumaitis-Kennedy, 1981). The synthesis and translation of the mRNA for a soluble protein initially called "g" (Thompson & Lane, 1980) and later called "germin" (Grzelczak & Lane, 1984) was concomitant with the initiation of growth in germinating wheat embryos (Rahman et al., 1988). The appearance of germin mRNA was coincident with the secondary water uptake phase and the accumulation of the germin protein reached its highest level between 24-48 hours postimbibition (about ~ 40ng/embryo in 40 hour germinated embryos (Grzelczak et al., 1985).

Germin was first detected in germinating cereals, but subsequently, germin-like proteins were also identified in a protist (Lane et al., 1991), dicotyledonous angiosperms (Michalowski & Bohnert, 1992; Hofte et al., 1993; Delseny et al., 1994; Heintzen et al., 1994), and gymnosperms (Domon et al., 1995). Wheat germin is a relatively rare water-soluble glycoprotein (less than 0.1% of the mass of soluble proteins in germinating wheat embryos) which in homogenates exists as an oligomeric complex even which does not dissociate when analyzed by SDS-polyacrylamide gel electrophoresis (SDS-PAGE) unless boiled in the presence of the detergent (Grzelczak & Lane, 1984). Three isoforms of germin have been defined. These are G, G' and ΨG (pseudogermin). The G and G' oligomers are water soluble, resistant to digestion by pepsin and to dissociation in aqueous SDS solutions at room temperature (Grzelczak & Lane, 1983, 1984). Amino acid composition and sequencing indicate that G and G' have the same apoprotein and differ only in that G has two further N-acetylglucosamine units attached to the basic core of its N-glycans. (Jaikaran et al., 1990). Both forms of germin were defined as glycoproteins based on their positive Schiff reaction, which is a characteristic of glycoproteins as well as by the incorporation of radioactive glucosamine, mannose and fucose into the germin oligomer following labelling, *in vivo*. The carbohydrate content of germin is about 10% by weight (Lane et al., 1987; Jaikaran et al., 1990).

One of the most striking characteristics of germin is its resistance to proteolysis: it was found to be resistant to digestion by a broad range of proteases, including pepsin. By exploiting this property, milligram quantities of highly purified protein may be prepared from pepsinized supernatants of an homogenate, or simply from an aqueous wash fraction of germinated embryos, in a very short time (Grzelczak & Lane, 1984). It seems possible that the oligosaccharide substituents in germin play a role in the remarkable stability of protein towards protease (Lane et al., 1987). In addition to the germination related germin isoforms (G and G′), an antigenically related homotetrameric form of germin (pseudogermin - ΨG) has also been detected, particularly in the cell walls of immature wheat embryos at a time when the maximum cellular enlargement associated with embryogenesis and maturation is occurring (20-25 dpa) (Caliskan et al., 2004). Unlike germin, pseudogermin is thermostable; the oligomer remains undissociated even when boiled in the presence of SDS, so long as non-reducing conditions (*e.g.* in the absence of mercaptoethanol) are maintained (Lane et al., 1992). Physicochemical studies indicated that germination-related germin (G) was a homopentamer and its oligomeric mass was calculated to be ~ 130 KDal when measured by three different methods. Therefore a molecular mass of 130/5 = 26 KDal was calculated for the monomer (McCubbin et al., 1987).

A virtually full length germin cDNA was isolated (Rahman et al., 1988) and its polynucleotide sequence was determined (Dratewka-Kos et al., 1989). This germin cDNA has been used as a probe to indicate that germin is encoded by a multigene family which has ~5 copies on chromosome 4A, ~3 copies on chromosome 4B and ~9 copies on chromosome 4D in hexaploid wheat. This cDNA was also used to screen a genomic wheat DNA library and the nucleotide sequences of a 2.8 kbp fragment (gf-2.8) from one genomic clone, and of 3.8 kbp fragment (gf-3.8) from another clone were determined. The protein coding-regions of these two genes are intronless and 87% identical (Lane et al., 1991). The association of germin isoforms with apoplast cause speculation that germin may have a role in embryo-specific desiccation/hydration processes (Lane et al. 1991). It was observed that even if germin was purified free of other proteins, there was a selective association between germin and the highly substituted glucuronogalactoarabinoxylans (HS-GGAX) (Jaikaran et al., 1990) whose synthesis was reported to be closely associated with cell wall extension in cereals and grasses (Gibeaut & Carpita, 1991). Upon this observation, it was suggested that germin-like oxalate oxidase might play a crucial role in early plant development by controlling integration of cell wall extension, for example by transporting extending wall material (*e.g.* HS-GGAX) into the cell wall to support extension and at the same time promoting cross-linking between wall polymers to restrict extension (Lane, 1994).

In gymnosperms, three germin-like proteins were discovered among the extracellular proteins produced by cells grown in liquid tissue culture. These proteins were found to be present in cultures which retained embryogenic potential, but to be absent in non-embryogenic cell lines. These proteins had high N-terminal amino acid sequence homology with other germins and were immunologically cross-reactive with an antiserum raised against the *apo*-protein component of cereal germin. Their molecular weight was about ~ 26 KDal. Pine germin-like proteins have not yet been shown to have oxalate oxidase activity (Domon et al., 1995). It was also demonstrated that wheat embryo callus induction by auxin is associated with a rapid accumulation of germin-like oxalate oxidase (Caliskan et al., 2004). Photoperiodic treatments of the short-day plant (SDP) *Pharbitis nil* seedling resulted in synthesis of a germin-like protein during darkness-induced flowering. This germin-like protein had a molecular mass of 22 KDal in SDS-PAGE analyses, and it reached the highest level of accumulation after the critical length of the dark period (~ 10 hours after the light was turned off). The cotyledons and leaves, known to be the two major organs that perceive the photoperiod and produce the floral stimulus, were the only expression sites for this protein. Sequence analysis showed that the *Pharbitis nil* germin-like protein shared the highest homology a germin-like protein in another dicotyledonous plant, *Sinapus alba* (white mustard). No oxalate oxidase activity was found for *Pharbitis nil* germin-like protein (Ono et al., 1996).

The similarities between wheat germin and the barley germin-like polypeptides may imply that both of these proteins have some protective function during early plant development. In wheat embryos, germin increases significantly during seed germination. Based on the increase in carbohydrate synthesis during germination and the presence of adventitious arabinoxylans in wheat germin preparations this has prompted speculation that germin could have a function in cell wall expansion (Jaikaran et al., 1990). The identification of cereal germins as oxalate oxidases (Dumas et al., 1993; Lane et al., 1993) has caused us to modify this view (Caliskan & Cuming, 1998). The principal product of the degradation of oxalic acid is hydrogen peroxide which is a highly reactive compound known to be involved in several metabolic processes in higher plants. The discovery of its oxalate oxidase activity instantly suggested several specific ways in which germin might function. Specifically, linkage of the developmental appearance of cell wall bound germin (Lane et al., 1992) to oxalate degradation suggests that germin might have a role in cell wall reinforcement by producing Ca^{++} and H_2O_2 for pectic cross-linking and peroxidase mediated cross-linking of cell wall polymers respectively. Proteins could be cross-linked through tyrosine side-chains, lignin via -OH groups and carbohydrates via -COOH groups (Cassab & Varner, 1988; Showalter, 1993). The anchorage of at least a proportion of the germin-like oxalate oxidase in the cell wall with GGAX oligomers also implicates it in the cross-linking of these components within the cell wall matrix; notably, arabinoxylan comprises a substantial fraction of monocotyledonous cell walls, in which the oxalate oxidase activity of germin-like proteins is characteristic. Although germin-like proteins have been identified in dicotyedonous plants, these, for the most part, have not been shown to possess oxalate oxidase activity. An exception to this is the oxalate oxidase of *Beta vulgaris*, which has been found to be germin-like in its properties (a protease-resistant SDS-stable oligomer): notably, the cell walls of this species, and other members of the *Chenopodiaceae*, are more "cereal-like" than "typically dicotyledonous", being enriched in arabinoxylan content (Bacic et al., 1988). Germin synthesis also appears to be auxin responsive (Berna & Bernier, 1997). Typically, auxins stimulate cell wall loosening and bring about cell wall expansion. On the other hand, germin (oxalate oxidase activity) produces hydrogen peroxide which is believed to be required for the peroxidase mediated cross-linking reactions in the cell wall. Thus, it was suggested that germin synthesis might be associated with both initiation and termination of cell wall expansion in early plant development (Lane, 1994).

Germin-like oxalate oxidase is also a pathogen inducible enzyme (Zhang et al., 1995; Dumas et al. 1995; Hurkman & Tanaka, 1996b). The production of H_2O_2 (the "oxidative burst") is the primary response of some higher plants to pathogen (*Erisyphe graminis)* infection. H_2O_2 can be implicated in the development, differentiation, vascularization, defence and signalling processes of higher plants. The action of oxalate oxidase, in generating H_2O_2, could be an especially potent defence mechanism. The pathogen-responsive oxalate oxidase was different from both the commercially available barley root oxalate oxidase, and barley salt stress-induced root oxalate oxidase in its molecular weight. Commercial oxalate oxidase was reported, variably, to have a molecular mass 80 KDal in 8% SDS-PAGE (or, Dumas et al. (1995) report 100 KDal) whereas the pathogen-related oxalate oxidase is 100 KDal - 95 KDal. This may therefore represent a specific, individual member of the germin gene family. In a parallel study of transcripts induced in response to *Erisyphe* infection, two novel germin-like sequences were identified as pathogen-responsive (Zhang et al., 1995).

Dumas et al., (1995) demonstrated an increase in the activity of germin-like oxalate oxidase in association with the response of barley to *Erisyphe graminis*. In normally growing seedlings, oxalate oxidase activity was detected in a tissue specific manner in 3 day germinated seed (in roots) and 10 day germinated seedlings (in the residual coleorhiza). These activities were demonstrated to be associated with a germin-like protein by immunoblotting using anti-germin antibody. In 10 day old seedling coleoptiles a 26 KDal polypeptide reacted with anti-germin serum but appeared to lack oxalate oxidase activity. This suggested that an inactive form of oxalate oxidase could accumulate. On infection by the fungus, oxalate oxidase was induced in barley leaves (5 day), especially along the vascular bundles. This suggested that oxalate oxidase belonged to a class of proteins that responds to pathogen attack. The production of H_2O_2 by oxalate oxidase might be significant in two ways: (1) By further enhancement of the defence response (through induction of further plant defence responses as a signalling molecule), (2) Through H_2O_2 being used as a substrate by peroxidase to mediate cross-linking of cell wall polymers (e.g. lignin) in the hypersensitive response, causing the sealing of infected lesions (Dumas et al., 1995).

Similarly, in wheat, germin mRNA, germin and oxalate oxidase activity were induced in leaves of wheat upon infection with *Erisyphe* (6 day old seedling +2, 4, 6, 8 days after inoculation). The control leaves at the same age gave negative results. An increase in expression of peroxidase was also detected. These results reinforce the suggestions that germin activity has a role in plant defence through the local production of H_2O_2 for the hypersensitive defence response (Hurkman & Tanaka, 1996b). This suggests that the genes encoding germin-like oxalate oxidase might have potential in transgenic approaches to plant defence. Another peroxide generating enzyme, a fungal glucose oxidase, has been shown to enhance the resistance of transformed plants to fungal infection, when introduced as a transgene (Wu et al., 1995). In experiments designed to protect *Brassica napus* plants from the oxalate secreting fungus, *Sclerotinia* transgenic oilseed rape plants, transformed with barley oxalate oxidase, were found to express a 25 kDal protein reactive with anti-germin antiserum and to express oxalate oxidase activity which protected plants against potentially toxic applications of oxalic acid (Thompson et al., 1995). In a current study, 36 expressed sequence tags (ESTs) encoding GLPs from peanut (*Arachis hypogaea* L.) were identified. The purified AhGLP2 has displayed superoxide dismutase (SOD) activity in enzymatic assay, but not oxalate oxidase activity. It was reported that the SOD activity of AhGLP2 was stable up to 70A °C and resistant to hydrogen peroxide, suggesting that AhGLP2 might be a manganese-containing SOD and likely protects peanut plants from reactive oxygen metabolites (Chen et al., 2011). The rice germin-like protein (OsGLP1) being a cell wall-associated protein involved in disease resistance also revealed to possess superoxide dismutase (SOD) activity as recognized by heterologous expression in tobacco (Banerjee et al., 2010).

It was discovered that both germins and spherulins had statistically significant sequence similarity with plant seed storage globulin domains. The germins were clearly related both to seed globulins (Baumlein et al., 1995) and spherulins (Lane et al., 1991). On the basis of these similarities, it has been proposed there is a superfamily (groups of gene families encoding structurally related but functionally distinct proteins) of related genes encoding vicilins, legumins, SBPs (sucrose-binding proteins), germins and spherulins. SBPs are proteins associated with the plasma membrane and which have a role in sucrose transport, and it is known that spherulins have a function in the cellular desiccation process, including osmotic regulation. It has been suggested that sucrose may serve as one of the principal agents in the acquisition of desiccation tolerance in seeds and other plant tissues, where the role of disaccharides in the assumption of a "glassy state" by the cytosol has been inferred (Leopold et al., 1992). The legumin-like 11S and the vicilin-like 7S seed proteins are synthesized and accumulated during seed maturation, and are stored in protein bodies in mature seeds. In the timing of their accumulation, and their regulation by agents such as abscisic acid, they are at least associated with the acquisition of desiccation tolerance by seeds, which occur during the maturation phase of seed development (Braun et al., 1996).

2.2 The enzymatic activity of germins: oxalate oxidase

Oxalic acid is one of the strongest organic acids with pKa values of 1.3 and 4.3 (Lane, 1994). Oxalic acid has a wide variety of industrial and household applications for instance it is used as an analytical reagent. Oxalic acid is also a constituent of cleaning solutions for removing paint, varnish, rust and ink stains as well as being used extensively in laundries as a scouring agent to remove excess alkalinity remaining in washed fabrics. It is also used for cleaning or bleaching wood and straw, as a chrome stripper and as a bleach in leather manufacture.

In plants, the highest oxalate concentrations commonly occur in the leaves and the lowest in roots. Meanwhile, the oxalate content of plants can vary according to their age, the season, the climate and the type of soil. Wide variations can occur in the oxalate content of plants. In some plants, such as rhubarb, oxalate content tends to increase as the plants mature, whereas, in other plants, e.g. spinach, sugar beet leaves, oranges, bananas, there is a large increase in oxalate content during the early stages of development, followed by a decrease as the plants mature.

Although the origin of oxalic acid in plants is controversy, the early studies on plant organic acids, including oxalic acid, yielded evidence that the production of these acids was related to photosynthesis and carbohydrate metabolism. Myers (1947) noted that oxalate concentration in rhubarb leaves increased in parallel with the growing seasons, being correlated with the seasons of most active photosynthesis. Later, experiments on rhubarb and Begonia indicated that oxalic acid was not a direct product of photosynthesis but it was synthesized from precursors synthesized in the photosynthetic pathway (Stuta & Burris, 1951; Tavant, 1967).

Now, it is well known that oxalic acid is synthesized via several major pathways. Although glyoxylate and L-ascorbic acid appear to be the major precursors of oxalic acid in plants (Davies & Asker, 1983; Yang & Loewus, 1975), some other possible pathways have been reported. Glucose, acetate and some acids of the tricarboxylic acid cycle were determined to be involved in oxalate biosynthesis in red beet roots and young spinach leaves (Chang & Beevers, 1968). Moreover, glycolic and isocitric acids (Millerd et al., 1963), oxaloacetic acid (Chang & Beevers, 1968) are known to donate carbon to oxalic acid in plants. The relative significance of these metabolites as precursors of oxalic acid has not been established (Wagner, 1981).

The enzymes involved in the synthesis and degradation of oxalic acid are of interest. In lettuce, two enzymes identified to have a role in the oxidation of glycolate and glyoxylate to oxalic acid were lactate dehydrogenase and glycolate oxidase (Davies & Asker, 1983). Since oxalic acid could be formed from several precursors, there is no clear evidence of its synthesis and deposition site. However, it was observed that the primary site of deposition of oxalic acid formed from L-ascorbic acid was the vacuole in barley which is a low level oxalic acid accumulator (Wagner, 1981).

Most of the plants and animals produce oxalic acid, and it is of interest that they share some common pathways of oxalic acid synthesis. Oxalate may be present in the tissues as the free oxalic acid, as soluble sodium and potassium salts or as insoluble calcium oxalate crystals. Calcium oxalate crystal formation in animals is generally considered to be pathological. The pathological role of oxalic acid in the formation of urinary stones in animals and humans has been known since the early 18th century. For example, excess consumption of oxalate-rich foods leads to hyperoxaluria which is recognized as a key risk factor for calcium oxalate stone formation (Sharma et al., 1991). Furthermore, there is evidence that high ascorbic acid intake increases urinary oxalate levels which could lead to the formation of calcium oxalate stones in the kidneys and other regions of the urinary system (Roth & Breitenfield, 1977). Oxalic acid is regarded as an undesirable component of our food not only because it raises the risk of urinary stones but also because it sequesters calcium, which is one of the essential ions, as insoluble calcium oxalate. The toxicology of oxalic acid in humans was reviewed by several researchers (Polson & Tattersall, 1959; Hodgkinson, 1977). In medicine, the knowledge of the oxalate concentration in blood and other body fluids can be very important in certain clinical situations such as primary hyperoxaluria (Pertrarulo et al., 1990). Various methods have been formulated for the assay of oxalate from different sources, for instance, gas-liquid chromatography, ion chromatography, high performance liquid chromatography, mass spectrometry and enzymatic determination methods have all been developed. Of these, the oxalate oxidase activity-based determination of oxalate has become very popular and is used widely because of its simplicity, specificity and sensitivity (Pundir et al., 1985).

Oxalic acid has usually been seen as an inert end product of metabolism and only plants have been reported to be able to metabolize oxalic acid and oxalates. However, recent studies indicate that the levels of oxalate are too high for the substance to only be an end-product of metabolism in animals (Emsley, 1994). Therefore, it has been suggested that there could be an oxalate oxidase pathway in animals which uses oxalate to produce H_2O_2, which could than be used to promote a "burst" of phagocytes: cells that engulf and break down foreign particles, cell debris and disease-producing microorganisms. Unlike animals, plants are highly tolerant of oxalic acid and oxalates. Oxalic acid and oxalates have been detected in varying quantities in all parts of most plants' leaves, leaf stalks, flowers, tubers and roots (Srivastava & Krishnan, 1959). It is well established that plants are capable of metabolizing oxalate by observing fluctuations in oxalate concentrations under certain conditions (Vincent & Harry, 1980), and enzymes degrading oxalic acid have been detected in numerous plants. Several functions have been proposed for the presence of oxalic acid in plants. It has been implied that oxalic acid might be related to ionic balance, since it can combine with various plant ions to form soluble or insoluble compounds. It was suggested that oxalate synthesis occurred to balance the excess of inorganic cations (represented by K^+, Na^+, NH_4^+, Ca^{++} and Mg^{++}) over anions (represented by NO_3^-, Cl^-, $H_2PO_4^-$, SO_4^{-2}) normally present in the plant - the ability of nitrate ions and chloride ions to inhibit oxalic acid oxidase activity in *Beta vulgaris* results in an accumulation of oxalate.

Calcium oxalate crystals were amongst the first objects observed in plants in the early days of light microscopy in the late 17th century (Lane, 1994). Insoluble calcium oxalate formation enables plants to control the concentration both of ionically active oxalic acid and calcium. Both of these molecules might have a toxic effect when accumulating in excess quantities. Thus plants could induce calcium oxalate crystal formation to remove excess oxalic acid or calcium. Although calcium is essential to biological growth and development, free calcium at high concentrations is toxic to cells. So it was suggested that calcium oxalate precipitation serves to sequester excess calcium and remove it from active metabolism (Webb et al., 1995). The rapid induction of calcium oxalate crystal formation by calcium in Lemna plants suggests that the crystals may serve as a storage form for calcium for future needs (Hepher & Wayne, 1985). Further support for this came from the observation that in some plants the crystals appear to be dissolved during calcium deficient conditions, presumably to supply calcium for growth and cell maintenance (Franceschi, 1989). Calcium is required for the activation and/or stabilization of certain enzymes; for example plant cells need calcium to release peroxidases which are related to the control of cell elongation since they can rigidify walls by their cross-linking activity and their ability to participate in the formation of lignin. Thus, for this role they are under the control of cellular calcium levels (Sticher et al., 1981). One of the major roles of calcium in plant cells is its action in the formation of the middle lamella where Ca^{2+} ions form stabilising ionic bridges between pectin chains. Although it has been suggested that calcium oxalate crystals are a means of detoxifying excess oxalic acid, the fact that many plants are able to retain high concentrations of soluble and free oxalic acid within their vacuoles indicates that oxalic acid may not be particularly toxic to plant tissues. However, accumulation of oxalic acid may have some toxic consequences (for example in causing osmotic problems and destabilization of cells), unless it is readily metabolized (Raven & Smith, 1976).

It is well known that certain plant pathogenic fungi secrete oxalic acid as part of the process for invasion of plant tissues. For example, *Sclerotium rolfsii* Sacc., a fungus, causes diseases of plants in nearly 100 plant families. Considerable quantities of oxalate were detected in infected but not in healthy tissues. Oxalic acid produced by this pathogenic fungus played an essential role in its pathogenic capabilities (Maxwell & Bateman, 1968). Another pathogenic fungus infecting a wide range of plant species is *Sclerotinia sclerotiorum*. During infection, the fungus produces high levels of a necrosis phytotoxin identified as oxalic acid (Noyes & Hancock, 1981). The role of oxalic acid in the pathogenicity process is still unclear. However, oxalic acid may have a number of functions in the infection process including chelating calcium from the cell wall thus making the pectic fraction more available to fungal hydrolases, and providing an acid pH needed for maximum activity of the wall degrading enzymes released by the pathogenic fungus (Keates et al., 1996). Furthermore, it was suggested that oxalic acid produced by fungi played a key role in lignin biodegradation through its stimulation of lignin-degrading enzymatic activities (e.g. Mn-peroxidase activity) (Kuan & Tien, 1993). One part of the plant defence response to this may be the production of oxalate oxidase which is an oxalic acid degrading enzyme. A second response would be the induction of deposition of oxalic acid in the form of soluble or insoluble salt.

Oxalic acid and its salts, oxalates, are widely distributed within the cells and cell walls of plants and probably they play an important role in tissue metabolism. There are two possible enzymatic reactions for the degradation of oxalate in plants: (1) decarboxylation by oxalate decarboxylase (oxalate carboxy-lyase, EC 4.1.1.2) which catalyzes conversion of one mole of oxalate to one mole of CO_2 and formate, (2) oxidation by oxalate oxidase. Oxalate oxidation has been found to occur in fungi, mosses and higher plants, but the biological significance of oxalate oxidation is not yet clear. The discovery that germin, a protein marker of early plant development, is an oxalate oxidase suggested that oxalate oxidase, the enzymatic formation of H_2O_2, and Ca^{++} release from poorly soluble calcium oxalate might play an important role in metabolic regulation, particularly in cell wall modification during germination and seedling development.

Oxalate oxidase is an oxidoreductase (oxalate: oxygen oxidoreductase EC 1.2.3.4) which catalyzes the formation of one mole of H_2O_2 and two moles of CO_2 from one mole of oxalate and aerobic oxygen. Oxalate oxidase has been found in Pseudomonas sp. OX-53 (Koyama, 1988), *Tilletia controversa* (Vaisey et al., 1961), banana peel (Raghwan & Devasagayam, 1985), mosses (Laker et al., 1980), *Bougainvillea* leaves (Srivastava & Krishnan, 1962), barley seedlings and roots (Chiriboga, 1966; Pietta et al., 1982), spinach, beet stem and leaves (Obzansky & Richardson, 1983; Leek et al., 1972), and leaves and roots of CSH-1 and CSH-5 varieties of grain sorghum (Pundir & Nath, 1984; Pundir & Kuchhal, 1989).

An oxalate oxidase was purified from sorghum leaves with a molecular weight of 62 kDal and pH optimum of 4.3. This enzyme had maximum activity at 40 °C and it was insensitive to Na^+ . This property made it well suited for medical diagnostic use in the detection of oxalate in urine - the principal cause of kidney stone formation. The enzyme was suggested to be a flavoprotein on the basis of stimulation of its activity by FAD. However, neither the barley oxalate oxidase nor that of Pseudomonas sp. OX-53 was classified as flavoproteins (Pundir, 1991).

Of the other oxalate oxidases purified from plant sources, that from grain sorghum leaves (*Sorghum vulgare* L hybrid CSH-5) (Kuchhal et al., 1993) was active in a range of pH from 4-6, with an optimum of pH: 5 and optimal temperature of 40° C. The optimum pH for barley oxalate oxidase was pH: 3.2, and for mosses pH: 4, whilst the bacterial (Pseudomonas) enzyme was maximally active at pH: 4.8. The enzyme purified from grain sorghum leaves was heat sensitive, as it lost 80% of its activity when heated at 80 °C for 3 min whilst the inhibition of the enzyme by EDTA, and its stimulation by metal ions (Cu^{++}, Mg^{++}), led it to be subsequently classified as a metalloprotein and not a flavoprotein. Both the enzymatic activity and oxalate content showed progressive decrease with the advance of germination in sorghum. These data suggested that the high enzyme activity during initial growth was required for degradation of endogenous oxalate to form H_2O_2 which might play an important role in cellular regulation, such as glucose transport, glucose incorporation into glycogen, lipid synthesis, and release of Ca^{++} (Kuchhal et al., 1993).

The oxalate oxidase purified from leaves of grain sorghum hybrid CSH-5 was reported to have a molecular mass of 120 KDal and to be composed of 62 KDal monomeric subunits, unlike the enzymes purified from moss, barley, and banana peel, it was unaffected by Cl^-, in the physiological concentration range, which made it particularly suitable for urinary diagnostic tests, since the removal of Cl^- from urine, prior to oxalate assays was not required (Satyapal & Pundir, 1993). More recently, an oxalate oxidase has been purified from beet stems (*Beta vulgaris* L.) (Azarashvili et al., 1995). When the activity of this enzyme was determined as a function of pH and temperature, the optimal conditions were found to be at pH: 4 and 30 °C. This enzyme has been incorporated into an automated method for determination of urinary oxalate - a rapid (3 hours) and reliable procedure based on the measurement of H_2O_2 produced by oxidation of oxalate. The chromogen MBTH-DMA was used to measure the hydrogen peroxide produced (Obzansky & Richardson, 1983). Nitrate ions (at concentrations as low as 5x10-5 M) was identified as the sole factor in crude beet extracts responsible for the inhibition of oxalate oxidase activity in vitro (Meeuse & Campbell, 1959). This would lead to accumulation of oxalate in cells.

Barley seedling oxalate oxidase was purified by Sugiura et al. (1979). It was originally reported to have a monomeric molecular mass of 75 KDal and to assemble as a homodimer of 150 KDal. The optimal pH and temperature were found to be 3.2 and 37 °C, respectively. The enzyme was found to be extremely stable at temperatures up to 70 °C and to be strongly inhibited by 2-mercaptoethanol and halogen ions (Cl^-, F^-). This enzyme appeared to differ in its properties from the sorghum leaf enzyme, which in addition to its different pH and temperature optima, and molecular mass, was not inhibited by halogen ions (Satyapal & Pundir, 1993).

Oxalate oxidase activity in plants may vary between species, and differ between organs and tissues of a plant even though these may have similar oxalate contents. For example, a close examination of germinated cereal embryos indicated that there are only modest differences between the oxalate contents of different varieties. However, there may be gross differences between their oxalate oxidase activities. For example, cold-tolerant maize (C0255) contains ~ 20 fold higher levels of oxalate oxidase activity than cold-sensitive maize (C0286), whilst in 6 day germinated wheat seedlings, the coleoptile and leaf have the similar oxalate content but the coleoptile has ~ 16 times higher levels of oxalate oxidase activity than the leaves (Lane, 2000).

Analysis of beet shows that the oxalate content of this dicotyledonous plant is ~ 10-fold greater than that determined in wheat, but that the oxalate oxidase/oxalate ratio is 100-fold smaller than for hexaploid wheat. Beet oxalate oxidase shares several properties with the wheat enzyme: typically the "germin-like" characteristic of resistance to SDS-denaturation, allowing its activity to be determined directly in SDS-PAGE gels. However it was sufficiently immunologically similar to be recognized by wheat anti-germin in Western-blots. By contrast, the sorghum oxalate oxidase, although derived from a monocotyledonous source did not have "germin-like" stability, but it did react weakly with the anti-serum raised against wheat germin (Lane 2000).

2.3 The expression of germin genes under salt stress

Although there are plenty of organisms need salt to carry out their life (Ozcan et al., 2006, 2007) excessive salt is a stress factor for most of the living beings. Salt stress is also an important agricultural problem, particularly since the majority of crop plants have low salt tolerance. The response of plants to salt stress is a complex phenomenon that involves biochemical and physiological processes as well as morphological and developmental changes (Flowers et al., 1977; Greenway & Munns, 1980). The identification of genes whose expression enables plants to adapt to or tolerate to salt stress is essential for breeding programs, but little is known about the genetic mechanisms for salt tolerance. One approach in clarifying the molecular mechanisms involved in salt stress is to identify the genes whose levels change as a result of salt stress.

Isolation and examination of the two germin genomic clones (Rahman et al., 1988) and the determination of the predicted amino acid sequences have revealed high homology with spherulin 1a/1b proteins of the slime mould *Physarum polycephalum* (Lane et al., 1991). The synthesis of these proteins occurs during spherulation: a transition leading to developmental arrest imposed by environmental conditions such as osmotic stress and starvation (Bernier *et al.*, 1987), this similarity led to the suggestion that another possible function for germin might be related to the changing osmotic properties of cells. In support of this notion, synthesis of germin-like proteins was discovered to be altered upon salt stress in barley (Hurkman et al., 1991) and in the halophytic "ice plant", *Mesembryanthemum crystallinum*, (Michalowski & Bohnert, 1992). In salt-stressed barley, these proteins, like wheat germin, were resistant to protease and were glycosylated and heat stable. They were detected in barley roots (but not in tips) and coleoptiles but not leaf of 6 day seedlings. Their synthesis increased in roots upon salt stress, but decreased in coleoptiles. On the other hand, it was observed that the synthesis of a germin-like protein in the ice plant declined after salt stress (Michalowski & Bohnert, 1992). Thus, these different studies implied that germin might represent a family of proteins of which individual members may have different biochemical functions related to changes of the osmotic properties of the cell.

Experiments based on the extraction of mRNA from different organs suggest that wheat and barley have different spatial distributions of germin mRNA expression (Hurkman & Tanaka, 1996a). The vascular transition region was reported to contain the highest levels of germin mRNA in wheat, whereas roots displayed the highest germin expression levels in barley seedling. It was additionally observed that salt stress caused an increase in germin mRNA in roots at an early developmental stage (3 days on 0.2M NaCl), whereas, in the whole seedling, salt stress cause the "normal" expression of germin to be prolonged, relative to that in control seedlings. Overall, it was concluded that germin gene expression in barley seedlings was developmentally regulated in a tissue-specific manner, and also, potentially, by various plant hormones (IAA, ABA). Interestingly, it was discovered that a decline in germin expression occurred after 3 days in control seedlings and after 4 days in stressed seedlings - in both case seedlings were at a similar developmental stage, as assessed by seedling weight (approximately 25 mg in each treatment). Thus one interpretation of the experimental data is that NaCl treatment prolongs germin gene expression for an additional 1 day indirectly, through a slowing of seedling growth (Hurkman & Tanaka 1996a).

Among the salt stress responsive gene products, germin and germin-like proteins (GLP) suggested to be a member of "superfamily" which comprises various growth-related genes were identified (Caliskan, 1997; Hurkman et al., 1989). Cereal germin proteins have strong oxalate oxidase activity (Lane et al., 1993), an activity that produces one mole of H_2O_2 and two moles of CO_2 from degradation of oxalic acid. It is reported that H_2O_2 might act as a signaling molecule at low concentration (Luthell, 1993) or a component of cell wall modifications at high concentrations (Showalter, 1993). Another germin-like protein isolated from the cells of a moss, *Barbula unguiculata*, was shown to have manganese superoxide dismutase activity (Yamahara et al., 1999). Germin genes and proteins have been shown to be associated with various aspects of plant development (Caliskan, 2000; Lane, 2002) such as defense system (Berna and Bernier, 1999; Donaldson et al. 2001), embryonic development (Caliskan, 2001), photoperiodic oscillations (Ono et al. 1996), and hormonal stimuli (Berna and Bernier, 1997). The accumulation of germin gene products in wheat and barley seedlings in the presence of NaCl was analyzed previously, but little is known about the possible role of germin gene products during salt stress (Hurkman et al., 1991; 1994; Hurkman & Tanaka, 1996a; Berna & Bernier, 1999).

Germin and germin-like proteins are suggested to be salt-responsive gene products and their response to salt stress seems to be various. For example, accumulation of germin mRNA is up-regulated during the growth of germinating barley seedlings in the presence of NaCl (Hurkman & Tanaka, 1996a). In contrast, it is reported that in ice plant (*Mesembryanthemum crystallinum*) the synthesis of GLPs declined after salt stress (Michalowski and Bohnert, 1992). On the other hand, it was reported that in wheat seedlings germin synthesis was remained unchanged in the presence of NaCl (Berna & Bernier, 1999; Caliskan, 2009). The addition of NaCl to the cells of moss, *Barbula unguiculata*, during the logarithmic phase increased both the *BuGLP* mRNA levels and total SOD activity of BuGLP, but decreased the SOD activity bound to the cell wall due to release of most of the SOD activity into the medium. On the other hand, the addition of NaCl to the cells during the stationary phase hardly affected *BuGLP* mRNA levels or SOD activity levels bound to the cell wall. These results suggest that the induction of *BuGLP* gene by salt stress is caused by dissociation of BuGLP protein from the cell wall into the medium in the cells during the logarithmic phase (Nakata et al., 2002).

In situ RNA hybridization is one of the most powerful techniques developed for localizing the expression site of particular gene products at the cell, tissue and organ levels. This method is particularly useful in understanding the function of specific gene products in particular tissues and the relation between tissue function and its localization in the whole structure of an organ (Ranjhan et al., 1992). This technique was employed for analyzing the possible germin functions and it was shown that germin mRNAs synthesized in the cells of coleorrhiza in wheat seedlings and it was considered that the enzymatic activity of germin, oxalate oxidase, might play an important role in metabolic regulation, particularly in cell wall modification during germination and seedling development (Caliskan & Cuming, 1998). It is well known that stress factors alter the synthesis of gene expression. Indeed, upon salt stress the localization pattern of germin gene expression was changed (Caliskan, 2009). It is shown that although the water grown embryos and salt stress grown embryos accumulate the similar amount of germin mRNAs, the synthesis site of germin mRNAs are completely different from each other. In germinating wheat embryos the salt stress somehow caused germin mRNAs to be synthesized in coleoptile instead of coleorrhiza cells (Caliskan, 2009). It is possible to envisage that an explanation for the reasons of this shift will help us to have a better understanding of stress physiology of plants.

3. Conclusion

Although germin and germin-like proteins have been studied extensively since the 1980s, their biological importance and functions remain confusing. Furthermore, germin, germin-like proteins and oxalate oxidase have been found in a broad range of various plant species under different circumstances, and related with different aspects of plant development. For example, germin-like protein and oxalate oxidase enzyme activity have been identified in plants in relation to salt stress, pathogen infection, photoperiodic oscillations, germination and embryogenesis. It has also reported that germin and germin-like proteins and oxalate oxidase are responsive to various plant growth regulators such as auxin and abscisic acid. Germin-like oxalate oxidase therefore seems to fulfil some crucial functions in plants. In addition to these properties, recently it was suggested that germins and various other proteins which are involved in early plant development might belong to an evolutionarily ancient superfamily: cupin superfamily (Dunwell et al., 2000; 2008).

4. Acknowledgment

I would like to thank for the contribution of Turkish High Education Board, University of Leeds (UK) and Mustafa Kemal University (Turkey) during my studies.

5. References

Azarashvili, T.S., Evtodienko, V., Yu. & Kudzina, L., Yu (1996). Isolation, purification, and kinetic properties of oxalate oxidase (EC 1.2.3.4) from beet shoots. *Russian J. Plant Physiol.* 43(2), 169-173

Bacic, A., Harris, P.J. & Stone, B.A. (1988). Structure and function of plant cell walls. In: *The biochemistry of plants, a comprehensive treatise*, J. Preiss (Ed.), volume 14, 297-358, Academic Press Inc. London

Banerjee, J., Das, N., Dey, P. & Maiti M.K. (2010). Transgenically expressed rice germin-like protein1 in tobacco causes hyper-accumulation of H_2O_2 and reinforcement of the cell wall components. *Biochem. Biophys. Res. Commun.* 402, 637–643

Baumlein, H., Braun, H., Kakhovskaya, I.A. & Shutov, A.D. (1995). Seed storage proteins of spermatophytes share a common ancestor with desiccation proteins of fungi. *J. Mol. Evol.* 41, 1070-1075

Bernier, F., Lemieux, G. & Pallotta, D. (1987). Gene families encode the major encystment-specific proteins of Physarum polycephalum plasmodia. *Gene* 59, 265-277

Berna, A. & Bernier, F. (1997). Regulated expression of a wheat germin gene in tobacco: oxalate oxidase activity and apoplastic localization of the heterologous protein. *Plant Mol. Biol.* 33, 417-429

Berna, A. & Bernier, F. (1999). Regulation by biotic and abiotic stress of a wheat germin gene encoding oxalate oxidase, a H_2O_2-producing enzyme. *Plant Mol. Biol.* 39, 539-549

Bewley, J.D. & Marcus, A. (1990). Gene expression in seed development and germination. *Prog. Nucl. Acid Res.* 38, 165-193

Bewley, J.D. (1997). Seed germination and dormancy. *The Plant Cell* 9, 1055-1066

Braun, H., Czihal, A., Shutov, A.D. & Baumlein, H. (1996). A vicilin-like seed protein of cycads: similarity to sucrose-binding proteins. *Plant Mol. Biology* 31, 35-44, 1996

Bryant, J.A. & Cuming, A.C. (1993). Molecular control of development. In: Plant *Biochemistry and Molecular Biology*, P.J. Lea & R.C. Leegood (Eds.), 241-273. John Wiley & Sons Ltd. England

Butler, W.M. & Cuming, A.C. (1993). Differential molecular responses to abscisic acid and osmatic stress in viviparous maize embryos. *Planta* 189, 47-54

Caliskan, M. (1997). Temporal and spatial analysis of germin synthesis. University of Leeds, PhD Thesis, Leeds UK

Caliskan, M. & Cuming A.C. (1998). Spatial specificity of H_2O_2-generating oxalate oxidase gene expression during wheat embryo development. *Plant J.* 15, 165-171

Caliskan, M. (2000). Germin, an oxalate oxidase, has a function in many aspects of plant life. *Tr. J. Biol.* 24, 717-724

Caliskan, M., Bashiardes, S., Ozcan, B. & Cuming, A.C. (2003). Isolation and localization of new germination-related sequences from wheat embryos. *J. Biochem. Mol. Bio.* 36(6), 580-585

Caliskan, M., Turet-Sayin, M., Turan, C. & Cuming, A. (2004). Identification of germin isoforms in wheat callus. *Cereal Res. Com.* 32, 355-361

Caliskan, M., Turet, M. & Cuming, A.C. (2004). Formation of wheat (Triticum aestivum L.) embryogenic callus involves peroxide-generating germin-like oxalate oxidase. *Planta* 219, 132-140

Caliskan, M. (2009). Salt stress causes a shift in the localization pattern of germin gene expression. *Gen. Mol. Res.* 8, 1250-1256

Cassab, G.I. & Varner, J.E. (1989). Tissue printing on nitrocellulose paper: a new method for immunolocalization of proteins, localization of enzyme activities and anatomical analysis. *Cell Biol. Inter. Rep.* 13, 147-152

Chang, C.C. & Beevers, H. (1968). Biogenesis of oxalate in plant tissues. *Plant Physiol.* 43, 1821-1828

Chen, X., Wang, M.L., Holbrook, C., Culbreath, A., Liang, X., Brenneman, T. & Guo, B. (2011). Identification and characterization of a multigene family encoding germin-like proteins in cultivated peanut (Arachis hypogaea L.). *Plant Mol. Biol. Rep.* 29, 389–403

Chiriboga, J. (1966). Some properties of an oxalate oxidase purified from barley seedlings. *Biochem. Biophys. Res. Commun.* 116, 566-573

Cuming, A.C. (1984). Developmental regulation of gene expression in wheat embryos: Molecular clonong of a DNA sequence encoding the early-methionine-labelled (Em) polypeptide. *Eur. J. Biochem.* 145, 351-357

Datta, K., Parker, H., Averyhart-Fullard, V., Schmidt, A. & Marcus, A. (1987). Gene expression in the soybean seed axis during germination and early seedling growth. *Planta* 170, 209-216

Davies, D.D. & Asker, H. (1983). Synthesis of oxalic acid by enzymes from lettuce leaves. *Plant Physiol.* 72, 134-138

Delseny, M., Gaubier, P., Hull, G., Saez-Vasquez, J., Gallois, P., Raynal, M., Cooke, R. & Grellet, F. (1994). Nuclear genes expressed during seed desiccation: Relationship with responses to stress. In: *Stress-Induced Gene Expression in Plants*, Basra, A.S. (Ed), 25-59. Harwood Academic Publishers

De Vries, S.C., Harmsen, M.C., Kuiper, M.T.R., Dons, H.J.M. & Wessels, J.G.H. (1983). Molecular clonong of pea mRNAs encoding a shoot-specific polypeptide and light - induced polypeptides. *Plant Molecular Biology* 2, 295-303

Domon, J.M., Dumas, B., Laine, E., Meyer, Y., David, A. & David, H. (1995). Three glycosylated polypeptides secreted by several embryogenic cell cultures of pine show highly specific serological affinity to antibodies directed against the wheat germin apoprotein monomer. *Plant Physiol.* 108, 141-148

Donaldson, P.A., Anderson, T., Lane, B.G., Davidson, A.L. & Simmonds, D.H. (2001). Soybean plants expressing an active oligomeric oxalate oxidase from the wheat gf-2.8 (germin) gene are resistant to the oxalate-secreting pathogen *Sclerotina sclerotiorum*. *Physiol. Mol. Plant Pathol.* 59, 297-307

Dratewka-Kos, E., Rahman, S., Grzelczak, Z.F., Kennedy, T.D., Murray, R.K. & Lane, B.G. (1989). Polypeptide Structure of Germin as Deduced from cDNA Sequencing. *J. Biol. Chem.* 264, 4896-4900

Dumas, B., Sailland, A., Cheviet, J-P., Freyssinet, G. & Pallett, K. (1993). Identification of barley oxalate oxidase as a germin-like protein. *C. R. Acad. Sci. Paris, Sciences de la vie/Life sciences* 316, 793-798

Dumas, B., Freyssinet, G. & Pallett, K.E. (1995). Tissue-specific expression of germin-like oxalate oxidase during development and fungal infection of barley seedling. *Plant Physiol.* 107: 1091-1096

Dunwell, J.M., Khuri, S. & Gane, P.J. (2000). Microbial relatives of the seed storage proteins of higher plants: Conservation of structure and diversification of function during evolution of the cupin superfamily. *Microbiol. Mol. Biol. Rev.* 64, 153-179

Dunwell, J.M., Gibbings, J.G., Mahmood, T. & S.M. Saqlan Naqvi, S.M.S. (2008). Germin and germin-like proteins: evolution, structure, and function. *Crit. Rev. Plant Sciences* 27, 342-375

Dure III, L. (1985). Embryogenesis and gene expression during seed formation. Oxford Surv. *Plant Mol. Cell Biol.* 2, 179-197

Emsley, J. (1994). Rhubarb toxin found in blood. *New Scientist*, pp. 17

Fincher, G.B. (1989). Molecular and cellular biology associated with endosperm mobilization in germinating cereal grains. Ann. Rev. Plant Physiol. *Plant Mol. Biol.* 40, 305-346

Flowers, T.J., Troke, P.F. & Yeo, A.R. (1977). The mechanism of salt tolerance in halophytes. *Ann. Rev. Plant Phsiol.* 28, 89-121

Franceschi, V.R. (1989). Calcium oxalate formation is a rapid and reversible process in *Lemna minor* L. *Protoplasma* 148, 130-137

Galau, G.A., Jakobsen, K.S. & Hughes, D.W. (1991). The controls of late dicot embryogenesis and early germination. *Physiol. Plant.* 81, 280-288

Gibeaut, D.M. & Carpita, N.C. (1991). Tracing cell wall biogenesis in intact cells and plants: selective turnover and alteration of soluble and cell wall polysaccharides in grasses. *Plant Physiol.* 97, 551-561

Goldberg, R.B., Barker, S.J. & Perez-Grau, L. (1989) Regulation of gene expression during plant embryogenesis. *Cell* 56, 149-160

Greenway, H. & Munns, R. (1980). Mechanisms of salt tolerance in nonhalophytes *Ann. Rev. Plant Physiol.* 31, 149-190

Grzelczak, Z.F., Sattolo, M.H., Hanley-Browdoin, L.K., Kennedy, T.D. & Lane, B.G. (1982). Synthesis and turnover of proteins and mRNA in germinating wheat embryos. *Can. J.Biochem.* 60, 389-397

Grzelczak, Z.F. & Lane, B.G. (1983). The growth-related 28-kilodalton protein in germinating wheat. Use of peptide mapping to identify cryptic forms in cell-free extracts and protein-synthesizing systems. *Can. J. Biochem. Cell Biol.* 61, 1233-1243

Grzelczak, Z.F. & Lane, B.G. (1984). Signal resistance of a soluble protein to enzymic proteolysis. An unorthodox approach to the isolation and purification of germin , a rare growth-related protein. *Can. J. Biochem. Cell Biol.* 62, 145-149

Grzelczak, Z.F., Rahman, S., Kennedy, T.D. & Lane, B.G. (1985). Germin. Compartmentation of the protein, its translatable mRNA, and its biosynthesis among roots, stems, and leaves of wheat seedling. *Can. J. Biochem. Cell Biol.* 63, 1003-1013

Heintzen, C., Melzer, S., Fischer, R., Kappeler, S., Apel, K. & Staiger, D. (1994). A light- and temperature-entrained circadian clock controls expression of transcripts encoding nuclear proteins with homology to RNA-binding proteins in meristematic tissue. *The Plant Journal* 5(6), 799-813

Hepler, P.K. & Wayne, R.O. (1985). Calcium and plant development. *Ann. Rev. Plant Physiol.* 36, 397-439

Hodgkinson, A. (1977). Oxalic acid in biology and medicine. Academic Press London

Hofte, H., Desprez, T., Amselem, J., Chiapello, H., Caboche, M., Moisan, A., Jourjon, M.F., Charpenteau, J.L., Berthomieu, P., Guerrier, D., Girauda, J., Quigley, F., Thomas, F., Yu, D-Y., Mache, R., Raynal, M., Cooke, R., Grellet, F., Delseny, M., Parmentier, Y., Marcillac, G.D., Gigot, C., Fleck, J., Philipps, G., Axelos, M., Bardet, C., Tremousaygue, D. & Lescure, B. (1993). An inventory of 1152 expressed sequence tags obtained by partial sequencing of cDNAs from Arabidopsis thaliana. *The Plant Journal* 4(6), 1051-1061

Hurkman, W.J., Fornari, C.S. & Tanaka, C.K. (1989). A comparison of the effect of salt on polypeptides and translatable mRNAs in roots of a salt-tolerant and a salt-sensitive cultivar of barley. *Plant Physiol.* 90, 1444-1456

Hurkman, W.J., Tao, H.P. & Tanaka, C.K. (1991). Germin-like polypeptides increase in barley roots during salt stress. *Plant Physiol.* 97, 366-374

Hurkman, W.J., Lane, B.G. & Tanaka, C.K. (1994). Nucleotide Sequence of a Transcript Encoding a Germin-Like Protein That Is Present in Salt-Stressed Barley (*Hordeum vulgare* L.) Roots. *Plant Physiol.* 104, 803-804

Hurkman, W.J. & Tanaka, C.K. (1996a). Effect of salt stress on germin gene expression in barley roots. *Plant Physiol.* 110, 971-977

Hurkman, W.J. & Tanaka, C.K. (1996b). Germin gene expression is induced in wheat leaves by powdery mildew infection. *Plant Physiol.* 111, 735-739

Jaikaran, A.S.I., Kennedy, T.D., Dratewka-Kos, E. & Lane, B.G. (1990). Covalently Bonded and Adventitious Glycans in Germin. *J. Biol. Chem.* 265, 12503-12512

Jendrisak, J. (1980). The use of α-amanitin to inhibit in vivo RNA synthesis and germination in wheat embryos. *J. Biol. Chem.* 255, 8529-8533

Keates, S.A., Zhang, D., Loewus, F.A. & Franceschi, V.R. (1996). Oxalate oxidase is synthesized and secreted from bean leaf cells in response to fungal infection. *Plant Physiol.* 111, 311-311

Konomi, K., Abe, H. & Furuya, M. (1987). Changes in the content of phytochrome I and II apoproteins in embryonic axes of pea seeds during imbibition. *Plant Cell Physiol.* 28(8), 1443-1451

Koyama, H. (1988). Purification and characterization of oxalate oxidase from Pseudomonas sp. OX-53. *Agric. Biol. Chem.* 52, 743-748

Kuan, I.C. & Tien, M. (1993). Stimulation of Mn peroxidase activity: a possible role of oxalate in lignin biodegradation. *Proc. Natl. Acad. Sci. USA* 90, 1242-1246

Kuchhal, N.K., Satyapal & Pundir, C.S. (1993). Partial purification, some properties and possible physiological role of an oxalate oxidase from grain sorghum leaves. *Indian J. Plant Physiol.* 36(3), 159-162

Laker, M.F., Hofmann, A.F. & Meeuse, B.J. (1980). Spectrophotometric determination of urinary oxalate with oxalate oxidase prepared from moss. *Clin. Chem.* 26, 827-830

Lane, B.G. & Tumaitis-Kennedy, T.D. (1981). Comparative study of levels of secondary processing in bulk mRNA from dry and germinating wheat embryos. *Eur. J. Biochem.* 114, 457-463

Lane, B.G., Grzelczak, Z.F., Kennedy, T.D., Hew, C. & Joshi, S. (1987). Preparation and analysis of mass amounts of germin: demonstration that the protein which signals the onset of growth in germinating wheat is a glycoprotein. *Biochem. Cell Biol.* 65, 354-362

Lane, B.G. (1988). The wheat embryo, then and now. In: *The roots of modern biochemistry*, Kleinkauf, H., Von Dohren, H. & Jaenicke, L. (Eds), 457-476, Walter de Gruyter and co., New York

Lane, B.G. (1991). Cellular desiccation and hydration: developmentally regulated proteins, and the maturation and germination of seed embryos. *FASEB J.* 5, 2893-2901

Lane, B.G., Bernier, F., Dratewka-Kos, E., Shafai, R., Kennedy, T.D., Pyne, C., Munro, J.R., Vaughan, T., Walters, D. & Altomare, F. (1991). Homologies between members of the germin gene family in hexaploid wheat and similarities between these wheat germins and certain *Physarum* spherulins. *J. Biol. Chem.* 266, 10461-10469

Lane, B.G., Cuming, A.C., Fregeau, J., Carpita, N.C., Hurkman, W.J., Bernier, F., Dratewka-Kos, E., & Kennedy, T.D. (1992). Germin isoforms are discrete temporal markers of wheat development: pseudogermin is uniquely thermostable water-soluble oligomeric protein in ungerminated embryos and like germin in germinated embryos, it is incorporated into cell walls. *Eur. J. Biol. Chem.* 209, 961-969

Lane, B.G., Dunwell, J.M., Ray, J.A., Schmitt, M.R. & Cuming, A.C. (1993). Germin, a protein marker of early plant development, is an oxalate oxidase. *J. Biol. Chem.* 268: 12239-12242

Lane, B.G. (1994). Oxalate, germin, and the extracellular matrix of higher plants. *FASEB J.*, 8, 294-301

Lane, B.G. (2000). Oxalate oxidases and differentiating surface structure in wheat: germin. *Biochem J.* 349, 309–321

Lane, B.G. (2002). Oxalate, Germins, and Higher-Plant Pathogens. *IUBMB Life* 53, 67-75

Leek, A.E., Halliwell, B. & Butt, V.S. (1972). Oxidation of formate and oxalate in peroxisomal preparations from leaves of spinach beet (Beta vulgaris L.). *Biochim. Biophys. Acta* 286, 299-311

Leopold, A.C., Bruni, F. & Williams, R.J. (1992). Water in dry organisms. In: *Water and life*, G. Somero, C.B. Osmond & R.J. Williams (Eds), 161-169, Springer-Verlag, Berlin

Luttrel, B.M. (1993). The biological relevance of the binding of calcium ions by inositol phosphates. *J. Biol. Chem.* 268, 1521-1524

Marcus, A., (1969). Seed germination and the capacity for protein synthesis. *Symp. Soc. Exp. Biol.* 23, 143-160

McCubbin, W.C., Kay, C.M., Kennedy, T.D. & Lane, B.G. (1987). Germin: physicochemical properties of the glycoprotein which signals onset of growth in the germinating wheat embryo. *Biochem. Cell Biol.* 65, 1039-1048

Maxwell, D.P. & Bateman, D.F. (1968). Influence of carbon source and pH on oxalate accumulation in culture filtrates of

Sclerotium rolfsii. *Phytopathology* 58, 1351-1355

Meeuse, B.J.D. & Campbell, J.M. (1959). An inhibitor of oxalic acid oxidase in beet extracts. *Plant Physiol.* 34, 583-586

Michalowski, C.B. & Bohnert, H.J. (1992). Nucleotide sequence of a root-specific transcript encoding a germin-like protein from the halophyte *Mesembryanthemum crystallinum*. *Plant Physiol.* 100, 537-538

Millerd, A., Morton, R.K. & Wells, J.R.E. (1963). Oxalic acid synthesis in shoots of Oxalis pes-caprae L. *Biochem. J.* 86, 57-62

Moreland, D.E., Hussey, G.G., Shriner, C.R., & Farmer, F.S. (1974). Adenosine phosphates in germinating radish (*Raphanus sativus* L.) seeds. *Plant Physiol.* 54, 560-563

Myers, A.T. (1947). Seasonal changes in total and soluble oxalates in leaf blades and petioles of rhubarb. *J. Agric. Res.* 74, 33-47

Nakata, M., Shiono, T., Watanabe, Y. & Satoh, T. (2002). Salt stress-induced dissociation from cells of a Germin-Like Protein with Mn-SOD activity and an increase in its mRNA in a moss, *Barbula unguiculata. Plant Cell Physiol.* 43, 1568-1574

Noyes, R.D. & Hancock, J.G. (1981). Role of oxalic acid in the Sclerotinia wilt of sunflower. *Physiol. Plant Pathol.* 18, 123-132

Obzansky, D.M. & Richardson, K.E. (1983). Quantification of urinary oxalate with oxalate oxidase from beet stems. *Clin. Chem.* 29, 1815-1819

Ono, M., Sage-Ono, K., Inoue, M., Kamada, H. & Harada, H. (1996). Transient increase in the level of mRNA for a germin-like protein in leaves of the short-day plant *Pharbitis nil* during the photoperiodic induction of flowering. *Plant Cell Physiol.* 37, 855-861

Ozcan, B., Cokmus, C., Coleri, A. & Caliskan, M. (2006). Characterization of extremely halophilic archaea isolated from saline environment in different parts of Turkey. *Microbiology* 75, 739–746

Ozcan, B., Ozcengiz, G., Coleri, A., Cokmus, C. (2007). Diversity of halophilic archaea from six distinct parts of Turkey. *J. Microbiol. Biotechnol.* 17, 985–992

Petrarulo, I.S., Bianco, O., Marangella, M., Pellegrino, S. & Linari, F. (1990). Ion chromotographic determination of plasma oxalate in healthy subjects, in patients with chronic renal failure and in cases of hyperoxaluric syndromes. *J. Chromatography* 511, 285-291

Pietta, P.G., Calatroni, A., Agnellini, D. & Pace, M. (1982). Improved purification protocol for oxalate oxidase from barley roots. *Preparative Biochem.* 12, 341-353

Polson, C.J. & Tattersall, R.N. (1959). Clinical Toxicology. English Univ. Press London

Pundir, C.S. & Nath, R. (1984). Occurrence of an oxalate oxidase in Sorghum leaves. *Phytochemistry* 23, 1871-1874

Pundir, C.S., Nath, R. & Thind, S.K. (1985). Enzymatic assay of oxalate oxidase from sorghum leaves. In: *Urolithiasis and Related Clinical Research*, P.O. Schwille, L.H. Smith, W.G. Robertson and W. Vahlensieck (Eds), 657-660, Plenum Press New York

Pundir, C.S. & Kuchhal, N.K. (1989). Detection of an oxalate oxidase in grain Sorghum leaves. *Phytochemistry* 28, 2909-2912

Pundir, C.S. (1991). Purification and properties of oxalate oxidase from Sorghum leaves. *Phytochemistry* 30, 1065-1067

Raghwan, K.G. & Devasagayam, T.P.A. (1985). Oxalate oxidase from banana peel for determination of urinary oxalate. *Clin. Chem.* 31, 649-652

Rahman, S., Grzelczak, Z.F., Kennedy, T.D. & Lane, B.G. (1988). Germin. Molecular cloning of cDNA that selects germin mRNA from bulk wheat mRNA. *Biochem. Cell Biol.* 66, 100-106

Ranjhan, S., Karrer, E.E. & Rodriguez, R.L. (1992). Localizing α-amylase gene expression in germinating rice grains. *Plant Cell Physiol.* 33, 73-79

Raven, J.A. & Smith, F.A. (1976). Nitrogen assimilation and transport in vascular land plants in relation to intracellular pH regulation. *New Phytol.* 76, 415-431

Rogers, S.O. & Quatrano, R.S. (1983). Morphological staging of wheat caryopsis development. *Amer. J. Bot.* 70(2), 308-311

Roth, D.A. & Breitenfield, R.V. (1977). Vitamin C and oxalate stones. *J. Am Med. Ass.* 237, 768-768

Satyapal & Pundir, C.S. (1993). Purification and properties of an oxalate oxidase from leaves of grain sorghum hybrid CSH-5. *Biochim. Biophys. Acta* 1161, 1-5

Sharma, S., Vaidyanathan, S., Nath, R. & Thind, S.K. (1991). Advances in pathophysiology of calcium oxalate stone disease. *Ind. J. Urol.* 8, 25-37

Showalter, A.M. (1993). Structure and function of plant cell wall proteins. *The Plant Cell* 5, 9-23

Srivastava, S.K. & Krishnan, P.S. (1959). Oxalate content of plant tissues. *J. Sci. Industr. Res.* 18C, 146-148

Srivastava, S.K. & Krishnan, P.S. (1962). An oxalic acid oxidase in leaves of Bougainvillea spectabilis. *Biochem. J.* 85, 33-38

Sticher, L., Penel, C. & Greppin, H. (1981). Calcium requirement for the secretion of peroxidases by plant cell suspensions. *J. Cell Sci.* 48, 345-353

Stutz, R.E. & Burris, R.H. (1951). Photosynthesis and metabolism of organic acids in higher plants. *Plant Physiol.* 26, 226-243

Sugiura, M., Yamamura, H., Hirano, K., Sasaki, M., Morikawa, M. & Tsuboi, M. (1979). Purification and properties of oxalate oxidase from barley seedlings. *Chem. Pharm. Bull.* 27, 2003-2007

Tavant, H. (1967). Fixation de $^{14}CO_2$ et absorption de glucose-U-^{14}C par des feuilles de Begonia semperflorens Link et otto. Etude des conditions de la genese de l'acide oxalique. *Physiol. Veg.* 5, 57-69

Thompson, F.B. & Lane, B.G. (1980). Relation of Protein Synthesis in Imbibing Wheat Embryos to the Cell-free Translational Capacities of Bulk mRNA from dry and Imbibing Embryos. *J. Biol. Chem.* 255, 5965-5970

Thompson, C., Dunwell, J.M., Johnstone, C.E., Lay, V., Schmitt, M., Watson, H. & Nisbet, G. (1995). Degradation of oxalic acid by transgenic oilseed rape plants expressing oxalate oxidase. *Euphytica* 85, 169-172

Vaisey, E.B., Cheldelin, V.H. & Newburgh, R.W. (1961). Oxalate oxidation by an obligately parasitic fungus Tilletia controversa. *Arch. Biochem. Biophys.* 95, 66-69

Vincent, R.F. & Harry, T.H. JR. (1980). Calcium oxalate crystals in plants. *Bot. Rev.* 46, 361-427

Wagner, G.J. (1981). Vacuolar deposition of ascorbate-derived oxalic acid in barley. *Plant Physiol.* 67, 591-593

Webb, M.A., Cavaletto, J.M., Carpita, N.C., Lopez, L.E. & Arnott, H.J. (1995). The intravacoular organic matrix associated with calcium oxalate crystals in leaves of Vitis. *The Plant Journal* 7(4), 633-648

Wu, G., Shortt, B.J., Lawrence, E.B., Levine, E.B., Fitzsimmons, K.C. & Shah, D.M. (1995). Disease resistance conferred by expression of a gene encoding H2O2-generating glucose oxidase in transgenic potato plants. *The Plant Cell* 7, 1357-1368

Yamahara, T., Shiono, T., Suzuki, T., Tanaka, K., Takio, S., Sato, K., Yamazaki, S. & Satoh, T. (1999). Isolation of a germin-like protein with manganese superoxide dismutase activity from cells of a moss, Barbula unguiculata. *J. Biol. Chem.* 274, 33274-33278

Yang, J.C. & Loewus, F.A. (1975). Metabolic conversion of Lascorbic acid to oxalic acid in oxalate-accumulating plants. *Plant Physiol.* 56, 283-285

Zhang, Z., Collinge, D.B. & Thordal-Christensen, H. (1995). Germin-like oxalate oxidase, a H_2O_2-producing enzyme, accumulates in barley attacked by the powdery mildew fungus. *The Plant Journal* 8(1), 139-145

Part 4

Adaptation and Tolerance

12

Does Environmentally Contingent Variation in the Level of Molecular Chaperones Mirror a Biochemical Adaptation to Abiotic Stress?

Branka Tucić, Sanja Manitašević Jovanović and Ana Vuleta
University of Belgrade, Siniša Stanković Institute for Biological Research
Department of Evolutionary Biology
Serbia

1. Introduction

As Hochachka and Somero emphasized in their seminal book "Biochemical Adaptation: Mechanism and Process in Physiological Evolution", the key question to be posed in the study of biochemical adaptation is: "How have living systems, which are based on a common set of biochemical structures and processes and subject to a common set of physical-chemical lows, been able to adapt to the enormously wide spectrum of environmental conditions found in the biosphere?" (Hochachka & Somero, 2002). Given that the biosphere encompasses habitats with tremendously diverse combinations of physical, chemical, and biotic environmental factors, it seems reasonably to believe that the diversity of life forms that are observable in these habitats is the outcome of *adaptations* which have evolved to permit organisms the exploitation of nearly all land and water areas around the globe (Hochachka & Somero, 2002).

1.1 Conceptual approaches to study adaptation

Evolutionary biologists are well conscious that the term adaptation refers to both a process and its product; however, there still exists the controversy about the perception of that biological phenomenon. Amudson pointed out that "Natural selection, the mechanism universally regarded as the primary causal influence on phenotypic evolutionary change, is first and foremost an explanation of adaptation"(Amudson, 1996). In this context, adaptation is a process, which creates the *generic state* of adaptedness, as well as *specific traits* that - because arising by adaptive modifications - are referred as adaptations. However, irrespective of being a process, a generic state, or an individual trait, adaptation is a relational concept. Thus, if it is viewed as a process, adaptation signifies how an entity fits to the other one. The generic state reflects the functional efficiency of a given trait in an environment adapted to it, whereas *an* adaptation is a modified part of an organism that fulfills a biological function for that organism. Notably, in all of these definitions, natural selection is the principal mechanism that simultaneously generates both adaptation and adaptation(s) (Amudson, 1996).

Gould & Vrba (1982) called attention to the presence of another two distinct concepts of adaptation, one designated as *historical* and one as *nonhistorical*. According to the historical

definition of adaptation, an existing trait is an adaptation for a current function only if it was *improved by natural selection* for that function. On the nonhistorical definition, however, a trait is an adaptation just in the case if it *provides the current fitness benefits.* As proponents of the historical concept of adaptation, they introduced the term "aptation" to modify the nonhistorical definition of adaptation. Thus, aptation is a trait that confers fitness advantage irrespective of its origin. They also contrasted the concept of adaptation with the term "exaptations", which refers to aptations that are nonadaptations. In other words, exaptations are traits currently showing beneficial effects for which they were not selected.
Vermeij (1996) advocated a *comparative* concept of adaptation, defining adaptation as a "heritable attribute of an entity that confers advantages in survival and reproduction of that entity in a given environment" (Vermeij, 1996). His definition implicates the superiority of one trait over the alternative ones. Yet, since its advantage is context dependent, it implies that under different environmental conditions an alternative trait will be adaptive. As was previously pointed out, the term adaptation refers to both the beneficial trait and the process that generates and maintains it. Vermeij's concept of adaptation assumes that *adaptive traits have to be heritable*, as well as that *there must be a selective factor*, which discriminates among heritable alternatives, favoring those with greater reproduction or survival. At the organismal level, inheritance occurs by means of transmission of nuclear genes, although maternal effects, repeated infection, etc. may influence it. Since the mechanisms of inheritance in ancestral organisms cannot be precisely determined, Vermeij (1996) has recommended to make a difference between *functional benefit* or adaptation and the mechanism of *transmission* of the trait in question. Effectively, there are two possibilities for doing such comparisons. One of them is to contrast co-occurring individuals that bear a hypothetical adaptation with those individuals that possess alternative traits. Accordingly, any difference in the performance among compared individuals is the evidence of the presence or the absence of that adaptation. However, when some traits influence the fitness of populations, species or higher clades, it is better to compare the sister-groups instead of co-occurring individuals (Lauder, 1981).

1.1.1 Biochemical adaptation

Any biological entity that satisfies three necessary conditions for natural selection to operate: has the ability to vary, has continuity (heritability), and differs in its success relative to another co-occurring entity, can be adapted and can produce adaptations (Frank, 1996; Vermeij, 1996). In line with this statement, a frequently addressed question is whether the amazing adaptive diversity discovered in morphology, habitat preferences, mode of life and other attributes of contemporary organisms is comparable with the degree of their adaptive differentiation at the biochemical level. Unfortunately, the literature data indicate that the answer to that question is ambivalent (i.e., a mixture of "yes" and "no").
Because natural habitats occupied by different life forms exhibit a great diversity in physicochemical parameters, biochemical systems must confront with numerous environmental challenges during the process of adaptation. All of these challenges are focused exclusively to the two most important sources: (i) the essential biochemical constituents of every organism, including nucleic acids, enzymatic and structural proteins, and lipoprotein structures, and (ii) the competence of cells to maintain an adequate level of energy turnover to sustain life. Given that the core biochemical structures (and the interactions among them) are highly susceptible to direct perturbations from external

physical and chemical factors, each of them must remain a delicate balance between stability and instability to preserve its functional uniqueness, even in the face of environmental forces that may danger their integrity. However, many environmental factors can damage the cell indirectly by influencing its ability to maintain sufficient rates of energy turnover. In such cases, cells can modulate activities of the existing biochemical systems by redirecting metabolic flux in the manner that compensates for particular environmental insult (Hochachka & Somero, 2002). The unity of cellular biochemical design was found in all organisms and in all environments, presumably as the result of adaptive processes that allow conservation of the core set of structures and processes that form their biochemical architecture. Moreover, there is consensus that genes encoding the core components of cellular function and structure are highly conserved in most species as well, including genes for direct sensing the extracellular environmental changes, and those ones for transdusing these informations to corresponding intraracellular targets. However, the latter kind of genes is fundamental for the evolution of physiological diversity and species specificity. In this context, Hochachka and Somero pointed out "Underlying biochemical unity is preserved at the same time that diversity is generated"(Hochachka & Somero, 2002).

One of the classical examples of biochemical adaptation defined in the restrictive sense of the word (Gould & Vrba, 1982) is molecular chaperones. Molecular chaperones are a highly conserved group of proteins that occur universally in all prokaryotes and eukaryotes (Pearl & Prodromou, 2006). Chaperones assist in various cellular functions, such as *de novo* folding, refolding of stress-inactivated proteins, oligomeric assembly, protein transport within cell and proteolityc degradation, but without becoming part of a protein final structure (Frydman, 2001; Hartl & Hayer-Hartl, 2009; Rutheford, 2003; Wegele et al., 2004). Although essential under physiological conditions, protein chaperones are indispensable for survival under environmental conditions that interfere with protein folding, including extreme temperature or other environmental stresses (Frydman, 2001). Importantly, protein-folding chaperones do not modify genotype; instead, they influence "target-protein activity by allowing structurally unstable mutant protein sequence to fold into active configuration" (Rutherford, 2003). Because highly induced under conditions of conformational stress, a large number of structurally unrelated chaperones are denoted as stress proteins or "heat shock proteins" (Parsell & Lindquist, 1993); that is, in the case when the amount of aggregation-prone folding intermediates amplifies in the cell (Hartl & Hayer-Hartl, 2009). Molecular chaperones form different families according to their molecular weight (Hsp40, Hsp60, Hsp70, Hsp90, Hsp100 and so-called small Hsp proteins). Since their current beneficial properties in all organisms are very likely the same as the benefits that initially favored their evolution, there is no doubt that protein-folding chaperones arose and are maintained by natural selection (Feder & Hofmann, 1999). Here, we test this statement by exploring spatial and temporal variation in the cellular level of the two heat shock proteins, Hsp70 and Hsp90, which exhibit chaperone properties.

The 70-kDa heat shock proteins (Hsp70s) are a large family of chaperones that plays diverse role in the cell. In addition to the folding of *de novo* translated and recovery of stress-denatured proteins, they assist the post-translational unfolding and translocation of nuclear-encoded proteins through the lipid bilayers of organelles (Bukau & Horwich, 1998; Rutheford, 2003). The Hsp70 proteins are present in all eukaryotic cells, in eubacteria, as well as in many archaea (Macario et al., 1999). Recently, Albanèse at al. (2006) have applied a system biology approach that combines genomic and functional analyses to examine whether there is a difference between chaperone-mediated *de novo* folding and polypeptide refolding following stress, in yeast *Saccharomyces cerevisiae*. Their global analyses, together with previous studies investigating the functional organization of individual cytosolic chaperones, suggest that the eukaryotic Hsp70 chaperone machinery consists of two robust networks with highly specialized functions. One of them, denoted as the Hsp network, facilitates the protection of the cellular proteome from environmental stress by refolding stress-denatured proteins or by directing misfolded proteins to the ubiquitin-proteosome system for degradation (Parsell & Lindquist, 1993). The other network, called the CLIPS (Chaperones Linked to Protein Synthesis) network, is transcriptionally coregulated with the translational apparatus and functions co- and post-translationally to mediate *de novo* folding (Albanèse at al., 2006). Yet, whereas the stress-inducible chaperones dedicated to refold stress-denatured proteins reside within the bulk cytosol (Thulasiraman et al., 1999), the stress-repressed chaperones devoted to folding of nascent polypeptides occur in a sequestered environment, which is physically linked to the translational apparatus (Frydman & Hartl, 1996; Thulasiraman et al., 1999). It has also been revealed that both chaperone networks consist of distinct chaperones, which exhibit different but partially overlapping functions. For example, a small number of Hsp70 chaperones, including *SSA1* and *SSE1*, that are upregulated during highly proteotoxic heat shock, while repressed by most other stresses, are categorized into the CLIPS subset, because performed phenotypically, and bound newly made polypeptides as did the CLIPS proteins. According to Albanèse at al. (2006), "these chaperones function at the interface of both CLIPS- and HSP networks by fulfilling a dual function: normally assisting *de novo* folding but also contributing to the rescue and/or quality control of stress denatured proteins". This stress-induced subset of cytosolic chaperones is also indispensable for termotolerance (Parsell & Lindquist 1993).

In addition to Hsp70s, the 90-kDa heat shock proteins (Hsp90s) are also essential components of the chaperone system in all eukarya and eubacteria, with exception of the archaea, were they are no detected (Pearl & Prodromou, 2006; Wegele et al., 2004). In contrast to Hsp70, which is less selective to its substrates, Hsp90 is restricted to the conformational regulation of highly specific cell-cycle- and developmental regulators, including cyclin-dependent kinases, tyrosine kinases, steroid hormone receptors, transcription factors or mitochondrial membrane components (Buchner, 1999; McClellan et al., 2007; Picard, 2006; Rutherford, 2003; Young et al., 2004). It is likely that Hsp90 recognizes its client proteins on the basis of their hydrophobic surface features that are found on nearly mature proteins in normal condition (Richter & Bucher, 2001) or at the initial stage of unfolding on proteins damaged by stress (Freeman & Morimoto, 1996). Although each of the two Hsps is indispensable for viability, they fulfill non-overlapping functions in the cell (Frydman, 2001; Young et al., 2004). Moreover, regardless of an elevated content of both Hsps during environmental stress, there is no data to demonstrate that the Hsp90 chaperone is required for thermotolerance and/or disaggregation of heat-denatured proteins, as was found for Hsp70.

To elucidate the functions of Hsp90 more deeply, McClellan et al. (2007) have applied a genome-wide chemical-genetic screen in *S. cerevisae* combined with bioinformatic analyses. They revealed a number of new Hsp90-dependent cellular function under physiological (normal) conditions and in response to environmental stress. In general, at stress-induced high temperatures, Hsp90 is essential for cell-cycle progression, meiosis, and cytokinesis, whereas at normal growth temperatures, Hsp90 facilitates protein trafficking and assembly or stabilization of oligomeric complexes (Frydman, 2001; McClellan et al., 2007; Young et al., 2004). Because several important cellular processes, such as transcription factor activity or mitochondrial function, are similarly expressed at 30°C and 37°C, it suggests that Hsp90 plays a housekeeping role in daily activities of the cell as well (McClellan et al., 2007). There is growing evidence that the functionally active Hsp90 is an evolutionary conserved capacitor of hidden genetic and epigenetic variation that accumulated in wild-type phenotypes (Manitašević et al., 2007; Queitsch et al., 2002; Rutherford & Lindquist, 1998; Salathia & Queitsch, 2007; Sangster et al., 2004). However, if the function of Hsp90 is impaired either by environmental stress or by genetic or pharmaceutical treatments, this variation usually releases, providing the raw material for natural selection to operate (Salathia & Queitsch, 2007).

2. Testing hypotheses about local adaptation to environmental conditions

The phrase "*local adaptation*" refers to patterns and processes detected across conspecific populations associated by gene flow (Kawecki & Ebert, 2004). The intensity of natural selection within local populations is usually specific to habitat conditions, leading to genotype x environment interaction for individual fitness. If there are no other forces or constraints, such spatially heterogeneous or *divergent selection* affects each population to develop traits which are functionally advantageous under specific composition of environmental factors prevailing in its own habitat, not considering the effects of these traits for fitness in other habitats. Consequently, local genotypes in each population would express the elevateted mean fitness relative to genotypes stemming from the other habitats. According to Williams (1966), the process resulting in such a pattern is defined as *local adaptation*. In this context, local adaptation would be the outcome of two evolutionary factors that operate antagonistically to each other: the spatially heterogeneous natural selection with potentially differentiating effect and the homogenizing effect of gene flow. This conception regards populations either as discrete entities within fixed environmental patches or random sampling units in a continuous species range. Similarly, the spatial environmental variation can be also disconnected and composed of a number of different habitat types, or, as in the case of a continuous environmental gradient, a "habitat" can represent an arbitrary point along this gradient with a specific combination of environmental factors (Kawecki & Ebert, 2004).

2.1 Reciprocal transplant experiments

Experimental biologists interested in testing hypotheses about local adaptation to native environments have used reciprocal transplant experiments for more than a half-century (see Emms & Arnold, 1997; O'Hara et al., 2004; Schluter, 2000). The method requires reciprocally transplanting two or more populations among their natural habitats in the wild and, within a single generation, comparing the relative performance of a sample of genotypes from the "local" population and those one from the "foreign" population(s) under the same environmental conditions, *i.e.*, in the same habitat. An alternative approach involves comparing the performance of the same genotype among contrasting habitats; that is, "at home" and "away". Currently, there is a controversy over which of the two criteria can be used as diagnostic to local adaptation. The "local *vs.* foreign" criterion accentuates the contrast between populations within habitats, predicting that in each habitat local populations should exhibit greater fitness relative to populations from the foreign habitats. Conversely, the "home *vs.* away" criterion gives emphasis to the contrast between a population's fitness among alternative habitats. In this case, local adaptation is expected to occur if each population exhibits a higher performance in its native habitat ("at home") in comparison to other habitats ("away") (Kawecki & Ebert, 2004).

2.2 Chaperone Hsps as biochemical adaptation

There is growing evidence on adaptive variation in thermal resistance in natural populations, indicating that stressful environmental conditions select for adaptations in natural populations (Hoffmann et al., 2003). For example, in laboratory experiments, very small heat induction of Hsps results in measurable effects on development, life span, fecundity, and stress resistance in plants and animals (Queitsch et al., 2002; Rutherford & Lindquist, 1998). In the field, adaptive changes in the Hsp level over days (Nguyen et al., 1994) and/or over season (Hoffmann & Somero, 1995; Manitašević et al., 2007) appeared to be the ecologically significant for natural populations as well. Hovewer, in contrast to laboratory conditions where the importance of Hsps for survival following heat shock is evident, under field conditions their ecological significance is less explicit and rarely directly explored (Gehring & Wehner, 1995; Manitašević et al., 2007). Recently, the ecological significance of Hsps for adaptation in natural populations have been confirmed using data from latitudinal and climatic clines (Frydenberg et al., 2003; Hemmer-Hansen et al. 2007; Sørensen et al., 2009), which documented that natural selection affects *hsp* variation, which was found to be very low in the coding regions of the *hsp* genes (Sørensen, 2010)

To this end, the goals of this study were to corroborate the statements: (i) that variation in the endogenous relative level of Hsp70 and Hsp90 chaperones is a biochemical adaptation to fluctuating environmental conditions, and (ii) that the amount of plasticity in the relative level of these two chaperones is protein-specific and dependent on the habitat type that the examined plants were derived from. The model-organism used was *Iris pumila*, a perennial herb, naturally growing in lowlands of central and southeastern Europe, including the Deliblato Sands in northern Serbia (Randolph, 1955).

3. Material and methods

3.1 Studied are and species

The Deliblato Sands is an isolated district of sand masses located between the western Carpathian slopes and the Danube River in southern Banat (Serbia). The relief of this area is of eolian origin and therefore has an undulating dune shape. The sand dunes extend in a straight south-east-north-west direction, as does the complete sandy district (44° 47' 39'' N / 21° 20' 00'' E to 45 ° 13' 10'' N / 28 ° 26' 08"E). Surface water streams are completely absent in the Deliblato Sands. The uniqueness of its relief combined with shortage of ground water generated the specific ecological conditions and, as a result, the diversity of habitats and wildlife therein.

Iris pumila L. (Iridaceae) is a rhizomatous perennial monocot, which is very abundant in the dune system at the Deliblato Sands (Gajić, 1983). Within its natural habitats, the species forms circle-chapped clones that differ in size, depending on the clone age (Tucić et al., 1989). *Iris* clones result from the development of horizontally growing and tightly packed rhizome segments that spread radially from the center of each clone toward its margin. Populations of *I. pumila* are markedly polymorphic for flower colour due to segregation at several gene loci. Consequently, each of the flower colour variants commonly found in an *Iris* population represents a unique clonal genotype (Tucić et al., 1988).

3.2 Experimental setup

To detect local adaptation in the level of Hsps in natural populations of *I. pumila*, we conducted a reciprocal transplant experiment in the Deliblato Sands. For this experiment, we choose two populations of *I. pumila* naturally growing in contrasting light habitats. One at an open dune site, where plants experienced multiple abiotic stresses, and one in a more "friendly" environment under the canopy of a *Pinus silvestris* stand. In April 2001, samples of adult genotypes from each of the two populations (developed from rhizome modules raised under controlled ambient conditions in a growth-room from May 1998 to April 2001), we transplanted reciprocally between their original habitats. During 2007, when most transplants produced several new leaves, we cut one fully developed leaf from each genotype between 15:00 and 16:00 h, once in each of two seasons (spring and summer), for determining the endogenous level of Hsp70 and Hsp90 chaperones. After cutting leaves were immediately frozen in liquid nitrogen, transported to the laboratory, and stored at -70°C until preparation.

3.2.1 Leaf extracts preparation

We pulverized frozen leaf tissue under liquid nitrogen and resuspended in two volumes (w/v) of extraction buffer (0.1 M Tris, pH 7.6, 1 mM EDTA, 1 mM DTT and 0.1 mM PMSF). After sonification for 3 x 15 s on ice, at 1A and 50/60 Hz, with 30% amplitude (Hielscher Ultrasound Processor), the homogenates were centrifuged twice at 12000g at 4 °C, for 15 min and the supernatants analyzed for total protein content by the method of Spector (1978) with bovine serum albumin (BSA) used as a standard.

3.2.2 SDS-polyacrilamide gel electrophoresis and Western blot analysis

Tissue extracts, mixed with equal volumes of 2X SDS-sample buffer (0.125 mM Tris-HCl, pH 6.8, 4% SDS, 20% glycerol, 10% 2-mercaptoethanol), were boiled for 5 min at 100°C. Samples containing 40 µg proteins were loaded onto 7.5% SDS-PA gels (Laemmli, 1970) and separated by electrophoresis at 120 V and 4 °C, using Mini-Protean II Electrophoresis Cell (Bio-Rad Laboratories, Hercules, CA). We applied a Page Ruler Prestained Protein Ladder (Fermentas International Inc., Canada) for precise molecular weight determinations. After electrophoresis, the separated proteins were transferred from the gels to nitrocellulose membranes (Hybond-C, Amersham) and electroblotted overnight at 135 mA and 4°C in 25 mM Tris buffer, pH 8.3, containing 192 mM glycine and 20% (v/v) methanol, using a Mini Trans-Blot Electrophoretic Transfer Cell (Bio-Rad Laboratories, Hercules, CA). Unbound sites on the membranes we blocked by incubation in PBS (1.5 mM KH_2PO_4, 6.5 mM Na_2HPO_4, 2.7 mM KCl, 0.14 M NaCl, pH 7.2) containing 1% nonfat dry milk (GE Healthcare Bio-Sciences) for 1.5 h at room temperature. Proteins of interests were detected using appropriate monoclonal antibodies; Hsp70 by SPA-820 (1:1000; StressGen, Canada) and Hsp90 by SPA-830 (1:2000; StressGen, Canada). After washing with PBS containing 0.1% Tween 20, we incubated the membranes for 1h with alkaline phosphatase-conjugated secondary antibody (1:20000). The immunoreactive bands we visualized and quantified by enhanced chemifluorescence (ECF) detection system using STORM Imager and ImageQuant image analysis software (Amersham Biosciences Limited, UK). To make quantitative comparisons between multiple immunoblots reliable, i.e. to exclude the inter-gel variation, an internal reference sample consisting of a mixture of all samples was run simultaneously on each gel. Prior to any comparison, the intensity of each analyzed immunospecific band we normalized to the intensity of the respective internal reference band on the same blot. The representative immunoblots display the presence of one immunospecific band for Hsp70 and two different bends for Hsp90, in all leaf samples of *I. pumila* plants from an exposed and a shaded habitat across spring and summer (Figure 1).

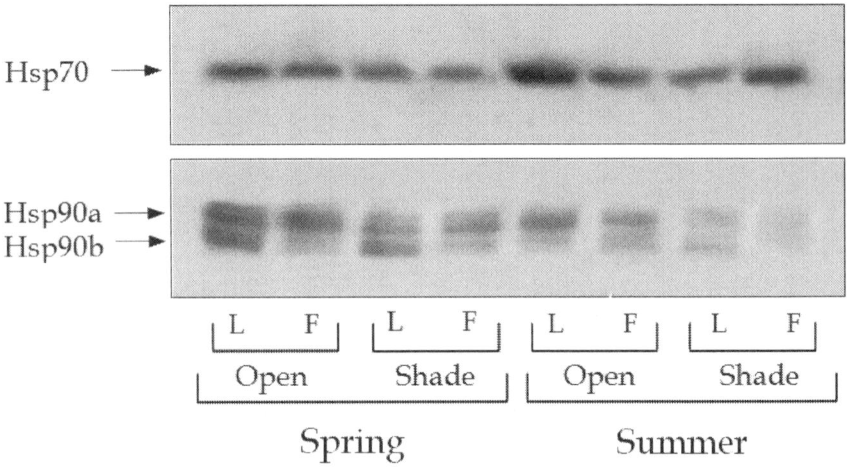

Fig. 1. Immunoblot detection of Hsp70 and Hsp90 in the foliage of *Iris pumila* genotypes from an exposed (Dune) and a shaded (Woods) habitat over spring and summer. L = local habitat; F = foreign habitat. Proteins from whole cell extracts (40 µg) were separated by 7.5% sodium dodecyl sufate-polyacrylamide gel electrophoresis and transferred to nitrocellulose membrane. Hsp70 was detected by N27F3-4, whereas Hsp90 isoforms (Hsp90a and Hsp90b) were revealed by AC88 monoclonal antibody. Immunoreactive bands were visualized by enhanced chemifluorescence reaction.

4. Statistical analyses

All analyses were conducted using 22 *Iris* genotypes: 11 from the Dune and 11 from the Woods population. The reaction norm plots were depicted for each Hsp genotype using its individual value expressed within each habitat and season. To examine the effect of environmental conditions in different light habitats as well as within the same habitat over seasons, a factorial ANOVA was implemented using GLM procedure in SAS (SAS Institute 2003). The full ANOVA model included the following sources of variation: habitat (amount of phenotypic plasticity between habitats), season (amount of phenotypic plasticity between seasons), population (genetic variation in trait means between populations), habitat-by-population interaction (genetic variation for plasticity to habitat conditions between populations), season-by-population interaction (genetic variation for plasticity to seasonal changes between populations), and habitat-by-season-by-population (season-dependent genetic variation for plasticity to habitat conditions between populations). In these analyses, we specified light habitat, season and population as fixed factors. Since all traits (i.e. Hsps relative level) were normally distributed and had equal variances, all ANOVAs were computed on row data. Apart from producing environmentally contingent differences between populations in the mean value of a trait, natural selection can influence the trait plasticity between different habitats and/or within the same habitat over season. In this study, we estimated the amount of plasticity of a clonal genotype for the relative level of each Hsp by calculating an index of plasticity, PI_C (Valladares et al., 2006):

$$PI_C = |(X_1 - X_2)| / |(X_1 + X_2)|,$$

where X_1 is the Hsp level of a clone (c) in the open habitat, or in the spring within that (open) habitat, whereas X_2 refers to the Hsp level of the same clone in the shaded habitat, or in the summer within that (shaded) habitat. The index of plasticity, PI_C, measures the changes in a trait induced by spatial and/or temporal variation in the environmental states. Wilcoxon 2-sample test (SAS Institute, 2003) was used as a nonparametric procedure to compare the mean plasticity to seasonal variation in abiotic factors between the local and foreign genotypes within the same habitats for different Hsps. The same test we used to determine whether the plasticity to contrasting light conditions in the sample of Dune genotypes varied for a given Hsp between the sample of the Woods genotypes.

5. Results

5.1 Habitat and seasonal variation in the mean relative level of Hsp70 and Hsp90 chaperones

In the sun-exposed habitat during both seasons, the mean relative level of all tree foliar chaperones, Hsp70, Hsp90a and Hsp90b, was generally greater in the local (Dune) genotypes relative to their foreign (Woods) counterparts (Table 1).

Considering individual Hsps, in spring, Hsp70 exhibited the lowest (0.63 in the Dune, and 0.29 in the Woods population), while Hsp90a had the highest mean value (0.91 in the Dune, and 0.53 in the Woods population) among all chaperones studied. Yet, when Hsp90a and Hsp90b were compared one to another, the relative level of Hsp90a was greater than that of Hsp90b in both populations of *I. pumila* (0.91 *vs.* 0.78 in the Dune, and 0.53 *vs.* 0.50 in the Woods population). In summer, however, the mean level of Hsp70 was more than twofold greater compared for its spring value (1.46 *vs.* 0.63 in the Dune, and 0.96 *vs.* 0.29 in the Woods population). Conversely, the amount of the chaperones Hsp90a and Hsp90b appeared to be twofold lower than that observed during spring (Table 1).

In the shaded habitat during spring, the mean relative level of all tree foliar chaperones, Hsp70, Hsp90a and Hsp90b, was generally greater in the local (Woods) genotypes as compared to the foreign (Dune) genotypes (Table 1). Again, the mean relative level of Hsp70 chaperone appeared to be the lowest in the Dune genotypes (0.33). In contrast to the open habitat, the average relative level of Hsp90b chaperone in the Woods genotypes was appeared to be the greatest (0.79). In the summer, the foreign (Dune) genotypes produced a higher relative amount of all heat shock proteins (Hsp70, Hsp90a and Hsp90b) than the local (Woods) genotypes.

	OPEN HABITAT					
	Spring			Summer		
Trait	Mean	SD	CV%	Mean	SD	CV%
DUNE (n=11)						
Hsp70	0.63	0.39	62.29	1.46	0.32	21.71
Hsp90a	0.91	0.13	14.06	0.56	0.18	32.82
Hsp90b	0.78	0.03	4.46	0.39	0.06	14.77
WOODS (n=11)						
Hsp70	0.29	0.12	39.46	0.96	0.45	46.94
Hsp90a	0.53	0.15	27.81	0.30	0.04	13.28
Hsp90b	0.50	0.12	23.98	0.24	0.05	20.42
	SHADED HABITAT					
	Spring			Summer		
Trait	Mean	SD	CV%	Mean	SD	CV%
DUNE (n=11)						
Hsp70	0.33	0.21	63.56	0.56	0.10	17.93
Hsp90a	0.58	0.10	17.92	0.65	0.22	34.22
Hsp90b	0.68	0.11	16.28	0.54	0.20	37.08
WOODS (n=11)						
Hsp70	0.34	0.23	67.45	0.50	0.20	39.41
Hsp90a	0.67	0.24	36.70	0.51	0.17	33.59
Hsp90b	0.79	0.26	33.21	0.40	0.07	17.57

Table 1. Sample size (n), mean value (in AU = arbitrary units), standard deviation (SD) and coefficient of variation (CV%) for the relative level of three heat shock proteins, Hsp70, Hsp90a, and Hsp90b, measured during spring and summer in different *Iris pumila* genotypes from a sun-exposed (Dune) and a shaded (Woods) natural populations transplantated in each of their alternative light habitats

The three-way ANOVA applied to each of the three Hsps revealed a highly significant main effects of abiotic environmental conditions on all but one of these traits, the mean relative level of Hsp90a between contrasting light habitats (Table 2). These results suggest that *I.pumila* genotypes possess the capability to alter the level of the two chaperones in accordance with abiotic environments conditions they have happened to experience. A significant main effect of population was obtain for all three chaperones (Hsp70, Hsp90a and Hsp90b), indicating that the exposed and shaded population of *I. pumila* are genetically differentiated one from the other for these biochemical traits. The two-way interaction between habitats and populations were highly statistically significant for all three chaperones studied (all $P < 0.001$), implying the presence of genetic differences between populations in plasticity to environmental conditions within contrasting light habitats. The season-by-population interaction appeared to be insignificant for the relative level of all Hsp chaperones. Conversely, the three-way interaction was highly significant for each of the Hsp (all $P < 0.001$), indicating that genetic differences in plasticity to habitat conditions between the Dune and the Woods population are dependent upon season (Table 2).

Source of variation		Hsp70		Hsp90a		Hsp90b	
	d.f.	*F*	*P*	*F*	*P*	*F*	*P*
Habitat	1	46.36	**0.0001**	0.55	0.4623	19.57	**0.0001**
Season	1	62.68	**0.0001**	21.57	**0.0001**	103.79	**0.0001**
Population	1	13.79	**0.0004**	24.13	**0.0001**	16.07	**0.0001**
H x P	1	11.39	**0.0011**	16.99	**0.0001**	11.42	**0.0011**
S x P	1	1.01	0.3183	0.74	0.3929	1.15	0.2871
H x S x P	2	10.89	**0.0001**	9.00	**0.0003**	5.86	**0.0042**

Table 2. Factorial ANOVA results for the relative level of three heat shock proteins, Hsp70, Hsp90a and Hsp90b, measured *in situ* during spring and summer in leaves of distinct *I. pumila* genotypes, stamming from a sun-exposed (Dune) and a shaded (Woods) population, which were reciprocitally transplanted between their local habitats in the Deliblato Sands. Significant *F* values are given in bold face

5.2 Individual variation (CV %) in the relative level of Hsp70 and Hsp90 between habitats and season

Apart from exhibiting seasonal and habitat-specific differences in the average relative level of Hsp70 and Hsp90, our study provides evidence that the individual variation among genotypes, expressed in term of a coefficient of variation (CV %), also changed over seasons and between the populations from which they originate. In general, Hsp70 expressed the greatest individual variation in both populations across both seasons, with only exception during summer (21.17%). In addition, the lowest level of CV% was observed for Hsp90b during spring (4.46%) in the Dune population, as well (Table 1).

The individual variation (CV%) in the mean relative level of the three analyzed chaperones in the shaded habitat displayed similar trend to that revealed at the open Dune site. However, the percentage of individual variation for each chaperone analyzed appeared to be greater for the local (Woods) genotypes compared to that revealed for the foreign (Dune) genotypes.

Again, in the summer, the Dune genotypes exhibited the lowest individual variation for the mean relative level of Hsp70 (17.93%), as was revealed for the individual variation of the mean relative level of Hsp90b in the Woods genotypes (17.57%) (Table 1).
At the open habitat, the univariate ANOVAs revealed a significant difference in the endogenous level of all Hsps (Hsp70, Hsp90a and Hsp90b) between the Dune and the Woods genotypes in both seasons (all $P < 0.001$; Table 3A). Conversely, in the shaded habitat, a significant difference appeared exclusively for Hsp90b chaperone in the summer ($P < 0.05$; Table 3B).

A.		OPEN HABITAT											
		Spring						Summer					
Source of variation		Hsp70		Hsp90a		Hsp90b		Hsp70		Hsp90a		Hsp90b	
	d.f.	*F*	*P*	*F*	*P*	*F*	*P*	*F*	*P*	*F*	*P*	*F*	*P*
POPULATION	1	7.43	0.013	41.85	0.000	55.27	0.000	9.17	0.007	21.43	0.000	43.48	0.000

B.		SHADED HABITAT											
		Spring						Summer					
Source of variation		Hsp70		Hsp90a		Hsp90b		Hsp70		Hsp90a		Hsp90b	
	d.f.	*F*	*P*	*F*	*P*	*F*	*P*	*F*	*P*	*F*	*P*	*F*	*P*
POPULATION	1	0.03	0.859	1.29	0.269	1.58	0.223	0.73	0.404	3.04	0.097	5.09	0.035

Table 3. ANOVA exploring the effect of population origin in a single generation during spring and summer on the relative level of Hsp70, Hsp90a and Hsp90b chaperones in leaves of *Iris pumila* genotypes grown at an open (A.) and a shaded habitat (B.). The *F*-value for each effect is reported

When the mean relative level of all three Hsps (Hsp70, Hsp90a and Hsp90b) in genotypes stemming from the two populations were compared between alternative light habitats over seasons, a univariate ANOVA revealed that the Dune genotypes expressed significantly different level of all these Hsps in both season (all $P < 0.05$), with only exception of Hsp90a in summer (Table 4A). An inverse trend exhibited the Woods genotypes. In summer, their mean relative level for all three Hsps differed significantly between alternative light habitats (all $P < 0.001$; Table 4B), in contrast to their spring counterparts that differed significantly only for the mean relative level of Hsp90b chaperones between the contrasting light habitats ($P < 0.001$; Table 4A).
We presented the reaction norm plots to habitat type for the level of three Hsps (Hsp 70, Hsp90a and Hsp90b) in leaves of *I. pumila* clonal genotypes from the Dune and the Woods population during spring (Fig. 2A and 2B) and summer (Fig. 2C and 2D).

A.		SPRING											
		Population Dune						Population Woods					
Source of variation		Hsp70		Hsp90a		Hsp90b		Hsp70		Hsp90a		Hsp90b	
	d.f.	*F*	*P*	*F*	*P*	*F*	*P*	*F*	*P*	*F*	*P*	*F*	*P*
HABITAT	1	5.15	0.034	44.77	0.000	6.94	0.016	0.40	0.533	2.62	0.121	11.41	0.003

B.		SUMMER											
		Population Dune						Population Woods					
Source of variation		Hsp70		Hsp90a		Hsp90b		Hsp70		Hsp90a		Hsp90b	
	d.f.	*F*	*P*	*F*	*P*	*F*	*P*	*F*	*P*	*F*	*P*	*F*	*P*
HABITAT	1	80.96	0.000	1.07	0.313	5.90	0.025	9.51	0.006	15.44	0.001	37.96	0.000

Table 4. ANOVA exploring the effect of habitat type in a single generation during spring and summer on the relative level of Hsp70, Hsp90a and Hsp90b chaperones in leaves of *Iris pumila* genotypes from the Dune (A.) and the Woods population (B.). The *F*-value for each effect is reported

The plots of reaction norms over seasons are shown for the level of three Hsps (Hsp 70, Hsp90a and Hsp90b) in leaves of *I. pumila* clonal genotypes from the Dune and the Woods population at an open (Fig. 3A and 3B) and a shaded habitat (Fig. 3C and 3D).

The mean reaction norms were steep for all Hsps measured (Figs. 2 and 3), suggesting a general ability of *I. pumila* clones for plastic adjustment of their leaf biochemistry to spatial and temporal variation of environmental conditions. Pattern of reaction norms was rather complex in both populations, with some genotypes exhibiting reversals of ranking in different seasons and/or habitat. Crossed-reaction norms indicate that there was genetic variation for plasticity in the leaf Hsp level within and between populations, corroborating the factorial ANOVA results (Table 2).

The plasticity means between seasons or habitats (average PI_C) appeared to be strongly trait-specific (Table 5). The relative level of Hsp70 chaperone was found to be the most plastic, whereas the relative level of Hsp90a appeared to be the least plastic in both populations of *I. pumila*. Plastic response of Hsp90b chaperone was intermediate in the magnitude.

A Wilcoxon 2-sample test revealed that the amount of plasticity for identical *Iris* genotypes was similar between distinct light habitats or between seasons within the same habitat for any of the three chaperones, in each of the two populations studied. Regarding contrasting light habitats, however, in the summer, the mean plasticity was significantly greater for Hsp70, while in the spring, Hsp90b expressed the lowest mean plasticity, when the Dune and the Woods genotypes were compared.

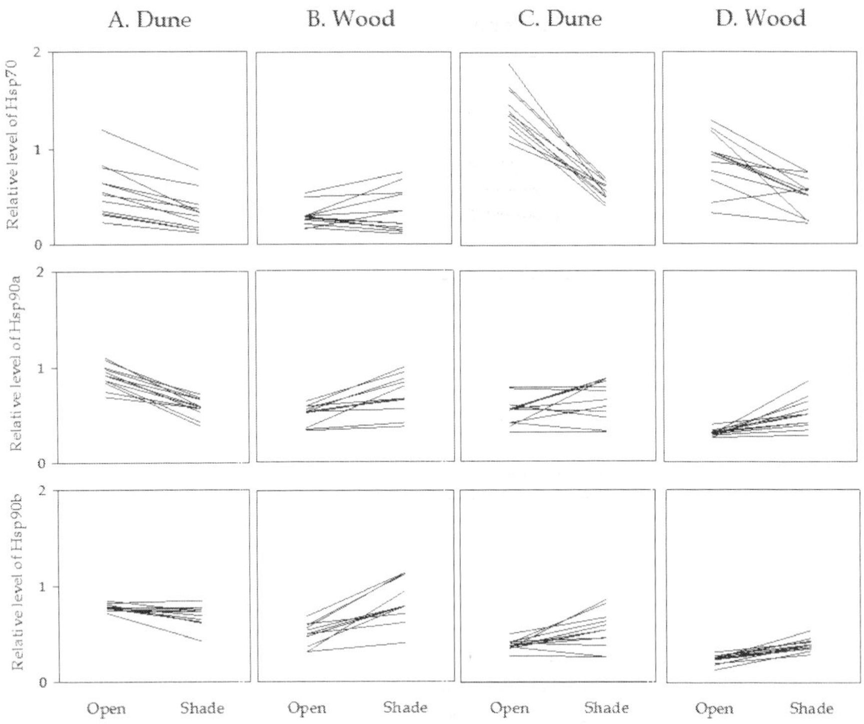

Fig. 2. Reaction norm plots for 22 *Iris pumila* clonal genotypes, native to an exposed (Dune) and a shaded (Woods) population, which were reciprocally transplanted between their original habitats. The relative level of leaf Hsp chaperones in 11 Dune genotypes and 11 Woods genotypes was observed at an open and a shaded habitat during spring (A and B) and summer (C and D)

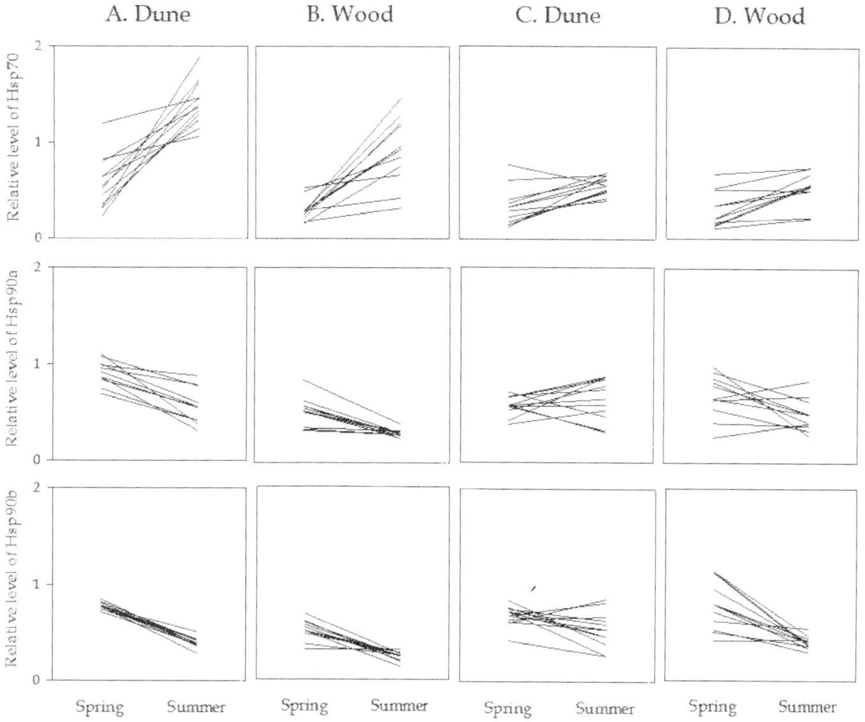

Fig. 3. Reaction norm plots for 22 *Iris pumila* clonal genotypes, native to an exposed (Dune) and a shaded (Woods) population, which were reciprocally transplanted between their original habitats. The relative level of leaf Hsp chaperones in 11 Dune genotypes and 11 Woods genotypes was observed during spring and summer at an open (A and B) and a shaded (C and D) habitat

	SPRING *VS.* SUMMER				OPEN HABITAT *VS.* SHADED HABITAT			
Trait (n = 11)	Open habitat		Shaded habitat		Spring		Summer	
DUNE	PI_C	SD	PI_C	SD	PI_C	SD	PI_C	SD
Hsp70	0.444	0.222	0.340	0.201	0.315	0.130	0.439	0.106
Hsp90a	0.249	0.138	0.185	0.105	0.225	0.093	0.122	0.129
Hsp90b	0.335	0.069	0.189	0.139	0.074	0.070	0.171	0.127
WOODS	PI_C	SD	PI_C	SD	PI_C	SD	PI_C	SD
Hsp70	0.484	0.224	0.324	0.223	0.224	0.106	0.315	0.178
Hsp90a	0.261	0.131	0.221	0.148	0.143	0.091	0.231	0.153
Hsp90b	0.337	0.182	0.307	0.168	0.213	0.136	0.250	0.097

Table 5. Sample size (n), mean plasticity index (PI_C), and standard deviation (SD) for the relative level of three heat shock proteins, Hsp70, Hsp90a and Hsp90b, measured during spring and summer in leaves of different *I. pumila* genotypes from a sun-exposed (Dune) and a shaded (Woods) populations transplanted in each of their alternative light habitats

6. Discussion

During the course of evolution, plants have evolved a variety of different biochemical mechanisms for preventing fitness reduction under adverse environmental conditions (Bazzaz, 1996; Lambers et al., 2008; Tucić et al., 2009; Vuleta et al., 2010). One of such mechanisms is molecular chaperones - a group of proteins that respond to sudden increase in temperature or exposure to other environmental stresses (Feder & Hofmann, 1999; Salathia & Queitsch 2007; Sangster et al., 2004; Sørensen et al., 2003). Molecular chaperones, particularly Hsp90, also restrict stochastic phenomena within cells, by minimizing developmental perturbations, thereby canalizing the organism's development (Samakovli et al., 2007). Results presented in this study provide evidence that in *Iris pumila* plants the relative level of heat shock proteins Hsp70 and Hsp90 (Hsp90a and Hsp90b isoforms), varied significantly in a single generation, between the samples of local and foreign genotypes reciprocally transplanted to their original light habitats ("local *vs.* foreign" approach), and in the same genotype grown under alternative light conditions ("at home *vs.* away" approach), or among different seasons in the same habitat ("spring *vs.* summer"). In general, local genotypes exhibited significantly higher relative amounts of all three chaperones (Hsp70, Hsp90a and Hsp90b) compared to the foreign genotypes. Similarly, the relative level of all the three Hsps in each genotype from both populations of *I. pumila* was greater "at home" (within its native habitat) than "away" (within the non-native habitat). Theoretically, a higher performance under native environmental conditions, and a lower performance under non-native environmental conditions can be interpreted as an evidence of local adaptation (Kawecki & Ebert, 2003). Based on the results obtained using the two experimental approaches, it seems reasonable to conclude that the analyzed populations of *I. pumila* are genetically differentiated for the mean relative level of the two Hsp chaperones, Hsp70 and Hsp 90, most probably due to divergent natural selection operating within alternative light habitats. Although the two criteria can frequently be simultaneously satisfied, Kawecki and Ebert have the preference to the "local *vs.* foreign" criterion as diagnostic for the pattern of local adaptation. They believe that "This criterion is directly relevant to the driving force of local adaptation – divergent natural selection - which acts on genetic differences in relative fitness within each habitat. In contrast, the "home *vs.* away" criterion confounds the effects of divergent selection with intrinsic differences in habitat quality" (Kawecki & Ebert, 2003). The term divergent selection is closely related to the phenomenon of ecological speciation – the process by which reproductive isolation between populations evolves as a results of ecologically based divergent selection (Rundle & Nosil, 2005). According to Schluter (2000), divergent selection can arise due to differences between populations in their environmental conditions, including habitat structure, climate, resources, and the predators or competitors present. Indeed, in the studied populations of *I. pumila,* the ambient air temperature and the instantaneous light intensity markedly differed between contrasting light habitats, as well as over seasons in the same habitat. For example, 2004 measurements, at an open habitat in the spring, the mean air temperature amounted 19.5 ± 0.5°C and light intensity 1797 ± 16 μmol m^{-2} s^{-1}, while in the summer, the average air temperature appeared to be 29.7 ± 0.6°C and light intensity 1378 ± 16 μmol m^{-2} s^{-1}. In the forest shade, however, the spring temperature mean was 21.6 ± 0.6 °C and the mean light intensity 136 ± 1.8 μmol m^{-2} s^{-1}, whereas in the summer, the average air temperature reached 19.5 ± 0.5°C and the mean light intensity 45 ± 3.1 μmol m^{-2} s^{--1} (Vuleta et al., 2010). It has been recently reported that the relative level of the two leaf Hsp chaperones, Hsp70 and Hsp90, varied significantly across seasons in the same clonal genotypes native to an exposed and a shaded population of *I. pumila,* and between different naturally growing genotypes in these *I. pumila* populations experiencing contrasting light conditions as well (Manitašević et al., 2007). Our study provides evidence that the average relative level of Hsp70 chaperone was significantly higher at an exposed site than under the forest understorey, reaching its maximum in the summer, especially in plants exposed to full sunlight. Of note, in the open habitat, the relative level of Hsp70 in the local (Dune) genotype appeared to be significantly greater in spring- and summer-collected leaves (0.63 *vs.* 1.46, respectively) than in their foreign (Woods) counterparts (0.29 *vs.* 0.96, respectively) (Table 1; Fig. 2), presumably due to ecologically based divergent selection.

6.1 The role of Hsp70 chaperones for adaptation

It is well known that molecular chaperones play a crucial role at two stages in the life of a protein: throughout *de novo* folding following translation, and during denaturation imposed by environmental stress (Morimoto et al., 1997; Parsell & Lindquist, 1993). Although many chaperones are highly elevated during stress, the investigations of eukaryotic cells revealed a significant difference between the folding of newly translated polypeptides and stress-denatured proteins. Regarding Hsp70 protein family, the former - stress-repressed chaperones are located in a sequestered cell environment close to the translational apparatus, while the latter - stress-induced chaperones occur in the bulk cytosol (Thulasiraman et al., 1999). Recently, Albanèse et al. (2006) have applied a systems biology approach to elucidate the functional organization of cytosolic chaperones in *Saccharomyces cerevisiae*. They revealed that the eukaryotic chaperone machinery comprises two networks with specialized functions. The first, stress-repressed network consists of chaperone linked to protein synthesis, and, therefore, is denoted as CLIPS, and the second, stress-inducible or the Hsp chaperone network, which includes components that either renature or clear misfolded proteins. These authors proposed that Hsp70 chaperones, particularly the Hsp70 Ss1/2p, plays a central role early during polypeptide synthesis, i.e., in *de novo* folding of newly made polypeptides, but also in response to stress.

Because the essential role of Hsp70 in stressful environments is to prevent aggregation, and to facilitate refolding and/or proteolytic degradation of nascent proteins (Wang et al., 2004), the elevated level of Hsp70 in sun-exposed plants, especially during summer, might be viewed as a kind of "anticipatory" phenotypic plasticity to increasing chances of heat stress in that habitat. In the forest understory, however, where thermal fluctuations are fewer and less frequent, a lower relative level of Hsp70 are sufficient to maintain native protein structure and occasional refolding of damaged proteins (Manitašević et al., 2007). Given that increased level of Hsp70 chaperone correlates well with greater thermotolerance in many plant species and that heat stress jointly occurs with high irradiance, the *I. pumila* genotypes naturally exposed to multiple abiotic stresses could be though as more stress-tolerant compared to those ones inhabiting a more "benign" vegetation shade.

6.2 The role of Hsp90 chaperones for adaptation

Contrary to Hsp70 proteins, which achieved their maximal relative level in the summer, the average level of both Hsp90 isoforms (inducible-Hsp90a and constitutive-Hsp90b) were highly suppressed in the summer compared to their spring counterpart, especially Hsp90b isoform (Table 1; Fig. 3). In the open habitat, genotypes from both populations differed significantly in the average level of the two Hsp90 isoforms (Fig. 3A and 3B). Conversely, in the shaded habitat, their relative level was similar between local and foreign genotypes in both seasons, with only exception of Hsp90b isoforms in the summer, which relative amount appeared to be greater in the Dune than in the Woods population (0.54 *vs.* 0.40, respectively; Table 1; Fig. 3C and 3D).

In the eukaryotic cytosol, Hsp70 and Hsp90 chaperones are each essential for cell viability under all growth conditions, implying that they fulfill non-overlapping function (Frydman, 2001; Young et al., 2004). Hsp90s are highly conserved group of molecular chaperones, which constitute about 1-2% of all cytosolic proteins in most cells under non-stress conditions (Parsell & Lindquist, 1993). Hsp90 is not a chaperone for newly synthesized proteins, but, instead, its cellular function is restricted to the conformational regulation of the limited group of substrates or "clients" (McClellan et al., 2007). In higher eukaryotes, Hsp90 work together with a large set of co-chaperones to mediate the conformational regulation of tyrosine kinases and steroid hormone receptors (Picard, 2006), but also to prevent phenotypic variation of these signaling molecules in the face of gene mutation (Sangster et al., 2004). The current understanding of Hsp90 function in tyrosine kinase and steroid hormone receptors maturation suggests that Hsp90 binds to "clients" that are substantially folded, facilitating their conformational remodeling.

Recently, McClellan et al. (2007) have used a genome-wide chemical-genetic screen combined with bioinformatic analyses to elucidate more deeply the Hsp90 functions. They identified several *unanticipated function* of Hsp90 under normal conditions and in response to stress. One of new informations obtained from these studies is the modular nature of the

Hsp90 interaction network. The Hsp90 network consists of two major functional modules, one dedicated to cellular trafficking and transport, and the other dedicated to the cell cycle regulation. Hsp90 functions in almost all aspects of the exocytic and endocytic secretory pathway through direct physiological interaction with their components. It was found that the Hsp90 targets tend to interact with each other, and, surprisingly, the average distance between them was found to be smaller than expected from random chance, as well as that each of these targets contains higher than expected number of hubs (proteins with more than 25 interaction partners) (McClellan et al., 2007). Under normal environmental conditions, Hsp90 is essential for vesicular transport and protein trafficking, which require the ordered assembly and disassembly of large multisubunit complexes. Hsp90 stabilizes or assists in the development of these oligomeric complexes by stabilizing their subunits prior to assembly or by assisting in their conformational transition. During environmental stress, however, Hsp90 appears to stabilize unstable conformations of many proteins, and has a key role in the continued function of the cell cycle machinery (McClellan et al., 2007).

Our study provides evidence that at the sun-exposed habitat the relative level of inducible and constitutive Hsp90 isoforms of Hsp90 chaperone was lower in the summer-collected leaf samples from both local and foreign *Iris* genotypes, compared for their spring value, and, as a rule, was greater in the former than in latter ones. The same trend was detected in the shaded habitat, but more conspicuously during summer, in both populations studied. According to Sørensen (2010), it is not easy to decide "when the level of constitutive and inducible HSP expression should be interpreted as reflecting the capacity or ability to mount a strong defense (i.e. as a benefit) or when it should be interpreted as reflecting the need to mount a strong response as the organism is stressed (i.e. as a cost)" (Sørensen, 2010). Since at the open habitat, the local genotypes had an increased level of constitutive Hsp90b over seasons in general than the foreign genotypes, it suggests that this chaperone might be important for adaptation to usually higher temperature prevailing there. The finding that the relative level of both Hsp90 isoforms was lower in summer than in spring ought to have special attentions because it opposed the general prediction that the amount of Hsps is elevated by stressful stimuli. It is well known that heat shock response is energetically costly, since both the protein production and chaperoning activity of Hsps require energy supply. The costs may also arise from stress-related destruction of normal cellular functions, extensive use of energy by antioxidants, and as well as from negative influence of Hsps on fitness (Feder & Hofmann, 1999; Heckathorn et al., 1996; Manitašević et al., 2007; Sørensen, 2010). Contrary to Hsp90, the Hsp70 dramatically increased in the summer, but about 50% more in the Dune than in the Woods populations, suggesting than this Hsp might be most important during rare and unpredictable environmental stress episodes and not for continuous or regularly occurring stress exposure (Sørensen, 2010).

Apart from producing environmentally contingent differences between populations in the mean value of a trait, natural selection can influence the trait plasticity within the same habitat as well. We quantified the degree of plasticity in the level of Hsps in *I. pumila* using an index of plasticity, PI_C (Valladares et al., 2006). Among the three chaperones analyzed, Hsp70 exhibited the greatest amount of plasticity, regardless of the habitat type and/or season. However, a statistically significant difference in plasticity to habitat light conditions was observed between the Dune and the Woods populations for the relative level of Hsp70 (PI_C = 0.437 and 0.315, respectively) in the summer, and for Hsp90b (PI_C = 0.074 and 0.213, respectively) in the spring (Table 5). Unexpectedly, the degree of plasticity to seasonal variation in abiotic environment conditions for the relative level of all three Hsps (Hsp70, Hsp90a and Hso90b) appeared to be similar between the two *I. pumila* populations, in each of their native habitats. The observed results could be interpreted as the outcome of divergent selection on Hsp plasticity; one, operating within the thermally unstable sun-exposed dune sites, and the other, generated by more variable light conditions prevailing under the forest understory. In addition, our results corroborate the hypothesis that thermal variation occurring within a generation time scale likely selected for increased Hsp chaperone level and, consequently, for greater inducible thermotolerance.

7. Conclusions

There is a consensus among evolutionary biologists that the phenomenon of adaptation has dual meaning: as a process and as a product of that process, which mechanism is natural selection. In this context, any biological entity that satisfies three necessary conditions for natural selection to operate: has the ability to vary, has continuity (heritability), and differs in its success relative to another co-occurring entity, can be adapted and can produce adaptations. When defined in a restrictive sense of the word, the term "adaptation" refers to a "trait (i) that enhances the fitness of an organism, and (ii) whose current beneficial characteristics reflect the selective advantage of the trait at its time of origin" (Hochachka & Somero, 2002). Molecular chaperones are a highly conserved set of functionally defined proteins that are involved in the folding and degradation of stress-demaged proteins. Because the role they play in all contemporary organisms is the benefit, which is very likely to be the same as the benefit that initially favoured the evolutionary development of these proteins, molecular chaperones are viewed as "adaptations". Among the molecular chaperones, the heat shock proteins Hsp70s are involved in the folding of newly translated and stress-denatured proteins. In addition to stress-inducible chaperone networks, eukaryotes contain a stress-repressed chaperone network that is dedicated to protein biogenesis. Although the Hsp90 molecular chaperones are highly abundant under normal conditions, they are restricted on a limited set of nearly mature, but inherently unstable, signaling proteins. Thus, under normal conditions, Hsp90 plays a key role in various aspects of the secretory pathway and cellular transport, while during environmental stress, Hsp90 is necessary for the cell cycle, meiosis and cytokinesis. In this study local adaptations for the relative level of three heat shock proteins, Hsp70, Hsp90a and Hsp90b in leaves of *Iris pumila* genotype native to contrasting light habitats were tested using a reciprocal transplant experiment conducted in the wild. Two experimental approaches, "local *vs* foreign" and "at home *vs* away", were applied to find out which of them can be used as diagnostic for local adaptation (Kawecki & Ebert, 2004). At a sun-exposed site, local genotypes were found to produce a higher amount of all three Hsps than did the foreign genotypes transplanted from a shaded habitat. Similarly, each of the genotypes exhibited a greater level of all three Hsp chaperones "at home" than "away", indicating that both criteria are satisfied for testing local adaptations. The obtained results indicate that the revealed genetic differentiation between populations from the exposed and the shaded habitats could be ascribed to divergent selection operating within their natural habitats. Apart from producing environmentally contingent differences between populations in the mean value of a trait, natural selection can influence the trait plasticity within the same habitat as well. Among the three chaperones analyzed, Hsp70 exhibited the greatest amount of plasticity, regardless of the habitat type and/or season. However, a statistically significant difference in plasticity to habitat light conditions was observed between the Dune and the Woods populations for the relative level of Hsp70 in the summer, and for Hsp90b in the spring. The observed results could be interpreted as the outcome of divergent selection on Hsp plasticity; one, operating within the thermally unstable sun-exposed dune sites and the other, generated by more variable light conditions prevailing under the forest understory. In addition, our results corroborate the hypothesis that thermal variation occurring within a generation time scale likely selects for increased Hsp chaperone level and, consequently, for greater inducible thermotolerance.

8. Acknowledgments

The authors are grateful to Nikola Tucić for critical reading of the paper. This research was supported by a grant (No. 173007) to B. T. from the Ministry of Education and Science of the Republic of Serbia.

9. References

Albanèse, V.; Yen-Wen Yam, A.; Baughman, J.; Parnot, C. & Frydman, J. (2006). Systems analyses reveal two chaperone networks with distinct functions in eukariotyc cells, *Cell*, 124, 75-88, ISSN 0092-8674

Amudson, R. (1996). Historical Development of the Concept of Adaptation, In: *Adaptation*, Rose, M.R. & Lauder, G.V. (Eds.), 11-53, Academic Press, ISBN 0-12-596421-8, San Diego, USA

Bazzaz, F.A. (1996). *Plants in changing environments: linking physiological, population, and community ecology*, Cambridge University Press, ISBN 0-521-39843-6, Cambridge, UK

Buchner, J. (1999). Hsp90 & Co. - a holding for folding. *Trends in Biochemical Sciences*, 24, 136-141, ISSN 0968-0004

Bukau, B. & Horwich, A.L. (1998). The Hsp70 and Hsp60 chaperone machines. *Cell*, 92, 351-366, ISSN 0092-8674

Emms, S.K. & Arnold, M.L. (1997). The effect of habitat on parental and hybrid fitness transplant experiments with Louisiana irises. *Evolution*, 51, 1112-119, ISSN 0014-3820

Feder, M.E. & Hofmann, G.E. (1999). Heat-shock proteins, molecular chaperones and stress response: evolutionary and ecological physiology. *Annual Review of Physiology*, 61, 243-282, ISSN 0066-4278

Frank, S.A. (1996). The design of natural and artificial systems, In: *Adaptation*, Rose, M.R. & Lauder, G.V. (Eds.), 451–505, Academic Press, ISBN 0-12-596421-8, San Diego,USA

Freeman, B.C. & Morimoto, R.I. (1996). The human cytosolic molecular chaperones in hsp90, hsp70 (hsc70) and hdj-1 have distinct roles in recognition of a non-native protein and protein refolding. *EMBO Journal*, 15, 2969–2979, ISSN 0261-4189

Frydenberg, J.; Hoffmann, A.A. & Loeschcke, V. (2003). DNA sequence variation and latitudinal associations in *hsp23, hsp26* and *hsp27* from natural populations of *Drosophila melanogaster*. *Molecular Ecology*, 12, 2025–2032, ISSN 0962-1083

Frydman, J. (2001). Folding of newly translated proteins *in vivo*: the role of molecular chaperones. *Annual Review of Biochemistry*, 70, 603–647, ISSN 0066-4154

Frydman, J. & Hartl, F.U. (1996). Principles of chaperone-assisted protein folding: differences between *in vitro* and *in vivo* mechanisms, *Science*, 272, 1497-1502, ISSN 0036-8075

Gajić, M. (1983). *The Flora of the Deliblato Sand*, Parabucski, M.; Gajić, B.; Šajinović, B.; Stojakov, B. & Vlatković, S. (Eds.), 6-446, University of Novi Sad, Serbia

Gehring, W.J. & Wehner, R. (1995). Heat shock protein synthesis and thermotolerance in *Cataglyphis*, an ant from the Sahara desert. *Proceeding of National Academy of Sciences of USA*, 92, 2994-2998, ISSN 0027-8424

Gould, S.J. & Vrba, E.S. (1982). Exaptation - a missing term in the science of form. *Paleobiology*, 8, 4-15, ISSN 0094-8373

Hartl, F.U. & Hayer-Hartl, M. (2009). Converging concepts of protein folding *in vitro* and *in vivo*. *Nature Structural and Molecular Biology*, 16, 574-581, ISSN 1545-9985

Heckathorn, S.A.; Poeller, G.J.; Coleman, J.S & Hallberg, R.L. (1996). Nitrogen availability alters pattern of accumulation of heat stres-induced proteins in plants. *Oecologia*, 105, 413-418, ISSN 0029-8549

Hemmer-Hansen, J.; Nielsen, E.E.; Frydenberg, J. & Loeschcke, V. (2007). Adaptive divergence in a high gene flow environment: Hsc70 variation in the European flounder (*Platichthys flesus* L.). *Heredity*, 99, 592-600, ISSN 0018-067X

Hochachka, P.W. & Somero, G.N. (2002). *Biochemical Adaptations: Mechanism and Process in Physiological Evolution*, Oxford University Press, ISBN 0-19-511703-4, New York, USA

Hoffmann, G.E. & Somero, G.N. (1995). Evdence for protein damage at environmental temperatures - seasonal changes in levels of ubiquitin conjugates and Hsp70 in the intertidal mussel *Mytilus trossulus*. *Journal o Experimental Biology*, 198, 1509-1518, ISSN 0022-0949

Hoffmann, A.A.; Sørensen, J.G. & Loeschcke, V. (2003). Adaptation of *Drosophila* to temperature extremes: bringing together quantitative and molecular approaches. *Journal of Thermal Biology*, 28, 175-216, ISSN 0306-4565

Kawecki, T.J. & Ebert, D. (2004). Conceptual issues in local adaptation. *Ecology Letters*, 7, 1225–1241, ISSN 1461-023X

Laemmli, U.K. (1970). Cleavage of structural proteins during the assembly of the head of bacteriophage T4. *Nature*, 227, 680-685, ISSN 0028-0836

Lambers, H.; Chapin III, F.S. & Pons, T.L. (2008). *Plant Physiological Ecology* (2nd), Springer Science + Business Media, ISBN 978-0- 387-78340-6, New York, USA

Lauder, G.V. (1981). Form and function: structural analysis in evolutionary morphology. *Paleobiology*, 7, 430-442, ISSN 0094-8373

Macario, A.J.; Lange, M.; Ahring, B.K. & De Macario, E.C. (1999). Stress genes and proteins in the archaea. *Microbiological and Molecular Biological Reviews*, 63, 923-967, ISSN 1092-2172

Manitašević, S.; Dunđerski, J.; Matić, G. & Tucić, B. (2007). Seasonal variation in heat shock proteins Hsp70 and Hsp90 expression in an exposed and a shaded habitat of *Iris pumila*. *Plant Cell and Environment*, 30, 1-11, ISSN 0140-7791

McClellan, A.J.; Xia, Y.; Deutschbauer, A.M.; Davis, R.W; Gerstein, M. & Frydman, J. (2007). Diverese cellular functions of the Hsp90 molecular chaperone uncovered using systems approaches. *Cell*, 131, 121-135, ISSN 0092-8674

Morimoto, R.I.; Kline, P.M.; Bimston, D.N. & Cotto, J.J. (1997). The heat-shock response: regulation and function of heat-shock proteins and molecular chaperones. *Essays of Biochemistry*, 32, 17-29, ISSN 0071-1365

Nguyen, H.T. Joshi, C.P.; Klueva, N.; Weng, J.; Hendershot, K.L. & Blum, A. (1994). The heat-shock response and expression of heat shock proteins in wheat under diurnal heat stress and field conditions. *Australian Journal of Plant Physiology*, 21, 857-867, ISSN 0310-7841

O'Hara, R.J.; Hines, W.G.S. & Robinson, B.W. (2004). A new statistical test of fitness set data from reciprocal transplant experiments involving intermediate phenotypes. *American Naturalist*, 163, 97-104, ISSN 0003-0147

Parsell, D.A. & Lindquist, S. (1993). The function of heat shock proteins in stress tolerance: degradation and reactvation of dameged proteins. *Annual Review of Genetics*, 27, 437-496, ISSN 0066-4197

Pearl, L.H. & Prodromou, C. (2006). Structure and mechanism of the Hsp90 molecular chaperone machinery. *Annual Review of Biochemistry*, 75, 271–294, ISSN 0066-4154

Picard, D. (2006). Chaperoning steroid hormone action. *Trends in Endocrinology and Metabolism*, 17, 229–235, ISSN 1043-2760

Queitsch, C.; Sangster, T.A. & Lindquist, S. (2002). Hsp90 as a capacitor in phenotypic variation. *Nature*, 417, 618-624, ISSN 0028-0836

Randolph, L.F. (1955). The geographic distribution of European and eastern mediterranean species of bearded *Iris*, In: *Iris yearbook 1955*, 35-46

Richter, K. & Buchner, J. (2001). Hsp90: chaperoning signal transduction. *Journal of Cellular Physiology*, 188, 281–290, ISSN 0021-9541

Rundle, H. D. & Nosil, P. (2005). Ecological speciation. *Ecology Letters*, *8*, 336–352, ISSN 1461-023X

Rutheford, S.L. (2003). Between genotype and phenotype protein chaperones and evolvability. *Nature Reviews Genetics*, 4, 263-374, ISSN 1471-0056

Rutherford, S.L. & Lindquist, S. (1998). Hsp90 as a capacitor for morphological variation. *Nature*, 396, 336-342, ISSN 0028-0836

Salathia, N. & Queitsch, C. (2007). Molecular mechanisms of canalisaton: Hsp90 and beyond. *Journal of Biosciences*, 32, 457-463, ISSN 0250-5991

Samakovli, D. Thanou, A.; Valmas, C. & Hatzopoulos, P. (2007). Hsp90 canalizes developmental perturbation. *Journal of Experimental Botany*, 58, 3513-3524, ISSN 0022-0957

Sangster, T.A.; Lindquist, S. & Queitsch, C. (2004). Undercover: causes, effects and implications of Hsp90-mediated genetic capacitance. *Bioassays*, 26, 348-362

SAS Institute, Inc. (2003) SAS/STAT *User's Guide, Version 6.*, 4th edn. SAS Institute, Inc. Cary, NC, USA

Schluter, D. (2000). *The ecology of adaptive radiation*, Oxford University Press, ISBN 978-0-19-850522-8, Oxford, UK

Sørensen, J.G. (2010). Application of heat shock protein expression for detecting natural adaptation and exposure to stress in natural populations. *Current Zoology*, 56, 703-713, ISSN 1674-5507

Sørensen, J.G.; Kristensen, T.N. & Loeschcke, V. (2003). The evolutionary and ecological role of heat shock proteins. *Ecology Letters*, 6, 1025-1037, ISSN 1461-023X

Sørensen, J.G.; Pekkonen, M.; Lindgren, B.; Loeschcke, V.; Laurila, A. & Merilä J. (2009). Complex patterns of geographic variations in heat tolerance and hsp70 expression levels in the common frog *Rana temporaria*. *Journal of Thermal Biology*, 34, 49-54, ISSN 0306-4565

Spector, T. (1978). Refinement of the coomassie blue method of protein quantification. *Analytical Biochemistry*, 86, 142-146, ISSN 0003-2697

Thulasiraman, V.; Yang, C.F. & Frydman, J (1999). *In vivo* newly translated polipeptydes are sequestered in a protected folding environment. *EMBO Journal*, 18, 85-95, ISSN 0261-4189

Tucić, B.; Milojković, S.; Vujčić, S. & Tarasjev, A. (1988). Clonal diversity and dispersion in *Iris pumila*. *Acta Oecologica. Oecologia Plantarum*, 9, 211-219, ISSN 0243-7651

Tucić, B.; Tomić, V.; Avramov, S. & Pemac, D. (1989). Testing the adaptive plasticity of *Iris pumila* leaf traits to natural light conditions using phenotypic selection analysis. *Acta Oecologica*, 19, 473-481, ISSN 1146-609X

Tucić, B.; Vuleta, A. & Manitašević, S. (2009). Protective function of foliar anthocyanins: *in situ* experiments on a sun-exposed population of *Iris pumila* L. (Iridaceae). *Polish Journal of Ecology*, 57, 4, 779-783, ISSN 1505-2249

Valladares, F.; Sanchez-Gomez, D. & Zavala, M.A. (2006). Quantitative estimation of phenotypic plasticity: bridging the gap between the evolutionary concept and its ecological applications. *Journal of Ecology*, 94, 1103-1116, ISSN 0022-0477

Vermeij, GJ. (1996). Adaptations of Clades: Resistance and Response, In: *Adaptation*, Rose M.R. & Lauder G.V. (Eds.), 363-379, Academic Press, ISBN 0-12-596421-8, San Diego, USA

Vuleta, A.; Manitašević Jovanović, S.; Šešlija, D. & Tucić, B. (2010). Seasonal dynamic of foliar antioxidative enzymes and total anthocyanins in natural populations of *Iris pumila* L. *Journal of Plant Ecology*, 3, 1, 59-69, ISSN 1752-9921

Wang, W.; Vinocur, B.; Shoseyov, O. & Altman. A. (2004). Role of plant heat shock proteins and molecular chaperones in the abiotic stress response. *Trends in Plant Science*, 9, 246-252, ISSN 1360-1385

Wegele, H.; Müller, L.; & Buchner, J. (2004). Hsp70 and Hsp90—A relay team for protein folding. *Reviews of Physiology, Biochemistry and Pharmacology*, 151, 1–44, ISSN 0303-4240

Williams, G.C. (1966). *Adaptation and Natural Selection*. Princeton University Press, ISBN: 978-1-4008-2010-8, Princeton, USA

Young, J.C.; Agashe, V.R.; Siegers, K. & Hartl, F.U (2004). Pathways of chaperone mediated protein folding in the cytosol. *Nature Reviews: Molecular and Cell Biology*, 5, 781-791, ISSN 1471-0072